NUTRITIONAL AND TOXICOLOGICAL CONSEQUENCES OF FOOD PROCESSING

ADVANCES IN EXPERIMENTAL MEDICINE AND BIOLOGY

Recent Volumes in this Series

NUTRITIONAL AND TOXICOLOGICAL CONSEQUENCES OF FOOD PROCESSING

Edited by

Mendel Friedman

U.S. Department of Agriculture
Albany, California

PLENUM PRESS • NEW YORK AND LONDON

Library of Congress Cataloging-in-Publication Data

American Institute of Nutrition-Federation of American Societies
for Experimental Biology Symposium on Nutritional and Toxicological
Consequences of Food Processing (1990 . Washington, D.C.)
Nutritional and toxicological consequences of food processing /
edited by Mendel Friedman.
p. cm. -- (Advances in experimental medicine and biology ; v.
289)
"Proceedings of an American Institute of Nutrition-Federation
of American Societies for Experimental Biology Symposium on
Nutritional and Toxicological Consequences of Food Processing, held
April 1-5, 1990, in Washington, D.C."--T.p. verso.
Includes bibliographical references and index.
ISBN 0-306-43891-7
1. Food--Toxicology--Congresses. 2. Food industry and trade-
-Health aspects--Congresses. 3. Nutrition--Congresses.
I. Friedman, Mendel. II. American Institute of Nutrition.
III. Federation of American Societies for Experimental Biology.
IV. Title. V. Series.
[DNLM: 1. Food Analysis--congresses. 2. Food Contamination-
-congresses. 3. Food Handling--congresses. 4. Nutrition-
-congresses. W1 AD559 v. 289 / WA 695 A5115n 1990]
RA1258.A43 1990
615.9'54--dc20
DNLM/DLC
for Library of Congress

91-3546
CIP

Proceedings of the American Institute of Nutrition-Federation of
American Societies for Experimental Biology Symposium on
Nutritional and Toxicological Consequences of Food Processing,
held April 1-5, 1990, in Washington, D.C., with additional invited contributions

ISBN 0-306-43891-7

A Division of Plenum Publishing Corporation
233 Spring Street, New York, N.Y. 10013

Printed in the United States of America

I dedicate this book to my wife Dora for foregoing the pleasure of my company during many a weekend that went into this effort.

Mendel Friedman
Moraga, California
December, 1990

The earth is full of the fruit of Thy works.
Thou causest grass to spring up for cattle,
And herbs for the service of man.
Thou bringest forth bread out of the earth
To sustain man's life,
And wine to gladden his heart.

Psalm 104: 28 -33

The eyes of all look to you expectantly,
and you give them their food when it is due,
You give it openhandedly, feeding every
creature to its heart's content.

Psalm 145: 15 -16

Eat nothing that will prevent you from eating.

Ibn Tibbon, 1190

PREFACE

A variety of processing methods are used to make foods edible; to permit storage; to alter texture and flavor; to sterilize and pasteurize food; and to destroy microorganisms and other toxins. These methods include baking, broiling, cooking, freezing, frying, and roasting. Many such efforts have both beneficial and harmful effects. It is a paradox of nature that the processing of foods can improve nutrition, quality, safety, and taste, and yet occasionally lead to the formation of anti-nutritional and toxic compounds. These multifaceted consequences of food processing arise from molecular interactions among nutrients with each other and with other food ingredients.

Since beneficial and adverse effects of food processing are of increasing importance to food science, nutrition, and human health, and since many of the compounds formed have been shown to be potent carcinogens and growth inhibitors in animals, I organized a symposium broadly concerned with the nutritional and toxicological consequences of food processing. The symposium was sponsored by the American Institute of Nutrition (AIN) - Federation of American Societies for Experimental Biology (FASEB) for its annual meeting in Washington, D.C., April 1-5, 1990. Invited speakers were asked to develop at least one of the following topics:

1. Nutrient-nonnutrient interactions between amino acids, proteins, carbohydrates, lipids, minerals, vitamins, tannins, fiber, natural toxicants, etc.
2. Effects of radiation.
3. Thermally induced formation of dietary mutagens, antimutagens, carcinogens, anticarcinogens, antioxidants, and growth inhibitors.
4. Effects of pH on nutritional value and safety.
5. Effects of oxidizing agents on food quality and safety.
6. Effects of processing on food allergenicities.
7. Metabolic detoxification pathways.
8. Prevention of adverse effects of processing.
9. Beneficial effects of processing.
10. Risk assessment.

The most important function of a symposium, I believe, is dissemination of insights and exchange of ideas so as to catalyze progress by permitting synergistic interaction among related disciplines. I hope that the reports presented at the symposium fulfilled this purpose. In addition, a number of scientists who could not participate in the symposium accepted invitations to contribute manuscripts to this volume on the theme of the symposium. This book is, therefore, a hybrid between symposium proceedings and a collection of invited papers.

Brought together here are outstanding international authors from twelve countries who discuss the multidisciplinary aspects of nutritional and toxicological significance of the processing of foods. The major theme of this book is that a better understanding of the molecular changes during food processing is needed to optimize beneficial effects such as bioavailability, food quality and food safety, and to minimize the formation and facilitate inactivation and removal of deleterious mutagens, carcinogens, and other toxicants.

The described multidisciplinary studies reveal a complex interplay between chemistry, biochemistry, nutrition, physiology, pharmacology, and toxicology of food ingredients.

I am particularly grateful to all contributors for excellent cooperation, to Dr. R. G. Allison of the American Institute of Nutrition for helpful correspondence on the theme of the symposium, and to Lillie Davis for excellent secretarial assistance.

Plenum Press is publishing the papers under the title *Nutritional and Toxicological Consequences of Food Processing,* as a volume in the series Advances in Experimental Medicine and Biology. This book is intended to complement the following published volumes which I edited for the same series: *Protein - Metal Interactions (1974) - Vol. 40; Protein Crosslinking: Biochemical and Molecular Aspects (1977) - Vol. 86A; Protein Crosslinking: Nutritional and Medical Consequences (1977) - Vol. 86B; Nutritional Improvement of Food and Feed Proteins (1978) - Vol. 109; Nutritional and Toxicological Aspects of Food Safety (1984) - Vol. 177;* and *Nutritional and Toxicological Significance of Enzyme Inhibitors in Foods (1986) - Vol. 199.*

I very much hope that these and related monographs which I edited (*Protein Nutritional Quality of Foods and Feeds,* Marcel Dekker, 1975 and *Absorption and Utilization of Amino Acids*, CRC Press, 1989) will be a valuable resource for further progress in agriculture, food chemistry, food safety, animal and human nutrition, physiology, pharmacology, toxicology, and medicine; all areas in which there is an urgent world-wide need to better the human condition. If so, the effort will be most worthwhile.

Mendel Friedman.

CONTENTS

1

Food Safety Assurance: The European Perspective

E. Quattrucci and R. Walker

Istituto Naz. della Nutrizione, 00179 Rome, Italy

and University of Surrey, Guildford GU2 5XH, U.K.

The association between diet and health is not only widely recognized but more and more is considered as a basis for Government strategies. In the past, food availability, food hygiene and food safety were the major concerns whereas the nutritional aspect of the wholesomeness of food played only a minor role. Recently, the Organizations dealing with public health in Europe, such as the WHO Regional Office for Europe, and the European Community have been seeking a more comprehensive approach to the area of healthy eating.

Created in 1957 with the Treaty of Rome, the EEC now comprises twelve countries (Belgium, Denmark, France, Greece, Ireland, Italy, Luxembourg, the Netherlands, Portugal, Spain, the United Kingdom and West Germany). The 300 million citizens of the Community certainly have no problem with the availability of an adequate, nutritious diet. On the other hand, unbalanced nutrition, often overnutrition, arises mainly due to poor dietary habits. Therefore, recent Community actions are also in hand on nutrition policy, nutritional education and information and research programmes.

In order to give a perspective to Community initiatives in the area of food safety and applied nutrition for those who are unfamiliar with the subject, let us take an historical glance at the Universe of the EEC, its tasks and ideals. The EEC was created in order to develop a Common Market for the free circulation of goods, including food commodities among the various member countries. Permanent Institutions, the most important of which are the Commission, the Council of Ministers, the European Parliament and the Court of Justice, have been created and endowed with real powers. It is easy to imagine the extent to which these Institutions are kept occupied by the task of harmonizing a multitude of national laws related to 12 countries with different currencies, languages and traditions. Examples of the powers given to the EEC Institutions are the cases of controversies brought before the Court of Justice. The first of these was the case of the blackcurrant liqueur, Cassis de

Dijon, which was produced in France in accordance with French laws but which did not meet German legislative requirements. The Court ruled that Article 30 of the Treaty of Rome prohibited differences in food standards from hindering trade between EEC countries. A product legally manufactured and marketed in a member state should be freely sold in all the other EEC countries unless a consequent health hazard provides a valid reason for opposing its trade nationally (Article 36). After the Cassis de Dijon case, a series of judgements added emphasis to the need for reasoned arguments, the most recent cases involving German beer and Italian pasta. At present, Italy and Greece are still before the Court for not having allowed entrance to imports of Dutch cheese produced with the addition of nitrate, on the basis of the aforementioned Article 36.

The means of achieving harmonization of food law, as laid down in Article 100 of the Treaty of Rome, are Community "Directives". These might be "horizontal directives" dealing with general topics such as food additive categories (colouring matters, preservatives, emulsifiers etc.), or "vertical directives" setting up standards for specific commodities such as wine, soft drinks, etc. The presence of certain contaminants (e.g. packaging materials, mycotoxins) in specific foods may also be regulated by vertical directives (Haigh, 1978).

The Commission elaborates Directives taking into account the opinions expressed by its Scientific Committees, such as the Scientific Committee for Food (SCF). These Committees work on the basis of the available toxicological information and their opinions, published periodically as Commission of the European Community Reports, are also largely endorsed by Parliaments, Governments, Industry and Consumers (Haigh, 1985). The EC approach, from regulatory statements to the last stage of control, presents only few similarities with other systems such as those of the USA or Japan. An interesting comparative overview was presented by Malaspina (1987).

A number of directives have been published, revised or are in various stages of preparation. They include both specific food standards and very broad and complex issues. The technique of food irradiation and related chemical, nutritional and microbiological aspects, dietary foods and infant formulae, and the use of anabolic agents in farm animals are only a few examples. The banning of anabolic agents stemmed from a concoction of conflicting scientific opinions, economic and political choices and consumer pressure. Although, for several anabolic agents, scientific evidence suggested that it might be possible to establish acceptable residue limits based on "no hormonal effect" levels in biological tests and adequate withdrawal periods before slaughter, there was concern about the ability adequately to monitor and enforce these control measures.

Another area of current public concern relates to the use or abuse of pesticides. In this area, the Commission elaborates directives on Maximum Residue Limits (MRLs) based upon advice from the Scientific Committee on Pesticides. This Committee adopts an approach to safety evaluation which is similar in many respects to that of the Joint FAO/WHO Meeting on Pesticide Residues (JMPR) with Acceptable Daily Intakes being assessed on a toxicological basis and MRLs being allocated on the basis of Good Agricultural Practice (GAP). MRLs are set with a view to ensuring that residues are kept to the minimum consistent with efficacy and that ADIs are not exceeded; in many cases the MRLs achievable with GAP result in exposures which are very much lower than the ADI. Harmonization has been achieved on compulsory MRLs relating to cereals and products of animal origin; partial agreement has been reached on MRLs for fruits and vegetables with lower limits that may be increased in a member state if necessary.

With regard to food additives and some processing residues, such as extraction solvents, the SCF advises the Commission on safety issues and has published guidelines on toxicological evaluation of materials for food contact and for food additives. The philosophy of the SCF is broadly similar to that of the Joint FAO/WHO Expert Committee on Food Additives (JECFA) in that it allocates Acceptable (or Tolerable) Daily (or Weekly) intakes for specific additives or contaminants based on a toxicological evaluation in a battery of toxicity tests, including special studies on reproductive toxicity and teratogenicity, mutagenicity and carcinogenicity. The ADI would normally be based on the no-observable-adverse-effect level (NOEL) in the most sensitive test/species unless supporting metabolic and pharmacokinetic studies or human tolerance studies indicate otherwise e.g. where the effect in the most sensitive species is due to mechanisms which do not apply to man because of metabolic or physiological differences, or conversely, where there is evidence of serious idiosyncratic reactions in man. In calculating the ADI, generally a safety factor of 100 is applied to the NOEL unless the nature of the toxic response, the quality of the toxicological data or supporting human data indicate that it should be greater (or smaller). Member states may then use the ADI along with national food survey data and estimated potential intakes to frame national regulations to ensure safety-in-use.

Lists of approved food additives are contained in "horizontal directives" designated with a unique "E" number and there has been a good deal of progress in harmonization of permitted lists of food additives in member states. In some cases, temporary derogations have been given for member states to continue to use particular additives for specified purposes e.g. the food colours Red 2G and Brown FK for use in sausages and kippers respectively in the U.K. However, such additives would not be allocated an "E" prefix which is restricted to those additives which are considered generally acceptable on health grounds and for which there is a proven need.

There are a number of differences in approach between Europe and the United States in regulating food additives and contaminants which are worth highlighting. Firstly, there is no codified equivalent of the Delaney Clause in EC legislation, although ADIs would not usually be allocated to genotoxic carcinogens; this has made it easier to deal with non-genotoxic carcinogens like saccharin. To date, the SCF has eschewed the use of quantitative risk assessment and the calculation of "virtually safe doses"; instead, as indicated previously, it relies on threshold models for compounds other than genotoxic carcinogens with the application of appropriate safety factors.

Another difference of note is that there is no equivalent of the GRAS concept in the EEC although substances with very low toxicity may be allocated an ADI "not specified" i.e. their level of use is limited by good manufacturing practice rather than toxicological considerations. Otherwise, the Directives permit member states to set upper limits on the levels of use of food additives generally or related to specific applications, but subject to the general provisions of Article 30 of the Treaty. As in many other countries, the EEC is wrestling with the problems posed by food flavours by virtue of the number of substances used, the fact that many are natural or nature-identical, and the need to establish a rational basis for prioritisation for toxicity testing and the extent of testing required. No definitive solution has yet emerged.

Quite apart from toxicological issues, in considering submissions on additives the EEC has adopted a requirement for a "case for need" to be established. This implicitly introduces the concept of limiting the number of permitted food additives to the minimum which are needed to fulfil established needs and differs from the requirement only to establish safety and efficacy. The former approach potentially leads to higher levels of exposure to fewer substances while the latter could lead to exposure to a greater number of additives at individually lower levels.

Moving to a subject of major interest in the context of this Symposium, i.e. nutritional aspects, there has been an increased sensitivity towards these factors in recent EEC strategies. In moving from deficiency into surplus of some commodities, such as butter or meat, in the past the measures adopted to dispose of stockpiles (subsidized Christmas butter etc.) were often nutritionally undesirable. Fortunately, quota measures have almost eliminated surplus stocks. Agricultural policies throughout the world are moving toward the GATT Uruguay Round indications on reduction of economic distortions and of excessive support of agricultural production.

The EC programme to legislate all the issues concerning public health, fair trading and public control with required or proposed actions are indicated in Table 1. From these considerations it is clear that proper nutritional labelling may be accomplished only if the existing RDAs are harmonized across the EEC. This subject is now on the agenda of the SCF.

In accordance with this increased sensitivity towards nutritional safety of the diet, in the context of the programme "Europe against Cancer", the Commission has elaborated and promulgated a code which also contains advice on reduction of fat intakes and increase in consumption of fruits and vegetables.

Following the well-known 7 country study on diet and cardiovascular disease (Keys, 1970), in which several

European countries participated, a number of member states have taken initiatives to attempt to reduce the incidence of cardiovascular disease. In Italy, an ongoing epidemiological study is monitoring the incidence of cardiovascular disease (the MONICA project) with a view to assessing the long-term efficacy of advice to effect dietary change by reducing fat and salt intakes. In the U.K., the Committee on Medical Aspects of Food Policy (COMA) has issued similar dietary advice.

Moreover, in the new four-year programme which is being developed in the EC, nutrition will be given greater importance both in terms of time and financial support. Plans are under consideration to designate 1994 "*European Year of Nutrition Information*". The feasibility of a large prospective study on the relation between diet and cancer is also being investigated.

Table 1. EEC actions/proposals for action

For food safety and nutrition

food additives;
dietetic foods;
novel foods;
materials and articles in contact with foodstuffs;
solvents;
flavours;
food hygiene;
rapid alert system for food contamination (set up as a consequence of the Chernobyl accident).

For fair trading

labelling, advertising and presentation of foodstuffs;
forthcoming proposals for quantitative ingredient declarations;
forthcoming proposals on claims;
nutrition labelling.

For public controls

the control directive;
coordinated programmes of sampling and analysis;
quality assurance in control laboratories;
exchange visits of public controllers.

(Gray, 1990)

It will be noted that the actions listed in Table 1 include initiatives relating to food hygiene. Concern has been expressed that, where adequate reporting systems exist, the statistics indicate that the number of cases of microbiological food poisoning has continued to increase in recent years despite measures to combat this e.g. Hazard Analysis Critical Control Point systems. This is exemplified by the number of cases of food poisoning reported to the Public Census Office in the U.K. shown in Table 2, where it can be seen that the number of cases doubled between 1986 and 1988; preliminary figures for 1989 indicate a further but small increase in cases of about 5%.

Table 2. Notifications of cases of food poisoning reported to the United Kingdom Office of Population and Census Studies - 1984-1988

Year	Number of cases reported*
1984	20702
1985	19242
1986	20948
1987	29331
1988	41196

* It has been estimated that the reported number of cases may be only 10% or less of actual occurrence

The highest level of public concern and political activity has been directed at the incidence of Salmonella infections associated with poultry and eggs although, as can be seen from Table 3, the incidence of *Campylobacter jejuni* infections has been even higher in the U.K. over the last eight years. Nevertheless, public awareness about salmonellosis has been the dominant issue, directed particularly at intensive methods of producing broiler chickens and eggs. The situation in Italy with regard to outbreaks of salmonellosis is stated to differ in that there have not been significant changes in incidence of salmonellosis in the last ten years (Toti *et al.*, 1990), however, these authors acknowledge that an efficient surveillance programme does not exist for adults and the data are underestimates.

Table 3. Incidence of Salmonella and Campylobacter jejuni infections in the United Kingdom - 1980-1988

Year	Number of cases reported	
	Salmonella	Campylobacter jejuni
1980	11,000	9,600
1981	10,900	12,300
1982	12,000	12,800
1983	14,250	17,500
1984	14,270	21,000
1985	11,900	23,500
1986	15,000	24,500
1987	18,000	27,390
1988	24,120	28,710

In the U.K., attention was particularly focussed on *Salmonella enteritidis* in eggs in late 1988 by a somewhat inaccurate statement by the then Junior Health Minister that the bulk of egg production in the U.K. was infected. "At risk" groups, identified as young children, the elderly, the immuno-compromised and pregnant women, were advised not to eat eggs unless thoroughly cooked (e.g. hard boiled) and to avoid products such as home-made mayonnaise or egg-based sauces which were only lightly cooked. The consequent fall in consumption of eggs was estimated to have cost the industry something like $30 million. A policy of culling infected flocks has been implemented to restore public confidence. In Italy, as in other European countries, *Salmonella enteritidis* has been detected with higher frequency in recent years, largely derived from poultry, eggs and egg-based foods.

Problems have also arisen with other pathogens, most notably *Listeria monocytogenes*, which can multiply slowly even at refrigerator temperatures and this has damaged confidence in "cook-chill" foods and some soft cheeses. In the U.K. there was a significant rise in cases of listeriosis between 1984 and 1988 (Table 4) and particular concern centres around exposure of pregnant women where infection may lead to abortions or still-births. Although the number of cases known to be associated with food is small, both in Italy and the U.K. recommendations have been made to place stringent controls on the performance of chill cabinets in retail stores so that product temperatures are maintained below 5°C; again the cost to retailers of implementing these recommendations is likely to be millions of dollars.

Table 4. Laboratory Reports of Listeriosis in the U.K.

Year	Cases	Abortions	Deaths (including stillbirths)
1984	115	7	34
1985	149	8	56
1986	137	5	32
1987	259	18	59
1988	291	11	52

Another regulatory issue concerning food safety and nutrition is that of novel foods and processes, including the use of genetic engineering in producing foods/ingredients. Currently there is no compulsory system in the EC for handling submissions relating to such processes or products which, like the issue of food irradiation, tend to have been dealt with in an *ad hoc* manner. However, the U.K. has established a voluntary system operating through an Advisory Committee on Novel Foods and Processes (ACNFP). This Committee has devised a "decision tree" approach to determine the manner in which each novel food or ingredient will be handled in respect of the nature and extent of toxicity or other tests which would be required; this approach is being studied by the SCF as a starting point for discussions on a Community-wide system of regulation. The ACNFP recently approved for the first time for food use a genetically manipulated micro-organism, a strain of bakers' yeast with genes from a sister strain inserted to speed up dough fermentation. A separate committee exists in the U.K. to regulate environmental release of genetically engineered microorganisms and, since some live yeast may survive the baking process, or otherwise "escape" into the environment, this Committee was also involved in the decision. It is anticipated that a similar situation will emerge in the Community at large.

Support for Research in Food Safety

With regard to support for research related to food safety and nutrition, the EC, mainly through its Directorate for Science, Research and Development, promotes research programmes carried on by national Institutions under the "Shared Cost" or "Concerted Action" schemes. The Framework Programme for Community activities in R&D an action - Action 4: "Exploitation and Optimum Use of Biological Resources". Within Action 4 is a sub action 4.2. "Agro-Industrial Technologies" which in turn consists of two programmes supporting research. These are:

(i) ECLAIR: "European Collaborative Linkage of Agriculture and Industry through Research; and

(ii) FLAIR: "Food Linked Agro-Industrial Research.

The ECLAIR programme is at an advanced stage whereas the FLAIR, adopted on 20th June 1989, will run from January 1990 to December 1993.

FLAIR concentrates on the processing-distribution-consumer end of the food chain and aims to encourage R&D in three broad sectors:

1. Assessment and Enhancement of Food Quality and Diversity;
2. Food Hygiene, Safety and Toxicology;
3. Nutrition and Wholesomeness.

These sectors represent frames for 16 sub-sector or priority themes of research which are listed in Table 5.

Table 5. Priority Themes Of Research under FLAIR

Sector 1: "Quality"	
1.1.	Quantitative measures of food "quality";
1.2.	Sensory evaluation and relationships to "quality";
1.3.	"Freshness" of processed foods;
1.4.	Effects of raw materials on "quality";
1.5.	"New technologies which enhance quality.
Sector 2: "Safety"	
2.1.	Rapid screening tests for toxicity factors;
2.2.	Natural plant toxins;
2.3.	Tests for specific organisms;
2.4.	Food intolerances;
2.5.	Applications in commercial plants;
2.6.	"New" technologies and "safety".
Sector 3: "Nutrition"	
3.1.	Effects of new technologies on nutrition/health;
3.2.	Nutritional methodologies;
3.3.	Nutrient bioavailability;
3.4.	Nutritional value of foods for special groups;
3.5.	"New" technologies for improved nutritional value.

Related to the theme "Nutrition and Health" the Community is operating a concerted action programme of medical research, named EURONUT!. This programme, started in 1988, will run until 1991. The most important actions in this programme are:

(i) a study on the role of nutrition in cardiovascular

disease (in collaboration with WHO). Preliminary results have already been obtained and presented in 1989.

(ii) a study on dietary patterns in the elderly in relation to health and performance.

EEC Strategy on Nutrition, Food Safety and Health

To sum up, according to P. Gray, Head of Division III/B/2 of the Commission of the EC in Brussels (Gray, 1990), the broad lines of Community strategy in the fields of Nutrition, Food Safety and Health are as shown in Table 6.

Table 6. Community Strategies approaching 1993

1. Removal of economic distortions in the internal market which could encourage wrong nutritional choices;
2. Protection of the public against health hazards;
3. Strict public controls, with equivalent application in all Member States (at the moment public control systems are rather patchy in different Member States);
4. Sound scientific bases for safety and nutritional assessment, obtained also by encouraging joint research projects;
5. Increased coherent information on the composition and nutritional value of foods (vide nutritional labelling);
6. Broad lines for nutrition education and information of European consumers.

References

Gray, P. (1990) A food and nutrition strategy for the EC. International Meeting on "Food and Nutrition Policy in Mediterranean Europe", Rome, March 21-23, 1990.

Gray, P. (1988) European Food Law - a Turning Point and a Challenge. *Food Technology International Europe*, 2, 25-28.

Haigh, R. (1978) Harmonization of legislation on foodstuffs, food additives and contaminants in the European Economic Community. *J. Food Technol.*, 13, 255-264.

Haigh, R. (1985) Control of food additives and contaminants: the EEC situation. *In*: Food Toxicology; Real or Imaginary Problems?, G.G. Gibson and R. Walker (Eds.), Taylor & Francis: London & Philadelphia.

Keys, A. (1970) Circulation, 41, Suppl.1.

Malaspina, A. (1987) Regulatory aspects of food additives. *In*: Toxicological aspects of food, K. Miller (Ed.), Elsevier Applied Science Publishers: London & New York.

Toti, L., Gizzarelli, S. & Mazzotti, M. (1990) Present status of Salmonella problem in food hygiene in Italy. Report to WHO Consultation Meeting on Salmonellosis Control in Agriculture, Orvieto, 9-12th April, 1990.

2

WHOLESOMENESS AND SAFETY OF IRRADIATED FOODS

A. John Swallow

Cancer Research Campaign Department of Biophysical
Chemistry, Patterson Institute for Cancer Research
Christie Hospital & Holt Radium Institute
Manchester M20 9BX, ENGLAND

ABSTRACT

Irradiation with gamma-rays, X-rays or fast electrons can be used to change foodstuffs in beneficial ways or to destroy harmful organisms. Gamma rays do not induce radioactivity in foods, but X-rays and fast electrons can induce short lived radioactivity if sufficiently energetic. This imposes limitations on the energies which can be used, and a short wait between irradiation and consumption may be advisable. Irradiation produces chemical changes in foodstuffs, and some foods are unsuitable for irradiation. With appropriate foods, trials with animals and human volunteers generally show that the product is safe. Some loss in nutritional quality can take place, which could be significant for some individuals, but are unlikely to be important for those on a balanced diet. Irradiation does not eliminate all risk from microbial contamination. Foods to be irradiated should be good quality, and need to be kept under proper conditions after irradiation. Irradiated foods should be appropriately labelled. Tests for radiation would help to enforce necessary controls. If the process is properly carried out on appropriate foods, and all due precautions are taken, irradiated foods are wholesome and safe.

INTRODUCTION

Unlike many other processes applied to food, irradiation has become highly controversial. On the hand, it is seen as a benevolent development which can help provide a varied and plentiful diet of food which is both wholesome and safe, to the benefit of supplier and consumer alike. On the hand, it is seen as a dangerous procedure, unwanted by consumers and the food industry, designed to mislead the public into eating rejected food. Extensive evidence for the wholesomeness and safety of irradiated foos has been presented (Joint FAO/IAEA/WHO Expert Committee 1981; Brynjolfsson, 1985; Advisory Committee on Irradiated and Novel Foods, 1986; idem 1987; Wierbicki and others, 1986; Scientific Committee for Food, 1987; Thayer, 1987; Hawthorn, 1989; Diehl, 1990). Strong criticisms have been voiced (Piccioni, 1988; Murray, 1990;Webb and Lang, 1990). This paper reviews the evidence in the light of the criticisms and of new data. Although irradiation is a potentialy valuable process, it is necessary for it to be properly controlled to ensure its safety. Although it could be generally beneficial, it must not be forced on an unwilling public. Care must be taken to prevent the process being misused.

THE IRRADIATION PROCESS

Some of the applications of irradiation rely on the production of changes in the food. The biochemical mechanisms responsible for the sprouting of bulbs and tubers are quite sensitive to irradiation, and relatively small radiation doses (0.05 - 0.15 kGy, where 1 Gy = $1 Jkg^{-1}$) can inhibit the sprouting of potatoes, sweet potatoes, onions, garlic, shallots, and chestnuts. The post-harvest ripening of climacteric fruits like papayas, mangoes, and tomatoes becomes delayed at somewhat higher dose levels (0.2 - 0.5 kGy).

Other applications of the process depend on the effect of radiation on the organisms in foods. Insects in fruit and grain are effectively destroyed at about the same doses as are needed for delay of ripening, or slightly higher. Helminths in meat, and especially *Trichonella spiralis* in pork, are effectively destroyed at higher levels (up to about 2 kGy). The microbial load in fish and poultry can be reduced sufficiently to extend the shelf life of the product and non-spore forming pathogens like *Salmonella* can be effectively eliminated at doses up to about 8 KGy. The effective sterilization (decontamination) of herbs and spices requires 10 KGy or more. Most spores and viruses are quite resistant to radiation and cannot be effectively destroyed except at doses which produce serious damage to foods (20 kGy or more). Toxins in foods are even more resistant to irradiation, although they can readily be inactivated on irradiation in simple aqueous solution, where all of the irradiation attack becomes concentrated on the toxin instead of being distributed among all the components of the food. Effects of radiation on foods is shown in Figure 1.

The radiations to be used to treat food must be sufficiently penetrating to reach every part of the food. They must not produce significant radioactivity. Sources of radiation must be available commercially at prices which make the process economically viable. The only radiations which meet these criteria are gamma rays of nuclear origin produced by machines. Gamma-ray photons and X-ray photons are indistinguishable. Their principal action on matter is to produce fast electrons within it, and these are responsible for the overwhelming majority of the chemical and biological effects produced by gamma- and X-rays. Since in every case it is fast electron which produce the effects, all the types of radiation produce basically similar effects, providing the extent of irradiation as measured by the radiation dose, i. e. the amount of energy absorbed per gram, is the same.

More than 30 countries have now approved the irradiation of at least some foods using gamma-rays, X-rays or fast electrons. Authorizations by the Council of the European Communities (Official Journal of the European Communities, 1988) are listed in Table 1.

RADIOACTIVITY

Natural radioactivity is present in all foods. Two of the most important sources are carbon-14 and potassium-40, the specific activation being 220 Bq kg^{-1} and 3000 Bq kg^{-1} of element, respectively. Other sources of radioactivity are tritium and some trace elements. The level of natural radioactivity can readily be calculated from their elemental composition to be in the region of 100 Bq kg^{-1}.

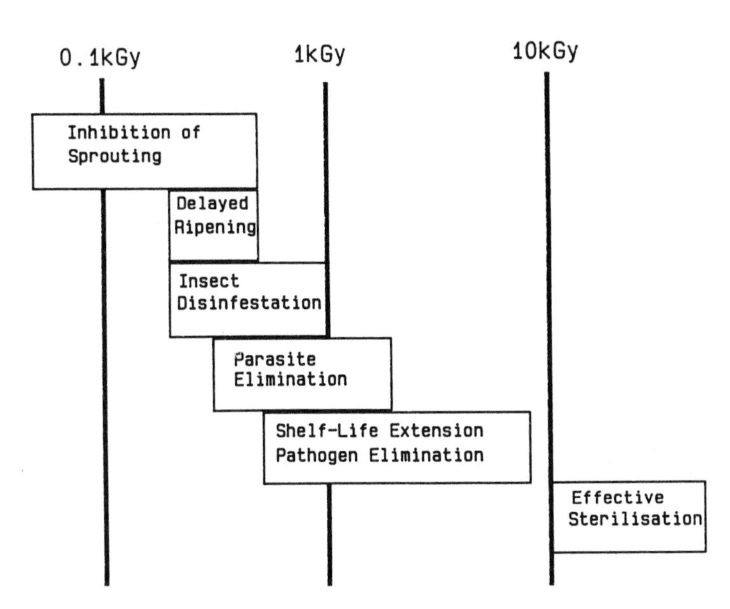

FIGURE 1. Effects of radiation on food at different dose levels. Note that a dose of 10 kGy would only raise the temperature of the food by about 2.5 degrees.

TABLE 1. Authorization of Radiation Treatment by the Council of the European Communities.

Foodstuff	Maximum overall average dose (kGy)
Strawberries, papayas, mangoes	2.0
Dried fruits	1.0
Pulses (legumes)	1.0
Dehydrated vegetables	10.0
Cereal flakes	1.0
Bulbs and tubers	0.2
Aromatic herbs, spcies, and vegetable seasonings	10.0
Shrimps and prawns	3.0
Poultry meat	7.0
Frog legs	5.0
Arabic gum	10.0

There are several processes by which high energy photons could produce additional radioactivity in foods:

1. Photoneutron Production

Irrespective of whether the process is caused by gamma- or X-rays, this process is abbrevirated as γ ,n. The neutrons may go on to produce further radioactivity. Photoneutron production does not happen at photon energies below 1.5 MeV, so that irradiation with the gamma-rays from cobalt-60 or caesium-137 does not give rise to photoneutrons. Irradiation with photons with energy 2.2 MeV can give rise to the raction D(γ,n)H from the deuterium atoms present in foods. Further photoneutron (γ,n) reactions become possible when photon energies are higher than 2.2 MeV.

As well as the photoneutron reactions taking place in the food itself, it is necessary to recognize that neutrons can be produced through materials which might be used in the construction of the irradiation plant. Lead-207 (22.6% abundance) has a photoneutron threshold of 6.7 MeV and should be avoided as a shielding material when using sources which produce photons of 6.7 MeV or greater (Wakeford et al., in preparation).

The neutrons from photoneutron reactions, whether generated in the food itself or in constructional materials, may give rise to radioactivity by neutron capture reactions (n,γ). Several common isotopes (23N, 37Cl) show strong radiative capture resonances with epithermal neutrons. Once thermalised, still further neutron capture reactions become possible.

2. Isomer Activation

In this process, a photon is absorbed by a nucleus and a photon of lower energy emitted, leaving the nucleus in a metastable state which reverts to the stable state by emitting one or more further photons. The half-lives of the metastable states are usually short, often less than a microsecond. This process is generally regarded as insignificant in foods, although attention has been drawn to the metastable states 107Ag and 109Ag half-lives 44 and 40 s respectively, which could be produced at a level up to 1 Bq kg^{-1} (Tuchscheerer, 1966). Isomer activation can be induced by the gamma-rays from caesium-137 or cobalt-60, but the activities produced are generally so trivial and the half-lives so short that, effectively, gamma-rays (as opposed to X-rays or high energy) do not induce radioactivity.

3. Other Processes

Two other processes are production of protons (γ,p) and tritons (γ,t) but these are even less important than isomer activation.

— Fast electrons can produce radioactivity via the production of X-rays (bremsstrahlung) followed by reactions of the photons. The efficiency of bremsstrahlung production rises with increasing electron energy and target atomic number. The low atomic number elements of which foodstuffs are comprised are inefficient at producing bremsstrahlung, but bremsstrahlung are produced more efficiently when fast electrons impinge on constructional materials, shielding, etc.

Numerous estimates have been made of the radioactivity induced in foodstuffs by irradiation with X-rays and gamma-rays and fast electrons. Some of the recent figures for a dose of 10kGy (National Radiological Protection Board, 1989) are shown in Table 2. Examination of the figures shows that in most cases the radioactivity induced by 10MeV X-rays is

TABLE 2. Induced Radioactivity (Bq kg^{-1}) after Irradiation with 10 MeV X-Rays.

	5 min	3h	12h	1d
Fish	190	44	24	12
Meat (pork/lamb/beef)	130	40	22	11
Poultry	120	47	32	22
Potatoes	180	53	39	16
Cereals	130	33	19	10
Mangoes	25	17	10	10

TABLE 3. Induced Radioactivity (Bq kg^{-1}) after Irradiation with 5 MeV X-Rays and 10 MeV Electrons.

	5 MeV X-rays		10 MeV electrons	
	5 min	1 d	5 min	1 d
Fish	1.9	0.13	5.7	0.38
Meat (pork/lamb/beef)	18	0.51	3.8	0.34
Poultry	1.2	0.22	3.6	0.66
Potatoes	2.7	0.15	5.5	0.48
Cereals	1.4	0.10	3.9	0.30
Mangoes	0.25	0.10	0.77	0.31

comparable to the natural radioactivity already present. By 24 hours the radioactivity had decayed to about a tenth of the natural level.

Measurements have recently been made of the radioactivity produced by irradiation of rice and a macerated meat product with the X-rays produced from 10 MeV electrons. After and X-ray dose of 15 kGy, radioactivity could easily be detected in manganese which was present as monitor for neutron production (55Mn (n,γ)56Mn, t 1/2 = 2.58 h). The radioactivity was orders of magnitude greater when the foods were artificially enriched with deuterium before irradiation so as to show up the consquences of the D(γ,n)H reaction. The experiments made it possible to calculate the maximum activity to be expected in ordinary foodstuffs. The activities determined in this way were similar to those estimated by calculations such as those whose results are given in Table 2. For example, beef was calculated to have an activity of at most 90 Bq kg^{-1} immediately after irradiation, falling to 10.7 Bq kg^{-1} after 24 hours (Wakeford et al., in preparation).

When used to irradiate foods directly, fast electrons produce about 30 times less radioactivity than when X-ray photons of the same energy are used to give the same dose.

Although the radioactivity in foods irradiated with 10 MeV X-rays is not large, it has been recommended on the basis of previous experiments and calculations (Meyer, 1966; Koch and Eisenhower, 1967) that the energy of photons to be used to treat foods should be restricted to 5 MeV rather than 10 MeV, but that the energy of electrons should be allowed to be up to 10 MeV (Joint FAO/IAEA/WHO Expert Committee, 1981). More recently, so as to allow short-lived activities to decay, it has been recommended as an additional precaution that irradiated food should not be consumed less than 24 hours after irradiation (Advisory Committee on Irradiated and Novel Foods, 1987; House of Lords Select Committee on the European Communities, 1989). Typical radioactivities to be expected under these conditions are included in Table 3. It should be pointed out that a 24 hour delay would not be necessary for foods irradiated with gamma-rays.

RADIOLYTIC PRODUCTS

The first step in the action of fast electrons on food (whether the electrons originate from gamma- or X-rays or are introduced directly) is to cause almost random excitation and, of much greater importance, ionizations along the electron track. Since water is a major component of most foods, the most abundant 'primary' action of radiation at the chemical level is to ionize water. Ionization of water results mainly in the formation of free hydroxyl radicals and hydrated electons:

$$H_2O \longrightarrow H_2O^+ + e^-$$

$$H_2O^+ + H_2O \longrightarrow H_3O^+ + {}^{\cdot}OH$$

$$e^- + {}_{aq} \longrightarrow e^-_{aq}$$

Smaller amounts of atomic and molecular hydrogen ($H^{\cdot}$ and H_2) and molecular hydrogen peroxide (H_2O_2) are also produced. The radiochemical yield of free radicals from water is about 0.6 umol radicals per J of radiation energy absorbed. As well as acting on water, fast electrons also produce radical from the organic compounds of foods.

The free radicals produced from the water have the opportunity of reacting with any of the organic components of the food. They will react rather indiscriminately, yielding radicals resembling (although not always identical with) those produced by direct action. From the known rate constants of the reactions of $^{\cdot}OH$, e^-_{aq} and $H^{\cdot}$, the half-life of the water radicals would be about 10^{-9} s. A great deal is known about the organic radicals and their reactions. Reactions in proteins (Friedman, 1973) are among those recently discussed (Swallow, in press). A typical half-life of organic radicals would be about 1 ms. After further reactions, the organic radicals give rise to radiolytic products (Swallow, 1977; idem, 1984). Since all the components in food are susceptbile to attack, the number of products must be very large and the quantity of each product must be very small. If the molecular weights of the products is typically 100, and two radicals are needed to give one molecule, the total amount of radiolytic products in foodstuffs given a dose of 10kGy can be calculated from the figures already given to be about 300 mg per kg. Radiolytic products are in general identical to compounds normally present in foods, and are present in similar amount to many of them (e. g. Merritt, 1972).

In frozen or dried foods and in hard parts of foods such as the pips of fruit or the bones of fish or meat, free radicals are physically unable to move and can be stable for periods greatly in excess of the shelf life of the food. Free radicals can readily be detected by electron spin resonance. A typical e.s.r. spectrum of the radicals in the scales of a fish is shown in Figure 2. The e. s. r. spectrum of radicals is the basis of a test for detecting whether foods have been irradiated (see below). Free radicals in frozen or dried foods react on thawing or adding water. The hard parts of foods are not generally eaten, but where they are (e. g. with strawberries or fish like whitebait) the radical would react on digestion yielding chemical products. In the case of fish bones and scales, the radical is the carboxylate anion radical, $CO_2^{\cdot -}$ (Dodd et al., 1988). If two such radicals were to react together they would yield oxalic acid. The yield of stable $CO_2^{\cdot -}$ in fish scales is about 0.1 nmol J^{-1}. If all of the radicals were to yield oxalic acid there would be 10^{-8} g oxalic acid per g of fish scales for a dose of 1.5 KGy. This quantity may be compared with the oxalic acid content of fruits, vegetables, and other foods. For example, a cup of tea contains about 20 mg oxalic acid, i.e. 2 million times as much as could be obtained by consuming the scales in a portion of irradiated fish.

In polyunsaturated fats the radiolytic free radicals can initiate lipid peroxidation, the reaction responsible for the normal development of rancidity. The reaction is a chain reaction so that oxidation is expected in relatively large yield, especially when the concentration of vitamin E or other antioxidants is low. The oxidation products are thought to be carcinogenic or mutagenic. Lipid oxidation products have also been implicated in the earlier phases of coronary heart disease (Addis and Park, 1989). Rancidity should be avoided, however caused. Irradiation and storage under anaerobic conditions will reduce the development of rancidity, but irradiation of fatty foods is often inadvisable on grounds of the flavour of the product, apart from the question of toxicity.

Strenuous efforts are now being made to discover chemicals which are present in irradiated foods, but not in unirradiated foods, and hence could be used to test whether foods have been irradiated or not (see also below). An early attempt was to seek *o*-tyrosine, which is known to be formed by attack of hydroxyl radicals on phenylalanine, and differs from the natural amino acid tyrosine. This compound has indeed been found in irradiated foods but has now also been found in unirradiated chicken, prawns, and strawberries (Hart et al., 1988). The difficulty of detecting characteistic differences between irradiated and unirradiated foods is part of the evidence that irradiated foods are as safe as unirradiated foods.

With sensitive modern methods of analysis it is hoped that unique irradiation products will indeed be found, but hitherto unsuspected products are likely to be found in unirradiated foods too. Gas chromatograms (flame ionization detector) of hexane extracts of irradiated (10 kGy) and unirradiated cinnamon illustrate this point (Figure 3). It can be seen that several of the compounents appear to be present at higher concentration in the irradiated sample than in the unirradiated. However, this could be connected with more efficient extraction. The chromatograms provide no proof of the formation of products which are unique to radiation.

Many of the components of foods are known to be toxic at sufficiently high concentrations, but are accepted as harmless when consumed as part of a normal diet.There is no reason based on the known radiation chemistry of food components (Elias and Cohen, 1977; idem 1983) to suppose that radiolytic products would be any more toxic at expected concentrations than are normally present in foods. However, the toxicity of compounds present in foods needs to be kept under review, whether or not the foods are treated, and if treated, whether treated by irradiation or by any other prcess.

Attention has recently been drawn to the possibility of toxicity arising from radiolytic effects on minor components of foods like permitted additives, contaminants, and pesticide residues. It is of course well established that antioxidants become consumed on irradiation, with obvious consequences. Effects on pesticides are likely to be small, but if anything would render them less harmful. It is difficult to see that there could be any unpredictable effect arising from the effect of radiation on additives, contaminants or pesticide residues, so it is not easy to see how further research on this topic could be justified, except as a check on expectations.

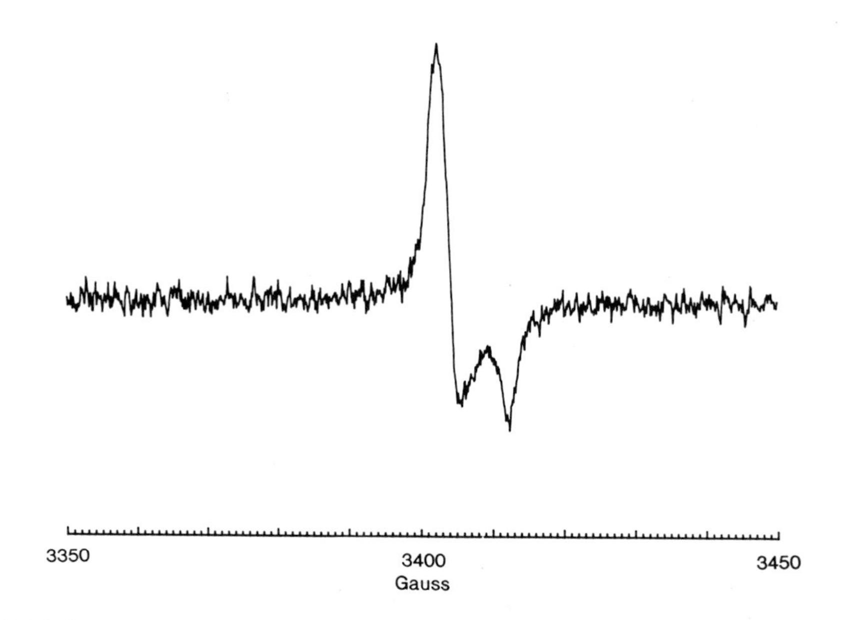

FIGURE 2. Electron spin resonance spectrum of irradiated fish scales. Irradiated bones and other calcified tissues also give this spectrum. The spectrum is not given by unirradiated samples

FEEDING TRIALS

Food additives are normally tested for toxicity by feeding to animals so as to establish levels which are without effect. A safety factor of 100 is then generally applied to calculate an accepted intake for humans. It is not possible to feed animals with irradiated foods at 100 times the normal level without causing nutritional imbalance. Neither can the radiation dose to foods be increased by a factor of 100 as this would render the diet unpalatable. Studies have, therefore, been conducted by feeding at nutritionally acceptable levels, using foods irradiated to doses comparable to, or higher than, those which would be used in practice. In some

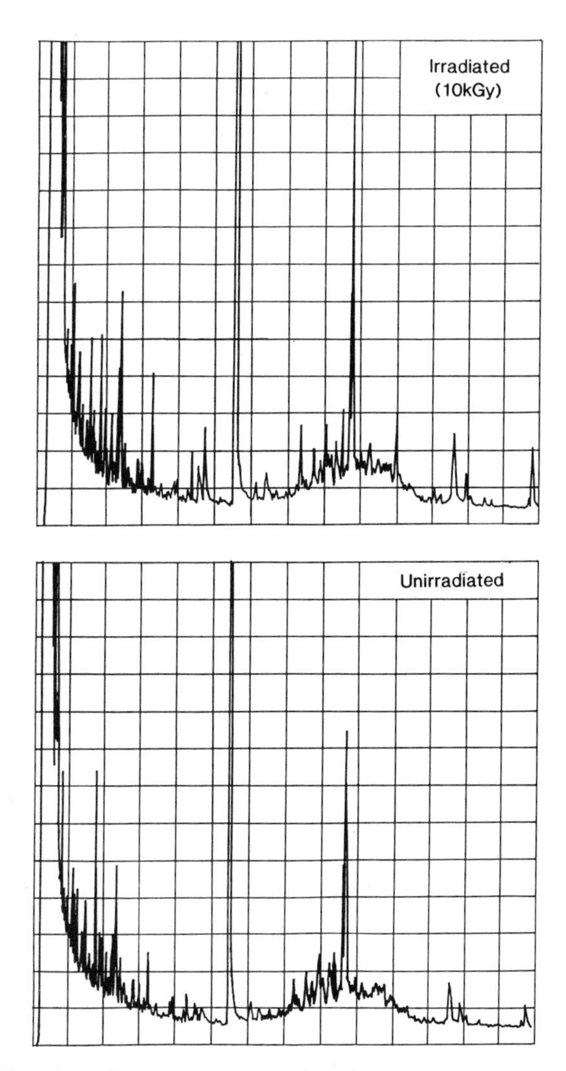

FIGURE 3. Gas Chromatograms of Solvent-Extracted Cinnamon.

experiments vitamin supplements have been incorporated in the diet so that any toxic effect would not be masked by effects caused by radiation-induced destruction of vitamins. Large numbers of feeding studies have now been carried out, backed up by other tests, on both animals and human volunteers. Among the most comprehensive is an eight-year nutritional, genetic, and toxicological study of irradiated chicken initiated by the U. S. Army (Thayer et al., 1987). Careful consideration of data on irradiated fruits, vegetables, cereals, pulses, spices, and condiments, miscellaneous foods, fish and fishery products, shell fish and meats (Scientific Committee for Food, 1987) indicates no adverse health effects from the consumption of irradiated food. It has even been concluded that there is no need for more animal feeding studies or toxicological testing, at least for foods given doses up to 10 kGy (Joint FAO/IAEA/WHO Expert Committee, 1981; Scientific Committee for Food, 1987).

Toxicological studies have also been reported on irradiated animal feeds, and on the basis of these studies, irradiated feeds are now used in normal laboratory practice. The process is in routine use in preference to autoclaving for pasteurization or sterilization of the whole diet of laboratory animals where it is necessary to maintain colonies with reduced bacterial load, although irradiation of high-fat diets is not recommended owing to the development of oxidative rancidity. During the last seven years in the Paterson Institute for example, about 2000 athymic nu/nu mice have routinely been fed diets irradiated to doses of 50 kGy. Over about 40 generations, the animals bred normally, showing no sign of genetic, teratogenic or any other abnormality. Morphological, bacteriological, parasitological, and serological screening has also shown the animals to be normal in every respect. Other laboratories have also found irradiated animal diets to be satisfactory. The experience of using radiation to treat animal diets is especially valuable as it would have been expected to reveal any long-term effects that might have been present. The experience with animals provides powerful grounds for believing that consumption of irradiated foods would be safe for human consumption, both in the short term and in the long term.

Research on irradiated wheat commissioned by the Indian Government is less clear-cut (Vijayalaxmi and Srikantia, 1989). In this work, irradiated wheat was used in feeding studies with several species of animals and with a small number of malnourished children. No difference was found between unirradiated wheat and irradiated wheat (0.75 kGy) which had been stored at 4° for 3 months before feeding. However, in experiments in which wheat was used for feeding within 20 days of irradiation, increased numbers of polyploid cells were reported to be present in the bone marrow of rats and mice and in peripheral blood lymphocytes of monkeys and children. Levels of polyploid cells returned almost to zero after ceasing to use freshly irradiated wheat in the preparation of the diet. Freshly irradiated wheat also appeared to cause a number of other abnormalities, whereas stored irradiated wheat did not. These experiments have been strongly criticised but have been confirmed by some independent experiments although not by others.

It is not easy to see why freshly irradiated wheat should differ from unirradiated wheat or from wheat which had been irradiated and stored. Moreover the significance of polyploidy for health is not clear. Nevertheless, further well-conducted work would seem to be required if the question of the safety of freshly irradiated wheat is to be resolved. In practice, irradiated wheat is likely to be stored before being used for food, so there should be no risk in consuming products made from irradiated wheat.

Despite the importance of wheat in the Indian diet, and frequent shotages of the commodity, the Indian Government does not yet appear to have approved the irradiation of wheat, although it has authorized the irradiation of potatoes and onions, and (for export only) of spices, frozen shrimps, and frog legs. More than ten other countries have also approved the irradiation of wheat or other grain, including the Soviet Union, in which country grain has been cleared for irradiation since 1959. In fact, more grain has now been irradiated than any other foodstuff, seemingly without damage to health.

Feeding trials of irradiated foods have been carried out in China between 1982 and 1985 with about 500 medical students and other human volunteers (Anon, 1987; Brynjolfsson, 1988; Yin Dai, 1989). In total, about forty kinds of food (including grain) irradiated with doses up to two thirds of the total diet, were fed to the volunteers over periods up to several months. Cytogenetic and numerous other tests were performed. In one of the studies in which food had been stored for at least 7 days and as much as 180 days or more, both those fed with irradiated and un-irradiated foods showed surprising increases in the content of polyploid cells in peripheral blood lymphocytes (Anon, 1987), although there was no significant difference between the two groups. No significant effect of irradiation was seen in any experiment. The experiments provide further indications of the safety of consuming irradiated foods. Of course, trials with human volunteers cannot easily prove the absence of long-term effects. If food irradiation comes to be widely practiced, it would be valuable to monitor the extent and pattern of the consumption of irradiated food as compared with untreated foods or foods treated in other ways to help to discover whether any of the treatments have produced favourable or unfavourable effects on human health.

NUTRITIONAL CONSIDERATIONS

For foodstuffs which make significant contribution to nutrition (but not for commodities such as spices) it is important to discuss to what extent the nutritional value of the food is affected by irradiation. All of the organic components of foodstuffs are susceptible to attack by hydroxyl radicals and/or hydrated electrons. Consequently, if isolated micro- or macro-nutrients are irradiated in dilute aqueous solution, the nutrients must become totally destroyed if the dose is sufficient. Similar considerations apply to nonaqueous solutions. The magnitude of the dose needed to destroy a high proportion of the nutrient will decrease as the concentration decreases. If other substances are present as well as the nutrient, some of the radicals would attack the other substances so that the nutrient would be protected from the radiation attack. On these grounds, nutrients in foods might be expected to be rather resistant to radiation. If all of the components in a foodstuff were to be equally susceptible to attack, then for a foodstuff containing 100 g organic matter per kg food, a dose of 10 kGy would destroy about 300×10^{-3} % = 0.3% of the food components. The proportion of nutrients destroyed would be the same, although it would be greater under conditions where chain oxidations are possible.

Although simplified radiochemical considerations alone might be thought to predict that irradiation would produce little effect on the nutritive value of foods, biochemical and other mechanisms can come into play in the real situation, so the effect of radiation on foods under conditions relevant to consumption needs to be examined. More work has now been done on the effects of irradiation than on the effect of other methods of food processing and preservation which are accepted for general use (e.g. Kraybill, 1982; Josephson et al., 1978).

The content of some vitamins in foods is found to be noticeably sensitive to radiation. Among the more recent studies, an examination has been made in Japan of the vitamin C content of potatoes. Figure 4 shows the change in total vitamin C content of raw potatoes of one variety as a function of time of storage at 5°C. It can be seen that radiation produced little immediate effect in this case, but that the decrease in vitamin C which took place over the first week or so after storage was amplified by irradiation. The unirradiated potatoes continued to lose vitam. C after the first few weeks of storage and after about 3 months the content of both irradiated and unirradiated poatoes became half the original value. In these experiments several different methods of cooking gave rise to further losses of vitamin C which were little dependent on whether the potatoes were irradiated or not. The greatest loss was when the potatoes were boiled. Losses of vitamin C have also been found in other experiments with different varieties of potato. Clearly the vitamin C content of potatoes can depend upon whether they have been irradiated or not, as well as on the duration and conditions of storage, the method of cooking, and the variety of potato.

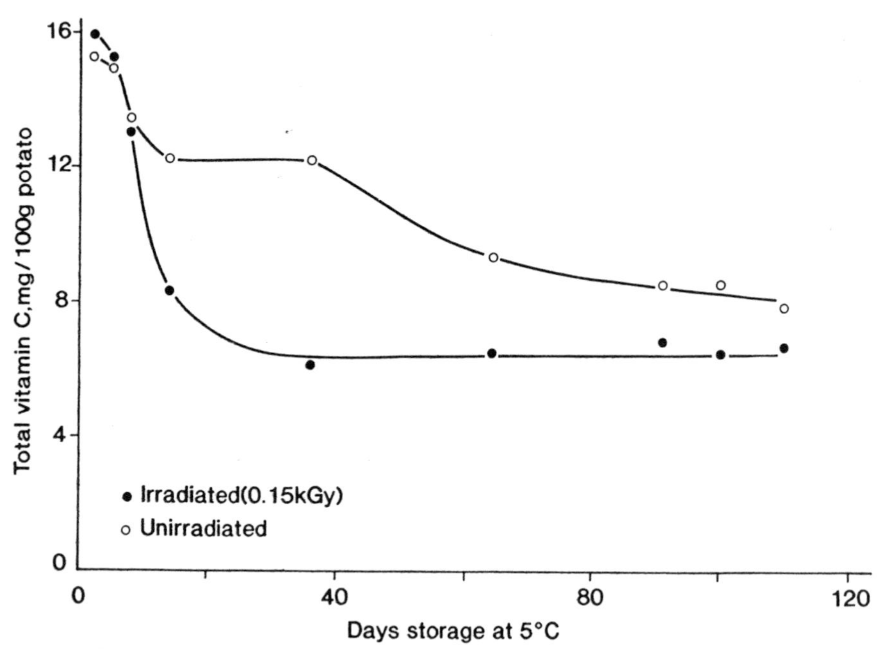

FIGURE 4. Effect of storage on vitamin C content of potatoes (based on data of Esashi, 1989).

Among other recent studies, it has been reported that the thiamine content of pork chops, as measured after cooking, decreases on irradiation, the decrease being more pronounced the higher the temperature of irradiation within the range -20 to 20° (Fox et al., 1989). For doses of 0.3 and 1.0 kGy (i. e. within the range suitable for control of Trichinella spiralis) the thiamine content decreased by 5.6 and 17.6%, respectively, when the temperature of irradiation was 0°. Pork in processed meat and as chops and roasts contributes about 15% of the thiamine in some diets, so if all the pork consumed were to be irradiated under the conditions described, and the contribution of pork to the diet remained unaltered, irradiation would cause a net reduction of between 1 and 3% in thiamine intake.

The vitamin E and polyunsaturated fatty acid contents of foods are also sensitive to irradiation, especially when oxygen is present during or after irradiation. It is generally accepted howeever that the nutritional quality of carbohydrate and protein constituents is little affected. Irradiation does not affect minerals. The intake of any nutrient by an individual depends on the extent of consumption of foods containing it, as well as on the nutritional value of each food item. The commercial introduction of irradiation may contribute to changing patterns of consumption, for example by leading to an increased availability of tropical fruit in temperate countries. The nutrient content of the diet needs to be kept under review, especially for individual with particular needs such as very young children or cigarette smokers. This is necessary in any case, whether irradiated foods are consumed or not.

MICROBIOLOGICAL CONSIDERATIONS

The irradiation of foods does not usually lead to the effective destruction of all the micro-organisms present. It is therefore necessary to consider whether any hazard arises from organisms which survive the irradiation treatment. With some foods, it is possible that spoilage organisms might be destroyed, leaving behind pathogenic organisms such as *Clostridium botulinum*. Even doses as high as 10 kGy, which would effectively destroy vegetative organisms, could leave spores behind. These might then revert to the vegetative form and grow to produce toxin without the food showing any sign of fermentation or putrefaction. This possibility is not unique to irradiation, but also needs to be considered when foods are subjected to other non-sterilising treatment such as salting, smoking, sulphiting, pasteurisation or vacuum packing. The irradiated foods must be subjected to the same controls as should be routine with other processes. In a few cases it might be possible to limit the hazard by controlling the dose so as to leave behind sufficient spoilage organisms to give detectable spoilage before pathogenicity could develop. In most cases it would be necessary at all times to keep the food under conditions which would prevent the growth of pathogens. *Listeria monocytogenes* can continue to grow slowly at temperatures which are commonly used to refrigerate foods so that storage at low temperatures could be necessary in cases such as poultry products irradiated to eliminate *Salmonella*, although some recent work indicates that *L. monocytogenes* would be effectively destroyed at doses which would be used for such a purpose (Patterson, 1989).

Another point that needs to be considered is whether irradiation could cause mutations in surviving micro-organisms leading to increased pathogenicity. Mutations can also be caused by sunlight, drying, heating, and certain chemicals. However, mutations normally lead to impairment of function and *loss* of pathogenicity. Mutations caused by irradiation are as unlikely to be dangerous as are those caused by other treatments.

Special attention has been drawn to the effect of irradiation on the production of mycotoxins, and especially aflatoxin, by *Aspergillus flavus* and other fungal moulds (Julius, undated). The production of aflatoxins is a serioud matter, because they are exceptionally carcinogenic. Levels need to be kept below about 10 ug per kg (10 ppb). Irradiated moulds are not particularly likely to mutate to forms with enhanced ability to produce aflatoxin. Moreover, irradiation will tend to decrease toxin production by moulds. Nevertheless, there is the possibility that irradiation might alter the food, e.g. by increasing the level of free fatty acids, so that surviving mould could produce toxin more effectively. Moreover, irradiation induced changes in the microflora of the food could lead to enhanced growth of the mould as a result of reduced competition from other microorganisms, Also, the increased storage life of the food could allow more time for the toxin to be produced. However, irradiation is not fundamentally different from any other treatment with regard to these matters. Where there is a risk of toxin production, foods need to be kept under cool, dry conditions. What is important is to ensure that aflatoxin levels are low, whether foods are irradiated, treated in other ways, or not treated at all.

Unlike many other treatment, irradiation does not provide any safeguard against re-infection. At-risk foods need to be packaged or otherwise protected to guard against contamination after irradiation. It is one of the advantages of irradiation treatments that foods can be treated while they are inside a protective package which will prevent infection after the irradiation has been completed.

OTHER ASPECTS OF SAFETY

When potatoes are exposed to light they produce glycoalkaloids, which at high concentrations are toxic and at lower concentrations can still rise to a bitter taste. Light also causes potatoes to become green owing to the formation of chlorophyll. The two processes are generally regarded as associated, so that a green colour should provide a warning of unpleasant flavour and possible toxicity. However, in experiments with Russett Burbank potatoes, glycoalkaloid production has been found to be unaffected by irradiation, whereas greening was reduced (Patil et l., 1971). In other experiments with potatoes, irradiation has been reported to decrease both the production of glycoalkaloid and of chlorophyll(for reference see Matsuyama and Umeda, 1983 and Thomas, 1983). In Hungarian experiments with two varieties of potato, the light induced production of the glycoalkaloid solanine was somewhat decreased on irradiation, although the dependence on irradiation was less important than the dependence on variety (Figure 5). In these experiments, the irradiated potatoes were reported to become more green than the unirradiated potatoes.

Potatoes containing excessive amounts of glycoalkaloid should not be eaten. Whether potatoes are irradiated or not, it is unwise to rely on the absence of a green colour as proof of a low glycoalkaloid level, although a green colour can provide a rough indication of excessive glycoalkaloid in many circumstances (cf. Friedman, 1991).

The materials to be used to package irradiated foods need to stand to irradiation without transmitting toxic or unpleasant flour to the food. They must also satisfy the requirements of any other packaging material, such as low permeability to moisture, oxygen or carbon dioxide. Even more is known about the effects of radiation on polymeric and other materials suitable for packaging than about the effect of radiation on foods. There is no fundamental difficulty in developing appropriate materials, and numerous materials have been approved for use, for example, by the FDA.

As well as the question of the safety of irradiated foods, there is also the question of the safety of the irradiation plant. The health and safety of employees and the general public will need to be ensured by adherence to the controls which apply to all premises containing sources of radiation. There is considerable experience with such premises, which include hospitals, industrial establishments employing radiation for sterilization of medical supplies, and factories irradiating plastic materials such as insulated wire and cable. All industrial processes need to be operated safely. Irradiation is no exception.

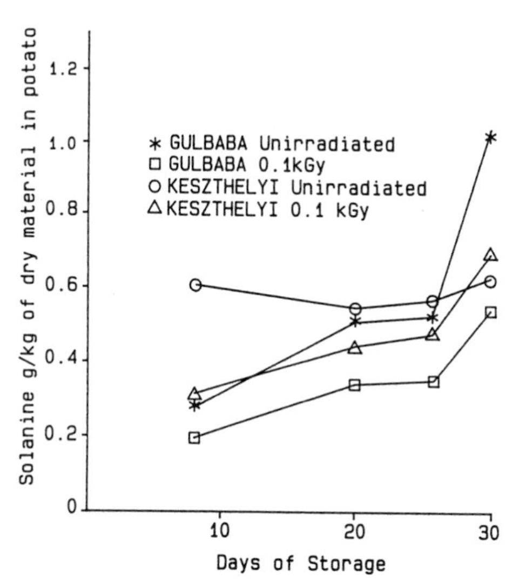

FIGURE 5. Production of the glycoalkaloid solanine on storage under illumination (based on data from Lindner and Szotyori, 1971).

IDENTIFICATION OF IRRADIATED FOODS

Apart from needing information about particulars such as recommended storage conditions, or vitamin content, many consumers will, for a variety of reasons, wish to know whether foods have been irradiated or not. This could be achieved if processors were to include with foods to be treated a simple indicator which would change colour on irradiation. Whether this is done or not, irradiated foods should be labelled in words, perhaps supplementing written information with symbol shown in Figure 6. It is also desirable to have tests which can be applied to the foods themselves so as to provide a check on any information provided, and, perhaps more important, to be able to show that unmarked foods have not been irradiated. It will not be easy to detect irradiation because if the process is carried out correctly, on appropriate foods, the appearance, flavour, and texture will not differ significantly from those of unirradiated foods and radiolytic effects such as the formation of radioactivity or chemical products, or perhaps in nutritional content or microbiological flora, will generally be masked by normal variability. To help in detecting the irradiation of

FIGURE 6. Widely recognized symbol for irradiated food, often green in colour.

TABLE 4. Some Methods of Detecting the Irradiation of Foods.

Method	Food	Status
Electron spin resonance	foods containing bone or calcified tissue, perhaps some fruit	successful method
Luminescence	spices, perhaps other foods containing mineral matter	successful method
Analysis of volatiles	foods containing fat	advanced development
Microbiological methods	uncooked foods	useful indication of irradiation
Conductance/impedance	potatoes	useful indication of irradiation
Changes in DNA	most foods	under development
Changes in protein	foods containing protein	under development
Viscosity	spices, dried vegetables	under development

foods, considerable effort has been devoted in recent years to searching for properties which are sufficiently characteristic of irradiated foods to enable them to be distinguished from unirradiated foods or from foods which have been treated in other ways (Delincee and Ehlermann, 1989; Bogl, 1990).

Some possibilities are listed in Table 4. The electron spin resonance method relies on detecting radiation-produced free radicals trapped in hard parts of foods (Dodd et al., 1985). The radicals in bone or other calcified tissues cannot be produced except by irradiation. They are stable for periods many times greater than any possible storage time. They are also little affected by cooking. The method is currently being tested by blind trials and should soon be suitable for routine use for testing poultry and related products. When correctly done with apropriate foods the method is claimed to provide 100% proof, either of irradiation at likely dose levels, or of the absence of such irradiation. With fruits such as strawberries, careful examination of electron spin resonance signals in the seeds may provide and indication of irradiation, but the presence of natural radical makes the results less easy to interpret in these cases.

The luminescence method relies on emission of the light associated with the release of radiation-produced centres of electron excess or deficiency, proably located in small particles of mineral matter associated with the food (Sanderson et al., 1989). The release can be achieved by heating the samples to temperatures of about 200-300° or in other ways. Most work has been with spices and herbs, etc., with many of which the method has been found to provide a good indication of irradiation or its absence, especially when the samples are examined within months after any irradiation has taken place.

Irradiation is unique among food treatments in that every molecule in the food has the possibility of being ionised to yield a radical cation in the first instance. From the point of view of detection, the randomness is unfortunate, because no single product can be expected in large yield. Worse, subsequent reactions generally yield ordinary compounds like those already present, as noted above. In the case of lipids, however, the radiolytic process might be expected to proceed through unique cleavages near the ester group, leading to hydrocarbon and other products, including alkylcyclobutanones, whose nature should depend on the fatty acid composition of the unirradiated food. Attempts are now being made to substantiate tests based on the analysis of volatiles arising in such ways (Nawar and Balboni, 1970).

Several ingenious microbiological methods have been proposed as tests for irradiation. In one of them, one technique, an aerobic plate count, is used to determine viable organisms in a food, while another technique, a direct epifluorescent filter technique,is used to determine the sum of the organisms which were viable before any irradiation and those still viable afterwards (Betts et al., 1988). Of course micro-organisms can be rendered non-viable by heat as well as by radiation, but if a sample contains a high proportion of non-viable organisms, and has obviously not been heated, the inference would be that it had been irradiated. A bonus of the method would be that it could also indicate the microbiological quality of a food before irradiation.

A simple method based on measuring electrical conductance or impedance shows promise as indication of the irradiation of potatoes. The potato is punctured with electrodes, and the AC impedance is measured. One version of the method measures the ratio of impedance at one frequency (e.g. 50 kHz) to that at another (5 kHz). The ratio is significantly lower for irradiated potatoes than for unirradiated potatoes (Hayashi, 1988). The method is not claimed to provide 100% proof of irradiation, but can provide a strong indication. Moreover, it is quick and easy to employ.

Amongst other possible methods, one possibility relies on detecting radiation-induced changes in the nucleic acid component of food. DNA is not an important constituent from the point of view of safety or wholesomeness, but is present in small amount in all except some processed foods. It can be extracted from the food, and examination by chronmatographic or other methods for radiolytic changes such as the formation of altered base unit. Another possibility is to detect changes in the protein constituent, for example the formation of abnormal amino acid units. Reference to o-tyrosine has been made above. Yet another possibility relies on changes in the molecular weight of macromolecules produced by radiation, whether directly or indirectly. This method could have a part to play in identifying the iradiation of foods such as some spices, where radiation-induced changes in the molecular weight of large molecules is irrelevant to the quality of the product. Viscosity of homogenates provides and adequate indication of molecular weight, and is easily measurable.

So far none of the proposed methods for detecting the irradiation of foods carry any implication for food safety.

CONCLUSIONS

Irradiation of foods with the wrong type of radiation, or with excessively energetic radiation of the right type, could render them unacceptably radioactive. Irradiation of inappropriate foods to excessive doses or under the wrong conditions, would produce unpleasant products which could even be dangerous if they were not so unpalatable that they would not be eaten. Even properly-irradiated foods cannot be relied upon to provide and adequate intake of nutrients if not consumed as part of a well-composed diet. Irradiation does not eliminate every risk from microbial contamination, and cannot be regarded as a substitute for good hygiene.

That a process can be used wrongly or has limitations does not mean that it does not have a useful part to play. For example, refrigeration is useful in extending storage times, even though it would be foolish to rely on it to keep foods indefinitely. Freezing damages foods such as lettuce, tomatoes, and eggs. Spices can be used to disguise off-flavours of bad meat, but sensibly employed can be used in the preparation of many safe and attractive dishes. Cooking can cause undesirable effects like, in some foods, the production of heterocyclic amines, and in others, serious destruction of vitamins, although cooking is generally both desirable and necessary.

The limitations of radiation are no more serious than those of other processes. Irradiated foods, if properly treated, are both wholesome and safe.

REFERENCES

Addis, P. B. and Park, S.-W. (1989). Role of lipid oxidation products in atherosclerosis. In "Food Toxicology: a Perspective on the Relative Risks", S. L. Taylor and R. A. Scanlan, eds., Marcel Dekker, New York, 297-300.

Advisory Committee on Irradiated and Novel Foods (1986). Report on the safety and wholesomeness of irradiated foods. HMSO, London, England.

Advisory Committee on Irradiated and Novel Foods (1987). The ACINF response to comments received on the "report on the safety and wholsomeness of irradiated foods". DHSS, London, England.

Anon (1987). Safety evaluation of 35 kinds of irradiated human foods. Chinese Medical J., 100, 715-718.

Betts, R. P., Farr, L., Bankes, P. and Stringer, M.F. (1988). The detection of irradiated foods using the direct epifluorescent filter technique. J. Applied Bacteriol., 64, 329-335.

Bogl, K. W. (1990). Methods for identification of irradiated food. Radiat. Phys. Chem., 35, 301-310.

Brynjolfsson, A. (1985). Wholesomeness of irradiated foods: a review. J. Fd. Safety, 7, 107-126.

Brynjolfsson, A. (1988). Results of feeding trials of irradiated diets in human volunteers: summary of the Chinese studies. In "Practical Application of Food Irradiation in Asia and the Pacific", IAEA, Vienna, Austria.

Delincee, H. and Ehlermann, D. A. E. (1989). Recent advances in the identification of irradiated food. Radiat. Phys. Chem., 34, 877-890.

Diehl, J. F. (1990). "Safety of Irradiated Foods". Marcel Dekker, New York.

Dodd, N. J. F., Lea, J. S. and Swallow, A. J. (1988). ESR detection of irradiated food. Nature, 334, 387.

Dodd, N. J. F., Swallow, A.J. and Ley, F. J. (1985). Use of ESR to identify irradiated food. Radiat. Phys. Chem., 26, 451-453.

Elias, P. S. and Cohen, A. J. (1977). "Radiation Chemistry of Major Food Components". Elsevier Scientific, Amsterdam, Holland.

Elias, P. S. and Cohen, A. J. (1983). "Recent Advances in Food Irradiation". Elsevier Biomedical, Amsterdam, Holland.

Esashi, T. (1989). Combined effects of irradiation, storage and cooking on the total vitamin C content of potato. Progress Report on Food Irradiation Research, Japan Radioisotope Association, Tokyo, 111-117.

Fox, J. B., jr., Thayer, D. W., Jenkins, R. K., Phillips, J. G., Ackerman, S. A. Beecher, G. R., Holden, J. M., Morrows, F. D. and Quirbach, D. M. (1989). Effect of gamma irradiation on the B vitamins of pork chops and chicken breasts. Int. J. Radiat. Biol. 55, 689-703.

Friedman, M. (1973). Radiation and hydrogen atom chemistry. In "The Chemistry and Biochemistry of the Sulfhydryl Group in Amino Acids, Peptides and Proteins", Pergamon Press, Oxford, England. Chapter 10.

Friedman, M. (1991). Composition and safety evaluation of potato berries, potato and tomato seeds, potatoes, and potato alkaloids. In "Evaluation of Food Safety", J. W. Finley and D. Armstrong, eds., American Chemical Society (ACS) Symposium Series, Washington, D. C.

Hart, R. J., White, J. A. and Reid, W. J. (1988). Technical note: occurrence of o-tyrosine in non-irradiated foods. Int. J. Fd. Sci. Technol. 23, 643-647.

Hawthorn, J. (1989). The wholesomeness of irradiated foods. In "Acceptance, Control of and Trade of Irradiated Food", IAEA, Vienna, Austria.

Hayashi, T. (1988). Identification of irradiated potatoes by impedometric methods. In "Health Impact, Identification, and Dosimetry of Irradiated Foods", K. W. Bogl, D. F. Regulla and M. J. Suess, eds., Institut fur Strahlenhygiene des Bundesgesundheitsamtes, Neuherberg by/Munchen, Germany, 432-452.

House of Lords Select Committee on the European Communities (1989). Irradiation of fodstuffs. HMSO, London, England.

Jospehson, E. S., Thomas, M. H. and Calhoun, W. K. (1978). Nutritional aspects of food irradiation: an overview. J. Fd. Processing and Preservation, 2, 299-313.

Joint FAO/IAEA/WHO Expert Committee (1981). Wholesomeness of irradiated food. WHO, Geneva, Switzerland.

Julius, H. (undated). Food irradiation and moulds: a time bomb. Submission to the Australian Consumers' Association and House of Representattives.

Koch, H. W. and Eisenhower, E. H. (1967). Radioactivity criteria for radiation prcessing of foods. ACS Series Advan. Chem., 65, 87-108.

Kraybill, H. F. (1982). Effect of processing on nutritive value of food: irradiation. In "Handbook of Nutritive Value of Processed Food", Vol. 1, M. Rechcigl, ed., CRC Press, Boca Raton, Florida, 181-208.

Lindner, K. and Szotyori, K. L. (1971). Solanine formation in potatoes treated with ionizing radiation to inhibit their germination. Elemiszervizsgalati Kozlemenyek, 17, (1-2), 25-28.

Matsuyama, A. and Umeda, K. (1983). Sprout inhibition in tubers and bulbs. In "Preservation of Food by Ionizing Radiation", Volume III, E. S. Josephson and M. S. Peterson, eds., CRC Press, Boca Raton, Florida, 159-213.

Merritt, C., jr., (1972). Qualitative and quantitative aspects of trace volatile components in irradiated foods and food substances. Radiat. Res. Rev. 3, 353-368.

Meyer, R. A. (1966). Induced radioactivity in food and electron sterilization. Health Physics, 12, 1027-1037.

Murray, D. R. (1990). Biology of food irradiation. Research Studies Press, Taunton, England.

National Radiological Protection Board (1989). In minutes of evidence to House of Lords Select Committee on the European Communities: Irradiation of Foodstuffs. HMSO, London, England, 112-122.

Nawar, W. W. and Balboni, J. J. (1970). Detection of irradiation treatment in foods. J. Res. Off. Anal. Chem., 53, 726-729.

Official Journal of the European Communities (1988). English edition. Information and Notices, 31, notice No 88/C 336/11.

Patil, B. C., Singh, B. and Salunkhe, D. K. (1971). Formation of chlorophyll and solanine in Irish potato (Solanum tuberosum l.) tubers and and their control by gamma radiation and CO_2 enriched packaging. Lebensm. Wiss. u. Technol., 4, 123-125.

Patterson, M. (1989). Sensitivity of Listeria monocytogenes to irradiation on poultry meat and in phosphate-buffered saline. Lett. Applied Microbiol. 8, 181-184.

Piccioni, R. (1988). Food irradiation: contaminating our food. The Ecologist, 18, 48-55.

Sanderson, D. C. W., Slater, C. and Cairns, K. J. (1989). Thermoluminescence of foods: origins and implications for detecting irradiation. Radiat. Phys. Chem., 34, 915-924.

Scientific Committee for Food (1987). Reports of the Scientific Committee for Food (Eighteenth series), Commission of the European Communities, Luxembourg, Belgium.

Swallow, A. J. (1977). Chemichal effects of irradiation. In "Radiation Chemistry of Major Food Components", P. S. Elias and A. J. Cohen, eds., Elsevier Scientific, Amsterdam, Holland, 5-20.

Swallow, A. J. (1984). Fundamental radiation chemistry of food components. In "Recent Advances in the Chemistry of Meat", A. J. Bailey, ed., Royal Society of Chemistry, London, England, 165-177.

Swallow, A. J. (in press). Effects of irradiation of food proteins. In "Developments in Food Proteins - 7", B. J. F. Hudson, ed., Elsevier Applied Science, Barking.

Thayer D. W. (1987). Assessment of the wholesomeness of irradiated food. In "Radiation Research", E. M.Fielden, J. F. Fowler, J. H. Hendry and D. Scott, eds., Taylor and Francis, Londong, England.

Thayer, D. W., Christopher, J. P., Campbell, L. A., Ronning, D. C., Dahlgren, R. R., Thompson, E. M. and Wierbicki, E. (1987). Toxicology studies of irradiated-sterilized chicken. J. Fd. Protection 50, 278-288.

Thomas, R. (1984). Radiation preservation of foods of plant origin. Part I. Potatoes and other tuber crops. CRC Crit. Rev. Food Sci. Nutr. 19, 327-379.

Tuchscheerer, T. (1966). Die bei der Strahlenkonservierung von Lebensmitteln in Energiebereich bis 10 MeV erzeugte kunstliche Radiaktivitat von Kernisomeren. Atomkernergie, 11, 333-338.

Vijayalaxmi and Srinkantia, S. G. (1989). A review of the studies on the wholesomeness of irradiated wheat, conducted at the National Institute of Nutrition, India. Radiat. Phys. Chem. 34, 941-952.

Wakeford, C.A., Blackburn, R. and Swallow, A. J., in preparation.

Webb, T. and Lang, T. (1990). Food irradiation: the myth and the reality. Thorsons, Wellingborough, England.

Wierbecki, E. and others (1986), Ionising energy in food processing and pest control: wholesomeness of food treated with ionizing energy. Council for Agricultural Science and Technology Report No 109. Ames, Iowa.

Yin Dai (1989). Safety evaluation of irradiated foods in China: a condensed report. Biomedical and Environmental Studies, 2, 1-6.

3

A LIGHT-INDUCED TRYPTOPHAN-RIBOFLAVIN BINDING: BIOLOGICAL IMPLICATIONS

Eduardo Silva[1*], Marta Salim-Hanna[1], Ana M. Edwards[1], M. Inés Becker[2] and Alfredo E. De Ioannes[2]

P. Universidad Católica de Chile, [1]Facultad de Química and [2]Facultad de Ciencias Biológicas
Casilla 6177, Santiago, Chile

ABSTRACT

We review here the covalent photo-binding induced by visible light between the essential amino acid trytophan and the vitamin riboflavin. We discuss the biological implications of this photoadduct in relation to the hepatotoxic and cytotoxic effect associated to parenteral nutrients and to culture media exposed to the action of light, respectively. We also analyze the formation of a photo-binding between riboflavin and the residues of tryptophan present in the proteins of the eye lens, a tissue which is permanently exposed to visible light.

INTRODUCTION

Most molecules of biological importance are relatively insensitive to the direct effects of visible light since they do not absorb radiation in such wavelength range. However, some of them are sensitized to damage and destruction by light in the presence of appropriate photosensitizers and molecular oxygen. These dye-sensitized photooxidation reactions are commonly termed photodynamic action (Tsai et al., 1985; Dubbelman et al., 1988; Roberts and Berns, 1989). The photochemical alteration induced in proteins by sensitized photooxidation is circumscribed to the specific modification of the side-chains of certain amino acid residues, with no alteration at the level of the peptide bonds (Spikes, 1988). Of the 20 types of amino acid residues that can be part of a protein, only the side chains of cysteine, histidine, methionine, tryptophan (Trp) and tyrosine (Tyr) residues are susceptible of being modified by sensitized photooxidation (Silva et al., 1974; Silva, 1987). Among these, the nutritionally essential amino acid Trp is exceptional in its diversity of biological functions (Friedman and Cuq, 1988).

Riboflavin (Rb) commonly known as vitamin B_2, is an essential ingredient of our physiological system. It is present in free form and as flavin mono-nucleotide (FMN) and di-nucleotide (FAD) conjugates, both in nature and in tissues. In the presence of light, riboflavin has photosensitizer characteristics (Griffin et al., 1981; Lozinova et al., 1986; Kanner and Fennema, 1987). Since Trp and Rb are common nutrients of our diet, it was of interest to describe the photochemical generation and the biological implications of a Trp-Rb adduct.

---

* To whom correspondence should be addressed.

A light-induced tryptophan-riboflavin binding

Exposure of Trp to visible light in the presence of molecular oxygen and a sensitizer results in the formation of photooxidation products. These products have been the subject of a number of studies (Benassi et al., 1967; Asquith and Rivett, 1971; Saito et al., 1977; Inoue et al., 1982; Creed, 1984). Three pathways involving the common intermediate hydroperoxyindolalanine have been proposed, for the transformation of Trp into its photo-products kynurenine and N-formylkynurenine (Saito et al., 1977; Nakagawa et al., 1981) whose fluorescent characteristics have been reported by Fukunaga et al (1982). In turn, Nakagawa et al. (1985) have shown that the pH not only exerts an influence on the Trp photooxidation rate, but that it also influences the formation of other photoproducts, such as 5-hydroxy-formylkynurenine at pH higher than 7.0 and a tricyclic hydroperoxide in the 3.6 to 7.1 pH range.

In 1977, it was reported that irradiation with visible light of lysozyme in the presence of Rb and molecular oxygen, not only produced the photooxidation of some amino acid residues of that enzyme, but also generated a binding of this sensitizer to the protein (Silva and Gaule, 1977). This photo-binding Rb-lysozyme was also obtained in an anaerobic atmosphere, thus avoiding photooxidative phenomena. In subsequent studies, after obtaining peptides from the lysozyme (Ferrer and Silva, 1981) it was verified that a Trp residue was specifically involved in the binding between Rb and the lysozyme (Ferrer and Silva, 1985). It is worth noting that a binding has also been obtained between Rb and the lyzozyme by the excitation of this vitamin through the transfer of energy from triplet acetone, generated enzymatically (Durán et al., 1983). Through the irradiation of this essential amino acid in its free form, in the presence of Rb, it was possible to isolate and characterize spectrophotometrically at least two types of photoadducts, according to the degree of modification of the flavin (Salim-Hanna et al., 1987).

Because of the modifications of Rb by the action of light (Cairns and Metzler, 1971; Heelis, 1982) and the complexity of the indole photochemistry (Creed, 1984), it has not yet been possible to elucidate the precise chemical nature of the binding between these compounds. At first, it was thought that the binding would occur between the N-1 of the indole and a photodecomposition product of Rb, of the formylmethylflavin or carboxymethylflavin type. This hypothesis was ruled out when it was found that lumiflavin, which lacks lateral functional groups, was also capable of binding with the Trp. With this in mind, it could be postulated that the binding would involve the isoalloxazine ring of the flavin.

Regarding the chemical behaviour of the indolic component during the formation of the photoadduct, it was thought that it could undergo the rupture of its lateral ring. In order to confirm or discard this possibility, different studies have been conducted with indolic derivatives such as indole-2-carboxylic acid, indole-3-carboxylic acid, indole-5-carboxylic acid and 5-methoxy-2-methyl-3-indoleacetic acid, which may undergo the rupture of the ring, as well as with DL-indoline-2-carboxylic acid which does not present this possibility. In all these cases, the adduct indole-flavin was found in approximately the same amount, which allows to rule out a mechanism involving a rupture of the indole lateral ring. This result suggested that we should orient our studies toward the reductive capacity of the N-1 group of the indole, since intermolecular photoreductions involving flavin have been previously reported (Heelis 1982). By using cyclic voltametry, it was shown that the oxidation potential values of all the compounds used in the previous study were consistent with this postulate. Moreover, it has been found that the process of formation of the indole-flavin photoadduct is quenched by $Fe(CN)_6^{-3}$, an electron acceptor (Rossi et al, 1981). With the available experimental data, and considering the literature related to photoreductions of flavins formed with selected substrates (Heelis, 1982) as well as the studies on the interaction between indole-flavin rings by X-ray structural studies (Inoue et al., 1983), it is possible to postulate a binding between the C-3 position of the indole and the N-5 position of the flavin (Figure 1).

Irradiation products of tryptophan and riboflavin solutions and hepatic dysfunctions

Self-administration of large doses of the amino acid Trp has become widespread for the induction of sleep in humans (Schmidt, 1983; Hartmann, 1983; Spinweber et al., 1983; Hartmann et al., 1983; Leathwood and Pollet, 1984; Hartmann and Greenwald, 1984; Schneider-Helmert, 1986; Rogemont et al., 1988). Despite the fact that Trp is an essential amino acid found in nearly all food substances containing proteins, the administration of Trp in high doses produces ultrastructural changes of the liver in rats (Trulson and Sampson, 1986). The presence of Trp and Rb has been related to hepatic dysfunctions produced by parenteral nutrition (Farrel et al., 1982; Vileisis et al., 1982). Moreover, hepatic dysfunctions in neonatal gerbils (Bhatia and Rassin, 1985) and a lethal effect to mammalian cells in culture (Wang, 1975; Nixon and Wang, 1977), are produced by photoproducts of Trp only in the presence of Rb. In all these works, the photooxidation products of Trp were postulated as give rise to for these anomalies. At present, it is known that the irradiation of Trp with visible light in the presence of Rb, not only produces the photooxidation of this amino acid (Silva, 1987) but also the formation of a photo-binding product between the sensitizer and Trp (Ferrer and Silva, 1985). The formation of a Trp-Rb photoadduct can be separated from the photooxidative process when working in an anaerobic atmosphere (Salim-Hanna et al., 1987).

As the presence of Rb and Trp has been associated to hepatic dysfunctions during parenteral nutrition, it was interesting to study the possible hepatotoxicity of the Trp-Rb adduct (Donoso et al., 1988). On this purpose, 3-week old female Wistar rats were randomly assigned to five groups. They received daily intraperitoneal injections over 12 days, as follows: Group I, a non light-exposed Trp and Rb solution (Trp+Rb-L); group II, a Trp-Rb solution irradiated for 48 h under N_2 atmosphere (Trp+Rb+N_2+L48h); group III, a Trp and Rb solution irradiated for 6 h under O_2 atmosphere (Trp+Rb+O_2+L6h); group IV, the same solution but irradiated for 48 h under O_2 atmosphere (Trp+Rb+O_2+L48h); group V (control group) received only saline. The aerobically irradiated solutions were injected for comparing the possible harmful effect of the Trp-Rb photobinding (anaerobically induced) with that of tryptophan photooxidation products (Bhatia and Rassin, 1985). When the solution was irradiated for 6 h in the presence of molecular oxygen, a spectrophotometric change was observed between 300 nm and 400 nm, similar to that obtained after 48 h of irradiation under N_2 atmosphere (Donoso et al., 1988). The purpose of the aerobic irradiation for 48 h was to obtain a maximum amount of Trp oxidation products (the Trp concentration was 100-fold higher than that of Rb). The final body weight and the gain in weight of control animals were significantly higher than those of the other groups that had been injected with different Trp and Rb solutions (Donoso et al. 1988). Table 1 shows the activity of serum γ-glutamyl transpeptidase (γ-GT) of the five groups of rats studied.

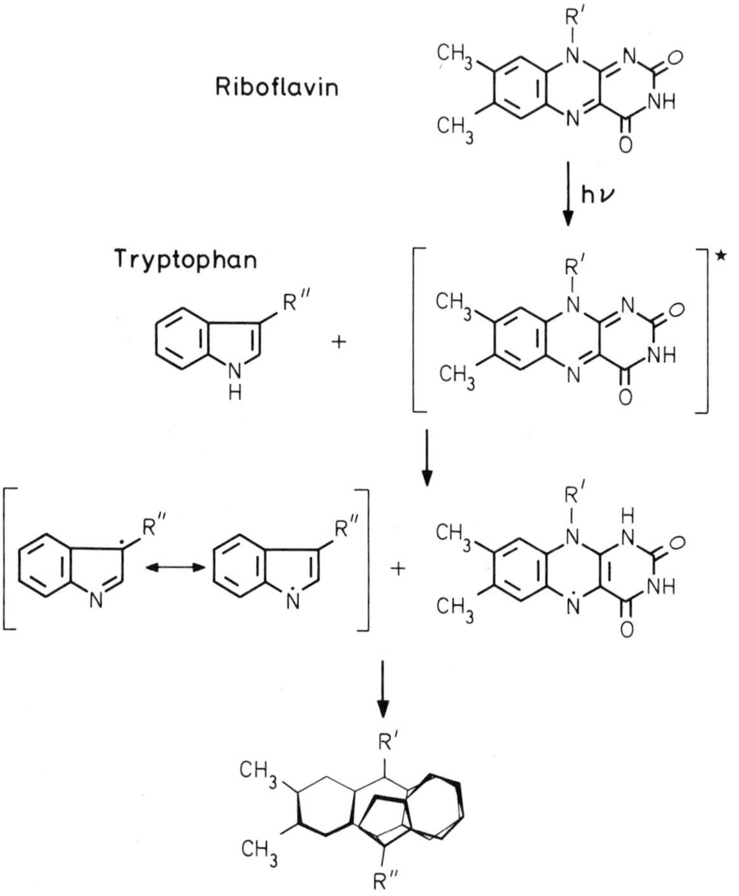

Riboflavin

Tryptophan

Figure 1. Postulated scheme for the formation of a photoinduced tryptophan-riboflavin adduct.

Table 1. Serum γ-Glutamyl transpeptidase (γ-GT) activity in liver (mean ± SD)

Group		γ-GT IU/l
I	: Trp+Rb-L	25.1 ± 3.3
II	: Trp+Rb+N_2+L48h	32.8 ± 3.3
III	: Trp+Rb+O_2+L 6h	25.1 ± 3.2
IV	: Trp+Rb+O_2+L48h	28.9 ± 1.2
V	: Control	14.5 ± 2.9
ANOVA		$p < 0.01$

All experiments were performed on groups of three rats, individual values of γ-GT activity were determined.
+L indicates irradiation of solutions prior injection.

The average activity of γ-GT, an indicator of hepatic dysfunction (Lum and Gambino, 1972; Whitfield et al., 1972), was significantly higher in all the groups injected with Trp and Rb solutions as compared to the controls. The highest activity of γ-GT were found in group II, injected with the Trp-Rb adduct, suggesting a role of this compound in the pathogenesis of an hepatic dysfunction when using rats as a model. Groups III and IV had higher γ-GT values than the control, though lower than those in group II, treated with the adduct. This suggests either that the photooxidation products of Trp induce hepatotoxic effects (which would be less harmful than those produced by the photoadduct) or that this increase of γ-GT is due solely to the effect of the photoadduct, generated in aerobic conditions, but to a smaller extent (Ferrer and Silva, 1985). The hepatic dysfunctions observed in rats injected with a non-light-exposed Trp and Rb solution (group I) indicates that the interaction of these compounds in the dark may induce the formation of somewhat toxic products. The formation of the Trp-Rb adduct has been also demonstrated using peroxidase-generated triplet acetone in the absence of light (Rojas and Silva, 1988). The biochemical generation of electronically excited species (photochemistry without light) is of wider biological significance, and accounts for a series of phenomena occurring at the cell level, that had not been explained so far (Cilento, 1980, 1984; Cilento and Adam, 1988). Polymorphonuclear leukocytes are rich in myeloperoxidase (Schultz and Kaminker, 1962) which, in the same way as horseradish peroxidase (Cilento and Adam, 1988) catalyzes the generation of excited species; therefore, a Trp-Rb photo-binding could be generated *in vivo* at the plasma level.

Cytotoxic effect of the Trp-Rb adduct on mammalian cell cultures in vitro

The potential toxic effect of daylight fluorescent lamps on mammalian cells, was reported by Stoien and Wang (1974). They demonstrated that mammalian cells in culture medium, exposed to visible light, suffered physiological damage and that proliferation was arrested. These authors provided evidence that the deleterious effect was due to the formation of toxic photoproducts resulting from a photodynamic reaction involving oxygen, Rb, Trp, and Tyr present in the culture medium (Stoien and Wang, 1974). Cell death does not occur when these components are withdrawn from the medium prior to irradiation (Nixon and Wang, 1977).

To find out which mechanisms may be involved in this lethal effect, we investigated whether or not the photoproducts had to be generated in situ, in order to damage the cells. At the conditions described by these authors, a Trp-Rb binding also occurs, therefore it was of interest to compare the cytotoxic effect of Trp-Rb solutions irradiated under either anaerobical or aerobical conditions. In the first case, the adduct Trp-Rb is preferentially generated, and in an oxygenated medium this Trp-Rb adduct, is accompanied by products of Trp photooxidation. To this purpose, we exposed to visible light solutions containing Trp and Rb under N_2 or O_2 atmosphere; the solutions were then added to F9-Teratocarcinoma cells (TC-F9), either at the time of seeding or in established cell cultures (Silva et al., 1988). The results showed that the Trp-Rb adduct generated under N_2 atmosphere was able to inhibit both the adhesion to the substrate and the ability of TC-F9 cells to proliferate, and a similar effect to that obtained when using a solution previously irradiated with visible light under an O_2 atmosphere (Figure 2). This effect was also found in preimplantation mouse embryos (Figure 3). Three main conclusions may be drawn from these findings: (1) the in situ generation of the adducts is not required; (2) adducts, generated in the absence of photo-oxidative processes are able to exert the cytotoxic effect in vitro; (3) these compounds alter the proliferation of both tumoral and normal cells, indicating that their mechanism of action is the restraint of a common function.

It is worth noting that preimplantation mouse embryos cultured in the presence of non irradiated Trp-Rb solutions arrest mitosis, suggesting an adduct generation in the dark (Rojas and Silva, 1988). A similar effect was observed in the study of hepatic dysfunctions in rats (Donoso et al., 1988) and has already been discussed here. This phenomenon cannot be discarded, since an enzymatically-generated binding between Trp and Rb, not requiring the presence of light, has been reported by Rojas and Silva (1988). This toxic effect of non irradiated Trp-Rb solutions was not observed in TC-F9 cell culture.

In the case of photooxidation products, a possible mechanism of action was suggested by the studies of Hoffmann and Meneghini (1979) who showed that when media containing Rb and Trp are irradiated with visible light, they form hydrogen peroxide. This product is responsible for the formation of single-strand breaks in DNA, and for chromosome damage in cultures of mammalian cells irradiated with visible light (Parshad et al., 1978).

Since the previous experiments were performed with a mixture of the adduct and the photodegradation products of Rb, we attempted the isolation of the adduct probably responsible for the cytotoxic effect. The irradiated solutions were fractionated on a Sephadex G-15 gel filtration column, and the elution profile of the column was monitored at 280 nm. The spectrophotometric properties of the fractions were analyzed by absorption and emission spectrophotometry. The comparison with the elution profiles of non irradiated solutions of Trp and Rb disclosed a small peak, that eluted after the elution volume of the dimeric form of Rb, which is not present in non irradiated solutions. According to the hypothetical properties of the adducts, this fraction might contain the molecule(s) responsible for the cytotoxic effect. The toxicity of all these fractions was tested on TC-F9 cell cultures, as described by Silva et al. (1988). It was found that, in fact, the isolated Trp-Rb adduct is responsible for the alterations observed in cell morphology and for the inhibition of cell proliferation. The other fractions of the elution pattern -non containing any adduct- showed no toxic effect and the cell appearance was similar to that of the control cells. The generation and

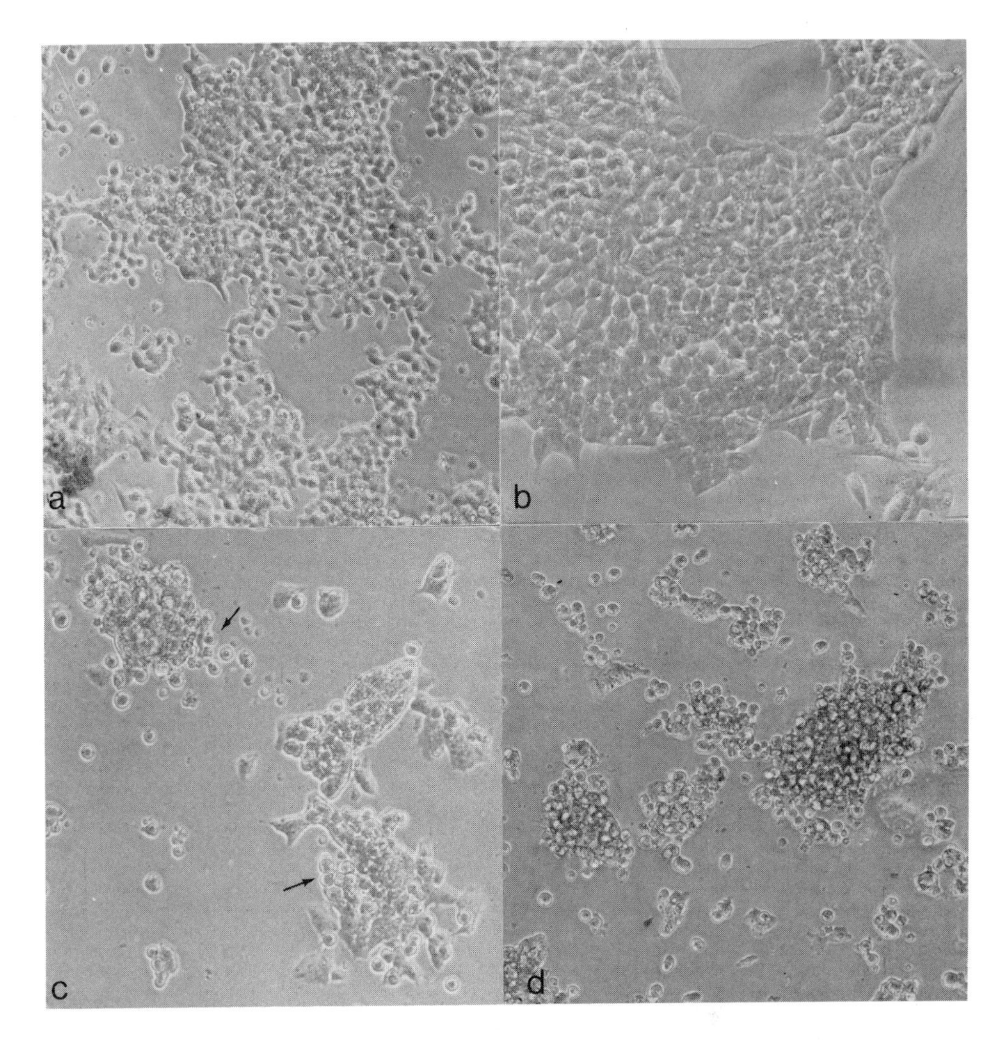

Figure 2. Effects of Trp and Rb irradiated solutions on established cultures of F9-Teratocarcinoma cells.

Controls, subconfluent culture, 24 h after seeding (a) and 48 h after seeding (b), cells show normal adherence to the substrate and their characteristic polygonal morphology; (c) subconfluent culture, 48 h after seeding, treated during the last 24 h with a Trp and Rb solution previously irradiated in an anaerobic atmosphere. Cells show a weaker adherence to the substrate than the control. Note the presence of many spheric cells, either alone or in small clumps (arrows); (d) subconfluent culture, 48 h after seeding, treated during the last 24 h with a Trp and Rb solution previously irradiated in an aerobic atmosphere. Cells did not adhere to the substrate and formed aggregates. Magnifications 450 X.

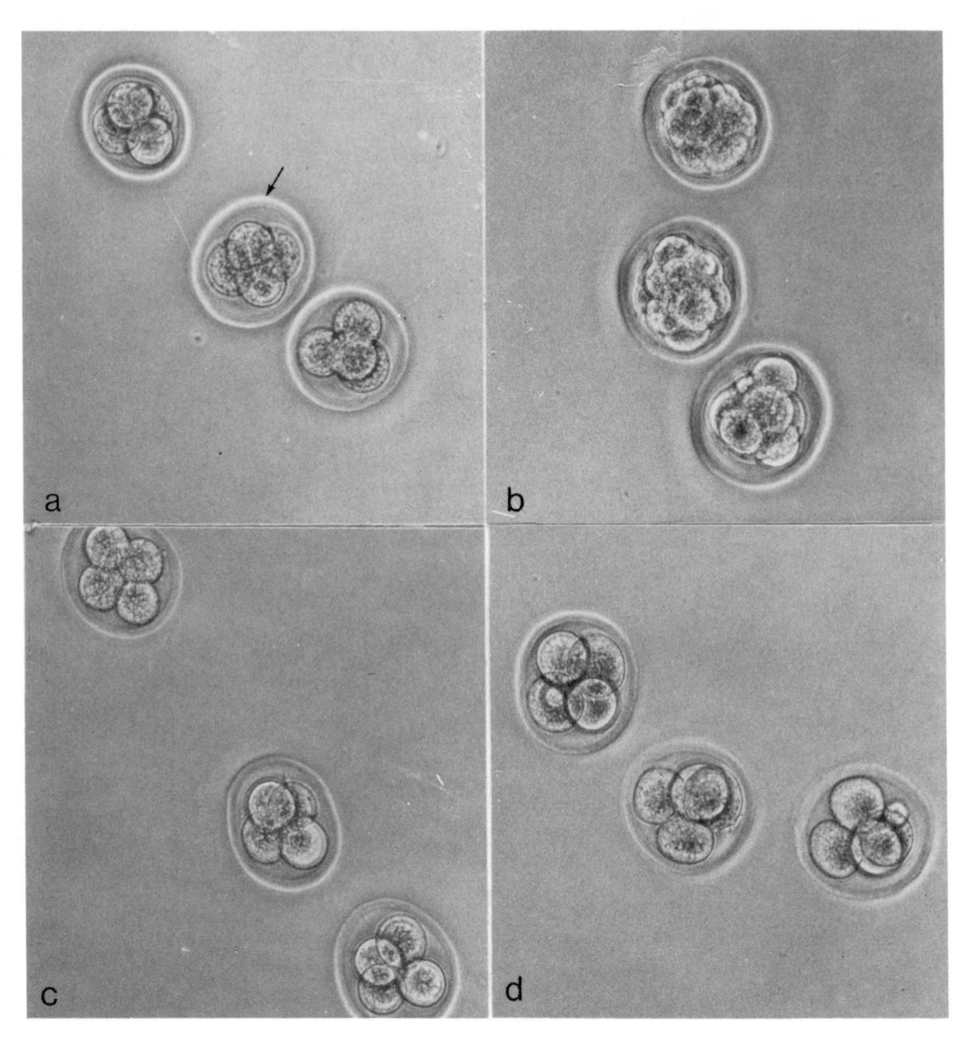

Figure 3. Effect of Trp and Rb irradiated solutions on preimplantation mouse embryos.

Normal development: (a) Four-cell mouse embryos in control medium, arrow shows the zona pellucida; (b) Morula stage after-14 h incubation of the 4-cell embryos in the control medium. Arrested development: (c) 4-cell mouse embryos cultured for 14 h in the presence of a Trp and Rb solution previously irradiated in an anaerobic atmosphere; (d) 4-cell mouse embryos cultured 14 h in the presence of a Trp and Rb solution previously irradiated in an aerobic atmosphere. Magnifications 400 X.

isolation of Trp-FMN and Trp-FAD photo-induced adducts was also studied. In both cases, an adduct was found, and the most deleterious effect was observed when TC-F9 cells were cultured in the presence of the Trp-FAD adduct (Silva et al., in preparation).

Our findings may be of great value for the treatment of tumors, when using photodynamic therapy. This procedure is based on the interaction of a photosensitizer retained in malignant tissue and photons of visible light, resulting largely from the formation of singlet oxygen, the lethal agent (Pandey and Dougherty, 1988; Spikes, 1988). This therapy has however a limitation: unlike ionizing radiation, its effects may last for a month or more. This is due to the low catabolism and excretion of the sensitizers. A more recent strategy to overcome this problem has been to develop sensitizers that are readily photo-bleached (Dillon et al., 1988). In this context, the sensitizer Rb has the advantage of being a natural compound, a vitamin, and of being easily photo-bleached (Moore and Baylor, 1969). Moreover, besides singlet oxygen, Rb generates superoxide-ion, hydroxyl radical and hydrogen peroxide (Joshi, 1989). These reactive species are more toxic than singlet oxygen (Borg et al., 1978; Greenstock and Ruddock, 1978; Khan, 1978). In the presence of Trp, Rb can also produce a Trp-Rb photo-adduct, whose toxic properties were discussed before. These findings point to Rb as a promising tool for the photodynamic treatment of tumors.

Photo-induced binding of Rb to the Trp residues of lens proteins

It would be interesting to explore the possible biological implications over time of the formation of a Trp-Rb adduct in the eye lens, a tissue permanently exposed to light. The lens grows through life. Anterior epithelial cells are displaced toward the lens equator, dividing and elongating into terminally differentiated fibre cells. New layers of fibre cells continually overlay their predecessors so that the central part of the lens, known as the nucleus, is finally formed by cells of embryonic origin. There is little, if any, protein turnover in the differentiated fibre cells, hence the proteins of the lens can be extremely old and may be exposed to bright light for decades (Wistow and Piatigorsky, 1988). Practically all the dry weight of the lens consists of proteins, and nearly 90% of these correspond to a specific structural type, known as crystallins (Tardieu and Delaye, 1988).

One of the many important processes associated to ageing is the appearance of cataracts in the eye lens. It has been proposed that some cataracts would occur as a consequence of protein chemical modifications. A covalent cross-linking between proteins and the accumulation of insoluble protein has been reported (Ziegler and Goosey, 1981a; Swamy and Abraham, 1987; Srivastava, 1988). Several authors have suggested that the formation of disulfide bonds would be an important factor in the formation of cataracts (Ziegler and Goosey, 1981a; Andley et al., 1982). Notwithstanding, recent works indicate that other factors may also be involved (Hum and Augusteyn, 1987a and 1987b; Bessems et al., 1987).

Several authors have attempted to associate photooxidation of lens proteins with the changes undergone by the lens crystallins because of ageing and/or cataractogenesis (Goosey et al., 1980; Ziegler and Goosey 1981a, 1981 b; Bose et al 1986). This type of process must be sensitized, since the lens proteins are typically transparent. Amongst the endogenous sensitizers, some researchers have postulated the vitamin Rb and products of the degradation of the essential amino acid Trp (Ziegler and Goosey, 1981b; Yegerov et al., 1987; Ichijima and Iwata, 1987; Zigman and Paxhia, 1988; Jernigan, 1985). In turn, other workers have studied the fixation of Rb in the rat eye lens (Hirano et al., 1983; Ono et al., 1986; Hirano et al., 1989).

Radioactivity was found in the lenses of Wistar-rats, when studying the physiological effects produced by intraperitoneal injection of a Trp and ^{14}C-Rb solution irradiated for 48 h under N_2 atmosphere and of a non irradiated solution. The results show that ^{14}C-Rb from either irradiated or non-irradiated solutions is distributed through the rat tissues and can be found in the eye lens. A decrease of the lens radioactivity with time has also been observed (Salim-Hanna et al., 1988). The level of ^{14}C-Rb detected in the lens may correspond to unaltered Rb, to ester forms of Rb, to photodecomposition products of Rb or to the Trp-Rb adduct (Salim-Hanna et al., 1987).

The natural presence of Rb in the lens led us to study the possible photoinduced binding of this vitamin to the lens proteins. In a first stage of this study, the lenses were removed and immersed for 48 h in a solution containing ^{14}C-Rb. They were then rinsed in saline and exposed to visible light (one lens per rat). Control lenses were kept in the dark. Irradiated and non-irradiated lenses were dialyzed separately against saline until the radioactivity in the solution became negligible. The preparations were then homogenized and centrifuged. Finally, radioactivity was assessed in the pellet and the supernatant. The results showed that irradiated and control lenses had a similar ^{14}C-Rb content in the soluble fraction, however the amount of ^{14}C-Rb from the insoluble fraction of irradiated lenses was significantly higher than that of controls. This suggests a photoinduced binding between Rb and the insoluble fraction, the most likely binding site for Rb being the Trp residues (Salim-Hanna et al., 1988).

Recent studies have dealt with the binding of Rb to the soluble protein of the eye lens in which the fractions α, β and γ crystallins are found.

When the lens homogenate and the soluble fraction were irradiated in the presence of Rb, photo-binding was obtained between this vitamin and the lens proteins. This photo-binding also occurs with the isolate lens protein fractions α, β, and γ-crystallins. Irradiation of the lens soluble proteins in the presence of Rb leads to a modification in the distribution of the lens protein fractions, with an important increase in the fraction of higher molecular weight, i.e., α-crystallin. In a study on ageing, it was found that the higher molecular crystallin fractions drastically increased with age while the lower molecular weight fractions decreased. Ageing also appears to produce an increase in the accessibility of the tryptophan residues of lens proteins, as has been established from a iodide quenching experiments. In the rat lens, endogenous Rb was detected at all ages (Ugarte et al., 1990).

FUTURE DIRECTIONS

The toxicity of the Trp-Rb photo-induced adduct at the hepatic level _in vivo_ and in tumoral cell culture -as well as modifications occurring at the lens protein level- point to an urgent need for further research. Such research should be directed to: (a) elucidate the adduct physiological mechanisms of action at the subcellular level; (b) to obtain antibodies to this compound which will serve for the adduct localization, characterization, quantification and purification in biological systems; and (c) to explore the possible formation of the adduct in light-exposed nutrients containing proteins and Rb (vitamin B_2) as well as its probable toxic effect on living beings.

ACKNOWLEDGEMENTS

We wish to thank Dr. Soledad Sepúlveda for careful reading of the manuscript. Ignacia Aguirre and Pedro Lira for assistance in preparing the manuscript, and María Teresa Pino for excellent typing. This work was supported by grants from FONDECYT (1051/86 and 389/89) and DIUC.

REFERENCES

Andley, V.P., Liang, J.N., and Chakrabarti, B. (1982). Spectroscopic investigations of bovine lens crystallins 2. Fluorescent probes for polar-apolar nature and sulfhydryl group accessibility. Biochemistry, 21, 1853-1858.

Asquith, R.S., and Rivett, D.E. (1971). Studies on the photooxidation of tryptophan. Biochim. Biophys. Acta, 252, 111-116.

Benassi, C.A., Scoffone, E., Galiazzo, G., and Jori, G. (1967). Proflavine-sensitized photooxidation of tryptophan and related peptides. Photochem. Photobiol., 6, 857-866.

Bessems, G.J.H., Rennen, H.J.J.M., and Hoenders, H.J. (1987). Lanthionine, a protein cross-link in cataractous human lenses. Exp. Eye Res., 44, 691-695.

Bhatia, J., and Rassin, D.K. (1985). Photosensitized oxidation of tryptophan and hepatic dysfunction in neonatal gerbils. JPEN, 9, 491-495.

Borg, D.C., Schaich, K.M., Elmore, J.J.,Jr., and Bell, J.A. (1978). Cytotoxic reactions of free radical species of oxygen. Photochem. Photobiol., 28 ,887-907.

Bose, S.K., Mandal, K., and Chakrabarti, B. (1986). Sensitizer-induced conformational changes in lens crystallin. II. Photodynamic action of riboflavin on bovine α-crystallin. Photochem. Photobiol., 43, 525-528.

Cairns, W.L., and Metzler, D.E. (1971). Photochemical degradation of flavins. VI. A new photoproduct and its use in studying the photolytic mechanism. J. Amer. Chem. Soc., 93, 2772-2777.

Cilento, G. (1980). Generation and transfer of triplet energy in enzymatic systems. Acc. Chem. Res., 13, 225-230.

Cilento, G. (1984). Generation of electronically excited triplet species in biochemical systems. Pure Appl. Chem., 56, 1179-1190.

Cilento, G., and Adam, W. (1988). Photochemistry and photobiology without light. Photochem. Photobiol., 48, 361-368.

Creed, D. (1984). The photophysics and photochemistry of the near-UV absorbing amino acids. I. Tryptophan and its simple derivatives. Photochem. Photobiol., 39, 537-562.

Dillon, J., Kennedy, J.C., Pottier, R.H., and Roberts, J.E. (1988). In vitro and in vivo protection against phototoxic side effects of photodynamic therapy by radioprotective agents WR-2721 and WR-77913. Photochem. Photobiol., 48, 235-238.

Donoso, M.N., Valenzuela, A., and Silva, E. (1988). Tryptophan riboflavin photo-induced adduct and hepatic dysfunction in rats. Nutr. Rep. Internat., 37, 599-606.

Dubbelman, T.M.A.R., Boegheim, J.P.J., and Van Steveninck, J. (1988). Mechanism of photodynamic damage induced in cellular system. In: "Light in Biology and Medicine", Douglas, R.H., Moan, J., and Dall'Acqua, F., eds., Plenum Press, New York, pp. 91-94.

Durán, N., Haun, M., De Toledo, S.M., Cilento, G., and Silva, E. (1983). Binding of riboflavin to lysozyme promoted by peroxidase-generated triplet acetone. Photochem. Photobiol., 37, 247-250.

Farrel, M.K., Balistreri, W.F., and Suchy, F.J. (1982). Serum sulfated lithocholate as an indicator of cholestasis during parenteral nutrition in infants and children. JPEN, 6, 30-33.

Ferrer, I., and Silva, E. (1981). Isolation and photo-oxidation of lysozyme fragments. Radiat. Environ. Biophys., 20, 67-77.

Ferrer, I., and Silva, E. (1985). Study of a photo-induced lysozyme-riboflavin bond. Radiat. Environ. Biophys., 24, 63-70.

Friedman, M., and Cuq, J.L. (1988). Chemistry, analysis, nutritional value, and toxicology of tryptophan in food. A Review. J. Agric. Food Chem., 36, 1079-1093.

Fukunaga, G., Katsuragi, Y., Izumi, T., and Sakiyama, F. (1982). Fluorescence characteristics of kynurenine and N'-formyl-kynurenine. Their use as reporters of the environment of tryptophan 62 in hen egg-white lysozyme. J. Biochem., 22, 129-141.

Goosey, J.D., Ziegler, J.S.Jr., and Kinoshita, J.H. (1980). Cross-linking of lens crystallins in a photodynamic system: a process mediated by singlet oxygen. Science, 208, 1278-1280.

Greenstock, C.L., and Ruddock, G.W. (1978). Radiation activation of carcinogens and the role of ·OH and O_{2^-}. Photochem. Photobiol., 28, 877-880.

Griffin, F.M., Ashland, G. and Capizzi, R.L. (1981). Kinetics of phototoxicity of Fischer's medium for L5178 Y leukemic cells. Cancer Res., 41, 2241-2248.

Hartmann, E. (1983). Effects of L-tryptophan on sleepiness and on sleep. J. Psychiatr. Res., 17, 107-113.

Hartmann, E., and Greenwald, D. (1984). Tryptophan and human sleep: an analysis of 43 studies. In: "Progress Tryptophan Serotonin Research", Schlossberger, ed., Hans Georg de Gruyter, Berlin, pp. 297-304.

Hartmann, E., Lindsey, J.G., and Spinweber, C. (1983). Chronic insomnia: effects of tryptophan, fluorazepan, fenogarbital, and placebo. Psychopharmacology, 80, 138-142.

Heelis, P.F. (1982). The photophysical and photochemical properties of flavins (isoalloxazines). Chem. Soc. Rev., 11, 15-39.

Hirano, H., Hamajima, S., Niitsu, Y., Oikawa, K., and Ono, S. (1983). Studies on the riboflavin-binding capacity of the rat lens. Internat. J. Vit. Nutr. Res., 53, 243-250.

Hirano, H., Horiuchi, S., and Ono, S. (1989). Effects of pH and light exposure on the riboflavin-binding capacity in the rat lens. Ophthalmic Res., 21, 93-96.

Hoffmann, M.E., and Meneghini, R. (1979). Action of hydrogen peroxide on human fibroblast in culture. Photochem. Photobiol., 30, 151-155.

Hum, T.P., and Augusteyn, R.C. (1987a). The state of sulfhydryl groups in proteins isolated from normal and cataractous human lenses. Curr. Eye Res., 6, 1091-1101.

Hum, T.P., and Augusteyn, R.C. (1987b). The nature of disulfide bonds in rat lens proteins. Curr. Eye Res., 6, 1103-1108.

Ichijima, H., and Iwata, S. (1987). Changes of lens crystallins photosensitized with tryptophan metabolites. Ophthalmic Res., 19, 157-163.

Inoue, K., Matsuura, T., and Saito, I. (1982). Mechanism of dye-sensitized photooxidation of tryptophan, tryptamine, and their derivatives. Singlet oxygen process in competition with type I process. Bull. Chem. Soc. Jpn., 55, 2959-2964.

Inoue, M., Okuda, Y., Ishida, T., and Nakagaki, (1983). On the interaction between flavin-adenine rings and between flavin-indole rings by X-ray structural studies. Arch. Biochem. Biophys., 227, 52-70.

Jernigan, H.M.Jr. (1985). Role of hydrogen peroxide in riboflavin-sensitized photodynamic damage to cultured rat lenses. Exp. Eye Res., 41, 121-129.

Joshi, P.C. (1989). Ultraviolet radiation-induced photodegradation and singlet oxygen, superoxide anion radical production by riboflavin, lumichrome and lumiflavin. Indian J. Biochem. Biophys., 26, 186-189.

Kanner, J.D., and Fennema, O. (1987). Photooxidation of tryptophan in the presence of riboflavin. J. Agric. Food Chem., 35, 71-76.

Khan, A.U. (1978). Activated oxygen: singlet molecular oxygen and superoxide anion. Photochem. Photobiol., 28, 615-627.

Leathwood, P., and Pollet, P. (1984). Tryptophan (500mg) decreases subjectively perceived sleep latency and increases sleep depth in man. In: "Progress Tryptophan Serotinin Research", Schlessberger, ed., Hans Georg de Gruyter, Berlin, pp. 311-314.

Lozinova, T.A., Nedelina, O.S., and Kayushin, L.P. (1986). Effect of adenosine diphosphate on light-induced oxygen consumption by flavins. Biofizika, 31, 10-15.

Lum, G., and Gambino, S.R. (1972). Serum gamma-glutamyl transpeptidase activity as an indicator of disease of liver, pancreas or bone. Clin. Chem., 18, 358-362.

Moore, W.M., and Baylor, C. Jr. (1969). The photochemistry of riboflavin. IV. The photobleaching of some nitrogen-9 substituted isoalloxazines and flavins. J. Amer. Chem. Soc. 91, 7170-7179.

Nakagawa, M., Kato, S., Nakano, K., and Hino, T. (1981). Dye-sensitized photo-oxygenation of tryptophan to give N'-formylkynurenine. J. Chem. Soc. Chem. Commun, 855-856.

Nakagawa, M., Yokoyama, Y., Kato, S., and Hino, T. (1985). Dye-sensitized photo-oxygenation of tryptophan. Tetrahedron, 41, 2125-2132.

Nixon, B.T., and Wang, R.J. (1977). Formation of photoproducts lethal for human cells in culture by daylight fluorescent light and bilirubin light. Photochem. Photobiol., 26, 589-593.

Ono, S., Oikawa, K., Hirano, H., and Obara, Y. (1986). Effects of ageing on the formation of ester forms of riboflavin in the rat lens. Internat. J. Vit. Nutr. Res., 56, 259-262.

Pandey, R.K., and Dougherty, T.J. (1988). Synthesis and photosensitizing activity of a di-porphyrin ether. Photochem. Photobiol., 47, 769-777.

Parshad, R., Sandford, K.K. Jones, G.M., and Tarone, R.E. (1978). Fluorescent light-induced chromosome damage and its prevention in mouse cells in culture. Proc. Natl. Acad. Sci. USA, 75, 1830-1833.

Roberts, W.G., and Berns, M.W. (1989). Cell biology and photochemistry of photodynamic sensitizers. Proc. SPIE-Int. Soc. Opt. Eng., 1065, 175-181.

Rogemont, C., Sarda, N., Gharib, A., and Pacheco, H. (1988). Changes in the rat sleep-wake cycle produced by D,L-β-(1-naphthyl) alanine, a tryptophan analogue. Neurosci. Lett., 93, 287-293.

Rojas, J., and Silva, E. (1988). Photochemical-like behaviour of riboflavin in the dark promoted by enzyme-generated triplet acetone. Photochem. Photobiol., 47, 467-470.

Rossi, E, Van de Vorst, A., and Jori, G. (1981). Competition between the singlet oxygen and electron transfer mechanisms in the porphyrin-sensitized photooxidation of L-tryptophan and tryptamine in aqueous micellar dispersions. Photochem. Photobiol., 34, 447-454.

Saito, I., Matsuura, T., Nakagawa, M., and Hino, T. (1977). Peroxidic intermediates in photosensitized oxygenation of tryptophan derivatives. Acc. Chem. Res., 10, 346-352.

Salim-Hanna, M., Edwards, A.M., and Silva, E. (1987). Obtention of a photo-induced adduct between a vitamin and an essential amino acid Binding of riboflavin to tryptophan. Internat. J. Vit. Nutr. Res., 57, 155-159.

Salim-Hanna, M., Valenzuela, A., and Silva, E. (1988). Riboflavin status and photo-induced riboflavin binding to the proteins of the rat ocular lens. Internat. J. Vit. Nutr. Res., 58, 61-65.

Schmidt, H.S. (1983). L-Tryptophan in the treatment of impaired respiration in sleep. Bull. Eur. Physiophathol. Respir., 19, 625-629.

Schneider-Helmert, D. (1986). Nutrition and sleeping behaviour. Nutr. Diet, 38, 87-93.

Schultz, J., and Kaminker, K. (1962). Myeloperoxidase of the leukocyte of normal human blood. I. Content and localization. Arch. Biochem. Biophys., 96, 465-467.

Silva, E. (1987). Sensitized photo-oxidation of amino acids in proteins. In: "Chemical Modification of Enzymes: Active Site Studies", Eyzaguirre, J., ed., Ellis Horwood Limited, Chichester, pp. 63-73.

Silva, E., and Gaule, J. (1977). Light-induced binding of riboflavin to lysozyme. Radiat. Environ. Biophys., 14, 303-310.

Silva, E., Risi, E., and Dose, K. (1974). Photooxidation of lysozyme at different wavelengths. Radiat. Environ. Biophys., 11, 111-124.

Silva, E., Salim-Hanna, M., Becker, M.I., and De Ioannes, A. (1988). Toxic effect of a photo-induced tryptophan-riboflavin adduct on F9 teratocarcinoma cells and preimplantation mouse embryos. Internat. J. Vit. Nutr. Res., 58, 394-401.

Spikes, J.D. (1988). Photochemotherapy: molecular and cellular processes involved. SPIE, 997, 92-100.

Spinweber, C.L., Ursin, R., Hilbert, R.P., and Hilderbrand, R.L. (1983). L-Tryptophan: effects on daytime sleep latency and the waking EEG. EEG Clin. Neurophysiol., 55, 652-661.

Srivastava, O.P. (1988). Age-related increase in concentration and aggregation of degraded polypeptides in human lenses. Exp. Eye Res., 47, 525-543.

Stoien, J.D., and Wang, R.J. (1974). Effect of near-ultraviolet and visible light on mammalian cells in culture. II. Formation of toxic photoproducts in tissue culture medium by black light. Proc. Natl. Acad. Sci. USA, 71, 3961-3965.

Swamy, M.S., and Abraham, E.C. (1987). Lens protein composition, glycation and high molecular weight aggregation in ageing rats. Invest. Ophthalmol. Vis. Sci., 28, 1693-1701.

Tardieu, A., and Delaye, M. (1988). Eye lens proteins and transparency: from light transmission theory to solution X-ray structural analysis. Ann. Rev. Biophys. Chem., 17, 47-70.

Trulson, M.E., and Sampson, H.W. (1986). Ultrastructural changes of the liver following L-tryptophan ingestion in rats. J. Nutr., 116, 1109-1115.

Tsai, C.S., Godin, J.R.P., and Wand, A.J. (1985). Dye-sensitized photooxidation of enzymes. Biochem. J., 225, 203-208.

Ugarte, R., Edwards, A.M., Diez, M., Valenzuela, A., and Silva, E. (1990). Anaerobic riboflavin photosensitized modification of rat lens proteins. A correlation with aged-related changes. (Submitted for publication).

Vileisis, R.A., Sorensen, K., González-Crussi, F., and Hunt, C.E. (1982). Liver malignancy after parenteral nutrition. J. Pediatr., 100, 88-90.

Wang, R.J. (1975). Lethal effect of "daylight" fluorescent light on human cells in tissue-culture medium. Photochem. Photobiol., 21, 373-375.

Whitfield, J.B., Pounder, R.E., Neale, G., and Moss, D.W. (1972). Serum -glutamyl transpeptidase activity in liver disease. Gut., 13, 702-708.

Wistow, G.J., and Piatigorski, J. (1988). Lens crystallins: the evolution and expression of proteins for a highly specialized tissue. Ann. Rev. Biochem., 57, 479-504.

Yegerov, S.Y., Babizhayew, M.A., Krasnowskii, A.A.Jr. and Shveeova, A.A. (1987). Photosensitized generation of singlet molecular oxygen by the endogenous substances of the crystalline lens. Biophysics, 32, 184-186.

Ziegler, J.S.Jr. and Goosey, J. (1981a). Ageing of protein molecules: lens crystallins as a model system. TIBS, 6, 133-136.

Ziegler, J.S.Jr. and Goosey, J.D. (1981b). Photosensitized oxidation in the ocular lens: evidence for photosensitizers endogenous to the human lens. Photochem. Photobiol., 33, 869-874.

Zigman, S., and Paxhia, T. (1988). The nature and properties of squirrel lens yellow pigment. Exp. Eye Res., 47, 819-824.

4

SYNTHESIS AND AVAILABILITY OF NIACIN IN ROASTED COFFEE

Jean Adrian and Régine Frangne

Chaire de Biochimie industrielle et agro-alimentaire
Conservatoire National des Arts et Métiers
292, rue Saint-Martin, 75003 Paris, France

ABSTRACT

The coffee bean contains about 1% of trigonelline that is demethylated at temperatures approaching 200° C ; it is partially converted into nicotinic acid. This operation is mainly proportional to the severity of dry heat treatment ; various other physico-chemical factors also influence the synthesis of niacin during the roasting. The niacin content of weakly roasted commercial coffee is about 10 mg/100 g (American coffee) and it reaches 40 mg in heavy roasted coffees, i.e. Italian coffee. Caffeine-free coffee is lower in niacin than the corresponding raw coffee. The drinking retains 85% of the niacin formed during roasting ; it is totally available for the organism and can constitute a noticeable part of the daily supply in niacin.

INTRODUCTION

Some observations attribute various deleterious physiological effects to high coffee consumption. These manifestations are due to the presence of particular constituents in coffee, such as chlorogenic acid (Teply et al., 1945 ; Chassevent, 1969 ; Challis and Bartlett, 1975) or caffeine (Feinberg et al., 1968 ; Naismith et al., 1969 ; Panigrahi and Rao, 1983 ; Pozniak, 1985 ; Stavric, 1988).

However, an unfavourable incidence of coffee on the glycemia, the lipidemia and the occurrence of cardiovascular pathology is not obvious for the healthy person who consumes coffee normally (Callahan et al., 1979 ; Querat and Heraud, 1987 ; Kitts and Mathieson, 1989).

Actually, the participation of coffee in the ethiology of cancer remains a point of question (Strubelt et al., 1973 ; Czok, 1977 ; Thomas, 1979 ; Sandler, 1983). The roasting process is responsible for the formation of characteristic molecules. Among them, such aromatic polycyclic hydrocarbons as the benzopyrene have a well established carcinogenic power (Kuratsune and Hueper, 1960 ; Fritz, 1969, 1975 ; Bracco, 1973 ; Soos and Fözy, 1974 ; Strobel, 1974) whereas others like harman and norharman (Adrian et al., 1985) are also likely to occur in the mechanism of the cancerization. These roasting components certainly represent the most probable risk bound to coffee consumption (Miller, 1983 ; Pozniak, 1985).

Nutritional and Toxicological Consequences of Food Processing
Edited by M. Friedman, Plenum Press, New York, 1991

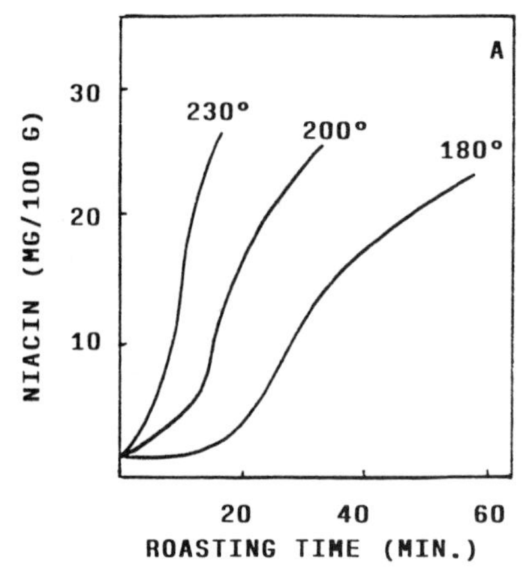

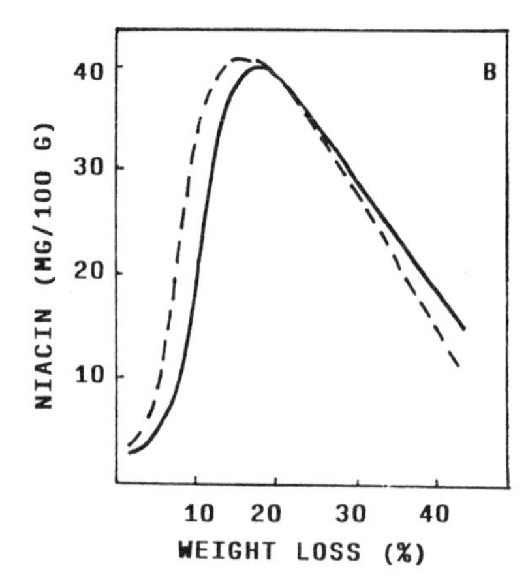

Figure 1. Niacin content in coffee according to the intensity of heat treatment. 1-A : Roasting measured by the time and temperature of treatment (Hughes and Smith, 1946) ; 1-B : Roasting measured by the weight loss of coffee (broken line : C. arabica ; heavy line : C. canephora var. robusta) (Adrian and Navellier, 1961).

However, roasting must not be considered as an operation with only negative consequences ; it develops the characteristic and enjoyed aroma of coffee and it provokes an abundant synthesis of nicotinic acid from trigonelline, which is its methylated precursor. This conversion during the roasting places coffee drinking among the most efficient food resources to satisfy the niacin requirement and to prevent or cure pellagra.

The mechanism of the synthesis of niacin from trigonelline and the efficiency of the formed nicotinic acid have been described by different teams whose conclusions all demonstrate the interest of roasted coffee as a contribution to the niacin supply. It would be useful to review these works, often old, because they show that, besides undesirable components, strong roasting can also have favorable consequences in an accurate vitamin field.

THE SYNTHESIS OF NICOTINIC ACID DURING COFFEE ROASTING

Trigonelline is the N-methyl betaine of nicotinic acid. This compound is widely present in vegetal produce : it can be found in all cereal grains, in potatoes, beetroots, tomatoes, mushrooms and sweet peppers at rates of betwwen 10 and 80 mg/100 g (Barbiroli, 1966). It is much more abundant in the coffee bean since C. arabica contains about 900 mg and C. canephora var. robusta, about 650 mg according to Hughes and Smith (1946). In a pure state, trigonelline heated in dry conditions is degraded in various derivated compounds, especially at temperatures ranging from 160°C to 230°C (Figure 1A) : about 5% are converted to nicotinic acid by demethylation. The same phenomenon occurs in some food technology processes. During experimental burnings, the nicotinic acid content increases in all the grains and vegetals containing trigonelline (Barbiroli, 1966). Under more usual modalities, an increase of the niacin content is observed as a result of an intense heating carried out in dry conditions : for example, the synthesis of niacin is noticeable in bread crust but not in crumb (Gassmann and Schneeweiss, 1959). Coffee roasting is nevertheless the operation most favorable provoking this synthesis from trigonelline : the bean contains a large quantity of trigonelline, its wetness is very weak and the heat treatment easily reaches 200°C during the process.

Incidentally, the nicotinic acid is one of the most stable vitamins when met with the customary conditions in food processing, storage, cooking, etc. (Adrian, 1959). In particular, it is remarkably thermostable, which allows it to resist efficiently during a strong heat treatment, at least as long as it does not reach extremes. That is noticed during the commercial torrefaction of coffee. In practice, the importance of the nicotinic acid synthesis and the niacin content in roasted coffee depend on many factors and parameters such as the intensity of treatment, the technological preparation of the bean, etc.

Table 1. Influence of roasting intensity on the niacin content in coffee (Teply and Prier, 1957)

	Niacin content	
	mg/100 g	relative value
green coffee	2.2	1
just before swelling	4.0	1.8
just swelling	8.3	3.8
New England roasting	13.0	5.9
French roasting	24.9	11.3
Italian roasting	41.6	18.9
Heavy roasting (excessive roasting)	43.6	19.9

Intensity of coffee roasting

The abundance of nicotinic acid in roasted coffee depends, first of all, on whether the intensity of torrefaction is estimated by the duration of heating, the browning of the bean or the weight loss (Hughes and Smith, 1946 ; Cravioto et al., 1955 ; Daum, 1955 ; Teply and Prier, 1957 ; Bressani and Navarrete, 1959 ; Adrian and Navellier, 1961 ; Bressani et al., 1962 ; Boddeker and Mishkin, 1963 ; Taguchi et al., 1985).

When the roasting is very heavy, the quantity of nicotinic acid always exceeds 40 mg/100 g of roasted coffee (Table 1, Figure 1B). This result corresponds to a value 20 to 30 times superior to that of the green bean, which is about 1.5 mg (Adamo, 1955 ; Cravioto et al., 1955 ; Daum, 1955 ; Teply and Prier, 1957 ; Bressani et al., 1959, 1962 ; Adrian and Navellier, 1961 ; Taguchi et al., 1985). According to Figure 1B, at the beginning of the roasting, the trigonelline is easily converted into nicotinic acid and the formed molecule remains stable in these modalities. It is only damaged by extremely severe burnings (Adrian and Navellier, 1961 ; Tchetche, 1979) ; this evolution is not observed in all torrefactions : Bressani et al. (1962) proceed to heatings reaching 42% of weight loss without the treatment being accompagnied by any decrease in the niacin content of the roasted produce.

Table 2. Niacin content of commercial roasted *C. arabica* and *C. canephora* types (**a**)

	Arabica	Robusta
Hughes and Smith (1946)	15.2 (4)	18.8 (4)
Bressani et al. (1961)	22.65 (29)	18.1 (2)
Adrian et al. (1967) (**b**)	18.45 (18)	15.4 (45)

(**a**) the values in brackets represent the number of samples
(**b**) the roasting time is 7.1 min. for Arabica and 8.2 min. for Robusta

In fact, coffee torrefaction primarily provokes a nicotinic acid synthesis and secondarily that of nicotinamide (Taguchi et al., 1985) :

in mg/100 g :	commercial roasting	heavy roasting
nicotinic acid	2.9 - 4.5	28.8
nicotinamide	0.48 - 0.84	2.3

The small proportion of nicotinamide is probably due to the partial degradation of aminoacids with releasing of amino groups, which react on nicotinic acid.

Influence of geographical and botanical factors

The Arabica and Robusta coffees constitute the two largest species of cultivated coffees. The former tends to contain more trigonelline but requires a less intense torrefaction than the Robusta types. Moreover, during heating, the possibilities of conversion can be compared in the both species. That is why the commercially roasted products have similar amounts of niacin (Figure 1B, Table 2).

On the other hand, within the species *C. canephora*, particularities appear according to the geographical area of cultivation. In Angola, a notion of "local variety" has been established to describe specific characteristics of Robusta cultivated in distinct regions. When samples of these "local varieties" are submitted to roasting, the influence of geographical data becomes especially evident : a series of 9 samples roasted under standardized modalities shows the following results : Ambriz "variety" hold 19.85 mg of niacin/100 g, Cazengo "variety" 15.35 mg and Amboim 13.15 mg (Adrian et al., 1967). These differences are attributed to the combination of geographical, pedologic and climatic factors on the development and the physiology of the coffee tree. They reveal a more important influence on the niacin synthesis than the differences observed between *C. arabica* and *C. canephora*. On the other hand, the year of the crop is not a determining factor to the rate of niacin in roasted coffees (Adrian et al., 1969).

Table 3. Influence of the bean size and of the swelling during the roasting on the niacin content (Adrian et al., 1967)

	Number of samples	Apparent swelling (%)	Niacin content (mg/100 g)
coarse beans (oversize of sifter 17/64")	21	69.3 (100)	14.60 (100)
middle beans (oversize of sifter 15/64")	21	74.0 (107)	16.43 (113)
small beans	21	80.8 (117)	17.40 (119)

Influence of cultural and technological factors

Technological factors involving in the conversion of trigonelline can be divised into two main groups : those which facilitate the heat transfer within the bean and are favorable to vitamin synthesis, and those which modify the concentration of trigonelline.

The smallness of the bean and its swelling capacity during roasting increase the capacity for heat transfer and, consequently, raise the synthesis rate in niacin. The data of Table 3 correspond to reproducible roastings of moderate intensity like that of American torrefaction. The differences observed would probably be unlike in the case of very intense treatments, of the Italian type. It must be pointed out that the part of swelling is difficult to define : the amount of niacin does not bear a direct relation to the importance of swelling in all the studies. However, the swelling makes the grain texture less dense and should have a positive effect on the penetration of heat during the process.

Some modalities seem to decrease the concentration of trigonelline in the bean ; they are the cause of lower niacin synthesis. First, most coffee plantations are established under high trees which supply shade and increase the productivity of the cultivation. According to Carvalho (1962), the shaded culture produce a roasted coffee holding 15% less niacin than unshaded coffee trees, which let us suppose that the synthesis of trigonelline is controlled by a photosynthetic mechanism.

In a general way, the technological preparation of the bean can indirectly modify the rate of niacin in roasted coffee : modalities which avoid a damping or soaking of the grain preserve its potential of trigonelline and result in a coffee richer in niacin. Thus, the cherries dried in parchment produce a coffee with more niacin (+19%) than the sun-dried cherries (Carvalho, 1962). Bressani et al. (1961) examine some technics of demucilagination (depulping) by either wet or dry procedure, the dry treatment allows more important syntheses during the roasting. This observation is confirmed by Adrian et al. (1967) who note a superiority of 15% with roasted coffee demucilaginated by a dry process.

Table 4. Niacin content of French commercial coffee, raw or caffeine-free samples (Adrian and Navellier, 1962)

	Raw coffee	Caffeine-free coffee
Coffee in bean		
number of samples	8	8
Total niacin (mg/100 g) :		
mean	24.0 ± 2.55	16.75 ± 6.0
extremes	19.8 - 28.4	6.6 - 26.4
Percentage of free niacin	93	94
Instant coffee		
number of samples	2	2
Total niacin (mg/100 g)	31.5 - 55.5	52.8 - 64.5
Percentage of free niacine	100	97

The most detrimental operation to the niacin production is the decaffeination : the caffeine-free products contain much less niacin than the raw coffees (Teply et al., 1945 ; Adrian and Navellier, 1962). This inferiority can be easily imputed to a partial solubilization of trigonelline when the caffeine is extracted (Trugo et al., 1983). Moreover, the niacin rates are more variable in caffeine-free samples : the amount varies in a proportion of 1 to 4 in French coffees (Table 4).

Finally, the commercial quality (estimated by the number of defects, color and smell of the green bean) has no repercussion on the richness in niacin in roasted coffees (Adrian et al., 1967).

NIACIN CONTENT IN COMMERCIAL COFFEES

Some countries, such as the U.S. and Portugal prefer slightly roasted coffee, while others such as France and Italy prefer a more heavily roasted coffee. As the intensity of torrefaction holds the main responsible for the niacin responsible for the niacin synthesis, the vitamin rate of commercial coffees varies in notable proportions according to consumer habits (Table 1). Possibly, apart from Italian coffees, torrefaction is stopped before producing the maximal quantity of nicotinic acid.

In Anglo-american countries, the most usual niacin content approaches 10 mg/100 g in U.S. (Teply et al., 1945 ; Teply and Prier, 1957), although Barton-Wright (1944) as well as Hughes and Smith (1946) indicate values between 13 and 18 mg for English products. In Latin America, the coffees are roasted more and have from 22 to 34 mg/100 g (Daum, 1955 ; Bressani et al., 1961). In Europa, the torrefaction is very often high and niacin is abundant in roasted goods. In France, coffee supplies between 19.8 and 26.0 mg of niacin (Table 4). Italian coffees, known for their strength, contain 32 to 50 mg, with an average of 42 mg (Adamo, 1955). Schematically, Italian products provide 4 times more niacin than American coffees.

Table 5. Niacin content of instant coffee from various origin

		Niacin content (mg/100 g)
U.S.	: instant caffeine-free coffee (Baker et al., 1976)	7.0
U.K.	: instant raw coffee (Trugo et al., 1985)	20.6 - 46.8
Japan	: instant raw coffee (Taguchi et al., 1985)	13.3 - 70.6
France	: instant raw coffee	35.5 - 55.5
	instant caffeine-free coffee (Adrian and Navellier, 1962)	52.8 - 64.5

The situation is very different for caffeine-free coffee, the niacin content remaining always very inferior to that of raw products, whatever the modality of the torrefaction :

	Raw coffees (mg/100 g)(a)	Caffeine-free coffees (mg/100 g)(a)	Difference (%)
American coffees (Teply et al., 1945)	9.5 (3)	4.5 (1)	- 53
French coffees Adrian et al., 1962)	24.0 (8)	16.75 (8)	- 30

(**a**) the values in brackets indicate the number of samples

The American caffeine-free products becoming at once weaker in trigonelline and undergoing a moderate heat treatment, the niacin content is inevitably very weak. The decaffeination may be less detrimental when beans are heavily roasted, like in Italian technology. This hypothesis seems to follow from the analyses of French samples.

Instant coffee seems to have a high content in niacin (Table 5). Only, the American instant caffeine-free product is very poor in niacin because of the decaffeination process and of weak roasting. All the other instant coffees contain a high niacin concentration, that shows the solubility of the vitamin synthetized during the torrefaction. But the technological modalities of production contribute to determine the richness of instant coffee ; this offers a wide variability of data (Table 5). Lastly, the quantities used for drinking preparations are weak and variable, consequently these products have a hardly calculable nutritional interest.

BIOLOGICAL EFFICIENCY OF NIACIN COFFEE

Almost all the determinations of niacin in coffees are carried out by microbiological method using *Lactobacillus arabinosus (L. plantarum)*. Only the nicotinic acid, the nicotinamide and the nicotinuric acid are measured by this procedure (*in* Adrian, 1959). Therefore, the vitamin potential estimated by L. arabinosus is likely to correspond to the available amount of niacin. This hypothesis has been checked with many methodologies. First, it must be noted that the niacin formed during roasting is in a free state and that a very high proportion is found in the drink. According to the results of numerous studies (Hughes and Smith, 1946 ; Daum, 1955 ; Cravioto et al., 1955 ; Bressani et al., 1961 ; Adrian and Navellier, 1961), it is admitted that an average of 85% of the niacin contained in roasted products are found in drinking, either raw or caffeine-free coffee. The nutritional value of a cup of coffee can be estimated thus : 10 g of coffee (sometimes 12 g) are used and 85% of the vitamin content are extracted by simmering water. Under these conditions, a cup of American coffee must provide about 0.85 mg of niacin, a cup of English product about 1.5 mg and a cup of Italian coffee about 3.1 mg. If the niacin requirement is assessed at 15 niacin-equivalent* per day, these rates supply respectively 7, 10 and 21 per cent of the daily need.

Table 6. Effect of coffee consumption on pellagra symptomatology (28 adult humans [**a**], 4 cups of coffee per day [**b**], 2 months) (Adrian et al., 1969)

	prevalence before treatment (%)	disappearing after treatment (%)
Cutaneous symptoms :		
hyperpigmentation	97	48
depigmentation	75	71
hyperkeratosis	61	59
Casal's collar	61	82
erythema	60	75
membranous desquamation	57	81
Merk's seam	25	86
Buccal symptoms :		
glossitis	54	32
fissured lip	21	68
scrotal tongue	18	0
cheilitis	18	60
Intestinal symptoms :		
diarrhea	36	100
intestinal colic	32	89
Psycho-nervous symptoms :		
asthenia	82	100
cephalgia	64	100
insomnia	47	100
buzzing in the ears	47	100

[**a**] customary diet : 8500 kJ/day, with 55 g of protein, the 4/5 of which are derived from corn.
[**b**] C. arabica (24 mg of niacin/100 g) whose 4 cups supply 11 mg/day.

Various studies have been undertaken on animals and on humans to reveal the efficiency of coffee niacin. Teply and Prier (1957) with rat and Bressani et al. (1962) with chicken demonstrate the overall efficiency of this vitamin. Goldsmith et al. (1959) confirm its availability for the human by establishing the rates of urinary outputs of niacin metabolites. It seems obvious that coffee could be a means of prevention or recovery against pellagra,

---

* 1 niacin-equivalent corresponds to a supplying of 1 mg of nicotinic acid or a furnishing of 60 mg of tryptophan, which are converted into 1 mg of nicotinic acid by metabolic way.

as was already supposed by Teply and Prier (1957), Goldsmith et al. (1959) as well as Bressani and Naverrete (1959). This has been demonstrated in the Central Angola, where corn is the main basis of the customary diet.

Table 7. Effect of coffee consumption on blood niacin content on pellagrous humans (72 adult humans, 4 cups of coffee per day, 3 months) (**a**) (Adrian et al., 1971)

	number of subjects	blood serum niacin (μg/ml) before treatment	after treatment
primary pellagra	34	0.120	0.267
chronic pellagra	38	0.118	0.255
control subjects	13	0.260	0.230

(**a**) **See** experimental procedure in Table 6.

Pellagra has been cured by distributing 2 cups of coffee in the morning and 2 in the evening to humans revealing clinical symptoms of deficiency. This treatment lasts for periods of 1 month, interspersed with one week without coffee. The coffee used was *C. arabica* because of its lower percentage in caffeine (Gounelle de Pontanel and Astier-Dumas, 1969). The 4 daily cups supplied 11 mg of niacin and only 0.4 g of caffeine. The pellagra symptoms were recorded before the cure and after 2 months of coffee consumption (Table 6). Most specific signs of deficiency (Casal's collar, Merk's seam, diarrhea and intestinal colic) have been resorbed in a large part. These favorable results do not yet constitute formal proof because the symptomatology of pellagra undergoes a spontaneous annual cycle, with a resorption of lesions during the winter period. That is why a second biochemical experiment has dealt with the restoration of the blood content in niacin in pellagrous adults : within 3 months, the blood concentration has risen to 120% (Table 7). It even became slightly superior to those of control subjects, in good health.

Thus, the niacin of roasted coffee is fully available and this drink could play an useful part in the prevention or cure of the pellagra. It is due to eating habits that Bressani and Navarrete (1959) attribute the absence of pellagra in Central America, where the consumption of coffee is high : they have noticed that in many regions of Guatemala people drank on average 3 cups of coffee per day and that concurrently pellagra remains non-existent in spite of the predominant part of corn in the diet. The results obtained with living organisms entirely confirm the above hypothesis.

The richness of niacin in coffee is particularly noticeable in Africa and Central and Latin America because of the frequent proximity of corn cultivations and of coffee plantations. It would be justified to incite local consumption of coffee in the producting countries to prevent any pellagra risk (Davis, 1978).

ACKNOWLEDGEMENTS

We thank deeply M. Joaquim Xabregas and The Instituto do Café de Angola for their help and keen interest in the work related to pellagra.

REFERENCES

Adamo, G. (1955). Il contenuto in acido nicotinico nel caffe. *Boll. Soc. Ital. Biol. Sper.*, *31*, 79-82.

Adrian, J. (1959). Le dosage microbiologique des vitamines du groupe B. Cahier technique du CNERNA, 183 p., CNRS, Paris.

Adrian, J. and Navellier, P. (1961). Intérêt nutritionnel du café comme source de vitamine PP. Café, Cacao, Thé, 5, 263-268.

Adrian, J. and Navellier, P. (1962). Teneur en vitamine PP de différents échantillons de café du commerce. Café, Cacao, Thé, 6, 224-227.

Adrian, J. (1963). Synthèse de la niacine au cours de la torréfaction du café et son efficacité biologique. Café, Cacao, Thé, 7, 359-365.

Adrian, J., Frangne, R., Xabregas, J., and Corte dos Santos, A. (1967). Teneur en vitamine PP des cafés grillés de l'Angola : rôle des facteurs botaniques et technologiques. 3ème Coll. Chimie Café, Trieste, 427-435.

Adrian, J., Pena, J., Morais de Carvalho, J., Miranda, A., Xabregas, J., and Corte dos Santos, A. (1969). La boisson de café dans le traitement de la pellagre humaine. 4ème Coll. Chimie Café, Amsterdam, 232-242 ; instalment Instituto do café de Angola, Luanda, 20 p.

Adrian, J., Xabregas, J., Pena, J., Morais de Carvalho, J., and Gomes, N. (1971). La restauration en vitamine PP par la consommation de café. Etude chez le pellagreux. 5ème Colloque Chimie Café, Lisboa, 371-374 ; Adrian, J. (1972). La consommation de café et la pellagre. Med. et Nutr., 8, 71-80.

Adrian, J., Guillaume, J.L., and Rabache, M. (1985). Occurrence of aminoacid pyrolysis products in roasted coffees and overgrilled cereal products. Sci. Aliments, 5, hors série, 199-203.

Baker, D.H., Yen, J.T., Jensen, A.H., Teeter, R.G., Michel, E.N., and Burns, J.H. (1976). Niacin activity in niacin amide and coffee. Nutr. Repts. Int., 14, 115-120.

Barbiroli, G. (1966). Ricerche sperimentali sulla tostatura degli alimenti : la trasformazione della trigonellina in acido nicotinico. Univ. Messina, atti del V convegno della qualita, 10-12 septembre.

Barton-Wright, E.C. (1944). The microbiological assay of nicotinic acid in cereal and other products. Biochem. J., 38, 314-319.

Boddeker, H. and Mishkin, A.R. (1963). Determination of nicotinic acid in coffee by paper chromatography. Anal. Chem., 35, 1662-1663.

Bracco, U. (1973). Détermination des hydrocarbures polycycliques aromatiques : technique et application aux huiles de café. Riv. Ital. Sost. Grasse., 50, 166-176.

Bressani, R. and Navarrete, D.A. (1959). Niacin content of coffee in Central America. Food Res., 24, 344-351.

Bressani, R., Fiester, D., Navarrete, D.A., and Scrimshaw, N.S. (1961). Effect of processing method and variety on niacin and ether extract content of green and roasted coffee. Food Technol., 15, 306-308.

Bressani, R., Gomez-Brenes, R., and Conde, R. (1962). Cambios de la composicion quimica del grano y de la pulpa del café durante el proceso de tostacion, y actividad biologica de la niacina del café. Arch. Venez. Nutr., 12, 93-104.

Callahan, M.M., Rohovsky, M.W., Robertson, R.S., and Yesair, D.W. (1979). The effect of coffee consumption on plasma lipids, lipoproteins and the development of aortic atherosclerosis in rhesus monkey fed an atherogenic diet. Am. J. Clin. Nutr., 32, 834-845.

Carvalho, A. (1962). Variability of the niacin content in coffee. Nature, 194, 1096.

Challis, B.C. and Bartlett, C.D. (1975). Possible cocarcinogenic effects of coffee constituents. Nature, 254, 532-533.

Chassevent, F. (1969). L'acide chlorogénique : ses actions physiologiques et pharmacologiques. Ann. Nutr. Alim., 23, 1-14.

Cravioto, R.O., Guzman, J.G., and Suarez, M.L.S. (1955). Increments del contenido de niacina durante la torrefaccion del café y su significado. Ciencia, 15, 24-26.

Czok, G. (1977). Kaffee und Gesundheit. Z. Ernährungsw., 16, 248-255.

Daum, M.G. (1955). La niacina en el café y su importancia nutricional en Venezuela. Arch. Venez. Nutr., 6, 61-70.

Davis, R.G. (1978). Increased bitter taste detection thresholds in Yucatan inhabitants related to coffee as a dietary source of niacin. Chem. senses flavor, 3, 423-429.

Feinberg, L.J., Sanberg, H., de Castro, O., and Bellet, S. (1976). Effects of coffee ingestion on oral glucose tolerance curves in normal human subjects. Metabolism, 17, 916-922.

Fritz, W. (1969). Zum Lösungsverhalten der Polyaromate beim Kochen von Kaffee-Ersatzstoffen und Bohnenkaffee. Deut. Lebensm. Rdsch., 65, 83-85.

Fritz, W. (1975). Entstehen bei der Zubereitung von Lebensmitteln krebserzeugende Stoffe ? Ernährungsforschung. Nahrung, 18, 83-87.

Gassmann, B. and Schneeweiss, R. (1959). The vitamins B1, B2 and PP in the profile of normally baked breads and those baked with infrared radiation. Nahrung, 3, 42-54.

Goldsmith, G.A., Miller, O.N., Unglaub, W.G., and Kercheval, K. (1959). Human studies of biological availability of niacin in coffee. Proc. soc. Exp. Biol. Med., 102, 579-580.

Gounelle de Pontanel, H. and Astier-Dumas, M. (1969). Café et Santé : la place du café en diététique. Bull. Acad. Nat. Médecine, 153, 628-640.

Hughes, E.B. and Smith, R.F. (1946). The nicotinic acid content of coffee. J. Soc. Chem. Ind., 65, 284-286.

Kitts, D.D. and Mathieson, R. (1989). Effects of caffeine on ovine maternal glucose-insulin response and fetal metabolite levels. Nutr. Repts. Int., 40, 673-684.

Kuratsune, M. and Hueper, W.C. (1960). Polycyclic aromatic hydrocarbons in roasted coffee. J. Nat. Cancer Inst., 24, 463-469.

Miller, A.B. (1983). Coffee and Cancer. Carcin. Mutag. Environm., 3, 13-20.

Naismith, D.J., Akinyanju, P.A., and Yudkin, J. (1969). Influence of caffeine-containing beverages on the growth, food utilization and plasma lipids of the rat. J. Nutr., 97, 375-381.

Panigrahi, G.B. and Rao, A.R. (1983). Influence of caffeine on arecoline-induced SCE in mouse bone-marrow cells in vivo. Mutation Res., 122, 347-353.

Pozniak, P.C. (1988). The carcinogenicity of caffeine and coffee. J. Am. Diet. Assoc., 85, 1125-1133.

Querat, F. and Heraud, G. (1987). Pathologie du café. Gaz. Médicale, 94, n° 33, 37-40.

Sandler, R.S. (1983). Diet and cancer : food additives, coffee and alcohol. Nutr. Cancer, 4, 273-279.

Slattery, M.L., West, D.W., and Robinson, L.M. (1988). Fluid intake and bladder cancer in Utah. Intern. J. Cancer, 42, 17-22.

Soos, K. and Fözy, I. (1974). The content of polyaromatic hydrocarbon carcinogens in various coffee types. Edisipar., 25, 7-10 ; 37-40 ; 65-69.

Stavric, B. (1988). Methylxanthines : toxicity to human. 2 Caffeine. Food Chem. Toxic., 26, 645-662.

Strobel, R.G.K. (1974). The determination of 3-4 benzopyrene of coffee products. 6eme Coll. Chimie Café, Bogota, 128-134.

Strubelt, O., Siegers, C.P., Breining, H., and Steffen, J. (1973). Tierexperimentelle Untersuchungen zur chronischen Toxizität von Kaffee und Coffein. Z. Ernährungsw., 12, 252-260.

Taguchi, H., Sakaguchi, M., and Shimabayashi, Y. (1985). Trigonelline content in coffee beans and the thermal conversion of trigonelline into nicotinic acid during the roasting of coffee bean. Agric. Biol. Chem., 49, 3467-3471.

Tchetche, A.G. (1979). Dosage quantitatif de la vitamine PP dans le Coffea canephora var. robusta par une méthode microbiologique utilisant Lactobacillus arabinosus. 8th Intern. Coll. on Coffee, 147-152.

Teply, L.J., Krehl, W.A., and Elvehjem, C.A. (1945). Studies on the nicotinic acid content of coffee. Arch. biochem., 6, 139-1.

Teply, L.J. and Prier, R.F. (1957). Nutritional evaluation of coffee including niacin bioassay. J. Agric. Food Chem., 5, 375-7.

Thomas, H.E.jr (1979). The relationship of coffee drinking to death and cardiovascular disease. 8th Intern. Coll. on Coffee, 305-310.

Trugo, L.C., Macrae, R., and Dick, J. (1983). Determination of purine alkaloids and trigonelline in instant coffee and other beverages using high performance liquid chromatography. J. Agric. Food Chem., 34, 300-306.

Trugo, L.C., Macrae, R., and Trugo, N.M.F. (1985). Determination of nicotinic acid in instant coffee using high-performance liquid chromatography. J. Micronutrient anal., 1, 55-63.

5

INTERACTION BETWEEN CASEIN AND VITAMIN A DURING FOOD PROCESSING

Annie Poiffait and Jean Adrian

Chaire de Biochimie Industrielle et Agro-alimentaire
Conservatoire National des Arts et Métiers
292, rue Saint-Martin, 75003 Paris, France

ABSTRACT

A particular relation exists between casein micelle and vitamin A.

In vitro, large amounts of retinol are fixed to acid casein and its different fractions by hydrophobic bindings. The binding on the hydrophobic amino-acid residues (Trp, Phe) is greatly facilitated by a configurational change in the molecule exposed to physico-chemical parameters : alcalinity and heat treatments. Another amount of retinol is even more strongly fixed by a binding which can only be broken by saponification.

Casein plays an important role in stabilizing retinol which does not degrade over time or during heat treatments.

In vivo, when the retinol availability is measured by the vitamin A content in the liver, acid casein somewhat increases the retinol efficiency and, even more so, the retinal efficiency. Nevertheless, this favourable action is only observed under particular conditions : balanced diet, casein from milk, etc.

For the moment, the animal data cannot be directly linked with the properties of binding appearing *in vitro*.

INTRODUCTION

The hydrophobic and phosphorylated character of casein influence many of the physico-chemical properties of this protein matter. This has had many important repercussions in the technological (Cheftel et al., 1985 ; Lorient et al., 1988) and nutritional fields (Adrian et Poiffait, 1987 ; Poiffait et Adrian, 1988). Owing to its high hydrophobicity (Bigelow, 1967), casein can be bound to various compounds such as detergents (Cheeseman et Knight, 1970 ; Marshall et Green, 1980), biliary acids (Lanzini et al., 1987), lecithins (Barratt et Raynen, 1972 ; Barratt et al., 1974), fatty acids (Kato et Nakaï, 1980 ; Kanazawa et al., 1987), pesticides (Rakotovelo et al., 1981), dyes (Aizawa et Takeyama, 1969 ; Lorient et Alais, 1974), etc.

A liposoluble substance such as retinol can also be bound to casein (Raica et al., 1959), probably by a mechanism like unsatured fatty acids (linolenic acid, parinaric acid) bound to the micelle.

Nutritional and Toxicological Consequences of Food Processing
Edited by M. Friedman, Plenum Press, New York, 1991

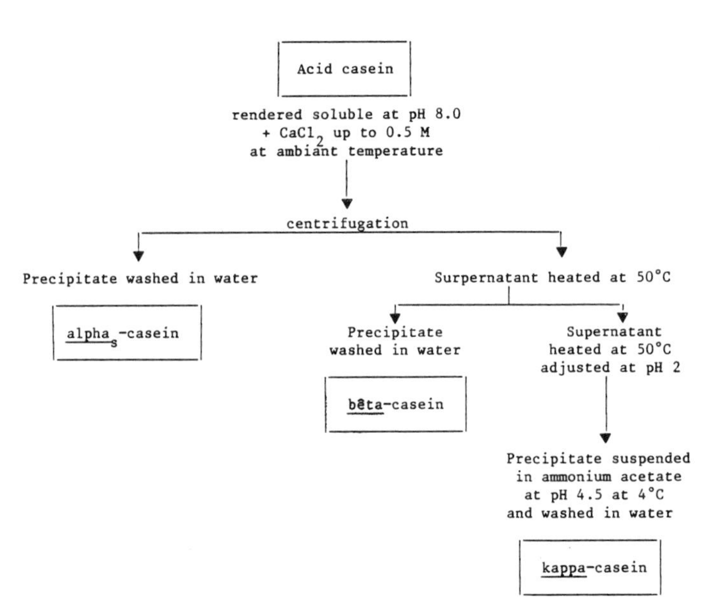

Figure 1. Preparation of casein fractions from acid casein

According to some nutritional results, casein tends to increase the efficiency of such trace nutrients as iron (Bron et al., 1967 ; Aschkenasy, 1967 ; Nelson et Potter, 1980), calcium (Wong et Lacroix, 1980), as well as vitamin A (Fraps, 1946 ; James et El Gindi, 1953 ; Raica et al., 1959 ; Olsen et al., 1959 ; Berger et al., 1962 ; Gronowska-Senger et Wolf, 1970 ; Kamath et Arnich, 1973 ; Underwood et al., 1979 ; Sharma et Misra, 1987 ; Wahid et Gerber, 1989).

Furthermore, we have observed that the reinforcement of vitamin A efficiency is a function of the spatial configuration of casein : the micelle insolubilization in the form of acid casein or rennet casein promotes nutritional activity in retinol, which is not observed with milk proteins in the native state (Adrian et al., 1984). In the same vein, the conversion of casein into caseinate seems to counteract this favourable action on vitamin A efficiency (Faruque et Walker, 1970 ; Adrian et al., 1984). Bêta-lactoglobulin, another hydrophobic milk protein (Kinsella, 1982) seems to possess the same properties as shown by Said et al. (1989) : it increases the intestinal absorption of retinol. This is attributed to the hydrophobic binding of protein and retinol (Hemley et al., 1979).

This paper summarizes some of our findings relevant to the binding *in vitro* between caseins and vitamin A as well as observations on laboratory rats showing an increase in retinol availability in dietary casein. However, it has not been shown that the mechanism observed *in vitro* has a direct relation to the increased nutritional efficiency found in the animals.

EXPERIMENTAL SECTION

MATERIAL AND PRODUCTS

The acid casein, rennet casein, caseinate, bêta-lactoglobulin, ovalbumin, gelatin, soybean isolate (Soyamin 90) used were commercial products. The different forms of vitamin A (all-trans retinol, all-trans retinal, retinol palmitate) used were supplied by Sigma Chemical Co. Casein micelle was prepared from crude milk according to the Van Hooydonk method (1981). The $alpha_s$, bêta and kappa-caseins were obtained according to the Girdhar et Hansen procedure (1978) modified as shown in **Figure 1.**

Table 1. Chemical composition of used caseins

	acid casein	$alpha_s$-casein	bêta-casein	kappa-casein
In % of dry matter				
-crude ash	0.80	1.34	4.78	11.89
-phosphorus	1.15	0.62	0.49	0.27
-crude protein	96.46	99.64	91.69	87.62
In % of protein				
-$alpha_s$-casein		78	90	0
-bêta-casein		16	0	0
-kappa-casein		traces	traces	100

Their purity was compared to commercial products (Sigma) by chemical analysis and high-performance size-exclusion chromatography on a Shodex WS 802.5 column. The results appear in **Table 1.**

VITAMIN A BINDING ASSAY

The investigated protein (1.25 10^{-3}M) was suspended during 12 hours in an appropriate buffer (citrate, phosphate or borax) at a determinated pH. Alcoholic solutions of retinol (0.875 10^{-4}M) and of BHT (0.227 10^{-3}M) were added ; the final concentration in ethanol was 20%. This mixture was immediatly shaken for 20 min. Then, the protein was centrifuged or ultrafiltrated according to its stage of hydrodispersibility ; it was washed and suspended in distilled water. Nitrogen and vitamin A were analyzed in the soluble and insoluble fractions to determine the binding rate of vitamin A. All the operations were carried out in nitrogen, sheltered from UV rays.

VITAMIN A DETERMINATION

After being precipitated in alcohol, the collected proteins were either saponified or not. The retinol was then extracted with hexane (3 extractions) and analyzed by HPLC (Rabache et Adrian, 1981).

PROTEIN DETERMINATION

The crude protein was obtained by the Kjeldahl method. The result was converted into protein by the specific factor of 6.38 for caseins and milk proteins, of 5.71 for soybean and of 6.25 for other products.

MEASURE OF U.V. ABSORBANCE

In order to characterize the binding of vitamin A to protein, the UV spectra were first established for vitamin A and casein, and then for the mixtures before and after extraction by hexane at 76° C. Kappa-casein was suspended (1.25 10^{-5}M) in a pH 7 - ethanol buffer solution (80:20) with retinol (0.55 10^{-5}M) or with retinal (0.905 10^{-5}M).

MEASURE OF FLUORESCENCE

The fluorescent intensity of hydrophobic pockets of casein was determined with a Kontron Fluorescence Spectrophotometer (Model SFM 25) according to the Sklar et al. procedure (1976, 1977a, 1977b). The bêta-casein

Legend -

kappa casein : 1.25 x 10^{-5} M
(in buffer of pH 7.0)

retinol : 0.55 x 10^{-5} M

retinal : 0.805 x 10^{-5} M

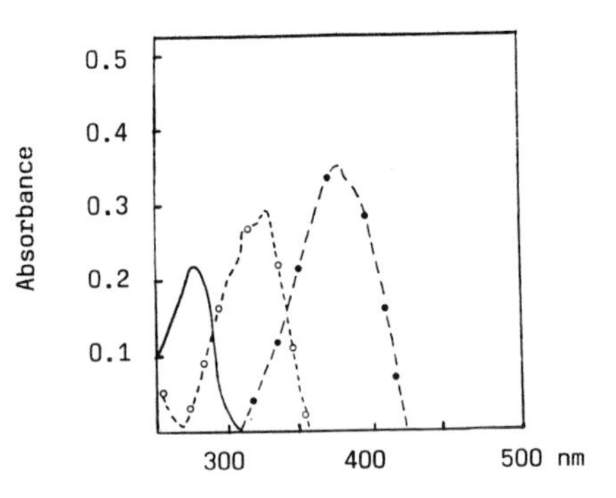

Figure 2. Absorption spectra of kappa casein (full line), retinol (white circle) and retinal (black circle).

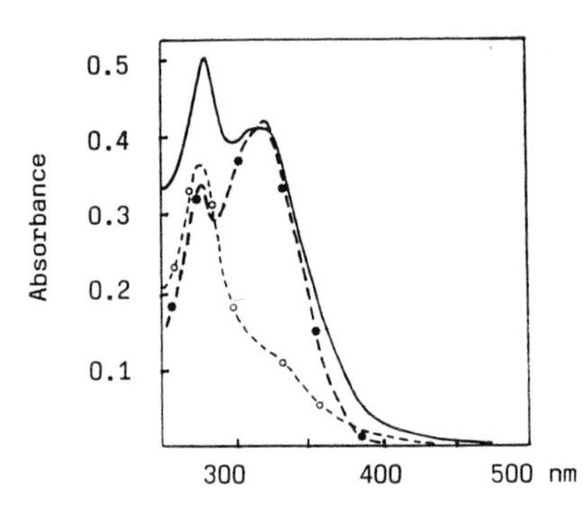

Figure 3. Absorption spectra of kappa casein - retinol mixture.

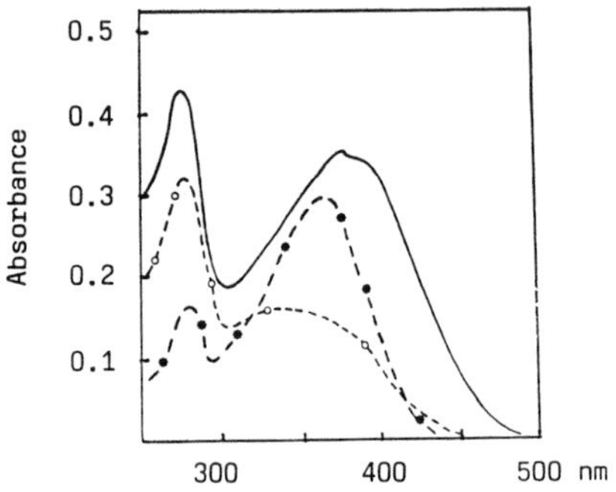

Figure 4. Absorption spectra of kappa casein - retinal mixture.

Legend -

full line : crude mixture

black circle : hexane phase after treatment with hexane at 76° C during 20 min.

white circle : residual aqueous phase, after hexane treatment.

(1.5 10^{-6}M) was suspended in the buffer solution (pH 7.0) in presence of retinol or retinal. Tryptophan was excited at 290 nm and its fluorescence emission was measured at 356 nm. The retinol excitation was produced at 348 nm and its emission was observed at 477 nm.

ANIMAL EXPERIMENTS

Weaned male Wistar rats received a maintenance diet containing no vitamin A, for a period of 15 to 18 days. At the end of this period, the liver contained 11.9 $\pm$ 1.9 μg of retinol per organ.

The animals were then divided into groups of 9 to 12 rats. They received a diet with a reduced lipid content (3% of soybean oil) and a variable level of protein and of vitamin A.

After 28-30 days, the animals were killed. The vitamin content in blood and in the liver was determined by HPLC and the total cholesterol by colorimetry (Crawford, 1958).

STATISTICAL ANALYSIS

Results were expressed as the mean ± SEM. Data were analyzed using the Student's "t" test and regression analysis.

RESULTS

NATURE OF VITAMIN A-CASEIN BINDING

Figure 2 shows the absorption spectra of kappa-casein (in buffer solution pH 7.0) of retinol and retinal (in alcohol solution). The absorption peaks are 278 nm for kappa-casein, 328 nm for retinol and 390 nm for retinal.

Figure 3 and 4 show the absorption spectra of the mixtures. The vitamin compounds create a hyperchrome effect on protein absorbance : the absorption peak at 278 nm increases from about 0.2 to 0.4 and 0.5 of absorbance.

Moreover, casein develops a bathochrome effect on retinal where the spectrum streches to 490 nm as opposed to 410 nm in the pure solution. The treatment of the two mixtures, casein-retinol and casein-retinal, with hexane at 76° C for 20 min., shows that the initial constituents are not dissociated totally from complexes : in the hydrocarbon phase, the characteristic peak of vitamin compounds are found along with a typical peak of casein. On the other hand, in the aqueous phase, the casein absorbance spectrum reveals a slight presence of retinol or retinal. This is particularly apparent with vitamin A aldehyde.

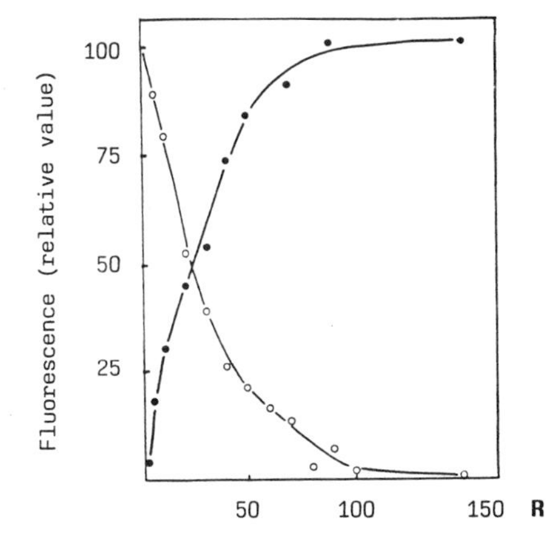

Figure 5. Fluorescence of bêta casein-retinol mixture : extinction of tryptophan (white circle) and emission of retinol bound to protein (black circle) in function of the ratio R = [retinol]/[bêta casein].

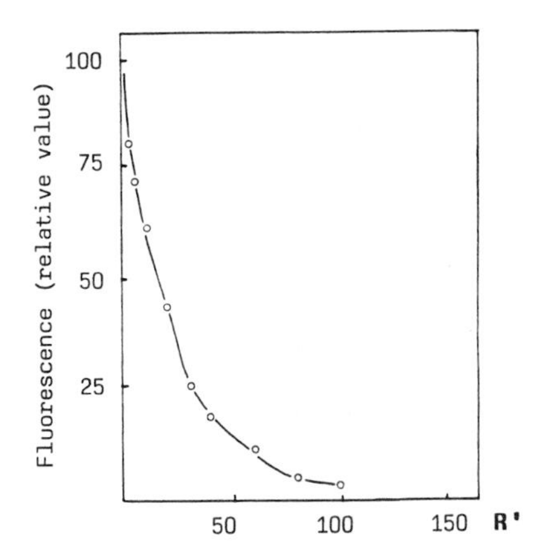

Figure 6. Fluorescence of bêta casein - retinal mixture : extinction of tryptophan (white circle) in function of the ratio R' = [retinal]/[bêta casein]

The absorbance spectra show that vitamin compounds are fixed on different sites of kappa-casein : vitamin A bound to the most hydrophobic fractions is extracted with hexane ; on the contrary, if binding is achieved with the hydrophilic fraction, vitamin A remains in its aqueous phase bound to the hydrophilic peptides, owing to the protein's polar properties.

This is confirmed by the saponification of the aqueous phase of the mixtures : after saponification, 12.2% of initial retinol and 14.2% of retinal are extracted by hexane. These amounts are linked to kappa-casein by binding which can only be broken down through saponification. There is therefore a possibility of strong binding with casein, whereas the retinol-bêta-lacto-globuline binding is totally broken down by extraction with hexane (Hemley et al., 1979). If these bindings are chemical, they can be different in the alcohol and aldehyde forms of the vitamin ; a retinol esterification can occur in the phosphoryl groups of casein, as well as a reaction between the aldehyde vitamin A and the amine function of the protein in the case of retinal.

However, the main vitamin A-casein binding is weak because it is broken down during hexane extraction. The changes in the fluorescence spectra of tryptophan in the presence of retinol or retinal (**Figures** 5 and **6**) indicate a hydrophobic bond near the tryptophyl residue sites and near the aromatic amino acid of casein.

The presence of these vitamin-compounds causes the progressive quenching of the amino acid : when the ratio of retinol/bêta casein reaches 100:1, the Trp fluorescence disappears completely and is simultaneously transferred to the retinol where the fluorescence emission reaches its maximum (**Figure** 5). The situation is similar with retinal except that no fluorescence is produced.

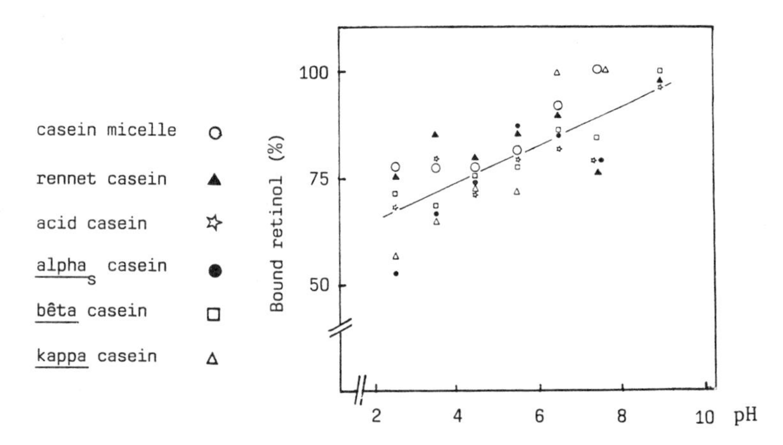

Figure 7. Relative capacity for binding retinol on caseins, according to the pH value.

The similarity of tryptophan-retinol and tryptophan-retinal bindings indicates that the mechanism cannot be dependent on the terminal alcohol or aldehyde function of vitamin A. Consequently, these substances demonstrate general properties like parinaric acid, a tetraunsaturated fatty acid which is bound to tryptophan by hydrophobic interactions (Sklar et al., 1976). It therefore can be concluded that tryptophan and casein aromatic amino acids are very capable of establishing hydrophobic bonds with vitamin A, as it has been observed with many water repellent compounds (fatty acids, pesticides, etc.).

CONDITIONS FOR VITAMIN A-CASEIN BINDING

Acid casein, rennet casein, caseinate, fractions of casein are able to fix vitamin A at the same ratio (**Figure 7**). This general property depends first on the pH of the protein suspension. On the whole, when the pH rises from 2.5 to 9.0, the retinol amounts are doubled : this progression is linear and is represented by the equation $Y = 4.43 X + 59.9$, where Y is the amount of bound retinol and X is the pH value. These two parameters correlate significantly ($r = + 0.974$). In fact, the rise in pH increases the negative charges of the protein progressively and provokes their spreading. Thus, the hydrophobic sites become more available and accessible for hydrophobic compounds. On the contrary, in the acid medium, the hydrophobic sites are established inside the nitrogenous structure and become much less accessible (Cheftel et al., 1985).

Table 2. Relation between the hydrodispersion of acid casein heated at pH 2.5 and its capacity for binding retinol (unpublished)

Heat treatment (30 min.)	Casein hydrodispersibility (%)	Bound retinol (%)
20° C	35	23
50° C	14	21
75° C	80	83
100° C	100	100
135° C	92	88

Table 3. Change of retinol bound to acid casein during storage at 4° C and in function of pH (unpublished)

Conservation at 4° C (in days)	% of recovered retinol				
	pH 4.5			**pH 9**	
	retinol in solution	retinol bound to casein : in aqueous phase	in lyophilized state	retinol in solution	retinol bound to casein, in aqueous phase
0	78 ± 13	70 ± 10	45.5 ± 4	69 ± 8	96 ± 2
1	34 ± 5	51 ± 9	39.2 ± 2	49.5 ± 7	100 ± 2
6	27.5 ± 3	25.5 ± 4	32 ± 2	51 ± 8	100 ± 2
15	27.5 ± 4	25 ± 5	25.5 ± 3	25.5 ± 3	92.5 ± 2
50	12 ± 1	11 ± 2	25.5 ± 2	19 ± 2	88 ± 2

The casein hydrodispersion is therefore an essential precondition for binding with a hydrophobic element like vitamin A. The other proteins studied (ß lactoglobulin, ovalbumin, gelatin, soybean) show a certain ability to fix retinol and the conditions under which they react are distinctly different ; in particular, their binding capacities don't seem proportional to the pH value (unpublished) : this, in fact, seems characteristic of the casein micelle and of its fractions.

The thermic treatments which cause casein hydrodispersion, allow for greater binding with retinol, confirming earlier observations. Thus, a 30 min. heating at pH 2.5 promotes the dispersion of acid casein and allows for retinol binding in relation to the level of hydrodispersibility obtained by the protein ; the two parameters are closely linked together (**Table 2**). These results also indicate that binding sites are very thermostable and that they don't affect the secondary functions of amino acids because most of them are degraded during high heatings (Adrian, 1975 ; Lorient, 1978). Indirectly, these data reinforce the hypothesis of hydrophobic bindings between retinol and casein.

CONSEQUENCES OF CASEIN-RETINOL BINDING

Consequences in the technological field

Under some conditions, the fixation of retinol on casein insures high vitamin A protection (**Table 3**). This fact was revealed by comparing the retinol stability in an alcoholic solution (A, D) either after fixation in casein suspension (B, E) or after the casein-retinol binding has been lyophilized (C). The experiments were made with acid casein at pH 4.5 (A, B, C) or 9.0 (D, E).

At pH 4.5, retinol bound to casein is not protected, at least while it is in an aqueous solution (B). On the other hand, if the retinol-casein mixture is lyophilized after binding, a twofold evolution is observed : the lyophilization provokes a significant retinol reduction probably because at - 75° C the low energy bindings are broken. After that, the retinol which remains inside the micelle demonstrates increased stability over time (C). In relation to the initial rates of retinol recovered (45.5% in lyophilized state and 70% in aqueous phase), a 50 day storage period results in a 43% loss of retinol in lyophilized casein versus a 84% loss after remaining in aqueous dispersion for the same period of time. At pH 4.5 casein contracts and folds together : this dense configuration can create a physical barrier between the oxygen in the air and the retinol maintained inside the protein structure of the lyophilized product.

Table 4. Change of retinol bound to acid casein at pH 7.5 during heat treatments (unpublished)

Heat treatment (30 min.)	% of recovered retinol	
	retinol in solution	retinol bound to casein, in aqueous phase
20° C	89	74
50° C	96	100
100° C	79	96
135° C	21	80

At pH 9.0, casein exerts a significant protection against retinol degradation (D, E) : after 50 days at 4° C, casein in suspension almost completely prevents all retinol break down. Under these conditions, it acts like antioxygen matter, as it has been observed with linoleic acid (Kanazawa et al., 1987). As a matter of fact, as soon as casein is in alcaline suspension, the unfolding of the molecule has a remarkable effect upon retinol stability. **Table 4** shows the protective effect of casein at pH 7.5 on retinol when heated up to 135° C. Even under these drastic thermic conditions, the binding on the unfolded casein may recover 80% of vitamin A after heat treatment.

All of these observations demonstrate the very important technological potential of casein when it is in an alcaline medium. Our results underline the dominant role of pH and elucidate casein's lowered protection action observed in milk (Lau et al., 1986 ; Mac Carthy et al., 1986) and in yoghurt (Ilic et Ashoor, 1988) owing to the acid nature of these products.

Consequences in the nutritional field

At present, casein's ability to fix retinol _in vitro_ cannot be directly connected to some observations on the increased nutritional efficiency of

Table 5. Role of acid casein on retinol and cholesterol content in the liver (Poiffait et al., 1985)

	Acid casein in diet (%)			
	5	10	15	25
retinol in % of ingested retinol	17.8	29.0	20.9	16.6
total cholesterol (m mol)	0.26	0.31	0.53	0.61

retinol in casein diets. However, we can associate the two phenomena even if the increase of retinol efficiency only becomes evident under limited conditions :

- it is efficient when casein is obtained chemically (acid casein) or enzymatically (rennet casein), but it is inefficient when it is introduced into milk or transformed into caseinate (Faruque et Walker, 1976 ; Adrian et al., 1984) ;
- its action is not directly proportional to its amount in the diet : it reaches maximal efficiency at a rate of 10% (**Table 5**) (Poiffait et al., 1985, 1988a). This means that the mechanism promoting the accumulation of retinol in the liver is different from that which supervises the free cholesterol content in liver (Poiffait et al., 1985) ;
- at a normal level in the diet (1.3 µg/kg) retinol's efficiency is clearly enhanced by acid casein ; however when excessive levels of retinol are present (6.5 µg/kg) acid casein becomes much less effective (Poiffait et al., 1988a). In this case, the liver vitamin A reflects only the retinol intake (Chew et Archer, 1984 ; Donoghue et al., 1983) ;

Table 6. Role of acid casein and isolated soybean on the nutritional efficiency of retinol or retinal*
(Poiffait et al., 1988b)

	Diets			
	Retinol		Retinal	
	Soybean	Casein	Soybean	Casein
Intake (g/d)	10.95	10.85	10.70	11.90
Liver retinol (µg)				
total	51.1 ± 14.4	88.7 ± 11.5	39.1 ± 10.5	137.0 ± 19.5
free	2.95 ± 0.75	6.0 ± 1.07	3.48 ± 1.37	15.45 ± 2.54
Serum retinol (µg/100 ml)	50.1 ± 5.3	62.1 ± 4.2	42.4 ± 2.8	49.2 ± 2.5

* all diets contain about 10% of protein and exactly 1 mg of vitamin A/kg.

- but acid casein's effects are not limited to retinol. Retinal efficiency and its rate of conversion into retinol are greatly increased in diets containing 10% of casein : vitamin A alcohol is responsible for a liver content of 88.7 mg of retinol ; along with vitamin A aldehyde, the content reaches 137 µg (**Table 6**).

Table 7. Retinol efficiency in presence of the whole casein or of its fractions (unpublished)

	Rate of P (%)	Ad libitum (28 days) Intake (g/d)	Liver retinol (μg/organ)	Pair fed (24 days) Intake (g/d)	Liver retinol (μg/organ)
whole casein	1.15	10.84	88.7 ± 11.5	5.24	131.4 ± 18.4
$alpha_s$-casein	0.62	-	-	5.05	77.0 ± 10.3*
kappa-casein	0.27	10.84	114.1 ± 25.0	-	-

* significant difference with diet containing the whole casein ($P < 0.05$).

Therefore, we cannot retain the hypothesis of an esterification reaction between retinol and the phosphoric function of casein which should assure better stability or increased absorption of vitamin A ;

- moreover, there is no relation between the phosphorus content of the casein fractions and the liver content in retinol (**Table 7**). With regard to whole casein, kappa-casein seems to increase the vitamin A content in the liver (insignificant) while $alpha_s$-casein lowers it ($P < 0.05$).

Briefly, these results suggest that acid casein favours or reinforces the vitamin A availability only when the ingesta furnish just sufficient amounts of protein and retinol to cover the nutritional requirements. **Table 6** illustrates very well the properties of acid casein on soybean isolate : in all experiments, acid casein promotes an important and significant increase of the retinol content in the liver in balanced diets.

ACKNOWLEDGMENTS

The authors would like to express their thanks to Kathryn Harper for her assistance in the revision of the article.

REFERENCES

Adrian, J. (1975). Les traitements thermiques appliqués aux produits laitiers et leurs consequences dans le domaine azoté. Le lait, 55, 24-40 and 182-206.

Adrian, J., Frangne, R., and Rabache M. (1984). Rôle des protéines laitières sur l'efficacité du rétinol. Sci. Aliments, n° Hors série III, 305-308.

Adrian, J., and Poiffait, A. (1987). La spécificité de la caséine. 1e partie : ses caractéristiques dans le domaine azoté. Med. et Nutr., 23, 377-384.

Aizawa, H., and Takeyama, I. (1969). (Interaction of food colours with protein. VIII. Binding capacity and free energy of binding on isoxanthene colour by protein). Eiyo to Shokuryo, 22, 235-239 (FSTA, 1971, 10T525).

Aschkenasy, A. (1967). Absorption gastrointestinale du 59-Fe et répartition tissulaire du radio-fer absorbé chez les rats carencés en protéines et restaurés avec des régimes comportant de la caséine ou divers mélanges d'acides aminés. Arch. Sci. Physiol., 21, 127-151.

Barratt, M. D., Austin, J. P., and Whitehurst, R. J. (1974). The influence of the alkyl chain length of lecithins and lysolecithins of their interaction with α_{s1}-casein. Biochim. Biophys. Acta, 348, 126-135.

Barratt, M. D., and Rayner, L. (1972). Lysolecithin casein interactions. I. Nuclear magnetic resonance and spin label studies. Biochim. Biophys. Acta, 255, 974-980.

Berger, S., Rechcigl, M., Loosli, J. K., and Williams, H. H. (1962). Protein quality and carotene utilization. J. Nutr., 77, 174-178.

Bigelow, C. (1967). On the average hydrophobicity of proteins and the relation between it and protein structure. J. Theoret. Biol., 16, 187-211.

Bron, C., Blanc, C., and Isliker, H. (1967). Hydrolyse trypsique de transferrine et de lactotransferrine humaines. Helv. physiol. pharmacol., 25, 337-352.

Cheeseman, G. C., and Knight, D. J. (1970). The interaction of bovine milk caseins with the detergent sodium dodecyl sulphate. II. The effect of detergent binding on spectral properties of caseins. J. Dairy Res., 37, 259-267.

Cheftel, J. C., Cuq, J. L., and Lorient, D. (1985). "Proteines alimentaires". Technique et Documentation, Lavoisier, Paris.

Chew, B. P., and Archer, R. G. (1983). Comparative role of vitamin A and ß-carotene on reproduction and neonate survival in rats. Theriogenology, 20, 459-472.

Crawford, N. (1958). An improved method for the determination of free and total cholesterol using the ferric chloride reaction. Clin. Chem. Acta, 3, 357-367.

Donoghue, S., Richardson, D. W., Sklan, D., and Kronfeld, D. S. (1985). Placental transport of retinol in ewes fed high intakes of vitamine A. J. Nutr., 115, 1562-1571.

Faruque, O., and Walker, D. M. (1970). Vitamin A and protein interrelationship in the milk fed lamb. Brit. J. Nutr., 24, 11-22.

Fraps, G. S. (1946). Effect of bulk, casein and fat in the ration on the utilization of carotenes by white rats. Arch. Biochem., 10, 485-489.

Girdhar, B. K., and Hansen, P. M. T. (1978). Production of -casein concentrate from commercial casein. J. Food Sci., 43, 397-406.

Gronowska-Senger, A., and Wolf, G. (1970). Effect of dietary protein on the enzyme from rat and human intestine which converts ß-carotene to retinal. J. Nutr., 100, 300-308.

Hemley, R., Kohler, B. E., and Siviski, P. (1979). Absorption spectra for the complexes formed from vitamin A and ß-lactoglobulin. Biophys. J., 28, 447-455.

Ilic, D. B., and Ashoor, S. H. (1988). Stability of vitamins A and C in fortified yogurt. J. Dairy Sci., 71, 1492-1498.

James, W. H., and El Gindi, I. M. (1953). The utilization of carotene. 1. As affected by certain proteins in the diet of growing albino-rats. J. Nutr., 51, 97-108.

Kamath, S. K., and Arnrich, L. (1973). Effect of dietary protein on the intestinal biosynthesis of retinol from ^{14}C-ß-carotene in rats. J. Nutr., 103, 202-206.

Kanazawa, K., Ashida, H., and Natake, M. (1987). Autoxidizing process interaction of linoleic acid with casein. J. Food Sci., 52, 475-478.

Kato, A., and Nakai, S. (1980). Hydrophobicity determined by a fluorescent probe method and its correlation with surface properties of proteins. Biochim. Biophys. Acta, 624, 13-20.

Kinsella, J. E. (1982). Relationship between structure and functional properties of food proteins. In "Food proteins", P.F. Box, and J.J. Condon, eds., Applied Science Publishers, London and New-York, p. 51-103.

Lanzini, A., Fitzpatrick, W. J. F., Pigozzi, M. G., and Northfield, T. C. (1987). Bile acid binding to dietary casein : a study in vitro and in vivo. Clin. Sci., 73, 343-350.

Lau, B. L. T., Kakuda, Y., and Arnott, D. R. (1986). Effect of milk fat on the stability of vitamin A in ultra-high temperature milk. J. Dairy Sci., 69, 2058-2059.

Lietaer, E., Poiffait, A., and Adrian, J. (1990). Nature et propriétés de la liaison vitamine A-caséine. Lebensm. Wiss. Technol. (soumis).

Lorient, D. (1978). Utilisation de protéines pures et de peptides pour l'étude des modifications chimiques et nutritionnelles subies par le lait au cours du chauffage. Ann. Nutr. Alim., 32, 391-406.

Lorient, D., and Alais, Ch. (1974). Dégradation thermique des caséines et ß de vache. II. Modifications des propriétés physicochimiques. α_s Biochimie, 56, 667-673.

Lorient, D., Colas, B., and Le Meste, M. (1988). "Propriétés fonctionnelles des macromolécules alimentaires". Les Cahiers de l'ENSBANA (Dijon), Technique et Documentation, Lavoisier, Paris.

Mac Carthy, D. A., Kakuda, Y., and Arnott, D. R. (1986). Vitamin A stability in ultra-high temperature process milk. J. Dairy Sci., 69, 2045-2051.

Marshall, R. J., and Green, M. L. (1980). The effect of the chemical structure of additives on the coagulation of casein micelle suspension by rennet. J. Dairy Res., 47, 359-369.

Nelson, K. J., and Potter, N. N. (1980). Iron availability from wheat gluten, soy isolate and casein complexes. J. Food Sci., 45, 52-55.

Olsen, E. M., Harvey, J. D., Hill, D. C., and Branion, H. D. (1959). Effect of dietary protein and energy levels on the utilization of vitamin A and carotene. Poultry Sci., 38, 942-949

Poiffait, A., and Adrian, J.(1988). La spécificité de la caséine. 2e partie : son action sur l'efficacité des micro-nutriments. Med. et Nutr., 24, 9-15.

Poiffait, A., Karisto, T., and Adrian, J. (1985). Influence de la caséine sur les quantités sanguines et hépatiques de rétinol et de cholestérol. Sci. Aliments, 5, n° Hors série V, 127-132.

Poiffait, A., Karisto, T., and Adrian, J. (1988a). Influence of vitamin A form and protein nature on the retinol status in the rat. Int. J. vit. nutr. res., 58, 33-36.

Poiffait, A., Moustaïzis-Carpelli, E., Karisto, T., and Adrian, J. (1988b). Effects of soybean and casein on the retinol status in the rat. Int. J. vit. nutr. res., 58, 27-31.

Rabache, M., and Adrian, J. (1981). Effets physiologiques des pigments d'Aspergillus Niger. 1. Propriétés des phénylpolyènes. Sci. Aliments, 1, 577-585.

Raica, N., Vavich, M. G., and Kemmerer, A. R. (1959). The effects of several milk components and similar compounds on the utilization of carotene by the rat. Arch. Biochem. Biophys., 83, 376-380.

Rakotovelo, V., Lhuguenot, J. C., and Lorient, D. (1982). Fixation des pesticides organo-phosphorés sur les constituants protéiques du lait. Le Lait, 62, 531-540.

Said, H. M., Ong, D. E., and Singleton, J. L. (1989). Intestinal uptake of retinol : enhancement by bovine milk ß-lactoglobulin. Am. J. Clin. Nutr., 49, 690-694.

Sharma, H. S., and Misra, U. K. (1987). Distribution of vitamin A in various organs of rats in relation to the quality and the quantity of dietary proteins. Zeitschrift für Ernährungswissenschaft, 26, 43-51.

Sklar, L. A., and Hudson, B. S. (1976). Conjugated polyene fatty acids as fluorescent membrane probes : Model system studies. J. Supramolecular Structure, 4, 449-465.

Sklar, L. A., Hudson, B. S., Petersen, M., and Diamond, J. (1977a). Conjugated polyene fatty acids on fluorescent probes : spectroscopic characterization. Biochemistry, 16, 813-819.

Sklar, L . A., Hudson, B. S., and Simoni, R. D. (1977b). Conjugated polyene fatty acids as fluorescent probes : binding to bovine serum albumine. Biochemistry, 16, 5100-5108.

Underwood, B. A., Loerch, J. D., and Lewis, K. C. (1979). Effects of dietary vitamin A deficiency, retinoic acid and protein quantity and quality on serially obtained plasma and liver levels of vitamin A in rats. J. Nutr., 109, 796-806.

Van Hooydonk, A. C. M., Hagedoorn, H. G., and Boerrigter, I. J. (1986). pH induced physico-chemical changes of casein micelles in milk and their effect on rennetting. 1. Effect of acidification on physico-chemical properties. Neth. Milk and Dairy J., 40, 281-296.

Wahid, A., and Gerber, L. E. (1989). Effects of varying protein intake on wound healing tissue bêta-carotene concentration in rats fed bêta-carotene. Nutr. Rep. Int., 40, 621-626.

Wong, N. P., and Lacroix, D. E. (1980). Biological availability of calcium in dairy products. Nutr. Rep. Int., 21, 673-680.

6

THERMAL DEGRADATION OF CAROTENES AND INFLUENCE ON THEIR PHYSIOLOGICAL FUNCTIONS

Lena Jonsson

SIK, The Swedish institute for food research
P.O. Box 5401, S-402 29 Göteborg, Sweden

ABSTRACT

Raw carrot juice contains a considerable amount of α-and β-carotene, which makes carrot an excellent source of vitamin A. Heat treatment of the juice at temperatures comparable to those at pasteurization and boiling does not change the carotenes, while heating at temperatures used during sterilization results in rearrangement of the carotene molecules and a decrease in total carotenes. The all-trans α- and β-carotenes appear partly as cis-isomers, especially the 13-cis--isomer. Isomerization of the carotenes leads to a decrease in their vitamin A activity. Carotenes also seem to be anticarcinogens but the extent to which this property is influenced by isomerization is still unknown.

INTRODUCTION

Plenty of carotenoids occur in nature, both in vegetable and animal tissue. They are of varying color, such as yellow, orange, red or pink, and they also seem to have antioxidative properties.

Nutritionally some carotenoids are known to exert vitamin A activity. The most potent vitamin A precursor is β-carotene.

Recently it has also been shown in epidemiological studies that diets rich in carotenoid-containing green-yellow vegetables protect humans against certain forms of cancer, particularly lung cancer. Similarly, a high level of β-carotene in serum reduces the risk of developing lung cancer. Long-term intervention trials with β-carotene supplement are in progress to confirm this findings. The mechanisms behind the protective action of carotenoids and the molecular shape of carotenoids most effective as anticarcinogens are still unknown.

Nutritional and Toxicological Consequences of Food Processing
Edited by M. Friedman, Plenum Press, New York, 1991

In vegetables, β-carotene primarily occurs as the all-trans isomer. This is also the most vitamin A active isomer, as shown by Bauernfeind (1972). Carotenoids are, however, reactive substances. They can oxidize and isomerize. When all-trans β-carotene is exposed to intense light or heat many cis-isomers are found.

During processing foods are often exposed to elevated temperatures, high or low pH levels, light, etc. - parameters which can accelerate the isomerization of carotenoids and decrease their vitamin A activity. The knowledge about the reaction rate of isomerization during common food preparation is insufficient, due to the lack of proper analytical methods. At SIK we have developed a rapid and reliable method to determine different carotenoids and their isomers.

This investigation was conducted to determine the effect of heat treatment on isomerization of α- and β-carotene in carrot juice and to calculate the influence on the pro-vitamin A activity of the juice.

EXPERIMENTAL SECTION

Materials and reagents. Mature carrots of the Duke variety were harvested in July and stored at -4°C for one month before juice preparation.

The α- and β-carotene standards were obtained from Sigma Chemical Company (St. Louis, MO). All chemicals used were either of analytical grade or of HPLC grade.

Preparation of carrot juice. The carrots were washed and processed in a raw juice centrifuge. The carrot juice was bottled in 1.5 ml brown, glass vials. The vials were stoppered with polyethylene caps and stored at -40°C up to six months before treatment.

Treatments. The vials with frozen carrot juice were placed at +4°C for 16 hours before heating.

Heat treatment was performed in a thermostatically controlled oil bath. The samples were heated to 80, 100, 121, 128 and 135°C. For each temperature three vials were removed at specific time intervals during 240 minutes. Immediately upon reaching the specified heating duration the vials were cooled in an ice bath for one minute. Extraction of carotenes was performed immediately.

Carotene extraction. The carotenes were extracted using equal amount of acetone and ethanol until the extraction solvent was colorless. The carotenes were then transferred to petroleum ether and evaporated to dryness using a rotary evaporator under vacuum and a water bath temperature of 35°C. The carotenes were resuspended in the HPLC mobile phase (hexane:acetone, 99.5:0.5, v:v) and then analyzed using adsorption HPLC. To minimize exposure to light the extraction was performed in dimmed (red) light with aluminum foil around all glasswares.

Chromatographic procedures. Adsorption HPLC with a slurry--packed calcium hydroxide (lime) column was used to separate cis/trans-

-isomers of α- and β-carotene. The mobile phase consisted of hexane: acetone, 99.5:0.5 (v:v). Tentative identification was made by measuring the retention time and UV/visible adsorption spectra obtained with a diode array detector. The adsorption was recorded simultaneously at 286, 347, 400, 442 and 450 nm. The carotenes were quantified by comparing peak areas of standards of known concentrations with those of the samples. The cis-isomers were presumed to have the same adsorption intensity as the corresponding trans-isomers. Detailed information of the HPLC procedure has been published elsewhere (Pettersson and Jonsson, 1990).

Evaluation of data. Calculations of pro-vitamin A activity of different carotenes were made using biopotency values for the major carotene stereo-isomers reported by Zeichmeister *et al.* (1962), see Table 1.

Table 1. Biopotency values of carotene stereo-isomers[1)]

Isomer	Biopotency
All-trans-β-carotene	100
9-cis- β-carotene	38
13-cis-β-carotene	53
All-trans-α-carotene	53
9-cis-α- carotene	13
13-cis-α-carotene	16

1) Values reported by Zeichmesiter (1962) based on all-trans-β-carotene as 100.

Small amounts of other carotenoids or stereo-isomers of undetermined vitamin A activity were also present. As these compounds make up less than 5% of the total carotene content in the carrot juices studied, they have not been included in the calculations.

RESULTS AND DISCUSSION

Carotenes in carrot juice

The raw carrot juice contained large amounts of α- and β-carotene, 21 and 88 μg/ml juice, respectively. Trace amounts of other carotenoids, such as phytoene and phytofluene were identified. The only stereo-isomer found in the raw juice was the all-trans form. Storage of the samples at -40°C before heat treatment did not change the carotene content.

Influence of heat treatment

Representative chromatograms of thermally treated and untreated juice samples are presented in Figure 1.

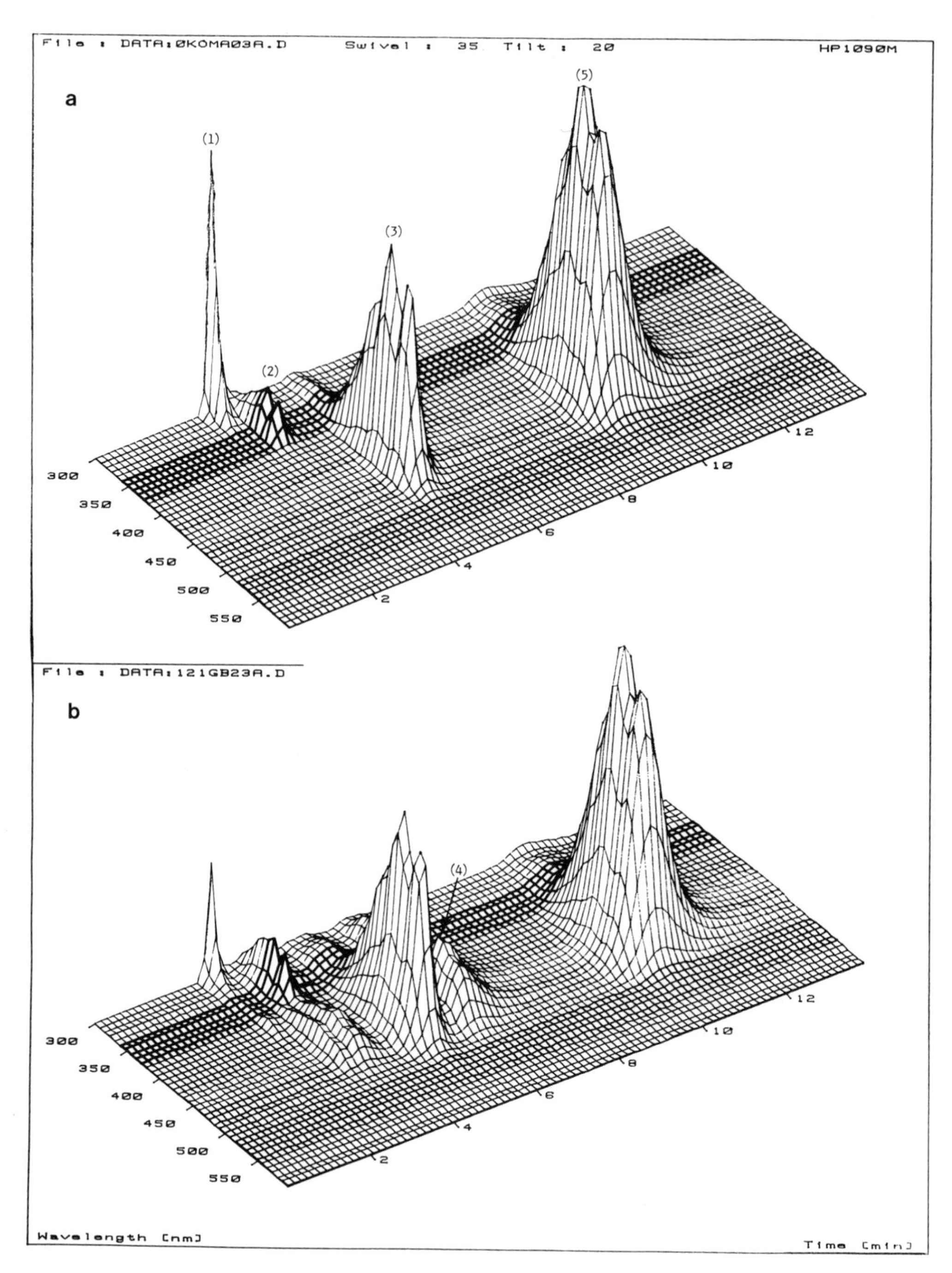

Figure 1. Representative chromatograms and spectra of carrot juice samples.
a) untreated sample
b) treated at 121 °C for 60 minutes.
(1) phytofluene; (2) phytoene; (3) all-trans-α-carotene; (4) 13-cis-β-carotene; (5) all-trans-β- carotene

As shown in Figure 1b, a considerable amount of the all-trans--carotenes are converted into cis-isomers during heating.

Figure 2 shows the effect of heating carrot juice at different times and temperatures.

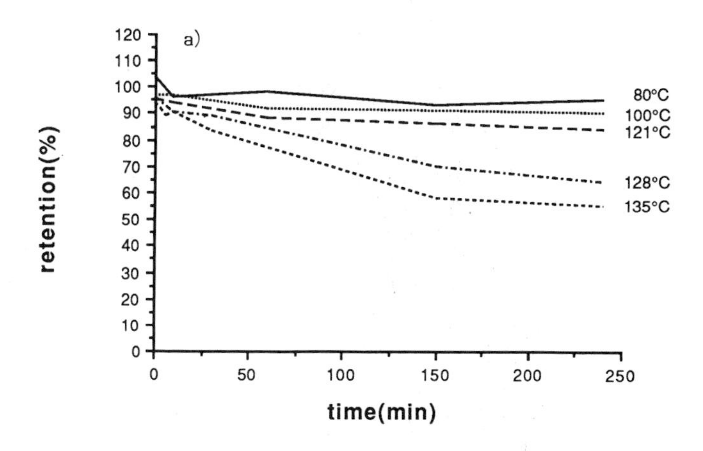

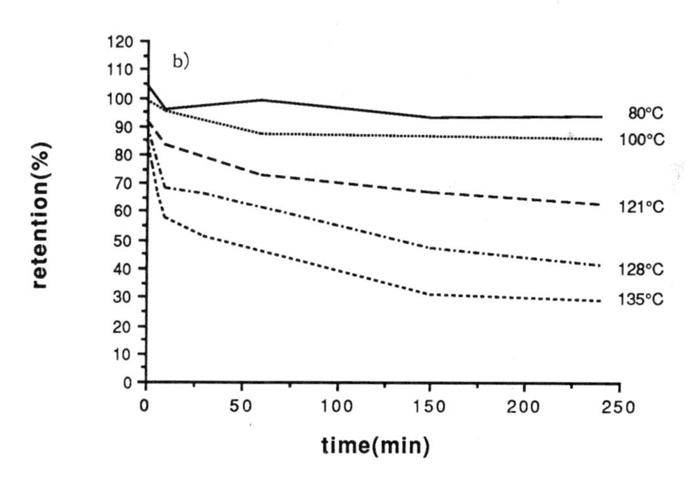

Figure 2. Effect of heat treatment on all-trans-α- and β-carotene in carrot juice.
a) all-trans-α-carotene
b) all-trans-β-carotene

During the first few minutes of treating the carrot juice at 80°C, a small increase in the content of carotenes was found. Similar results have been reported earlier by Hojilla et al. (1985), who thought that this increase was due either to a transformation of cis-isomers to the trans-form (they could not separate the different isomers) or perhaps to water-soluble compounds draining from the sample, thus increasing the concentration of water-insoluble compounds, such as carotenes. As we are able to follow the changes taking place in the isomers and because we extract the whole sample, we can reject the assumptions made by Hojilla and co-workers. We can not, however, explain the increase in carotene during mild heat treatment.

Further heat treatment of the carrot juice at 80°C resulted only in trace amounts of cis-isomers. Samples heated at 100°C during one hour lost about 10% of the initial amount of α- and β-all-trans-carotene. Some change into cis-isomers was found. The predominant isomer was the 13-cis-isomer. When calculated as having the same adsorption intensity as the all-trans-isomer, the 13-cis-β-carotene amounted to between 3 and 8% of all the β-carotene found. Even a short heating time at higher temperatures caused important loss of carotene. Ten minutes of heating at 121, 128, and 135°C resulted in 17, 32 and 42% loss of β-carotene, respectively. The α-carotene was much more stable during heating. Only about 10% of the α-carotene was destroyed during 10 minutes of heating at the temperatures studied. The rate of all-trans carotene destruction was highest during the first few minutes of heating.

As shown in Figure 1, many cis-isomers were found after heating at 121°C for 60 minutes. About 15% of β-carotene existed as the 13-cis-isomer.

Changes in vitamin A activity according to heat treatment

The vitamin A activity in the carrot juice was calculated using values presented by Bauernfeind (1972) and Zeichmeister (1949). According to these authors, one sixth of the all-trans-β-carotene is converted to active vitamin A.

Seventy-five per cent of the activity exerted by the all-trans-form is ascribed to the 13-cis-isomer. All-trans-α-carotene has about half as much vitamin A-activity as β-carotene. As the 9-cis-isomer of β--carotene and all the cis-isomers of α-carotene exert very little vitamin activity and, furthermore, occur in only small amounts they will each contribute with less than 0.1 µg retinolequivalents per milliliter heated juice. Therefore, we have decided to include only the all-trans-isomer of α-and β-carotene and the 13-cis-isomer of β-carotene in the calculation of vitamin A activity in the juice after different heat treatments.

The calculated vitamin A activity in the carrot juice is reported in Table 2. Mild heat treatment resulted in a decrease in vitamin A activity by a few percentage units only, due to loss of all-trans-β-carotene. More intense heat treatment increased the transformation of all-trans-β--carotene into its 13-cis isomer. This gave a rather small decrease in vitamin A activity.

Table 2. Effect of heat treatment on the content of carotenes and vitamin A activity of carrot juice

Heat treatment time min.	temp. °C	all-trans α-carotene	all-trans β-carotene	13-cis β-carotene	Vitamin A activity retinol eqv. μg/ml
		μg/ml juice			
0	0	21±0[1)]	88±2	ca 0	16.4
10	80	20±0	84±3	1±0	15.7
240	80	20±0	83±2	3±0	15.7
10	100	20±0	84±0	3±0	15.9
60	100	18±0	77±1	5±0	14.9
240	100	17±1	76±0	7±2	14.9
10	121	17±1	73±1	8±0	14.6
60	121	16±0	64±0	10±0	13.2
240	121	15±1	55±1	13±0	12.1
10	128	15±2	60±1	14±0	13.0
30	128	14±1	58±1	13±1	12.4
10	135	13±2	51±2	16±1	11.6
30	135	12±2	45±4	17±1	10.6

1) Each value represents the mean ± SD of 3 experiments.

Very high temperatures, however, resulted in a continuing increase in the amount of 13-cis-isomer, but the increase was not comparable to the decrease in all-trans-β-carotene. Heating at 135°C for 10 minutes caused a 30% decrease in vitamin A activity. The 13-cis-isomer accounted for almost one fifth of the vitamin activity. If calculation of vitamin A activity is based merely on all-trans-β-carotene, then the activity was only corresponded to 58% of the activity in raw carrot juice.

CONCLUSIONS

This study shows that low temperature heat treatment of relatively neutral foods, such as carrot, causes only insignificant changes in the carotenes and vitamin A activity. Examples of such heat treatment are cooking in water and pasteurization. However, when using high temperatures, as is done during heat sterilization, the loss is considerable. Conventional sterilization of cans at 121°C will result in a 20% reduction of the vitamin A activity. Very high temperatures, as those used during HTST treatment, enhance the need for proper control of the food process.

The results also show the importance of analyzing all occurring carotenes and their isomers to get a more reliable vitamin A value for foods exposed to heat treatment.

There is no information in the literature about the possible anticarcinogen effect of carotenes when they are changed into cis--isomers.

ACKNOWLEDGEMENTS

This work was supported in part by grants from the Volvo Research Foundation, Göteborg, Sweden. The author thanks Anders Pettersson for excellent technical assistance.

REFERENCES

Bauernfeind, J.C. (1972). Carotenoid vitamin A precursors and analogs in foods and feeds. J. Agr. Food Chem., 20 (3), 456-473.

Hojilla, M.P., Garcia, V.V. and Raymundo, L.C. (1985). Thermal degradation of β-carotene in carrot juice. ASEAN Food Journal, 1 (4), 157-161.

Pettersson, A. and Jonsson, L. (1990). Separation of cis-trans isomers of α-and β-carotene by adsorption HPLC and identification with diode array detection. Accepted for publ. in Journal of Micronutrient Analysis.

Zeichmeister, L. (1962). Cis-trans isomeric carotenoids. Academic Press, London, pp.251.

7

FORMATION OF MEAT MUTAGENS

Margaretha Jägerstad and Kerstin Skog

Department of Food Chemistry
Chemical Center, University of Lund, P.O. Box 124, S-221 00 Lund, Sweden

ABSTRACT

The formation of meat mutagens has been studied the last 10 years by carrying out modeling and meat cooking experiments in parallel. During this time, the list of meat mutagens has been growing, and continues to grow. The meat mutagens are usually produced in the crust of animal foods during frying, broiling, and baking. Another important source is meat extracts, consumed as gravies and meat bouillons. The formation of meat mutagens has been shown to depend physically on time, temperature, and water. Three major precursors have been identified: creatine or creatinine, certain amino acids, and monosaccharides or disaccharides. A requirement of sugar assumes a participation of the Maillard reaction, which also forms the basis for one of the major reaction mechanisms suggested. However, the meat mutagens are produced also in the absence of sugars, which means that other routes might be possible as well. Although the major precursors have been identified, more work needs to be done on the reaction mechanisms, the kinetics, and on the food constituents that might enhance or inhibit the formation of the meat mutagens. The results obtained to date point to several possibilities to control the formation of meat mutagens.

INTRODUCTION

Meat mutagens have been isolated from the crust of broiled, fried, and baked meat (beef, pork, ham, bacon, chicken, lamb), as well as from heat-treated meat extracts, such as gravies and bouillons (for review, see Sugimura and Sato, 1983; Felton et al., 1990). A certain minimum cooking temperature and time are required because boiled, microwave-treated, or stewed meat contains almost negligible amounts of mutagens (Felton et al., 1986). Studies so far have identified nearly a dozen mutagens of which those hitherto tested in long-term animal studies have been shown to be tumor-producing (Ohgaki et al., 1984,

Nutritional and Toxicological Consequences of Food Processing
Edited by M. Friedman, Plenum Press, New York, 1991

Table 1. Principal meat mutagens isolated from model systems and cooked meat products. (Jägerstad et al., 1989)

Compound	Short name	Structure
2-amino-3-methyl imidazo-[4,5-f] quinoline	IQ	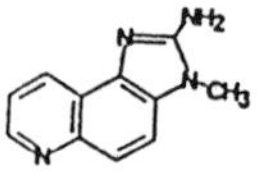
2-amino-3,4-dimethyl imidazo-[4,5-f] quinoline	MeIQ	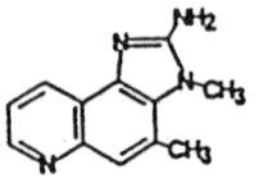
2-amino-3-methyl imidazo-[4,5-f] quinoline	IQx	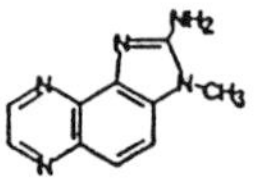
2-amino-3,8-dimethyl imidazo-[4,5-f] quinoxaline	MeIQx	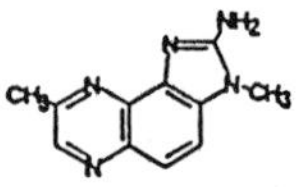
2-amino-3,4,8-trimethyl imidazo [4,5-f] quinoxaline	4,8-Di MeIQx	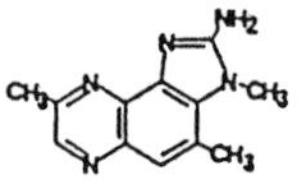
2-amino-3,7,8-trimethyl imidazo [4,5-f] quinoxaline	7,8-Di MeIQx	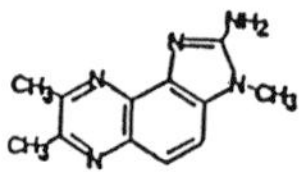
2-amino-1-methyl-6-phenyl imidazo [4,5-b] pyridine	PhIP	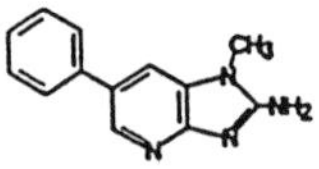
2-amino-n,n,n-trimethyl imidazo pyridine	TMIP	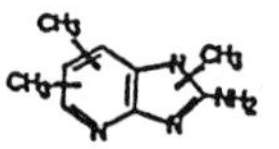
2-amino-n,n-dimethyl imidazo-pyridine	DMIP	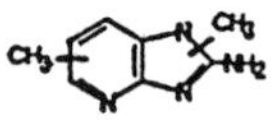
benzoxazines		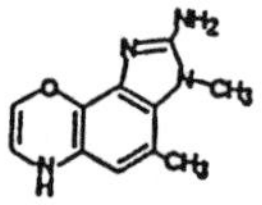

1986a,b, 1987; Tanaka et al., 1985; Kato et al., 1989). Analysis of fried or broiled meat shows that a portion of meat and gravy might contain everything from negligible amounts up to several micrograms of meat mutagens depending mainly on the cooking conditions and the meat composition. The risk of exposure to nanogram or microgram amounts of these meat mutagens is now a matter of great concern (Alexander et al., 1989).

Chemically, the meat mutagens have been classified as heterocyclic amines (Table 1). The major meat mutagens formed during normal cooking conditions (150-300°C) contain an imidazol part that is linked either to quinoline (IQ, MeIQ) (Kasai et al., 1980a,b), quinoxaline (IQx, MeIQx, DiMeIQx (Becher et al., 1988; Knize et al., 1988a; Kasai et al., 1981; Negishi et al., 1984, 1985; Grivas et al., 1985), or pyridine (PhIP and TMIP)(Felton et al., 1986). Recently, two other imidazo-containing compounds with a ring oxygen, the so-called benzoxazines, were isolated (Becher et al., 1988; Felton et al., 1990).

Studies on the formation of these meat mutagens have shown that creatine and certain amino acids are essential precursors (Jägerstad et al., 1990). In addition, glucose and other monosaccharides and disaccharides have been shown to be required, indicating a participation of the Maillard reaction (amino-carbonyl reactions or nonenzymatic browning reactions)(Jägerstad et al., 1990). There are, however, model studies that have reported high yields of the imidazo-containing meat mutagens in the absence of glucose or any other sugar (Taylor et al., 1987; Felton et al., 1990), indicating that the Maillard reaction route is not an obligatory pathway.

With the purpose to control the formation of meat mutagens, we have been working for several years with model systems and meat cooking experiments in parallel. The studies have focused on trying to understand the chemical aspects behind meat mutagen formation and the conditions that affect the yield.

POSTULATED ROUTE FOR FORMATION OF MEAT MUTAGENS

A possible reaction route for the formation of the imidazo-containing meat mutagens, the so-called IQ compounds (IQ, MeIQ, MeIQx) was presented at the Second International Maillard Meeting in Las Vegas in 1982 (Jägerstad et al., 1983a). Three precursors, all naturally occurring in beef, were assumed to participate: namely, creatine, certain amino acids, and sugars (Figure 1a). Creatine was postulated to form the 2-aminoimidazo part by cyclization and water elimination to creatinine, a reaction that easily takes place when the temperature is raised above 100°C. This imidazol part is a common moiety of the meat mutagens arising during normal cooking. The 2-aminoimidazo part is also responsible for the mutagenicity of these compounds, because without this part, and especially its 2-amino group, the mutagenicity of the

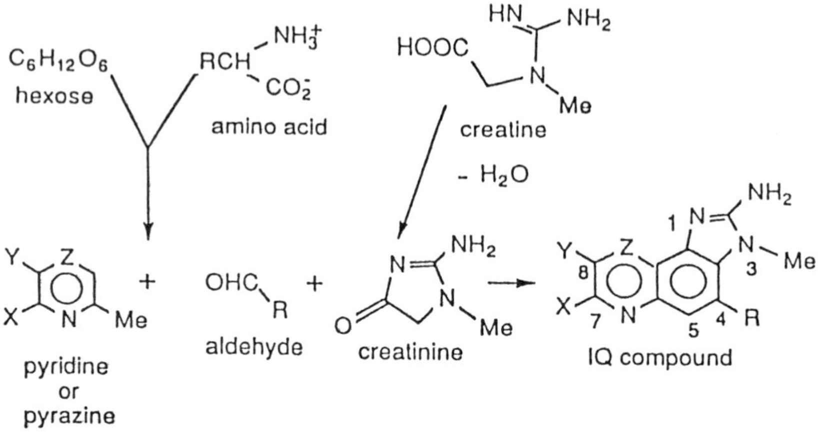

Figure 1a. Postulated reaction route for formation of IQ compounds. R, X, and Y may be H or Me; Z may be CH or N (Jägerstad et al., 1983a)

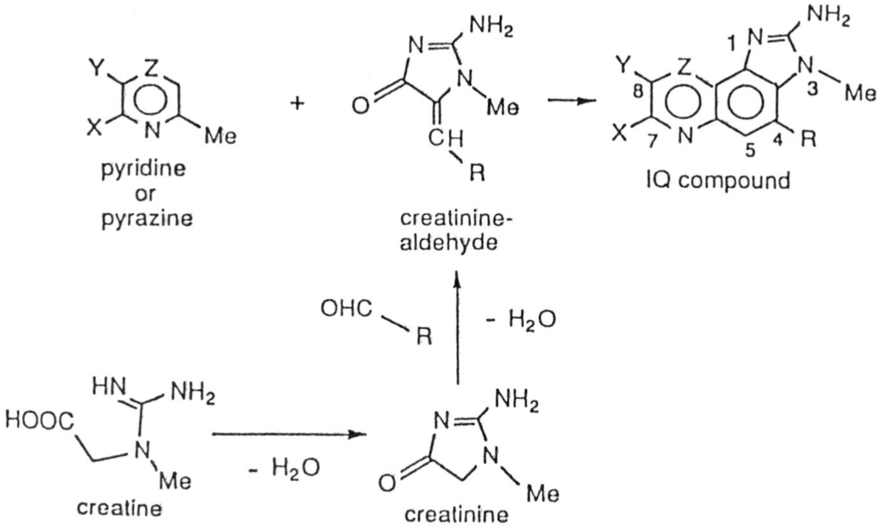

Figure 1b. Alternative route for formation of IQ compounds. R, X, and Y may be H or Me; Z may be CH or N (Nyhammar, 1986).

IQ compounds becomes almost negligible (Grivas and Jägerstad, 1984). The other two precursors, sugar and amino acids, were suggested to react according to the Maillard reaction; to produce typical Maillard reaction products, such as vinylpyrazines, vinylpyridines, and aldehydes, via Strecker degradation. By aldol condensations the quinoline or quinoxaline part of the IQ compounds were assumed to arise from vinylpyridines or vinylpyrazines and aldehydes.

According to our first hypothesis, vinylpyridines or vinyl-pyrazines condensed with aldehydes and then ring closed after condensation with creatinine as shown in Figure 1a. Another possibility equally likely was later outlined by Nyhammar (1986), who assumed that the condensation first occurred between aldehydes and creatinine, which then condensed with a vinylpyrazine or vinylpyridine according to Figure 1b.

RESULTS FROM MODEL SYSTEMS ON THE FORMATION OF MEAT MUTAGENS

Liquid Model System (Diethylene Glycol-H_2O; 5:1, v/v)

A special model system has been used to verify the postulated precursors and reaction route. Because the meat mutagens are produced at temperatures above 100°C, the model system was designed to operate at temperatures between 100 and 200°C. In a first series of experiments, the key reactants: creatine or creatinine, free amino acids and, sugars were dissolved in a mixture of water and diethylene glycol (1:5, v/v). This mixture was boiled under reflux for 2 hr at a temperature around 130°C. The molar ratio of the reactants was 1:1:0.5 for creatine, amino acid, and sugar, respectively. This model system produced high mutagenicity for TA 98 in the presence of S9 when all three groups of reactants were heated together. Heating the water-diethylene-glycol mixture alone or together with the reactants two by two produced only weak, if any, mutagenicity (Jägerstad et al., 1983a,b). The yield of mutagenicity varied greatly with the amino acid used; threonine being most active followed by glycine, alanine, etc. (Jägerstad et al., 1983b). An explanation of that phenomenon could be quantitative, as well as qualitative. It is known that the meat mutagens differ markedly in their specific activity in the Ames test. The variations in mutagenicity might therefore be due to the formation of different mixtures of the meat mutagens from each amino acid.

Purifications and fractionations have revealed the identity of the mutagenicity by using MS and ^1H-NMR with synthetic compounds as references. Reflux boiling a mixture of creatine, glycine, and glucose dissolved in diethylene glycol and water for 2 hr at 128°C produced MeIQx and 7,8-DiMeIQx (Jägerstad et al., 1984; Negishi et al., 1984). The latter product was a new mutagenic heterocyclic amine, and has to date only been demonstrated in cooked meat once (Turesky et al., 1988). When using threonine instead of glycine, MeIQx and another methyl derivative of MeIQx were produced, namely 4,8-DiMeIQx. This methyl derivative has also been demonstrated in fried hamburgers (Felton et al., 1986; Turesky et al., 1988; Becher et al. 1988). Almost simultaneously, Muramatsu and Matsushima (1985,1986), using the same model system, reported that alanine, and also lysine, produced MeIQx and 4,8-DiMeIQx irrespective of whether glucose or ribose was used. Grivas et al. (1985) isolated traces of MeIQ, together with 4,8-DiMeIQx, using the same model system and alanine as the amino acid and fructose instead of glucose. If alanine was replaced with glycine, MeIQx and small amounts of IQ were produced (Grivas et al., 1986). When using phenylalanine, PhIP was produced in this system (Shiyoa et al., 1987).

In a modification of the model system, the reactants creatine, glycine, and glucose were heated in the same proportions as before (1:1:0.5, molar basis) in the same diethylene glycol and water mixture (5:1, v/v) in open glass tubes at 180°C for 10-15 min. This modification comes closer to frying conditions as regards time and temperature than the previous system. Another similarity with cooking is that water is allowed to evaporate during the reaction. This modified system produced similar amounts of meat mutagens during 10 min as the previous ones after 2 hr. Although the mutagens produced are essentially the same as with the previous model system, we also found other mutagens as well. Heating creatine, glycine and glucose in the modified system produced MeIQx, 4,8-DiMeIQx, and 7,8-DiMeIQx, together with a not yet identified mutagen (Skog et al., 1990). Table 2 shows the imidazo-containing food mutagens isolated to date from model systems.

Dry-heating Model System

Dry heating is another model system that has been extensively used to study the formation of the meat mutagens. Yoshida et al. (1984) were the first group to report that heating 1 g each of proline and creatine for 1 hr at 150°C produced small amounts of IQ (0.4 nmol/mmol creatine). They also tried many other binary combinations of reactants, such as creatine/sugar, creatine/amino acids, amino acids/sugar; but none of these produced any IQ, which was the meat mutagen in search. A ternary system was never tried (Yoshida et al., 1980a,b, 1982, 1986).

IQ has also been isolated by dry heating (oven baking) equimolar amounts of creatine and phenylalanine in both the absence and the presence of glucose (Taylor et al., 1987). The yield of IQ increased about three times in the presence of glucose, while that of PhIP decreased from 735 nmoles to 560 nmole per mmole creatin(in)e.

Knize et al. (1988a) have isolated IQx after dry heating creatine and serine. Övervik et al. (1989) also dry-heated equimolar mixtures of creatine and amino acids. LC-MS analysis of the reversed phase HPLC separated amino acid-creatine mixture showed the presence of only one mutagenic compound per each amino acid-creatine

Table 2. Mutagenic meat compounds isolated from model systems of creatin(in)e, amino acid, and monosaccharide refluxed in diethylene glycol and water (5:1; v/v, 128°C/2 hr) (Jägerstad et al., 1990).

Precursors (mmols)	Isolated compounds (nmoles /mmoles creatin(in)e)		Ref
Crea/Glycine/Glucose (70:70:35)	MeIQx (4.2)	7,8-DiMeIQx	Jägerstad et al. (1984) Negishi et al. (1984)
Crea/Threonine/Glucose (70:70:35)	MeIQx	4,8-DiMeIQx	Negishi et al. (1985) Jägerstad et al. (1986)
Crea/Glycine/Fructose (35:35:17.5)	MeIQx (6.6)	IQ (1.0)	Grivas et al. (1986)
Crea/Alanine/Fructose (35:35:17.5)	MeIQ (<0.1)	4.8-DiMeIQx (2)	Grivas et al. (1985)
Crea/Alanine/Glucose or " / " /Ribose or " /Lysine /Ribose	MeIQx	4,8-DiMeIQx	Muramatsu and Matsushima (1985)
Crea/Phenylalanine/ Glucose (2.5:2,5:1.25)	PhIP (3.6)		Shioya et al. (1987)

combination. Thus, both phenylalanine/creatine and leucine/creatine produced PhIP. Threonine/creatine produced a mutagen having a mass number of 176, which is equivalent to that of TMIP. Serine, alanine and tyrosine heated one by one with creatine produced MeIQx.

MECHANISMS AND FACTORS AFFECTING THE YIELD

The yield of the imidazo-containing meat mutagens formed by using either the liquid model system or the dry-heating model amounts to only a few nanomoles per millimoles of each reactant except for PhIP, which was produced in much higher amounts (735 nmoles/mmole creatine) in the dry-heating system (Taylor et al., 1987). Such a low yield makes it difficult to verify the suggested route for the formation of the meat mutagens. The proposed participation of the Maillard reaction needs further substantiation. Heating of the suggested intermediates vinylpyrazines or vinylpyridines together with aldehydes and creatin(in)e has only been able to increase the formation of the imidazo-containing meat mutagens by about 50% (Jägerstad et al., 1983a,b).

The low yield is also a serious drawback for using isotopically labeled precursors. The high yield of PhIP in the dry-heating model system has made it possible to use specific carbon and nitrogen labeling in search of the precursors. Such studies have shown that all the carbon, as well as the nitrogen atoms in the ring structure of PhIP can be derived from creatine and phenylalanine (Felton et al., 1990).

The liquid model system has verified that the suggested precursors-creatine, monosaccharide, and certain amino acids-produce several of the imidazo-containing meat mutagens when heated together. There is obviously no doubt that creatine or creatinine is an essential precursor. Comparisons between creatine and creatinine show the latter to produce 20-30% higher yields (Jägerstad et al., 1983a,b; Skog et al., 1990).

There is also an obligatory role for amino acids in the formation of the imidazo-containing meat mutagens. Table 3 shows that certain amino acids are more active precursors than others. It has also been shown that each meat mutagen can be produced from several amino acids.

Table 3. Amino acids as precursors of mutagenicity and formation of meat mutagens (Jägerstad et al., 1990)

Amino acid	Rev/µmol I[1]	Meat Mutagens
Threonine	1068 ± 281	MeIQx; 4,8-DiMeIQx
Glycine	410 ± 59	IQ; MeIQx; 4,8-DiMeIQx;
Lysine	246 ± 108	7,8-DiMeIQx MeIQx; 4,8-DiMeIQx
Alanine	199 ± 32	MeIQ; MeIQx; 4,8-DiMeIQx
Serine	197 ± 85	IQ; IQx; MeIQx; 4.8-DiMeIQx
Leucine	161 ± 22	PhIP
Histidine	126 ± 33	
Arginine	101 ± 29	
Valine	91 ± 29	
Isoleucine	75 ± 26	
Asparagine	63 ± 23	
Tyrosine	56 ± 23	MeIQx
Aspartic acid	55 ± 17	
Phenylalanine	50 ± 26	IQ; PhIP
Tryptophan	50 ± 27	
Cysteine	40 ± 13	
Methionine	34 ± 17	
Glutamine	33 ± 12	
Proline	31 ± 22	IQ
Glutamic acid	30 ± 13	
Cystine	19 ± 8	

[1] Creatine, amino acid and glucose (1:1:0.5, molar ratio) boiled under reflux in DEG.H_2O for 2 hr at 128°C (Jägerstad et al., 1983b).

Thus, threonine, glycine, lysine, alanine, and serine have produced both MeIQx and DiMeIQx, often simultaneously. IQ has been produced from either proline, phenylalanine, glycine, or serine. Certain amino acids have also been reported to enhance the yield when added to meat exctract systems or when mixed in minced meat before cooking (Taylor et al., 1985; Ashoor et al., 1980; Övervik et al., 1989). On the other hand, studies by Jones and Weisburger (1988a,b,c) have showed certain amino acids e.g. tryptophan and proline to be inhibitory.

The role of monosaccharides in the formation of the meat mutagens is conflicting. While not necessary in the dry-heating model system, the monosaccharides are required in the liquid model system in certain concentrations. Fructose and ribose seem to produce higher yields compared with glucose (Matsushima et al., 1986). Sucrose and lactose are also able to produce the meat mutagens. Monosaccharides and disaccharides showed an optimum effect on the mutagenicity yield when present in amounts around half that of creatin(in)e or amino acid (Skog et al., 1990). By increasing the amount of glucose or fructose, the formation of mutagenic compounds decreased. At equimolar levels or in excess of the other reactants, the formation of mutagenic compounds was almost completely inhibited (Figure 2.) The mechanism behind the inhibitory effect of hexoses is unknown. It can only be speculated that by increasing the concentration by reducing sugars the Maillard reaction might favor the formation of other products that might interfere or compete with the reaction route that usually results in the formation of mutagenic heterocyclic amines. By monitoring the recovery of

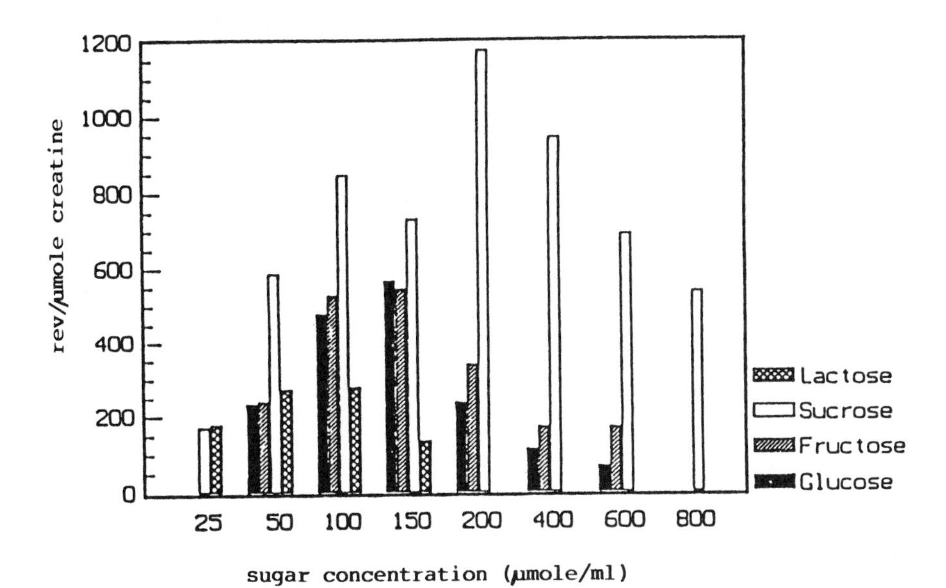

Figure 2. Effect of various sugar concentrations on mutagenicity formation in model systems. (Skog et al.,1990)

creatin(in)e during this reaction, Skog et al.(1990) found more than 90% unreacted creatine when glucose was added in half the molar concentration of that of creatine. With increasing amounts of glucose added, the recovery of creatin(in)e decreased. This effect was seen only in the presence of amino acid, which indicates that the Maillard reaction was involved and that some Maillard reaction products might have reacted with creatin(in)e to form non-mutagenic products.

The suggested pathway involving the Maillard reaction assumes amino acids and sugar to produce products like vinylpyrazines or vinylpyridines together with aldehydes that might condense to quinolines or quinoxalines. The observation that many of the imidazo-containing mutagenic meat mutagens also are produced without sugar indicates that other routes exist. However, it has been reported by Wang and Odell (1973) that dry heating of amino acids without sugar produced several derivatives of pyrazines, especially after heating hydroxy amino acids such as threonine and serine. The formation of pyrazines was markedly favored in the presence of sugar. This fact implies that sugar might have a catalytic effect on the formation of the necessary products that form quinolines or quinoxalines and in that case the suggested formation route given in Figures 1a and b could be valid both with and without the presence of sugar.

MEAT COOKING EXPERIMENTS

Meat Composition

Modeling has shown creatine, certain free amino acids, and sugar to produce meat mutagens when heated together simultaneously. These precursors are all naturally occurring in such muscle foods as beef, heart, and tongue. However, meat organs such as liver and kidney contain only traces of creatine, but more of free amino acids and glucose, especially the liver (Table 4a). Samples of these various bovine tissues were freed from visible fat, ground, and formed into approximately 10-mm thick patties each weighing 50 g. One patty at a time was panfried without adding any extra frying fat on a double-sided Teflon-coated plate for 3 min at 150°, 175° or 200°C. (Table 4b).

Only the patties made from beef, heart, and tongue produced significant mutagenicity. Neither the liver nor the kidney produced any significant mutagenicity. Such organs contain only traces of creatine, and this accords with results obtained in model systems showing creatine or creatinine to be an obligatory precursor.

When mixtures between muscle tissues and organs were fried in the same way at 200°C, only the mixed beef patties produced mutagenicity between 10,800 and 17,300 revertants per 100 gE. Multiple regression analysis showed creatine or creatinine to be the most important single factor to explain the variation in mutagenicity. Subsets of two variables showed that the sum of creatine and creatine

Table 4a. Chemical composition of raw bovine tissues. (Laser Reutersward et al., 1987b)

Component*	Concentration in				
	Meat	Heart	Tongue	Liver	Kidney
Water (%)	74.3	80.7	72.0	71.5	80.6
Fat (%)	3.3	1.5	11.7	2.2	1.4
Protein (%)	21.0	16.3	14.6	19.0	15.5
Ash (%)	0.8	1.1	0.9	1.4	1.1
Analytical residue (%)	0.7	0.4	0.8	5.9	1.4
Monosaccharides (μmol/g)	12	2.0	9.3	183	0.7
Glucose	8.2	1.6	7.4	183	0.45
Glucose-6-phosphate	3.0	0.25	1.4	0.30	0.27
Glucose-1-phosphate	0.15	0.07	0	0	0
Fructose-6-phosphate	0.75	0.07	0.47	0	0
Creatin(in)e** (μmol/g)	33	25	19	2.2	2.3
Creatine	31	23	17	1.7	1.8
Creatinine	2.0	1.9	2.0	0.50	0.47
Dipeptides (μmol/g)	23	0.4	2.2	0.8	0
Carnosine	21	0.4	1.5	0.8	0
Anserine	2.0	0	0.7	0	0
Total free amino acids (μmol/g)	30	37	46	101	71

* Concentrations of all components are calculated on the wet tissue.
** Creatin(in)e = creatine plus creatinine.

explained almost 90.4% of the mutagenic variation, whereas the corresponding figures for sugar and free amino acids were 80 %.

The essential role of creatine has also been shown by adding creatine or creatinine to meat. After cooking, the mutagenicity has increased dramatically (Nes, 1986; Becher et al., 1988; Knize et al., 1988b). Addition of certain amino acids to beef patties or meat extracts before heating has also been found to increase the mutagenicity (Taylor et al., 1985; Ashoor et al., 1980; Övervik et al., 1989). When Jones et al. (1988a,b,c) added either proline or tryptophan to meat during frying, they reported less mutagenicity.

Our first meat cooking experiment (Jägerstad et al., 1983a) showed that beef patties prepared from low-glucose beef obtained from animals that had developed strong stress before slaughter produced almost negligible mutagenicity as compared with beef patties prepared from meat containing normal amounts of glucose. The creatine level of both types of beef patties were similar. This supports the fact that frying, which is a typical high-temperature short-term process, requires sugar to be able to produce significant mutagenicity.

Table 4b. Creatine and creatinine levels and mutagenicity of the crust from bovine tissues fried as patties for 3 minutes at different temperatures on a double-sided Teflon-coated plate (Laser Reuterswärd et al, 1987b).

Sample origin	Temp. (°C)	Creatine (µmol/gdm)	Creatinine (µmol/gdm)	Creatin(in)e* (µmol(gdm)	Mutagenicity (reverants/100 gE)
Meat	Raw	120	7.9	128	-
	150	72.9	36.6	110	NS
	175	53.3	47.8	101	6,160
	200	45.1	70.9	116	10,703
Heart**	Raw	121	9.8	131	-
	150	54.6	28.5	83	5,859
	175	49.4	40.7	90	8,614
	200	47.5	45.0	92	15,520
Tongue**	Raw	62.1	7.0	69	-
	150	43.7	20.7	64	NS
	175	26.1	28.3	54	12,371
	200	15.8	26.6	42	19,603
Liver**	Raw	5.9	1.8	7.7	-
	150	1.8	2.4	4.2	NS
	175	2.3	2.1	4.4	NS
	200	2.1	2.4	4.5	NS
Kidney**	Raw	9.4	2.4	11.8	-
	150	4.5	3.1	7.6	NS
	175	4.0	4.5	8.5	NS
	200	3.7	4.4	8.1	NS

gdm = grams dry matter, gE = grams initial raw weight, NS = not significant * Creatin(in)e = creatine plus creatinine. ** Crust was not peeled off after frying.

Effect of Fat

Conflicting results exist concerning the role of fat content on the formation of mutagenic activity during cooking of meat products. Spingarn et al. (1981) demonstrated that mutagenicity of minced meat products reached a maximum at a fat level of 10%. They believed that fat was participating chemically. On the other hand, studies by Bjeldanes et al. (1983) indicate that fat does not take part chemically. Instead, along with the observation of Barnes et al. (1983), the role of fat was suggested to promote the heat transfer.

We designed a study with the purpose to check which explanation was most plausible (Holtz et al., 1985). Meat loaves were prepared from minced beef that was mixed with various amounts of back-fat (2.8-16.6%). The meat loaves were baked in an oven at 200°C until their surface temperature reached 140°C. As seen in Table 5, the baking time decreased with increasing fat content. The lowest fat level (2.8%) required almost 1 hr until the surface showed a temperature of 140°C, whereas the meat loaves containing 16.6% fat only required 40 min. The mutagenicity obtained

Table 5. Effect of fat on baking time and mutagenicity during the baking of meat loaves (Holtz et al., 1985).

Content of					Mutagenicity	
Recipe	fat (%)	water (%)	Final surface temp. (°C)	Baking time (min)	rev/g dry matter	rev/g fatfree dry matter
A	2.8	74.8	140	56	1,292 ± 267	1,494 ± 309
B	8.1	72.3	140	50	1,013 ± 203	1,460 ± 992
C	13.9	67.8	140	47	806 ± 166	1,360 ± 279
D	16.6	63.8	140	39	436 ± 95	820 ± 178
C	13.9	67.8	129	39	767 ± 254	1,327 ± 438
C	13.9	67.8	140	47	806 ± 166	1,460 ± 292
C	13.9	67.8	143	50	918 ± 178	1,549 ± 301
C	13.9	67.8	150	59	1,268 ± 278	2,060 ± 452
C	13.9	67.8	170	97	1,245 ± 209	2,014 ± 338

with the Salmonella typhimurium test (TA 98 + S9) showed that the mutagenicity decreased with increasing fat content (decreasing baking time) when calculated per gram dry material. When corrected for the varied fat content, the mutagenicity became almost equal in all the meat loaves. The results support the opinion that fat might act as a vehicle for heat transfer, but it is also clear that the water concentration decreases with increasing amounts of fat. Thus, less water has to be evaporated in the high-fat meat loaves facilitating the temperature to rise.

Later, Övervik et al. (1987) reported that addition of frying fat increased the mutagenicity of various types of fat during panfrying of pork at 250°C. The authors stated a similar conclusion as above; namely, that the addition of fat caused a more efficient heat transfer without excluding other factors.

Gravies and bouillons

Meat extracts, food-grade as well as for bacterial use (Difco beef extract) has been reported to induce high mutagenicity in the Ames Salmonella typhimurium test (Commoner et al., 1978; Munzner, 1981; Hargraves and Pariza, 1983, 1984; Laser Reuterswärd et al., 1987a). Beef broth, beef bouillon cubes and lyophilized gravies contain much less mutagenicities (for ref, see Hargraves and Pariza, 1984 and Laser Reuterswärd et al., 1987a). One reason for this could be a lower level of creatine(in)e in such products, as pointed out by Laser Reuterswärd et al., 1987a). Table 6 summarizes the mutageniciy and creatin(in)e concentrations of some commercial samples. In some countries, criteria for the total creatine content of bouillons and gravies exist; but for meat flavor samples, there are no such criteria. A commercial food-grade meat extract should have a maximum water content of 20%, maximum

Table 6. Creatinine and creatin(in)e levels, mutagenicity, and color of commercial samples of meat extracts, bouillons, and gravy (Laser Reuterswärd et al., 1987a)

Sample	Animal origin	Form	Creatinine (μmol/gdm)	Creatin(in)e* (μmol/gdm)	Creatine/creatin(in)e (%)	Mutagenicity** (revertants/gdm)	Color (absorbance/gdm)
Meat extract	Beef	Paste	301	593	51	9,800	41.0
Meat extract	Beef	Paste	465	849	55	6,900	54.7
Meat extract	Beef	Paste	149	719	21	2,500	57.4
Meat flavour, natural	Pork	Paste	5.1	5.1	100	NS	62.3
Consommé	Beef	Soup	24.8	43.4	57	122	22.1
Meat flavour, synth.	-	Powder	0	0	-	288	76.0
Meat bouillon	Beef	Cube	37.2	44.3	84	170	18.8***
Meat bouillon	-	Powder	14.2	27.5	52	106	10.6***
Meat bouillon	-	Cube	11.1	20.0	55	NS	10.9***
Meat bouillon	Beef	Cube	5.5	7.3	75	NS	34.7***
Meat gravy	Beef	Cube	3.6	8.5	42	NS	12.0***
Vegetable bouillon	-	Cube	2.5	3.7	68	NS	7.5***
Vegetable bouillon	-	Powder	1.8	2.0	90	42	6.4***
Difco Beef extract	Beef	Powder	-	-	-	2,500	-

gdm = Grams dry matter, NS = Not significant
* Sum of creatine and creatinine.
** To *Salmonella typhimurium*, TA 98, in the presence of S-9 mix.
*** Sugar color added.

fat content of 2%, and a total creatine content of 5-7%. As shown by Laser Reuterswärd and coworkers (1987a), such meat extracts containing creatine levels between 5 and 7% contained rather high mutagenicity (2,500-10,000 rev/g dm).

The amount of mutagenicity of meat extracts, beef stock, and bouillons depends on several factors, including the processing conditions, such as time, drying temperature, water content, and the amounts of precursors (Commoner et al., 1978; Dolara et al., 1979; Munzer, 1981; Aeschbacher, 1986, Laser Reuterswärd et al., 1987a).

During frying and baking, meat juice could leak out of the meat. Laser Reuterswärd et al. (1987a) collected meat juice drippings from round steaks of beef weighing around 450 g each and baked at various oven temperatures (115-245°C) until a center temperature of 70°C was obtained. The meat-juice drippings collected during baking showed a linear increase in mutagenicity with baking temperatures up to 180°C (48-828 revertants/100 gE) and a very sharp increase in mutagenicity for the gravy collected from the beef steak baked at 245°C (19,800 revertants/100 gE). This gravy was totally dried out and might therefore have reached higher temperatures than those meat-juice drippings collected at the lower baking temperatures. All the meat-juice drippings contained about five times higher concentrations of creatine than that of the crust of the beef steak (Table 7).

Table 7. Creatine and creatinine levels and mutagenicity of the crust and gravy from round steaks baked at various oven temperatures (Laser Reutersward et al., 1987a)

	Concentrations in steaks				Mutagenicity** (no. of revertants)	
	Oven temp. (°C)	Creatine (μmol/gdm)	Creatinine (μmol/gdm)	Creatin(in)e* (μmol/gdm)	No./gdm	No./100gE
Meat	Raw	138	7.7	146	-	-
Crust	115	111	11	122	NS	NS
	140	109	11	120	NS	NS
	170	103	12	115	NS	NS
	180	83	17	100	NS	NS
	245	68	29	97	68	135
Gravy	115	554	103	657	76	48
	140	536	194	730	1,146	435
	170	314	348	662	1,523	716
	180	-	-	-	1,035	828
	245	67	375	442	28,300	19,800

gdm = Grams dry matter, NS = Not significant
* Sum of creatine and creatinine.
** To Salmonella typhimurium in presence of S-9 mix.

Also during panfrying, meat juice leaks out that contributes considerably to the mutagenicicty in a meal based on fried meat with gravy. According to Övervik et al. (1987) as much as 60% of the mutagenicity might appear in the pan residue and 40% in the crust when frying meat pork at 200°C.

Effects of Time and Temperature

The formation of meat mutagens measured according to the Ames test is highly dependent on cooking time and temperature. Negligible mutagenicity is recorded at temperatures around 100°C, such as during boiling and microwave cooking or short-term frying when preparing stews and casseroles (Bjeldanes et al., 1983). Panfrying, broiling, and oven baking are the cooking methods that are most likely to produce mutagenicity.

The formation of mutagenicity as related to different temperatures is shown in a meat experiment (Figure 3). Here, minced beef was lyophilized to a dry powder and spread as a thin layer in a plastic bag, which was then wrapped in aluminum foil. These thin sheets of lyophilized meat were heated for various times between 110°C and 150°C. The mutagenicity was recorded with the Ames test using TA 98 and S9 activation. As seen, the formation of mutagenic activity increased with increasing temperature.

Panfrying exposes a rather big area of the meat to direct heating conducted through metallic material. Generally, panfrying is performed in the temperature range 175-250°C during a period from a few minutes up to at most 20 min. It has been shown that the mutagenicity is formed early during panfrying and then levels out after a while.

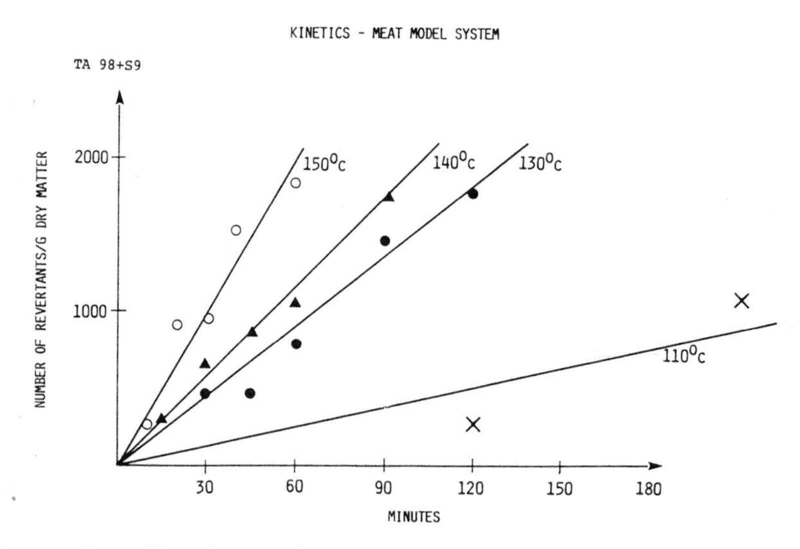

Figure 3. The formation of mutagenicity in lyophilized meat powder during heating. (Jägerstad et al., unpublished)

Most data reported in the literature are based on frying temperatures between 200°C and 300°C. Table 4 shows some panfrying data obtained from one previous study (Laser Reuterswärd et al., 1987b) carried out in the temperature range between 150° and 200°C. As seen, the mutagenicity increased with cooking time and was significant at 150°C only for patties made of minced bovine heart. Patties made of bovine muscle or tongue were not significantly mutagenic when fried at 150°C for 3 + 3 min. Oven baking has been shown to produce lower mutagenicity compared with panfrying (Bjeldanes et al., 1983; Laser Reuterswärd et al., 1987a). When round steaks weighing around 460 g were baked at various temperatures (115-245°C) until the temperature in the center reached 70°C, only the beaf steak baked at 245°C produced any mutagenicity in the crust, amounting to 135 rev 100 gE (Laser Reuterswärd et al., 1987a). This agrees with the data of Bjeldanes et al. (1982), who found 150 rev/100 gE in tests on a round beef steak baked at 176°C for 60 min, whereas after 90 min at 176°C, mutagenicity was increased considerably to 3,350 revertants/100 gE. In one unpublished experiment, we found 4,300 revertants/100 gE were induced in the crust of a beef steak initially weighing 1800 g and baked at 200°C for 130 min.

There are two major reasons why baking produces less mutagenicity as compared with frying. One reason is a lower heat transfer by air convection, compared with the more efficient metal conduction occurring during panfrying. Secondly, baking is often performed on round steaks or meat loaves, where the percentage of crust out of the total weight becomes much lower compared with panfrying. Baking might produce as little as 1-2% of the crust, whereas panfrying might result in between 5 and 25% (Laser Reuterswärd et al., 1987a,b).

Formation of mutagenicity during broiling when compared with other forms of cooking has only been reported by Bjeldanes et al. (1984). In this study, broiling produced mutagenicity less than that of pan-frying but more than baking. There is probably a wide variation in the mutagenicity produced depending on the existing times and temperatures used.

Color

Development of brown color during cooking is often associated with the Maillard reaction. Accordingly, the color intensity might be associated with the formation of mutagenicity. When such a relationship was investigated, it was found to exist if meat of comparable composition was cooked under different conditions (Jägerstad et al., 1983a, Holtz et al., 1985;), but not if the chemical composition has varied too much as regards the concentrations of carbohydrates and amino acids. It is mainly these two classes of compounds that produce colored Maillard reaction products. Therefore, browning increases with increasing amounts of reducing carbohydrates. Because sugars in increasing amounts might inhibit the formation of mutagenicity, there is no simple relationship between color and mutagenicity.

CONCLUSIONS

The formation of meat mutagens has been studied in model systems and experimental meat-cooking experiments. Both systems show the importance of physical parameters such as time, temperature, and water. Also of importantance is the availability of the right precursors in the right proportions. Certain food components might have enhancing or inhibitory effects on the formation of meat mutagens. Fortunately, the yield is much lower than what could be expected from the occurrence of the precursors. More work has to be done to elucidate the reaction mechanism in more detail; to study the kinetics considering the multifactorial effects from various physical and chemical parameters. From our present knowledge, there are encouraging results that indicate that it should be possible to further limit and control the formation of these meat mutagens. More rapid and less laborious chemical methods to extract and analyze the meat mutagens are also a desideratum.

ACKNOWLEDGEMENTS

We would like to thank Dr. A. Laser Reutersward, Swedish Meat Research Institute, Professor K. Olsson and his group at the Department of Chemistry, Swedish University of Agricultural Sciences, and Professor T. Sugimura's group at the National Cancer Research Center, Tokyo, for fruitful collaborative work within this field. The financial support from the Swedish Cancer Foundation and the Swedish Council for Forestry and Agricultural Research is gratefully acknowledged.

REFERENCES

Aeschbacher, H. (1986). Genetic toxicology of browning and caramelizing products. In: "Genetic Toxicology of the Diet", I. Knudsen, ed, Alan R Liss, NY, p 133-144.

Alexander, J., Becker, G., and Busk, L. (1989). Cooked food mutagens - a general overview. In "Risk assesment of cooked food mutagens", L. Busk, ed, Vår föda, 42, suppl 2, p 31-37.

Ashoor, S., Dietrich, R., Chu, F., and Pariza, M. (1980). Proline enhances mutagen formation in ground beef during frying. Life Sci., 26, 1801-1805.

Barnes, W.S., Maker, J.C. and Weisburger, J.H. (1983). High-pressure liqiud chromatographic method for the analysis of 2-amino-3-methyl imidazo (4.5-f) quinoline, a mutagen formed during the cooking of food. J. Agric. Food Chem. 31, 883-886.

Becher, G., Knize, M., Nes, I., and Felton, J. (1988). Isolation and identification of mutagens from fried Norwegian meat products. Carcinogenesis, 9, 247-253.

Bjeldanes, L.F., Morris, M.M., Felton, J.S., Healy, S., Stuermer, D., Berry, P., Timourian, H., and Hatch, F. (1982). Mutagens from the cooking of food. II. Survey by Ames/Salmonlla test of mutagen formation in the major protein-rich foods of the American diet. Food Chem. Toxic., 20, 357.

Bjeldanes, L.F., Morris, M.M., Timourian, H., and Hatch, F.T. (1983). Effects of meat composition and cooking conditions on mutagenicity of fried ground beef. J. Agric. Food Chem., 31, 18-21.

Commoner, B., Vithayathil, A.J., Dolara, P., Nair, S., Madyastha, P. and Cuca, G.C. (1978). Formation of mutagens in beef and beef extract during cooking. Science, 201, 913-916.

Dolara, P., Commoner, B., Vithayathil, A., Cuca, G., Tuley, F., Madyastha, P., Nair, S. and Kriebel, D. (1979). The effect of temperature on the formation of mutagens in heated beef stock and cooked ground beef. Mutation Res., 60, 231-237.

Felton, J., Knize, M., Shen, N., Andresen, B., Hatch, F., and Bjeldanes, L. (1986). Identification of the mutagens in cooked beef. Environ. Health Perspect., 67, 17-24.

Felton, J.S. and Knize, M.G. (1990). Heterocyclic-amine mutagens/carcinogens in foods. In: " Handbook of Experimental Pharmacology", C.S. Cooper and P.L. Grover eds, Springer-Verlag, Heidelberg, vol 94:1, pp 471-502.

Grivas, S., and Jägerstad, M. (1984). Mutagenicity of some synthetic quinolines and quinoxalines related to IQ in Ames test. Mutat. Res., 140, 55-59.

Grivas, S., Nyhammar, T., Olsson, K., and Jägerstad, M. (1985). Formation of a new mutagenic DiMeIQx compound in a model system by heating creatinine, alanine and fructose. Mutat. Res., 151, 177- 183.

Grivas, S., Nyhamma,T., Olsson, K., and Jägerstad, M. (1986). Isolation and identification of the food mutagens IQ and MeIQx from a heated model system of creatinine, glycine and glucose. Food Chem., 20, 127--136.

Hargraves, W.A., and Pariza, M.W. (1983). Purification and mass spectral characterization of bacterial mutagens from commercial beef extract. Cancer Res., 43, 1467.

Hargraves, W.A., and Pariza, M.W. (1984). Mutagens in cooked foods. J. Environ. Sci. Hlth, C2(1), 1.

Holtz, E., Skjöldebrand, C., Jägerstad, M., Laser Reuterswärd, A. and Isberg, P.E. (1985). Effects of recipes on crust formation and mutagenicity in meat producs during baking. J. Food Technol., 20, 57-66.

Jones, C., and Weisburger, J. (1988a). Inhibition of aminoimidazo- quinoxaline-type and aminoimidazol-4-one-type mutagen formation in liquid reflux models by 1-tryptophan and other selected indoles. Jpn. J. Cancer Res. (Gann), 79, 222-230.

Jones, C., and Weisburger, J. (1988b). L-tryptophan inhibits formtion of mutagens during cooking of meat and in laboratory models. Mutat. Res., 206, 343-349.

Jones, C., and Weisburger, J. (1988c). Inhibition of aminoimidazoquinoxaline-type and aminoimidazol-4-one type mutagen formation in liquid-reflux models by the amino acids 1-proline and/or 1-tryptophan. Environ. Mol. Mutagenesis, 11, 509-514.

Jägerstad, M., Laser Reuterswärd, A., Öste, R., Dahlqvist, A., Grivas, S., Olsson, K., and Nyhammar, T. (1983a). In: "The Maillard Reaction in Foods and Nutrition", G.R. Waller, and M.S. Feather, eds, Washington DC: ACS Symp. Ser., 215, p 507-514.

Jägerstad, M., Laser Reuterswärd, A., Olsson, R., Grivas, S., Nyhammar, T., Olsson, K., and Dahlqvist, A. (1983b). Creatin(in)e and Maillard reaction products as precursors of mutagenic compounds: Effects of various amino acids. Food Chem., 12, 255-264.

Jägerstad, M., Olsson, K., Grivas, S., Negishi, C., Wakabayashi, K., Tsuda, M., Sato, S., and Sugimura, T. (1984). Formation of 2-amino-3,8-dimethylimidazo[4,5-f])quinoxaline in a model system by heating creatinine, glycine and glucose. Mutat. Res., 126, 239- 244.

Jägerstad, M., Skog, K., Grivas, S., and Olsson, K. (1989). Mutagens from model systems. In: "Mutagens and Carcinogens in the Diet", M.W. Pariza, ed, Alan R. Liss, NY, in press., p.

Jägerstad, M., Skog, K., Grivas, S., and Olsson, K. (1990). Formation of heterolyctic amines using model systems. Mutation Research, Special Issue, in press,p.

Kasai, H., Yamaizumi, Z., Wakabayashi, K., Nagao, M., Sugimura, T., Yokoyama, S., Miyazawa, T., and Nishimura, S. (1980a). Structure and chemical synthesis of Me-IQ, a potent mutagen isolated from broiled fish. Chem. Lett., 1391-1394.

Kasai, H., Yamaizumi, Z., Wakabayashi, K., Nagao, M., Sugimura, T., Yokoyama, S., Miyazawa, T., Spingarn, N. E., Weisburger, J.H., and Nishimura, S. (1980b). Potent novel mutagens produced by broiling fish under normal conditions. Proc. Jpn. Acad., 56, 278-283.

Kasai, H., Yamaizumi, Z., Shiomi, T., Yokoyama, S., Miyazawa, T., Wakabayashi, K., Nagao, M., Sugimura, T., and Nishimura, S. (1981). Structure of a potent mutagen isolated from fried beef. Chem. Lett., 485-488.

Kato, T., Migita, H., Ohgaki, H., Sato, S., Takayama, S. and Sugimura, T. (1989). Induction of tumors in the Zymbal gland, oral cavity, colon, skin and mammary gland of F344 rats by a mutagenic compound, 2-amino-3,4-dimethylimidazo (4,5-f) quinoline. Carcinogenesis, 10(3), 601-603.

Knize, M.G., N.H. Shen, and Felton, J.S. (1988a). The production of mutagens in foods. Proc. Air. Pollution Control. Assoc., 88-130.

Knize, M., Shen, N., and Felton, J. (1988b). A comparison of mutagen production in fried-ground chicken and beef. Effect of supplemental creatine. Mutagenesis, 3(6), 506-508.

Laser Reuterswärd, A., Skog, K. and Jägerstad, M. (1987a). Effects of creatine and creatinine content on the mutagenic activity of meat extracts, bouillon and gravies from different sources. Food Chem. Toxicol. 25, 747-754.

Laser Reuterswärd, A., Skog, K., and Jägerstad, M. (1987b). Mutagenicity of pan-fried bovine tissues in relation to their content of creatine, creatinine, monosaccharides and free amino acids. Food Chem. Toxic., 25, 755-762.

Matsushima, T., and Muramatsu, M. (1986). Formation of MeIQx and 4,8- DiMeIQx by heating mixtures of creatinine, amino acids and monosaccharides. In: "Genetic Toxicology of the diet", I. Knudsen, ed, Alan R. Liss, NY, p 330 (abstract).

Muramatsu, M., and Matsushima, T. (1985). Formation of MeIQx and 4,8- DiMeIQx by heating mixtures of creatinine, amino acids and monoaccharides. Mutat. Res., 147, 266-267.

Münzner, R. (1981). Mutagenifetsprüfung von Fleischextrakten. Fleischwirtschaft, 61, 1586.

Negishi, C., Wakabayashi, K., Tsuda, M., Sato, S., Sugimura, T., Saito, H., Maeda, M., and Jägerstad, M. (1984). Formation of 2-amino-3,7, 8-trimethylimidazo[4,5-f])quinoxaline, a new mutagen, by heating a mixture of creatinine, glucose and glycine. Mutat. Res., 140, 55-59.

Negishi, C., Wakabayashi, K., Yamaizumi, J., Saito, H., Sato, S., and Jägerstad, M. (1985). Identification of 4,8-DiMeIQx, a new mutagen. Mutat. Res., 147, 267-268.

Nes, I.F. (1986). Mutagen formation in fried meat emulsion containing various amounts of creatine. Mutation Res., 175, 145-148.

Nyhammar, T., Grivas, S., Olsson, K. and Jägerstad, M. (1986). 4,8-DiMeIQx from the model system fructose, alanine and creatinine. Comparison with the isomeric 5,8- DiMeIQx. Mutat. Res., 174, 5-10.

Nyhammar, T. (1986). Studies on the Maillard reaction and its role in the formation of food mutagens, Doctoral Thesis.,Swedish University of Agricultural Sciences ISBN 91-576- 2658-8.

Ohgaki, H., Kusama, K., Matsukura, N., Morino, K., Hasegawa, H., Sato, S., Takayama, S. and Sugimura, T. (1984). Carcinogenicity in mice of a mutagenic compound, 2-amino-3-methylimidazo (4,5-f) quinoline, Carcinogenesis, 5, 921-924.

Ohgaki, H., Hasegawa, H., Kato, T., Suenaga, M., Ubukata, M., Sato, S., Takayama, S. and Sugimura, T. (1986a). Carcinogenicity in mice and rats of heterocyclic amines in cooked foods. Environ. Health Perspect., 67, 129-134.

Ohgaki, H., Hasegawa, H., Suenaga, M., Kato, T., Sato, S., Takayama, S. and Sugimura, T. (1986b). Induction of hepatocellular carcinoma and highly metastatic squamous cell carcinomas in the forestomach of mice by feeding 2-amino-3,4-dimethylimidazo (4,5-f) quinoline. Carcinogenesis, 7, 1889-1893.

Ohgaki, H., Hasegawa, H., Suenaga, M., Sato, S., Takayama, S. and Sugimura, T. (1987). Carcinogenicity in mice of a mutagenic compound 2-amino-3,8-dimethylimidazo (4,5-f) quinoxaline (MeIQx) from cooked foods. Carcinogenesis, 8, 665-668.

Övervik, E., Nilsson, L., Fredholm, L., Levin, Ö., Nord, C-E. and Gustafsson, J.Å. (1987). Mutagenicity of pan residues and gravy from fried meat. Mutation Res., 187, 47-55.

Övervik, E., Kleman,M., Berg, I., and Gustafsson, J-Å. (1989). Influence of creatine, amino acids and water on the formation of the mutagenic heterocyclic amines found in cooked meat. Carcinogenesis, 10, 2293-2301.

Shioya, M., Wakabayashi, K., Sato, S., Nagao, M., and Sugimura, T. (1987). Formation of a mutagen, 2-amino-1-methyl-6-phenylimidazo (4,5-b) pyridine(PhIP) in cooked beef, by heating a mixture containing creatinine, phenylalanine and glucose. Mutat. Res., 191, 133-138.

Skog, K., and Jägerstad, M. (1990). Effects of monosaccharides and disaccharides on the formation of food mutagens in model systems. Mutat. Res., in press.

Spingarn, N., Garvie-Gould, C., Vuolo, L., and Weisbarger, J.H. (1981). Formation of mutagens in fried beef patties. Cancer Letter, 13, 93-97.

Sugimura, T. and Sato, S. (1983). Mutagen-carcinogen in food. Cancer Res., (Suppl), 43, 2415s-2421s.

Sulser, H. (1978). Die Extraktstoffe des Fleisches. Wissenschaftliche Verlagsgesellschaft MBH, Stuttgart, p. 1-169.

Tanaka, T., Barnes, W.S, Williams, G.M. and Weisburger, J.H. (1985). Multipotential carcinogenicity of the fried food mutagen 2-amino-3-methylimidazo (4,5-f) quinoline in rats. Jpn. J. Cancer Res. (Gann), 76, 570-576.

Taylor, R., Fultz, E., and Knize, M. (1985). Mutagen formation in a model beef boiling system III. Purification and identification of three heterocyclic amine mutagens-carcinogens. J. Environ. Sci. Health, A20(2), 135-148.

Taylor, R., Fultz, E., Knize, M., and Felton, J. (1987). Formation of the fried ground beef mutagens 2-amino-3-methylimidazo- (4,5-f)quinoline (IQ) and 2-amino-1-methyl-6-phenylimidazo(4,5-b)pyridine (PhIP) from L-phenylalanine (Phe) + creatinine (Cre) (or creatine), Environmental Mutagenesis, 9(8), 276, (abstract)

Turesky, R.J., Bur, H., Huynh-Ba, T., Aeschbacher, H.U., and Milon, H., (1988). Analysis of mutagenic heterocyclic amines in cooked beef products by high-performance liquid chromatography in combination with mass spectrometry. Food Chem. Toxic., 26, 501-509

Wang, P-S., and Odell, G.V. (1973). Formation of pyrazines from thermal treatment of some amino-hydroxy compounds. J. Agric. Food Chem., 21, 868-870.

Yoshida, D., and Okamoto, H. (1980a). Formation of mutagens by heating creatine and glucose. Biochem. Biophys. Res. Comm., 96, 844-847.

Yoshida, D., and Okamoto, H. (1980b). Formation of mutagens by heating the aqueous solution of amino acids and some nitrogenous compounds with addition of glucose. Agr. Biol. Chem., 44(10), 2521- 2522.

Yoshida, D., and Fukuhara, Y. (1982). Formation of mutagens by heating creatine and amino acids. Agr. Biol. Chem., 46(4), 1069-1070

Yoshida, D., Saito, Y., and Mizusaki, S. (1984). Isolation of 2-amino-3-methyl-imidazo-(4,5-f) quinoline as mutagen from the heated product of a mixture of creatine and proline. Agr. Biol. Chem., 48(1), 241-243.

Yoshida, D., Matsumoto, T., Okamoto, H., Mizusaki, S., Kushi, A., and Fukuhara, Y. (1986). Formation of mutagens by heating foods and model systems. Environ. Health Perspect., 67, 55-58.

8

FORMATION OF HETEROCYCLIC AMINES DURING MEAT EXTRACT PROCESSING AND COOKING

H.U. Aeschbacher

Nestec Ltd.
Nestlé Research Centre, Vers-chez-les-Blanc
CH-1000 Lausanne 26 (Switzerland)

ABSTRACT

Standardized biological *in vitro* systems and in particular those used for cancer prediction are being used to monitor the development of food products in order to ensure the absence of potential mutagens or carcinogens. Maillard reactions occurring during meat extract production was followed in order to reduce the formation of heterocyclic amines. Possibilities to reduce the content of heterocyclic amines during meat extract processing have been proposed. However, several aspects, such as interaction with food-borne mutagen or carcinogen inhibitors, keeping quality, and organoleptic properties have also be taken into consideration. Whenever possible, food contaminants must be analytically determined and compared to total intake exposure and tolerated levels of other comparable food contaminants to establish realistic "tolerated" contamination levels.

INTRODUCTION

Deleterious effects on health such as allergies, obesity, diabetics, osteoporosis etc, have been associated with diet. However, the main worry today is the possible presence of food contaminants with mutagenic or carcinogenic potential (U.S. Dept. of Health 1988). Therefore toxicological screening tests (mutagenicity tests) are widely being used to evaluate food products. Such approaches in particular for *in vitro* tests require extraction and/or purification of complex substances such as food products. This intervention, however, can alter the outcome of the results by introducing artefacts (Aeschbacher *et al.*, 1983) or by loosing components with potential antimutagenic or anticarcinogenic property. Furthermore are *in vitro* results which only cover one aspect of risk assessment too often used to directly predict cancer risks. However, their cancer predictability is less appropriate than had previously been expected (Tennant *et al.*, 1987). Although this depends on the chemical class and their characteristics (Aeschbacher, 1990). The complexity of this aspect is therefore greater than generally is admitted. This is presently being illustrated with heat processing of meat products which generates exceedingly low amounts of mutagenic heterocyclic amines.

Nutritional and Toxicological Consequences of Food Processing
Edited by M. Friedman, Plenum Press, New York, 1991

MUTAGENICITY SCREENING

The most efficient tool to detect traces of the various heterocyclic amines in heated food is the Ames-Salmonella mutagenicity test (De Meester, 1989; Sugimura et al., 1986, 1990). Due to the lack of adequate and simple analytical methods this test was so far widely used to estimate levels of heterocyclic amines by semi-quantitative means.

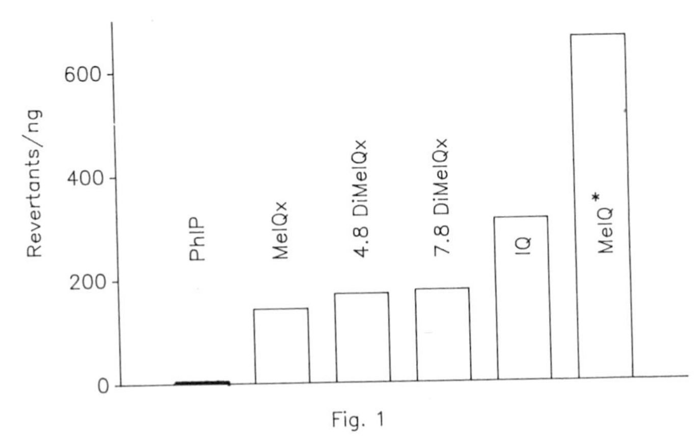

Fig. 1

*Mutagenic response to TA 98 by heterocyclic amines (Aminoimidazo Azaarenes). Some of the heterocyclic amines like protein pyrolysis products (Amino-carboline congeners) were shown to have even lower mutagenic responses in tester strain TA 98 e.g. 40 revertants/µg for Phe-P-1 (De Meester, 1989). *For MeIQ a very high reported value (De Meester, 1989), was not considered for calculating the mean number of revertants.*

Such a semi-quantitative approach, however, has serious drawbacks and therefore standardized and objective evaluation is required. In fact the actual mutagenic activity of the various heterocyclic amines observed in the Ames test varied considerably (Felton et al., 1988; De Meeser et al., 1990). Of the two major compounds in heated meat namely MeIQx and PhIP the difference in mutagenic response in Ames tester strain TA 98 is about a hundred fold. Hence an extracted heated meat sample of comparable mutagenic potential would contain about a hundred times less of one compound (MeIQx) than of the other (PhIP) (see fig. 1). Furthermore, considerable discrepancies between laboratories were observed for some compounds e.g. a 200 fold different response in TA 98 for MeIQ (De Meester, 1989).

For IQ compounds with strong mutagenic activity the "semi-quantitative" detection limit would hence lie far below 1pbb. Such a contamination level however is of no biological significance. In fact the heterocyclic amines with different mutagenic potentials in the Ames test showed comparable mutagenic or carcinogenic activity in mammalian tissues (Aeschbacher, 1986) as well as in rodents (Sugimura, 1990). Although the Ames test is a sensitive and valuable tool for identifying heterocyclic amines in heated food products this test is not suitable for quantification nor for "quantitative" cancer prediction.

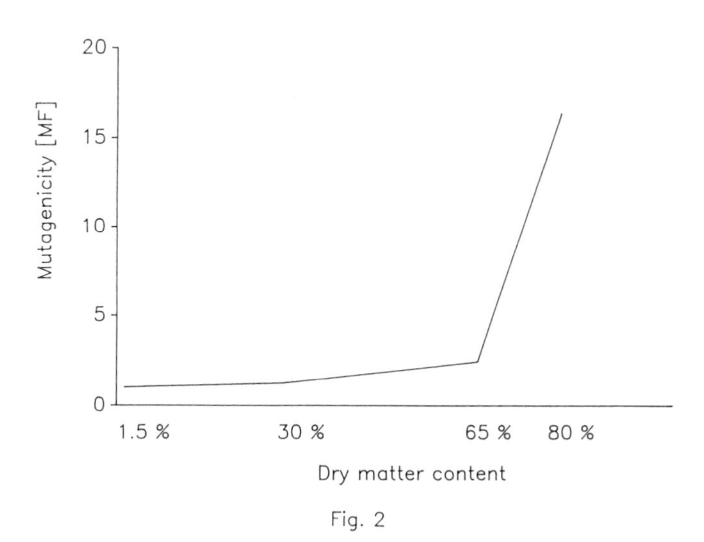

Fig. 2

Generation of mutagenic heterocyclic amines during meat extract processing

The above values were obtained when meat extract samples were purified according to Felton et al., *(1981) and evaluated with Ames tester strain TA 98.*

Analytical determination

As mentioned above, semi-quantitative evaluation based on mutagenic potential (Ames test) was used in the past for estimation of "contamination" levels of heterocyclic amines in heated food products because of lack of simple purification/quantification methods. So far, established sophisticated methods needed expensive equipment and tedious preparations which were not suitable for routine analysis (for review, see Felton and Knize, 1989). Simple cleanup methods e.g. monoclonal antibodies (Turesky et al., 1988) allowing HPLC quantification or derivatization methods employing GC (Murray et al., 1988) have recently become available. These methods however were restricted to detection of some heterocyclic amines only and are therefore suitable for studies where specific compounds can be used as markers.

A most recently developed method is now available for routine screening which is simple, rapid and yields highly purified elutions and hence only requires standard equipment e.g. HPLC, which allows quantification of the whole spectrum of heterocyclic amines (Gross et al., 1989; Gross and Aeschbacher, 1990). The detection limit of complex mixtures of all methods so far available lie within 1ppb, hence for routine purposes, quantification of heterocyclic amine "contamination" above 5-10 ppb would be feasible.

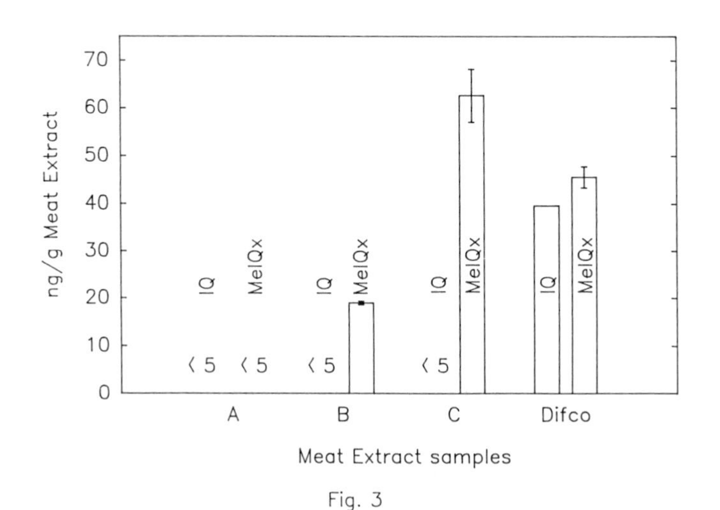

Fig. 3

Heterocyclic amine content in Meat Extracts

The above values were obtained with monoclonal antibody purified meat extract samples (Turesky et al., 1989) analyzed by HPLC. No detectable levels of PhIP were obtained in these commercial meat extract samples when analyzed according to the method of Gross et al., 1989.

Reduction of heterocyclic amine levels in heated meat

The mutagen forming reactions are dependent on factors which provide optimal conditions for Maillard reactions e.g. high temperature and dry matter content. Therefore the heterocyclic amine content was not only observed in fried, grilled or boiled meat at the ppb range (Felton and Knize, 1989) but also in the pan residues and the smoke formed during frying (Oevervik et al., 1987; Knize et al., 1988; Berg et al., 1988). It is therefore essential to find ways to reduce the exposure to heterocyclic amines formed during traditional cooking such as lower temperatures (Hatch et al., 1988). Similar ways have to be sought to avoid mutagen formation during commercial meat extract processing although levels are much lower than these of traditionally prepared meat because of the lower intake of meat extract e.g. a few grams per person and day.

Theoretically it would be possible to lower heterocyclic amine content in commercial meat extract below the detection limit (low ppb level) by partially avoiding Maillardisation during concentration "finishing stage".

Such an approach would necessitate not only changes in technical installation, but would lead to reduced keeping quality as Maillard reaction products e.g. melanoidins, have been shown to have considerable antioxidant property (Lingnert et al., 1986). New or modified meat extract processing might also alter their organoleptic aspects and hence influence "acceptability". Hence for decision making to reduce heterocyclic amines from heated meat products factors such as keeping quality, organoleptic changes, standardization of processing, legal aspects and technological feasibilities need to be considered.

"Tolerated" contamination levels

The evidence provided herein suggests that contamination levels should be based on analytically determined values and related to total exposure through human daily intake and carcinogenic potential in mammals. "Semi-quantitative" evaluation or extrapolation to a human risk from mutagenic potential of the Ames test must be avoided. Extrapolation of heterocyclic amine content from results of DIFCO meat extract, which generally contains much higher levels of IQ and MeIQ than commercial meat extract (fig. 3), is incorrect. An important factor for assessing the risk of meat extracts is the observation that commercial meat extracts contain mainly MeIQx but not PhIP as is the case in fried meat patties (Hatch et al., 1988). In particular since PhIP which is one of the most potent heterocyclic amines in mammalian cell mutagenicity tests (Felton et al., 1989) was found to occur in meat at much higher levels than MeIQx. Furthermore, it must also be remembered that heated meat products also contain constituents which might inhibit the carcinogenicity of heterocyclic amines. Such products include fatty acids (Ha et al., 1987) and melanoidins (brown pigments) which reduced the mutagenicity of IQ compounds (Kato et al., 1985).

And finally, estimated total daily intake of heterocyclic amine exposure through diet needs to be related to exposure levels from specific food products. The maximal daily human intake of 1 lt. beef bouillon containing about 1 g of meat extract results as a matter of fact in a maximum intake of only 100 ng of total heterocyclic amines whereas the total human intake is estimated at about $10^2 \mu g$ (Sugimura, 1986). The contribution to the heterocyclic amine exposure of industrially prepared soup products would hence represent 1/1000 or less of the estimated total daily intake. Therefore, in order to successfully reduce a possible cancer risk for man appropriate measures should be taken by considering all possible sources of heterocyclic amine exposure.

REFERENCES

Aeschbacher, H.U., Finot, P.A. and Wolleb, U. (1983). Interaction of histidine-containing test substances and extraction methods with the Ames mutagenicity test. Mutation Res., 113, 103-116.

Aeschbacher, H.U. (1986). Possible cancer risk of dietary heat reaction products. Proceedings of Euro. Food Tox. II Conf. (Inst. of Toxicol. ed). Schwerzenbach, Switzerland, pp 112-126.

Aeschbacher, H.U. (1990). Genetic toxicology of food products. In: Environment. Genotoxicity, Risk and Modulation. (M.L. Mendelson ed). Alan R. Liss Inc., New York, in press.

Berg, I., Oevervik, E., Nord, C.E. and Gustafsson, J.A. (1988). Mutagenic activity in smoke formed during broiling of lean pork. Mutation Res., 207, 199-204.

De Meester, C. (1989). Bacterial mutagenicity of heterocyclic amines found in heat processed food. Mutation Res., 221, 235-262.

Felton, J.S., Healy, S., Stuermer, D.H., Berry, C., Timurian, H., Hatch, F.T., Morris, M. and Bjeldanes, L.F. (1981). Mutagens from the cooking of food. Improved isolation and characterization of mutagenic fractions from cooked ground beef. Mutation Res., 88, 33-44.

Felton, J.S., Knize, M.G., Shen, N.H., Wu, R. and Becher, G. (1988). Mutagenic heterocyclic imidazoamines in cooked foods. In: Carcinogenic and Mutagenic Responses to Aromatic Amines and Nitroaarenes (C.M. King, L.J. Romano and D. Schutzle eds) Elseviers Sci; Publ., pp 73-85.

Felton, J.S. and Knize, M.G. (1989). Heterocyclic-amine mutagens/carcinogens in foods. In: Handbook of Experimental Pharmacology: Chemical Carcinogenesis and Mutagenesis, vol 94/7, pp 471-502.

Gross, G.A. Philippossian, G. and Aeschbacher, H.U. (1989). An efficient and convenient method for the purification of mutagenic heterocyclic amines in heated meat products. Carcinogenesis, 10, 1175-1182.

Gross, G. and Aeschbacher, H.U. (1990). Micromethod for the determination of heterocyclic amines. In: The Maillard Reaction in Food Processing, Human Nutrition and Physiology. (P.A. Finot, R.F. Hurrell, H.U. Aeschbacher, R. Liardon eds). Birkhäuser Verlag, Basle, in press.

Ha, Y.L., Grimm, N.K. and Pariza, M.W. (1987). Anticarcinogens from fried ground beef, heat altered derivatives of linoleic acid. Carcinogenesis, 8, 1881-1887.

Hatch, F.T., Felton, J.S. and Knize, M.G. (1988). Mutagens formed in food during cooking. ISI Atlas of Sci., 222-228.

Kato, H., Kim, S.B., Hayase, F., Nguyen, V.C. (1985). Desmutagenicity of melanoidins against mutagenic pyrolysates. Agric. Biol. Chem., 49, 3093-3095.

Knize, M.G., Shen, N.H. and Felton, J.S. (1988). A comparison of mutagen production in fried-ground chicken and beef: effect of supplemental creatine. Mutagenesis, 6, 503-508.

Lingnert, H. and Hall, G. (1986). Formation of antioxidative Maillard reaction products during food processing. In: Amino Carbonyl Reactions on Food and Biological Systems. (M. Fujimaki, M. Namiki, E. Kato eds). Elsevier, Tokyo, pp 273-279.

Murray, S., Gooderham, N.J., Boobis, A.R. and Davies, D.S. (1988). Measurements of MeIQx and DiMeIQx in fried beef by capillary column gas chromatography electron capture negative ion chemical ionisation mass spectrometry. Carcinogenesis, 9, 321-325.

Oevervik, E., Nilsson, L., Fredholm, L., Levin, O., Nord, C.E. and Gustafsson, J.A. (1987). Mutagenicity of pan residues and gravy from fried meat. Mutat. Res., 187, 47-53.

Skong, K., Jägerstad, M. and Olsson, K. (1990). Maillard reaction and the formation of genotoxic aminoimidazoazaarenes present in fried meat and fish. In: The Maillard Reaction in Food Processing, Human Nutrition and Physiology. (P.A. Finot, H.U. Aeschbacher, R.F. Hurrell, R. Liardon, eds). Birkhäuser Verlag, Basle, in press.

Sugimura, T. (1986). Past, present and future of mutagens in cooked food. Environ. Health Perspect., 67, 5-10.

Sugimura, T., Takayama, S., Ohgaki, H., Wakabayashi, K. and Nagao, M. (1990). Mutagens and carcinogens formed by cooking meat and fish; heterocyclic amines. In: The Maillard Reaction in Food Processing, Human Nutrition and Physiology. (P.A. Finot, H.U. Aeschbacher, R. Hurrell and R. Liardon eds). Birkhäuser Verlag, Basle, in press.

Tennant, R.W., Margolin, B.H., Shelby, M.D., Zeiger, E., Haseman, J.K., Spalding, J., Caspary, W., Resnick, M., Stasiewicz, S., Anderson, B. and Minor, R. (1987). Prediction of chemical carcinogenicity in rodents from in vitro genetic toxicology assays. Science, 236, 933-941.

Turesky, R.J., Forster, C.M., Aeschbacher, H.U., Würzner, H.P., Skipper, P.L., Trudel, L.J. and Tannenbaum, S.R. (1989). Purification of the food-borne carcinogens 2-amino-3-methylimidazo[4,5-f]quinoline and 2-amino-3,8-dimethylimidazo[4,5-f]quinoxaline in heated meat products by immunoaffinity chromatography. Carcinogenesis, 10, 151-156.

U.S. Dept. of Health and Human Services (1988). Cancer. In: The Surgeons General's Report on Nutrition and Health, DHHS Publication no 88-50210, Washington, pp. 177-247.

9

AN EXPERIMENTAL APPROACH TO IDENTIFYING THE GENOTOXIC RISK BY COOKED MEAT MUTAGENS

NICOLA LOPRIENO, GUIDO BONCRISTIANI and GREGORIO LOPRIENO

Dipartimento di Scienze dell'Ambiente e del Territorio dell'Università di Pisa Via S.Giuseppe 22, 56126 PISA, ITALY

SUMMARY

In order to define the toxicological risk for the human population derived from the chemical compounds formed during the process of cooking animal meat, which have been described to possess a mutagenic, genotoxic, and carcinogenic activity, an extensiye study has been developed on cooked meat extract and two cooked meat mutagens, IQ and MeIQx. The study has been based on toxicokinetics and mouse tissue distribution of the two chemicals, on *in vitro* and *in vivo* mutagenicity/genotoxicity analyses (gene mutation, chromosome aberration, micronuclea in mouse bone marrow cells, mice urine and faeces mutagenicity test), as well as *in vivo* protein and DNA binding.

The two chemicals have been found positive for the induction of gene mutation on *Salmonella,*but not in V-79 Chinese hamster cells; IQ only has been found positive for the induction of chromosome aberrations on CHO cells and cultured human lymphocytes. IQ and MeIQx were negative for the induction of micronuclea in mice treated with 40 mg/kg of the chemicals; the lowest effective administered dose to the mice which produced mutagenic urine was 0.4 mg/kg of IQ and 0.04 mg/kg of MeIQx. A dose of 40 mg/kg of IQ given by gavage to mice produced an excretion of 1-4% of the applied dose in the urine and 0.1-2% of the applied dose in the faeces, when evaluated chemically or mutagenically.

The DNA adducts for the liver were correlated with the dose of the IQ and MeIQx administered to the mice.

All the data have been used for defining a possible risk estimate derived to the human population as a consequence of a cooked meat diet.

INTRODUCTION

It has been widely demonstrated that during the cooking of the meat several mutagenic and carcinogenic chemicals are produced, as a result of a chemical reaction by aminoacids, creatine and carbohydrates, natural components of the animal muscular tissues (F.T. HATCH et al, 1982).

The aim of the present work has been that of developing *in vitro* and *in vivo* mutagenicity data, combined with toxicokinetics, tissues' distribution and DNA binding of two of the several chemicals identified among the cooking meat reaction compounds, namely 2-Amino-3-

Nutritional and Toxicological Consequences of Food Processing
Edited by M. Friedman, Plenum Press, New York, 1991

methylimidazo(4,5-*f*)quinoline (IQ) and 2-Amino-3,8-Dimethylimidazo(4,5-*f*)quinoxaline (MeIQx), in order to investigate, on the genotoxic/carcinogenic risk posed to humans by this class of food contaminants.

The present results have been compared with those reported in the literature, obtained in studies developed in different laboratories.

EXPERIMENTAL PART

Material and Methods

1 **Meat extract and Chemicals.** An industrial preparation of the meat extract (20 kg) was submitted to the extraction procedure according to J.S. FELTON et al. (1981), to obtain the basic and the acidic fractions. To prepare meat extract from home made meat broth, 1 kg of veal meat was cooked in 3.6 l of water for 4 hours at 100°C and then dried and extracted by means of the Felton's procedure. 2-Amino-3-methylimidazo (4,5-*f*)quinoline and 2-Amino-3,8-Dimethylimidazo(4,5-*f*)quinoxaline referred to, in this paper as IQ and MeIQx, were purchased from Dr. K.OLSSON, Swedish University of Agricultural Sciences, Uppsala, Sweden. ^{14}C-IQ and ^{14}C-MeIQx were also purchased from Dr. K.OLSSON. The specific activity of the two chemicals was the following: ^{14}C-IQ= 229 μC/mg; ^{14}C-MeIQx = 288 μC/mg.

2 **IQ and MeIQ determination**: the residues obtained from the extraction of meat extract or biological samples as a basic fraction were resuspended in 200 ml of a mixture of ethylacetate and eptafluorobutiric anhydride (3:1, v/v) and the samples were left to react for 30 min. at 70°C in a sand bath. The reaction mixture was brought to dryness under nitrogen and the residue redissolved in 100 ml of acetone. Aliquots of solutions obtained (1 ml) were injected in a Varian Mat 44 gas chromatograph-mass-spectrometer and quantitative determinations of IQ and MeIQx were performed focusing the instrument at 70 eV on the ions at m/z 225 (IQ) or at m/z 230 (MeIQx). The oven temperature was 240°C, the carrier gas (He) flow was 25 ml/min; the column was a SE-30 (3%) on Gas Chrom. Q (100-120 meh), 1 m long and 4 mm of internal diameter.

3 **Genotoxicity and Mutagenicity studies**: meat extracts, chemicals (IQ and MeIQx) and animal urine or fecal samples were evaluated for their mutagenicity activity on *Salmonella typhimurium* strain TA98, according to Ames's test. Basic Fraction of meat extract, IQ and MeIQx were also submitted to tests for the induction of gene mutation in the HPRT system of CH-V79 cells and for the induction of chromosome aberrations on CHO-cell line or on cultured human lymphocytes. IQ and MeIQx were also tested for the induction of micronuclea in bone marrow cells of mice treated orally with a single dose of the chemicals.

4 **Toxicokinetics studies**: CD-1 male mice have been treated orally with different doses of IQ and MeIQx, as reported in the results and analyzed for tissues' distribution and toxicokinetics in the blood. The following organs and liquids were evaluated for the chemical content: blood, urines, faeces, liver, lung, kidney, testes, bone marrow and brain.

5 **DNA binding studies**: the binding of IQ and MeIQx to proteins, RNA and DNA of mice treated with two doses of IQ or MeIQx has been determined according to K.SZCAWINSKA et al. (1981), and to C.PANTAROTTO and C.BLONDA (1984).

I. RESULTS (In vitro studies)

1 Mutagenic analysis on Salmonella (gene mutation). Fractions obtained by meat extract have been submitted to the *Salmonella* test in absence and in the presence of a S-9 mix obtained from Aroclor induced rats.

The results of typical experiments with basic fraction, IQ and MeIQx are reported in Table 1, 2 and 3: the specific mutagenic activity of the three samples, expressed as revertants/µg/plate has been the following: Basic Fraction=3,175; IQ=350,000; MeIQx=99,000. These values correspond to those reported in the literature (F.T. HATCH et al, 1982). The basic fraction employed in the *Salmonella* study, as well in the other studies contained 672,000 ppb of IQ and 2,656,000 ppb of MeIQx: the content of the two chemicals in the Meat Extract (industrial sample) employed for the present study was calculated to be 6.37 µg/kg for IQ and 25.25 µg/kg for MeIQx. The home made broth prepared as presented in Material and Methods contained 3.40 µg/kg of IQ and 5.23 µg/kg of MeIQx.

Table 1

Reverse Mutation in S.typhimurium TA98 by Basic Fraction of Meat Extract

Doses µg/pl	Without Met.Act. Mean ± S.E.	With Met.Act. Mean ± S.E.
Untreated	49 ± 10.6	47 ± 8.4
0.00	104 ± 31.9	61 ± 11.6
0.100	150 ± 27.4	251 ± 22.6
0.200	237 ± 25.1	365 ± 5.5
0.400	97 ± 34.8	355 ± 69.6
0.800	19 ± 10.1	355 ± 110.7
1.60	29 ± 15.7	278 ± 11.8
3.20		1030 ± 26.8

Positive Controls
2NF 2µg/pl(-S9)= 288±17.7
2AA 1µg/pl(+S9)=1109±68.0

Table 2

Reverse Mutation in S.typhimurium TA98 by IQ

Doses µg/pl	Without Met.Act. Mean ± S.E.	With Met.Act. Mean ± S.E.
Untreated	32 ± 2.6	39 ± 3.5
0.00	26 ± 3.1	41 ± 0.9
0.06	29 ± 1.5	243 ± 11.1
1.25	31 ± 1.5	350 ± 53.5
2.50	33 ± 2.3	795 ± 42.6
5.00	25 ± 1.7	1606 ± 68.0
10.00	28 ± 3.0	2276 ± 162.2

Positive Controls
2NF 2mg/pl(-S9)=210± 2.6
2AA 1mg/pl(+S9)=688±40.6

Table 3

Reverse Mutation in S.typhimurium TA98 by MeIQx

Doses µg/pl	With Met.Act. Mean ± S.E.
Untreated	48 ± 2.6
0.00	38 ± 4.4
0.62	112 ± 5.2
1.25	195 ± 10.9
2.50	336 ± 29.6
5.00	602 ± 48.8
10.00	1148 ± 34.7

Positive Controls
2NF 2µg/pl(-S9)= 107±2.0
2AA 1µg/pl(+S9)=787±19.5

2 **Mutagenic analysis on CH-V79 cell line (gene mutation).** For each chemical and for the basic fraction of meat extract two experiments have been performed with metabolic activation system (S-9 mix) obtained from livers of Aroclor induced rats. The induction of 6-Thioguanine resistant mutants have been evaluated at two expression times (6 and 9 days). The results are reported in Tables 4, 5 and 6: all three were negative.

Table 4

Gene Mutation in CH-V79 cells by Basic Fraction of Meat Extract
§ Positive control: NDMA nM

With Metabolic Activation				
	Experiment I		Experiment II	
Dose µg/ml	Day 6	Day 9	Day 6	Day 9
0.0	14.13	11.27	18.49	23.28
62.5	22.22	19.14	20.58	34.72
125.0	31.20	24.66	10.27	16.81
250.0	11.17	15.05	33.51	14.57
500.0	9.82	2.21	42.78	84.43
1000.0	15.09	2.17	54.46	45.67
5.00§	1088.57**	955.93**	910.97**	1092.23**
10.00§	1976.36**	1704.23**	1684.97**	1470.42**

Table 5

Gene Mutation in CH-V79 cells by IQ
§ Positive control: NDMA nM

With Metabolic Activation				
	Experiment I		Experiment II	
Dose µg/ml	Day 6	Day 9	Day 6	Day 9
0.00	1.72	10.13	9.24	13.50
0.80	10.03	2.36	31.10	59.48
1.60	4.05	4.96	27.91	27.81
3.20	3.72	2.03	15.06	18.08
6.40	6.28	15.22	42.62	13.90
12.80	5.74	8.28	18.18	17.25
25.60	3.54	2.45	5.83	19.39
5.00§	671.11**	443.60**	492.44**	572.21**
10.00§	1494.97**	615.60**	601.23**	543.40**

Table 6

Gene Mutation in CH-V79 cells by MeIQx
§ Positive control: NDMA nM

With Metabolic Activation				
	Experiment I		Experiment II	
Dose µg/ml	Day 6	Day 9	Day 6	Day 9
0.00	18.55	27.27	16.57	15.85
3.20	6.55	41.51	7.86	5.65
6.40	28.47	39.18	15.36	24.04
12.80	44.72	48.09	33.40	28.74
25.60	12.47	4.23	24.05	42.78
51.20	15.03	12.65	16.78	15.22
5.00§	358.66**	715.48**	652.56**	652.58**
10.00§	1120.75**	1287.80**	901.83**	901.62**

3 **In vitro cytogenetic analysis on CHO cell line and cultured human lymphocytes** The ability of the two chemicals and of the basic fraction of meat extract in inducing chromosome aberrations in mammalian cells treated in vitro has been evaluated by means of two test assays, the

Chinese hamster ovary cell line (CHO) and the cultured human lymphocytes. The treatment has been performed in the presence of S-9 mix induced rat liver metabolic activation system. For each treatment two cultures were considered. CHO cells were treated for three hours and then analysed after 12 and 24 hours; cultured human lymphocytes were treated for three hours in absence and in presence of S-9 mix and then incubated for further 25 hours before harvesting and scored for chromosome aberrations. The results of the analyses performed on CHO cells are reported in Tables 7, 8 and 9; those of analyses on cultured human lymphocytes are reported in Tables 10, 11 and 12: only IQ resulted positive for the induction of chromosome aberrations in the two types of assay.

Treatment	Doses	+S9 mix: 12 hr.		+S9 mix: 24 hr.	
	μg/ml	%CA	MI	%CA	MI
Untreated	-	4.5	20.9	2.5	11.2
Solvent	1%	1.5	17.9	1.0	12.1
B.Fraction	31.6	1.0	21.1	1.0	9.5
B.Fraction	100.0	1.0	22.2	0.0	9.9
B.Fraction	316.0	1.0	13.9	1.0	13.3
§ CPH	6.6	4.0	7.8	8.5*	11.9

Table 7

Chromosome aberrations in CHO cells by Basic Fraction of Meat Extract §Positive control: Cyclophosphamide (CPH)

Table 8

Chromosome aberrations in CHO cells by IQ §Positive control: Cyclophosphamide (CPH)

Treatment	Doses	+S9 mix: 12 hr.		+S9 mix: 24 hr.	
	μg/ml	%CA	MI	%CA	MI
Untreated	-	1.5	15.2	2.0	8.5
Solvent	1%	0.5	16.9	1.0	13.7
IQ	6.40	0.5	16.1	1.5	7.5
IQ	12.80	5.5**	5.9	4.5*	5.4
IQ	25.60	4.0*	3.9	5.5*	6.2
§ CPH	6.60	7.0***	9.2	12.0***	18.3

Treatment	Doses	+S9 mix: 12 hr.		+S9 mix: 24 hr.	
	μg/ml	%CA	MI	%CA	MI
Solvent	1%	2.0	31.2	3.0	17.2
Solvent	2%	3.5	30.0	7.5	18.9
MeIQx	17.20	7.0*	33.3	6.0	19.1
MeIQx	37.10	4.0	35.5	4.5	18.5
MeIQx	80.00	2.0	27.1	4.5	17.9
§ CPH	6.60	8.0**	18.5	23.0***	21.6
CPH	16.00	5.0	29.2	6.0	18.8

Table 9

Chromosome aberrations in CHO cells by MeIQx §Positive control: Cyclophosphamide (CPH)

Treatment	Doses µg/ml	-S9 mix: %CA	-S9 mix: MI	+S9 mix: %CA	+S9 mix: MI
Untreated (solvent)	-	3.0	4.7	5.0	5.3
B.Fraction	17.75	4.5	3.4	5.5	4.8
B.Fraction	35.50	1.5	4.2	1.5	4.5
B.Fraction	71.00	3.5	4.7	5.5	5.1
MMS	75.00	28.0**	4.7	-	-
CPH	6.60	-	-	30.0***	5.1

Table 10

Chromosome aberrations in cultured human lymphocytes by Basic Fraction of Meat Extract
§Positive control:
MMS: Methylmethanesulfonate
CPH: Cyclophosphamide

Table 11

Cromosome aberrations in cultured human lymphocytes by IQ
§Positive control:
MMS: Methylmethanesulfonate
CPH: Cyclophosphamide

Treatment	Doses µg/ml	-S9 mix: %CA	-S9 mix: MI	+S9 mix: %CA	+S9 mix: MI
Untreated (Solvent)	-	5.5	3.1	3.0	2.4
IQ	116.2	5.5	2.6	10.5**	1.4
IQ	178.8	5.5	2.1	11.0	1.8
IQ	275.0	8.5	2.4	14.0*	2.3
MMS	75.0	48.0***	2.4	-	-
§ CPH	12.5	-	-	54.0***	2.4

Treatment	Doses µg/ml	-S9 mix %CA	-S9 mix MI	+S9 mix: %CA	+S9 mix: MI
Untreated (solvent)	-	3.0	4.7	5.0	5.3
MeIQx	83.20	2.5	5.0	5.0	4.6
MeIQx	104.00	3.0	4.5	3.0	4.2
MeIQx	130.00	5.0	4.2	4.5	4.6
MMS	75.00	28.0**	4.2	-	-
CPH	12.50	-	-	30.0***	4.5

Table 12

Chromosome aberrations in cultured human lymphocytes by MeIQx
§Positive control:
MMS: Methylmethanesulfonate
CPH: Cyclophosphamide

4.4 Mutagenicity of IQ in the presence of Meat Extract Acidic Fraction. The possible inhibitory effect by Meat Extract's Acidic Fraction on the mutagenicity of IQ on *Salmonella* TA 98 strain (reverse gene mutations) and on CHO cell line grown in culture (chromosome aberrations) has been evaluated: the results are reported in Table 13 and 14: basic fraction was able to inhibit the mutagenicity of IQ on *Salmonella*, but not on CHO, or cultured human lymphocytes.

Table 13

Mutagenic activity of IQ on Salmonella TA98 in the presence of Meat Extracts' Acid Fraction

IQ ng/pl	ACID FRACTION mg/plate (IQ ng/plate)				IQ		PERCENTAGE OF IQ's MUTAGENICITY INHIBITION BY ACIDIC FRACTION			
	2000 (2574)	1000 (1287)	500 (643)	250 (322)	0	alone	2000	1000	500	250
Untreated	52	52	52	52	52					
0.0	95	93	82	72	39	-	-	-	-	-
0.625	108	125	151	270	199	0	45	37	24	0
1.25	124	159	204	290	411	0	70	61	50	29
2.50	131	200	345	521	831	0	84	76	58	37
5.00	176	313	580	862	1402	0	87	77	58	38
10.00	276	502	859	1365	2130	0	87	76	59	36

Table 14

Chromosome aberrations in CHO cells by IQ and in the presence of Meat Extract's Acid Fraction
CPA= Cyclophosphamide

Treatment	Dose level µg/ml	% Cells Aberrant	Mitotic Index
Untreated Control	-	2.0	12.5
Solvent Control	(1%)	3.5	16.0
Solvent Control	(1.14%)	5.5	15.9
Acid Fraction	1000	4.5	12.2
IQ	25.6	9.0*	5.2
IQ + Acid fraction	25.6 + 100	9.0*	5.3
CPA	6.60	9.0*	19.6
CPA + Acid Fraction	6.60 + 100	9.0***	10.5

II. RESULTS *(in vivo studies)*

1 **Mutagenic analyses of mice urines.** IQ and MeIQx were assayed *in vivo,* on mice, for the evaluation of their ability in producing urinary mutagens. Groups of three male CD-1 mice (24-31 g) were treated by oral gavage with IQ, MeIQx or the vehicle, 1% DMSO in 0.5% carboxymethylcellulose solution, at a dosage volume of 10 ml/kg. Treatment levels of 1.0, 0.4, 0.1, 0.04, 0.01 and 0.004 mg/kg were used for each test substance. Urine samples were collected from each group for the following 24 hours. The urine samples were tested in *S.typhimurium* TA98 with the preincubation procedure at 50 and 100 µl/plate. A dose-response relationship between the treatment level of the mice, and the numbers of revertant colonies induced by the urine sample was found for both test substances. Treatment with IQ at 0.4 mg/kg and above gave urine samples which induced increases in the numbers of revertant colonies which were greater than twice the values of control urine samples. MeIQx produced such increases at treatment levels of 0.04 mg/kg and above. The results of this study are reported in Fig.1.

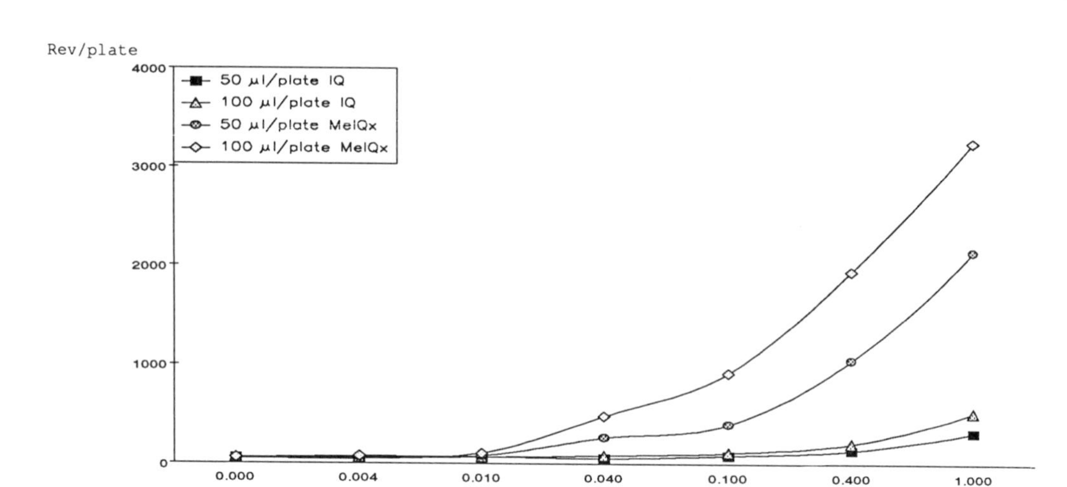

FIG. 1
REVERSE MUTATION IN Salmonella typhimurium TA 98 INDUCED BY 50-100 µl/plate OF URINES COLLECTED FROM MICE TREATED BY GAVAGE WITH IQ or MeIQx

2 Combined study with urinary analyses, micronuclea in the bone marrow and toxicokinetics in mice treated with IQ and MeIQx.
Males CD-1 (25-30 g) have been treated by gavage with 40 mg/kg of IQ in 1% DMSO in 0-5% of carboxymethylcellulose solution. Several effects have been evaluated: a) presence of IQ in the urines and faeces 12 and 24 hours after treatment; b) the induction of micronucleated cells in the bone marrow cells of the animals 12, 24 and 48 hours after the treatment with IQ or MeIQx; c) the IQ concentration in the peripheral blood of treated mice during the post-incubation time; d) the IQ concentration in liver, lung, kidneys, testes, bone marrow, and brain; e) the concentration of IQ in the urine and faeces 24 hours after the treatment. The urinary assay was performed in combination with *Salmonella typhimurium* analysis as described in the previous section. The bone marrow micronucleus test was performed according to the normal protocol. The toxicokinetics studies have been performed by extraction of blood and tissues samples with methylene chloride and by analyzing the extracts by gas-chromatography and mass-spectrometry. In Tables 15 and 16 the results of the micronuclea assay are reported; the toxicokinetics of IQ in the blood compartment is reported in Fig. 2; The results of tissues distribution are reported in Fig. 3 and the data on the excretion of IQ are reported in Table 17.

3 DNA binding of ^{14}C-IQ and ^{14}C-MeIQx in mice tissues treated via oral route. CD-1 Swiss mice have been treated orally with a single dose of ^{14}C-IQ (0.4 or 40 mg/kg), and ^{14}C-MeIQx (0.04 or 40 mg/kg). 6,12, and 48 hours after the treatment, groups of 3 or more animals have been killed and blood, brain, lung, heart, liver, kidneys, spleen, intestine and bone marrow have been taken and stored at -20°C until before being analysed. For animals treated with 40 mg/kg of either the chemicals only liver and bone marrow have been evaluated 48 hours after the treatment.

Table 15

Micronuclea induction in mice bone marrow treated by gavage with IQ

Treatment	Dose mg/kg	Sample Time	No.of mice	Cells Scored PCE	Cells Scored NCE	NCE/PCE Ratio	Incidence of MN/1000 cells Scored PCE MEAN ±SE	PCE Range	NCE MEAN ±SE	NCE Range
Vehicle	0.0	24 h	5	5024	4523	0.9	1.4±0.4	0-2	0.6±0.4	0-2.1
Test Substance	40.0	12 h	5	5012	4826	0.96	1.0±0.4	0-2	0.8±0.4	0-1.9
Test substance	40.0	24 h	5	4027	3850	0.96	1.2±0.5	0-2	1.4±0.8	0-3.7
Test substance	40.0	48 h	5	4041	4771	1.18	2.7±0.6	1-4	1.6±0.7	0-3.2
2-AAF	250.0	24 h	5	5035	5106	1.02	4.6±1.5**	0-9	2.1±0.3	8-2.7

Table 16

Micronuclea induction in mice bone marrow treated by gavage with MeIQx

Treatment	Dose mg/kg	Sample Time	No.of mice	Cells Scored PCE	Cells Scored NCE	NCE/PCE Ratio	Incidence of MN/1000 cells Scored PCE MEAN ±SE	PCE Range	NCE MEAN ±SE	NCE Range
Vehicle	0.(	24 h	5	5228	4894	0.94	2.1±0.7	0.9-0.4	0.9±0.4	0-2.3
Test Substance	40.0	12 h	5	5314	4338	0.81	2.8±0.6	0.9-3.8	1.6±1.0	0-4.0
Test substance	40.0	24 h	5	5208	6918	1.33	2.3±0.7	1.0-4.7	1.0±0.1	0.8-1.4
Test substance	40.0	48 h	5	5208	4505	0.86	2.5±0.9	0.9-5.7	0.4±0.2	0-1.0
Mitomycin C	250.0	24 h	5	5181	7048	1.36	42.5±5.6***	32.7-62.7	2.3±0.5	0.7-3.8

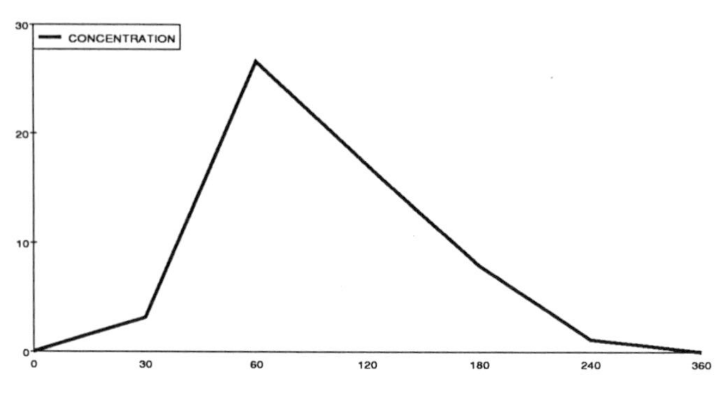

µg/ml

TIME (min.)

FIG.2

IQ CONCENTRATION IN THE BLOOD OF CD-1 MALE MICE
TREATED BY GAVAGE WITH 40 mg/Kg OF IQ

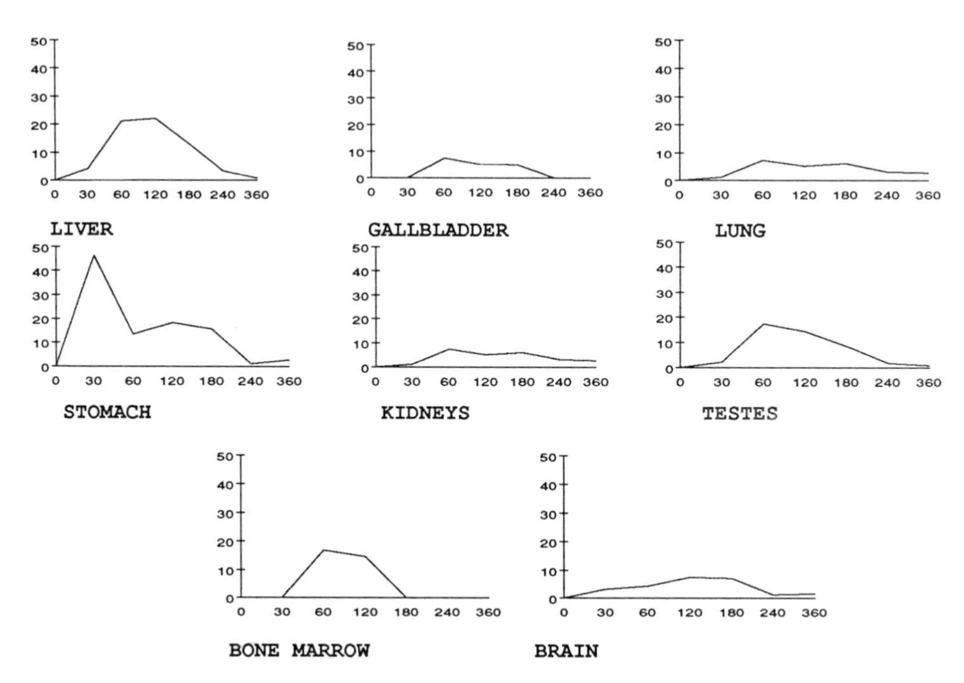

FIG.3

IQ CONCENTRATION IN DIFFERENT TISSUES OF CD-1 MALE MICE
TREATED BY GAVAGE WITH 40 mg/Kg OF IQ
(IQ µg/g/min)

Table 17

Excretion of IQ in the urine and faeces of CD-1 male mice treated by gavage with 40 mg/kg of IQ

SAMPLE	EXCRETION DURING 24 Hours (% of dose)
Urines	4.6
Faeces	2.7

The covalent binding to the proteins has been evaluated according to the procedure published by K.SZCZAWINSKA et al. (1981) and the DNA-binding according to the procedures of C.PANTAROTTO and L.BLONDA (1984). The results are reported in Tables 18, 19, 20 and 21.

Table 18

Radioactivity covalent binding to proteins (dpm/ml of blood or g of tissue): data after 48 hours in the treated mice with ^{14}C-IQ(0.4mg/kg) or ^{14}C-MeIQx (0.04 mg/kg)

Tissues	Chemicals	
	IQ	MeIQx
Blood	321 ± 48	52 ± 23
Brain	93 ± 17	7 ± 3
Lung	1,112 ± 374	123 ± 41
Heart	515 ± 126	41 ± 15
Liver	8,299 ± 1,426	1,122 ± 374
Kidney	2,936 ± 484	212 ± 56
Spleen	808 ± 179	48 ± 20
Intestine	1,427 ± 126	171 ± 32
Bone Marrow	1,116 ± 374	178 ± 41

Table 19

Covalent binding to proteins (dpm/g) after treatment of mice with 40 mg/kg of radiolabelled chemicals (average of three determination)

Tissues	IQ	MeIQx
Liver	1,274.351 ± 432.566	1,784.897 ± 323.910
Bone Marrow	175.666 ± 37.308	214.870 ± 41.117

Table 20

DNA covalent binding (dpm/mg) in different tissues of mice treated with ^{14}C-IQ (0.4 mg/kg and ^{14}C-MeIQx (0.04 mg/kg after 48 hours

Tissues	IQ	MeIQx
Brain	n.d.	n.d.
Lung	4 ± 2	n.d.
Heart	n.d.	n.d.
Liver	27 ± 8	5 ± 2
Kidney	8 ± 3	n.d.
Spleen	n.d.	n.d.
Intestine	4 ± 2	n.d.
Bone Marrow	6 ± 2	n.d.

Table 21

DNA covalent binding (dpm/g) in the liver with ^{14}C-IQ and ^{14}C-MeIQx (40 mg/kg)

Tissues	IQ	MeIQx
Liver	2,241 ± 376	8,423 ± 1,344
Bone Marrow	139 ± 57	221 ± 81

DISCUSSION

The extensive *in vitro* and *in vivo* (mice) mutagenicity/genotoxicity studies, the toxicokinetics and mice tissue distribution, the *in vivo* (mice) DNA and protein binding of IQ, MeIQx, and to some extent (in vitro mutagenicity data) of Basic Fraction obtained by Meat Extract are summarized in Fig. 4, and outlined here below:

1.1 Basic Fraction obtained from Meat Extract contains IQ and MeIQx; it is mutagenic for *Salmonella* TA98 strain in the presence of metabolic activation; it does not induce either gene-mutation in the HPRT genetic locus system of V-79 CH cell line, or chromosome aberrations in CHO cultured cells or cultured human lymphocytes, in the presence of metabolic activation.

1.2 The Acidic Fraction obtained from Meat Extract inhibits the mutagenic activity of IQ, when assayed in *Salmonella* TA98 strain, but not when assayed in the test for the induction of chromosome aberrations in CHO cultured cells.

2.1 IQ is mutagenic in *Salmonella* TA98 strain and in the assay for the induction of chromosome aberrations either in CHO cultured cells or in cultured human lymphocytes in the presence of metabolic activation. IQ does not induce gene mutation in the HPRT genetic locus system of V-79 CH cell line, when tested in the same conditions.

2.2 IQ, when administered orally to mice at different doses, produces mutagenic urines assayed on *Salmonella* TA98 strain down to a minimal oral dose of 0.4 mg/kg: the NOEML (the No Effective Mutagenic Level) of IQ on mice has been determined to correspond to a dose of 0.1 mg/kg.

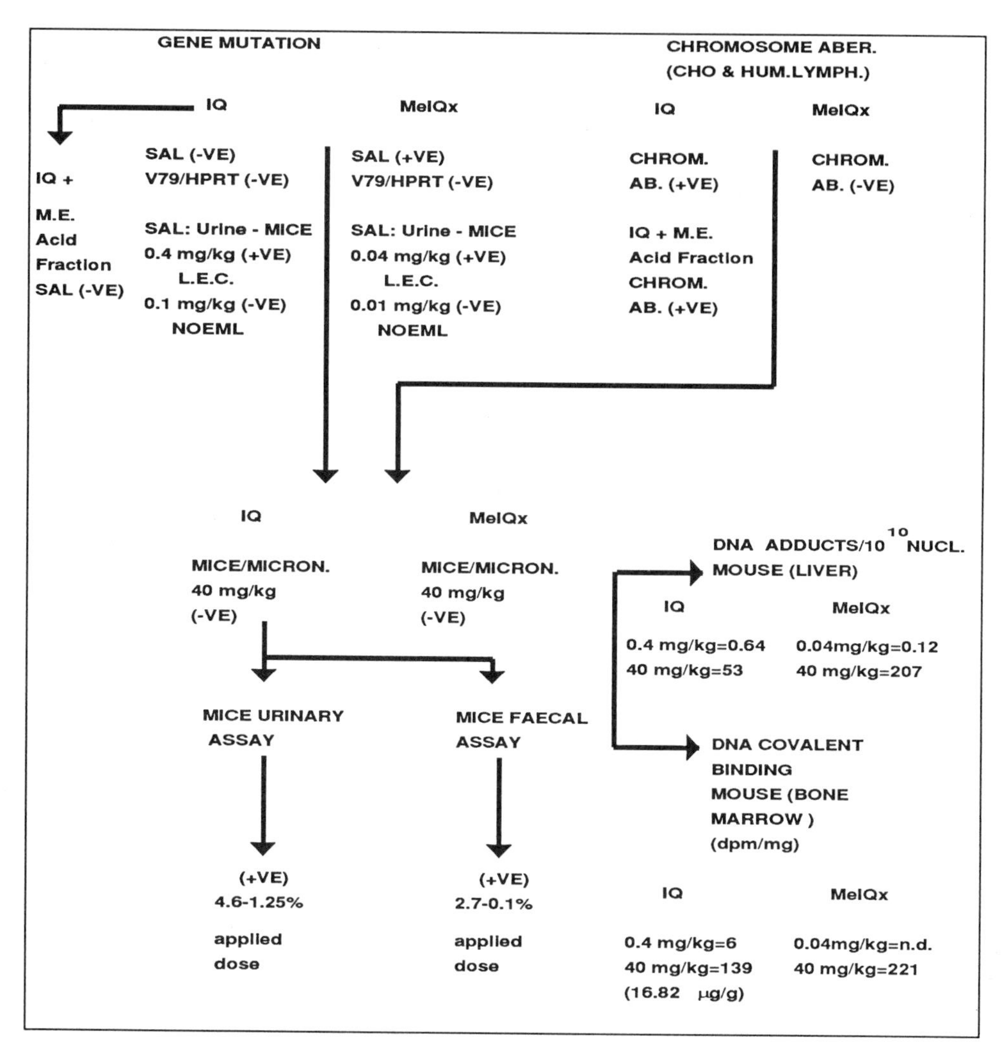

FIG.4

MUTAGENICITY / GENOTOXICITY OF IQ AND MEIQx

2.3 A dose of 40 mg/kg of IQ administered orally to mice reached a maximum concentration of 26.64 µg/ml in the blood 60 minutes after the treatment (Fig. 2); it has been found distributed in liver, lung, kidneys,testes, brain, bone marrow with a peak reached in 60-120 min (Fig. 3). In the bone marrow this treatment does not induce micronucleated cells, evaluated 12, 24, and 48 hours after treatment; the level of IQ in the bone marrow was evaluated to correspond to a concentration of 16.82 mg/kg, a dose found clastogenic *in vitro* (Table 8).

2.4 The presence of IQ in the urine of mice corresponded to a concentration of 4.6% of the applied dose (40 mg/kg), when it was evaluated by chemical analyses and to a value of 1.25%, when it was evaluated by *Salmonella* mutagenicity analysis; the corresponding values observed in the faeces were 2.7% and 0.1% of the applied dose.

2.5 DNA covalent binding analyses *in vivo* on mice have shown that IQ binds to liver DNA and it is dose related; DNA binding has been observed also in the bone marrow cells, at a level unable to produce chromosome aberrations.

3.1 The other cooked meat contaminant, MeIQx, when assayed in the same condition as for IQ, resulted mutagenic for *Salmonella*, but not for V-79 CH cultured cells, cultured human lymphocytes, and CHO cultured cells in the presence of metabolic activation.
3.2 In the *in vivo* studies on mice the production of mutagenic urine was observed when a minimal effective dose of 0.04 mg/kg of MeIQx was administered orally to the animals (NOEML=0.01 mg/kg).
3.3 A dose of 40 mg/kg of MeIQx orally administered to mice was not effective for the production of micronuclea (but the chemical did not induce this mutagenic end point even *in vitro:* see above 3.1), but it binds to liver DNA and it is dose-related.

Both positive and negative results in the genotoxic/mutagenic studies reported in the present study correspond to the data published in the literature: (D.WILD et al., (1985) have reported that IQ was unable to produce micronuclea in mice bone marrow, treated with 594 mg/kg i.p. or 198 mg/kg p.o.; J.D.TUCKER et al., (1989) have reported that MeIQx was unable to produce SCE in mice bone marrow, treated up to 50 mg/kg i.p.), thus confirming the finding that IQ and MeIQ are (1) mutagenic *in vitro,* but not *in vivo* and that (2) their mutagenic activity might be modulated by other substances naturally present in cooked meat or formed during the preparation of Meat Extract.
W.S. BARNES and J.H.WEISBURGER (1985) have found that only 1% of the dose of IQ applied to rats treated by intravenous injection with a dose up to 50 mg/kg was recovered in the urines; these were dose-related mutagenic on *Salmonella* in the presence of S-9 liver fraction. S.MURRAY et al. (1989) found that human urines, collected in the 48-60 h. period after a meal based on lean ground beef fried patties containing between 290 and 850 ng of MeIQx, contained a percentage of ingested heterocyclic amine in the range of 1.8 - 4.9.
By comparing the present data of the DNA binding studies with the data reported by ALDRICK & LUTZ (1989), the DNA binding index of IQ and MeIQx made possible their classification among the groups of weak carcinogens (Table 22), according to a previous evaluation made by W.K.LUTZ (1979). When a comparison is made between the data reported in the present paper for the mice treated with IQ and those obtained by E.G.SNYDERWINE et al. (1988) on rats and monkeys treated with the same compound, in regard to the relationship between the dose administered to the animals and the no. of radioactive adducts/10^{10} Nucleotides, the values reported in Table 23 are obtained: these data all togheter show a linear relation between oral dose (mg/kg) and No. of adducts/Nucleotides, (correlation factor of 0.94), indipendently from the animal species treated (Fig. 5). A linear dose-No. of DNA adducts relationship has been found also for MeIQx given to mice by J.S.FELTON et al. (1990).
The concentration of IQ present in the home made meat broth, evaluated in the present paper to be 3.40 µg/kg, allows us to attempt to evaluate the possible risk for man, due to this contaminant. The ingestion of 200 ml of broth/day by a person of 60 kg will allow an intake of IQ of the order of 11.3 ng/kg/day, a dose which is 35,000 times lower than that producing 0.64 DNA adducts/10^{10} Nucleotides (0.4 mg/kg): this value is less than 1 adduct/cell for man. To reach this dose a person should have meat broth for 96 years !
Another source of human exposure to chemicals like IQ is the use of industrial meat extract in the preparation of meat broth. The suggested use of industrial meat extract is a 25 g preparation of a commercial product containing 2% of meat extract: such a preparation is presently employed for 1 liter of broth, which will contain 0.5 g of meat extract. According to our analyses the industrial meat extract contains 6.37 µg/kg of IQ: the intake of 200 ml of broth prepared in this way will impose an exposure, for a person of 60 kg, of 0.01 ng/kg/day. This amount is 1,000 times lower than the previous value.

Table 22

Covalent binding index of different carcinogens

	CBI*
A. STRONG CARCINOGENS	
AFLATOXIN	17,000
DIMETHYLNITROSAMINE	6,000
B. MODERATE CARCINOGENS	
2-ACETYLAMINOFLUORENE	560
VINYL CHLORIDE	525
C. WEAK CARCINOGENS	
URETHANE	29-90
**MeIQx (Loprieno et al. 1990)	87.0
**IQ (Loprieno et al. 1990)	28.7
**MeIQx (Aldrick & Lutz, 1989)	9.0
p-AMINOBENZENE	2.0
D. NON CARCINOGENS	
SACCHARIN	<0.005

* µmol. adducts per mol. DNA/mmol. chemical applied per kg animal body weight

** liver DNA

Table 23

Species	DOSES mg/kg	No. of Adducts /10^{10} Nucleotides
MICE	0.4	0.64
	40.0	53.0
RATS (2)	330.0	103,000.0
MONKEYS (3)	200.0	107,000.0

(1) One doses

(2) Total dosage (2 weeks)

(3) Total dosage (3 weeks)

(2,3 from *E.G.SNYDERWINE et al.1988)*

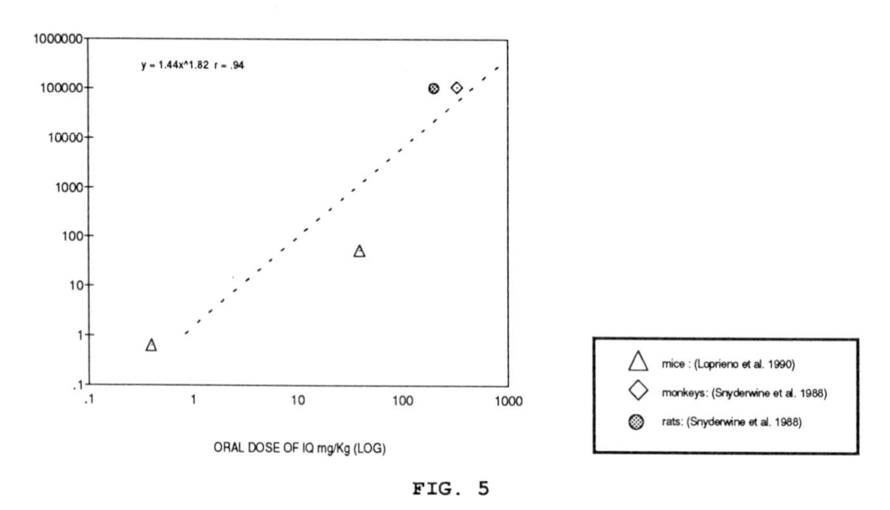

FIG. 5

Nr.DNA ADDUCTS/10^{10} NUCLEOTIDES (LOG)

T.SUGIMURA (1986) has calculated a TD50 value of heterocyclic amines of about 8 mg/kg/day, a dose which is extremely higher that those obtained by us for IQ in the DNA binding studies. According to T.SUGIMURA the intake of heterocyclic amines in human beings is of the order of 2×10^{-4} lower than the TD50, i.e. 0.4 µg/kg/day. This dose is 1,000 lower than that found by us to produce 0.64 DNA adducts/10^{10} Nucleotides.
The conclusion derived from the present study, which demonstrates the rather weak ability of the heterocyclic amines class as genotoxic agent *in vivo* allows us to support T.SUGIMURA's statement that "the heterocyclic amines may not be so serious for human cancer development"; however "it would be appropriate to establish a way to avoid (or to reduce) the formation of mutagens and carcinogens during cooking by improving heating conditions" (T.SUGIMURA, 1986). Thus the formation of such contaminants cannot be avoided, as the normal house cooking methods have demonstrated the presence of mutagenic contaminants in the meat broth (cfr. the present paper) and in the grilled lamb chops, sirloin and rump steaks, as well as in fried meats (P.J.BARRINGTON et al., 1990).

AKNOWLEDGEMENTS

The experimental work has been supported by the CONSIGLIO NAZIONALE DELLE RICERCHE, ITALY.

We aknowledge the cooperation made by C.PANTAROTTO, LSR Roma Toxicology Centre, Pomezia,Italy and Hazleton Microtest, York, UK.

REFERENCES

ALDRICH, A.J. and W.K. LUTZ: Covalent binding of [2-^{14}C] 2-amino-3,8-dimethylimidazo[4,5-*f*]-quinoxaline (MeIQx) to mouse DNA *in vivo*. Carcinogenesis *10*, 14919-1423, 1989.

BARNES, W.S. and J.H. WEISBURGER: Fate of food mutagen 2-amino-3-methylimidazo[4,5-*f*] quinoline (IQ) in Sprague-Dawley rats. I. Mutagens in the urine.
Mutation Res. *156*, 83-91, 1985.

BARRINGTON, P.J., R.S.U. BAKER, A.S. TRUSWELL, A.M. BONIN, A.J. RYAN, and A.P. PAULIN: Mutagenicity of basic fractions derived from lamb and beef cooked by common household methods.
Fd. Chem. Toxicol. *28,* 141-146, 1990

FELTON, J.S., M.K. NIZE, K.W. TURTELTAUB, M.H.BUONARATI, R.T.TAYLOR, M.VANDERLAAN, B.E. WATKINS, J.D. TUCKER , and L.H. THOMPSON: Mutagens and Carcinogens in cooked foods: concentration, potency, and risk.
The Faseb Journal *4,* A774, Abstract 2945, 1990.

FELTON, J.S., S. HEALY, D. STUEMER, C. BERRY, H. TUMOURIAN, F.T. HATCH, M. MORRIS, and L.F. BJELDANES: Mutagens from the cooking of the food. I. Improved extraction and characterization of the mutagenic fraction from cooked ground beef.
Mutat. Res. *88,* 33-44, 1981.

HATCH, F.T., J.S. FELTON, and L.F. BJELDANES: Mutagens from the coking of food: thermic mutagens in beef. In H.F. STICH, ed. "Carcinogens and mutagens in the environment" CRC Press, Inc. Boca Raton, Florida, USA, *1,* 147-163, 1982.

LUTZ, W.K.: In vivo covalent binding of organic chemicals to DNA as a quantitative indicator in the process of chemical carcinogens.
Mutat. Res. *65,* 289-356, 1979.

MURRAY, S., N.J. GOODERHAM, A.R. BOOBISM, and D.S. DAVIES: Detection and measurement of MeIQx in human urine after ingestion of a cooked meat meal.
Carcinogenesis *10,* 763-765, 1989.

PANTAROTTO, C. and C. BLONDA: Covalent binding to proteins as a mechanism of chemical toxicity.
Arch. Toxicol. 30, Suppl. 7, 208-218, 1984.

SNYDERWINE, E.G., K. YAMASHITA, R.H. ADAMSON, S. SATO, M. NAGAO, T. SUGIMURA, and S.S. TORGEIRSSON: Use of the ^{32}P-postlabeling method to detect DNA adducts of 2-amino-3-methylimidazolo[4,5-*f*] quinoline (IQ) in monkeys fed IQ: identification of N-(deoxyguanosin-8-yl)-IQ adduct.
Carcinogenesis *9,* 1739-1743, 1988.

SUGIMURA, T.: Past, Present, and Future of Mutagens in cooked foods.
Environmental Health Perspective *67,* 5-10, 1986.

SZCZAWINSKA, K., E. GINELLI, I. BARTOSEK, C. GAMBAZZA, and C.PANTAROTTO: Caffeine does not bind covalently to liver microsomes from different animal species and to proteins and DNA from perfused rat liver.
Chem. Biol. Interaction *34,* 345-354, 1981.

TUCKER, J.D., A.V. CARRANO, N.A. ALLEN, M.L., CHRISTENSEN, M.G. KNIZE, C.L. STROUT, and J.S. FELTON: In vivo cytogenetic effects of cooked food mutagens.
Mutation Res. *224,* 105-113, 1989.

WILD, D., E. GOCKE, D. HARNASH, G. KAISER, and M.T. KING: Differential mutagenic activity of IQ (2-Amino-3-methylimidazo[4,5-*f*] quinoline) in *Salmonella typhimurium* strains in vitro and in vivo, in Drosophila, and in mice.
Mutation Res. *156,* 93-102, 1985.

10

MUTAGENS AND CARCINOGENS IN COOKED FOODS: CONCENTRATION, POTENCY, AND RISK

J.S. Felton, M.K. Knize, K.W. Turteltaub, M.H. Buonarati, R.T. Taylor, M. Vanderlaan, B.E. Watkins, J.D. Tucker and L.H. Thompson

Biomedical Science Division, Lawrence Livermore National Laboratory, Livermore, CA 94550

The cooking of foods derived from muscle generates heterocyclic amines that are very potent bacterial mutagens and carcinogens in mice, rats, and monkeys. Presently 12 mutagenic compounds have been found in cooked foods derived from the Western Diet. Only 6 of these compounds have been definitely identified. Specifically designed monoclonal antibodies can detect nanogram amounts each of 2-amino-1-methyul-6-phenylimidazo[4,5-f]pyridine(PhIP), 2-amino-3-methylimidazo[4,5f] quinoline(IQ), and 2-amino-3,4-dimethylimidazo[4,5-f]quinoline (MeIQx) in the meat matrix, but specific quantitation requires a prior HPLC separation before use of antibodies for identification. Modeling of PhIP (20 ppb in beef) formation under dry heating conditions using heavy isotope incorporation coupled with MS and NMR shows that all the C and N atoms in the PhIP are contributed by creatin(ine) and phenylalanine. PhIP is both a potent frameshift mutagen in Salmonella bacteria and a potent inducer of mutations, sister chromatid exchange and chromosomal aberrations in Chinese Hamster Cells in culture, and of chromosome damage in mouse bone marrow. Using accelerator mass spectroscopy, we can measure in a linear fashion one MeIQz DNA adduct per 10^{11} nucleotides. N-OH-PhIP which is generated by both IA1 and IA2 forms of cytochrome P-450 appears to be a proximal mutagen.

Nutritional and Toxicological Consequences of Food Processing
Edited by M. Friedman, Plenum Press, New York, 1991

11

BEEF SUPERNATANT-FRACTION-BASED STUDIES OF HETEROCYCLIC AMINE-MUTAGEN GENERATION

R.T. Taylor, E. Fultz, M.G. Knize, and J.S. Felton

Biomedical Science Division, Lawrence Livermore National Laboratory
Livermore, CA 94550

To characterize the reaction conditions that generate frameshift mutagens in cooked meats, we have concentrated on a supernatant fraction (S_2) prepared from (1:1) aqueous homogenates of lean round steak. Soluble compounds <500 MW in S_2 are the sources of the S-9 dependent Salmonella TA1538 mutagenic activity in these homogenates, irrespective of how they are heated (100°C boiled, 200-300°C aqueous-pressure, or 200-300°C dry), as well as the outer surfaces (crust) of griddle-fried ground beef. Water is an important inhibitory reactant that influences not only the total TA1538 activity, but also the proportions of HPLC-polar, nitrite-resistant 2-amino-3-methyl-imidazo-type mutagens, as opposed to HPLC-nonpolar, nitrite-sensitive mutagens. Dry-state heating beef S_2 favors the former. It yields eight of the heterocylic amine-mutagens that have been identified in the surfaces of 100g beef patties fried at 250-300°C, including 2-amino-3-methylimidazo [4,5-f]quinoline (IQ), 2-amino-3-8-dimethylimidazo[4,5-f]quinoxaline (MeIQx), and the predominant mutagen 2-amino-1-methyl-6-phenylimidazo [4,5-b]pyridine (PhIP). Oven-baking the amounts of L-phenylalanine (Phe) and creatine (Cr) present in 100g (raw beef) equivalents of S_2 yields sufficient IQ and PhIP to accomodate the ppb quantities reported for high temperature fried beef patties. Dry-heating with heavy-isotope-labeled forms of Phe and Cr shows precisely how their C and N atoms are incorporated into PhIP. Our findings indicate that phIP and other 2-amino-3-methyl-imidazo-mutagens most likely arise in 250-300°C fried beef patties independent of Maillard reactions.

Nutritional and Toxicological Consequences of Food Processing
Edited by M. Friedman, Plenum Press, New York, 1991

12

CARCINOGENESIS IN OUR FOOD AND CANCER PREVENTION

J. H. Weisburger
American Health Foundation
Valhalla, New York 10595

ABSTRACT

Worldwide, locally prevailing nutritional traditions account for the occurrence of specific types of cancer. In the Orient, the custom of eating salted, pickled or smoked food parallels the risk of stomach cancer and hypertension-stroke. The underlying mechanisms and relevant carcinogens are partially known. In the Western world, the usual high-fat, low-fiber food is related to risk of cancer of the colon, pancreas, breast, prostate, ovary, and endometrium. The fat component translates to specific promoting mechanisms and fibers reduce risk of colon cancer through dilution of promoters. The associated genotoxic carcinogens may be the heterocyclic amines formed during cooking of meat. Methods have been developed to inhibit their formation. In all situations, a higher intake of vegetables and fruits has led to a lower risk for diverse types of cancer, through varied mechanisms. Based on current knowledge, more wholesome dietary traditions for chronic disease prevention in most countries can be developed.

INTRODUCTION

Much progress has been made in uncovering the causes and modifying factors associated with important types of cancer in man. Whereas the media and the public are under false impression that cancer stems from exposure to environmental and agricultural chemicals, pesticides, insecticeds, or food additives, factual information indicates that the major cancer types in any part of the world stem from the locally prevailing lifestyles. In the United States, tobacco use and particularly smoking accounts for about 30% of all cancer deaths. Because of effective health promotion by research groups, by voluntary associations, like the American Cancer Society and the Federal Government, through the Office of the Surgeon General and the National Institutes of Health (NIH), American males have progressively reduced the smoking habit (McGinnis, 1988-1989; Breslow, 1990). Therefore, since about 1984, lung cancer mortality in males has begun to decline. Unfortunately, women tend to smoke more, and lung cancer currently has greater mortality in women than the much feared breast cancer. More effective action in this country, and particularly anywhere else in the world where greater proportions of the adult population are still smoking, is needed to reduce the effect of this known human carcinogen. Smoking also sharply increases the rate of fatal heart attacks in the Western world.

Nutritional and Toxicological Consequences of Food Processing
Edited by M. Friedman, Plenum Press, New York, 1991

Nutritional traditions account for a large fraction of other types of of cancer (American Health Foundation, 1987). While more documentation and research is needed to fully underwrite the view that specific nutrients and methods of cooking relate to certain cancer types, the base of knowledge is adequate to begin making recommendations for alterations in nutritional habits as a means of health promotion and cancer prevention. This is the subject of this report. Prior to the development of this theme, we will review the major new acquisitions in the area of the mechanisms of cancer causation to provide a rational basis for the field of nutritional carcinogenesis.

MECHANISMS OF CARCINOGENESIS

Neoplastic Conversion

Historically, at least two distinct phases of carcinogenesis, namely an early stage and a late stage, or initiating agents and promoting agents were recognized (Berenblum, 1985). In the last 20 to 30 years, it has become apparent that cancer is the result of mutational events. Current major advances in molecular biology actually permit investigation at the molecular level of the precise nature of changes in the genome induced by carcinogens. The carcinogens leading to such mutational events are genotoxic (Williams and Weisburger, 1991). Such carcinogens have been found to be associated with cancers of environmental or occupational origin, or cancers related to lifestyle, because of use of tobacco products or exposure to tobacco smoke, or their presence in foods as a result of cooking, salting or pickling. Genotoxic carcinogens react with DNA, usually after host-mediated biochemical activation. They are thus converted to reactive electrophilic metabolite that react with nucleophilic centers in DNA, RNA, and proteins. There are new procedures involving sensitive immunological methods or biochemical approaches like 32P-postlabeling to determine to what extent such reactions have occurred (Williams, 1989). These markers also permit assessment of prior exposure of individuals to genotoxic chemicals (Bridges et al., 1982). Anticarcinogenesis involves efforts to trap reactive electrophiles with suitable nucleophiles (Wattenberg *in* Joossens et al., 1985; Hartman and Shankel, 1990).

In a qiescent cell population, the abnormal DNA can be repaired. However, during cell duplication, the carcinogen-modified DNA template forms the basis for the production of an altered DNA. The cells bearing such a changed DNA in its gene structure are considered typical of early neoplastic cells. Clearly, any endogenous event or exogenous chemical that increases the rate of DNA synthesis and mitosis may potentiate the effect of a carcinogen, or vice versa, attenuate it, if the rate is decreased (Newmark et al., 1990).

Overall, these series of events are called neoplastic conversion (Williams and Weisburger, 1991). This set of reactions at the level of DNA and the gene is effected by genotoxic chemicals, by specific tumor viruses, or directly or indirectly by radiation. In addition to altering DNA through translocation and gene amplification, such actions can also affect tumor suppressor genes, and thus provide a template typical of neoplasia for DNA synthesis (Weinberg, 1989; Kumar et al., 1990).

GROWTH AND DEVELOPMENT

The subsequent steps, namely the growth and development of early neoplastic cells, are subject to a totally distinct set of endogenous and exogenous growth controlling elements that operate through distinct mechanisms (Williams and Weisburger, 1991). This area of promotion and progression plays a key role in eventually leading to clinical, overt neoplasia. While additional alterations of the genome can take place during these stages, perhaps due to lack of fidelity of the DNA-polymerases in early cancer cells, the main actions involve growth and development. Early typical neoplastic cells are under the influence of growth controlling elements, transferred through intercellular gap junctions from surrounding normal cells. One role of promoters is to sever or constrict gap junctions. Neoplastic cells are thus subject to their intrinsic growth-enhancing controls. On the other hand, vitamin A can further intercellular transfer of growth controlling elements, and thus serve to decrease promoting actions (Hossein et al., 1989). A major role of dietary components like fat, fiber, or calcium ions is to influence the growth and development aspects of neoplastic cell systems through specific effector mechanisms. This is important since these actions are highly dose-dependent and are also reversible. Lowering the concentration of a growth-promoting substance by only 50% may exert a major delaying effect on the growth of neoplastic cells. Complete elimination of promoting substances can almost immediately and sharply decrease the development of neoplastic cells. Smoking cessation, for example, progressively lowers the risk of the appearance of clinical cancer (Weisburger, 1990). Based on these well documented properties inherent in the phenomenon of promotion, it is possible to delay or abolish the growth of abnormal cells. This is the foundation for dietary recommendations designed not only for the primary prevention of diet-related neoplasms but also to increase longevity in patients with diet-sensitive neoplasms that were surgically removed. The research-based deliberate decrease of promoting phenomena is designed to prevent, or at least delay recurrences, a sound means of adjuvant therapy.

CANCER OF THE ORAL CAVITY AND ESOPHAGUS

In the Western world, these diseases occur at a higher rate in individuals who chew tobacco or who smoke and also consume regularly alcoholic beverages (Garro and Lieber, 1990). This can be interpreted to mean an inadequate supply of protective elements such as vitamin A. It has been noted that exposure to toxic substances such as tobacco smoke increases the need for the vitamins, in part because tobacco smoke induces enzymes concerned with the metabolism of many substances including hormones and micronutrients (Conney, 1986).

In parts of China as well as the Near East, especially the Eastern part of Iran and Southern Soviet Union, cancer of the esophagus is seen in people who do not smoke or drink. Dietary elements, especially salted, pickled foods, have been associated with the risk of cancer of the esophagus, again in situation of poor intake of vegetables. The carcinogens have not yet been identified but may be specific nitrosamines (Wakabayashi et al., 1989; Bartsch et al., 1990) (Table 1).

CANCER OF THE STOMACH

This disease is still highly prevalent in Northern Japan, part of China, and Northern and Eastern Europe (Hayashi et al., 1986). It was a major type of cancer in all of the Western world including the United States. However, it has declined sharply in the last 60 years (Howson et al., 1986). The associated risk factor is the traditional intake of salted, pickled foods, or smoked meats or fish. In many parts of the world, therefore, gastric cancer is also associated with a higher incidence of hypertension related to dietary salt without protective elements like potassium and calcium salts (Joossens et al., 1985). The sharp decline of stomach cancer in the United States, and beginning elsewhere

Table 1. Geographic pathology of cancer and nutritional traditions

Organ	Lower-Risk Population	Lower-Risk Factors	Higher-Risk Population	Higher-Risk Factors
Colon	Japan	Low-fat diet	North America; Western Europe; New Zealand; Australia; Southern Scandinavia	High-fat and -cholesterol, low-fiber diets; fried foods
	Mormons	Higher-fiber diet		
	7th Day Adventists	Low or no fried food, higher-fiber diet		
	Finland	Higher-fiber diet, lower intake of fried food (calcium??)		
Breast	Japan	Low-fat diet	North America; Western Europe; New Zealand; Australia	High-fat diet, low fiber
Endometrium	Japan	Low-fat diet	USA(California); Europe	High-fat diet, obesity, estrogen use
Prostate	Japan	Low-fat diet	USA; Scandinavia; Western Europe	High-fat diet
Pancreas	Bombay, India		USA(California; blacks)	Early smoking habit, Western-style high-fat, high-cholesterol diet
Stomach	USA	Low pickled, salted foods, high fruits, salads, vitamins C,E	Japan, China, Chile, Columbia Eastern Europe	Salted, pickled foods, geochemical nitrate and low intake of vitamins A,C, and E.
Esophagus	Utah, USA; rural Norway	Low alcohol and smoking habits	France(Calvados Normandy); USA (lower socio-economic groups)	Alcohol and smoking
			India	Tobacco, chewing
			Eastern Iran, Southern Soviet Union, Central China	Low Vitamins A,C, and E; salted, pickled food?

in the world, is clearly linked to the introduction of commercial and especially home refrigerators as a means of food storage and preservation. This technical advance has negated the need for salting and pickling. For that reason, just as the incidence of gastric cancer has dropped, so has hypertension.

Salting and pickling has been reproduced in the laboratory by treating specific foods like fish with nitrite and sodium chloride at pH 3. This leads to the formation of a powerful direct-acting mutagen that in a pilot study has produced gastric cancer in rats. The structure of the carcinogen is not fully known but appears to be diazophenol (Weisburger, in Hayashi et al., 1986). In a pilot study, tyramine was reacted with nitrite at low pH, which gave the corresponding diazophenol. In rats, this chemical caused cancer of the oral cavity (Wakabayashi et al., 1989).

Clearly, head and neck, and gastric cancer can be prevented by omitting the habit of salting and pickling of foods, and consuming regimens with more vegetables and fruits. An associated benefit will be a much lower risk of high blood pressure.

WESTERN NUTRITIONALLY LINKED CANCERS

Important types of cancer prevalent in the Western World are those in the distal colon, postmenopausal breast, ovary, endometrium, pancreas, and prostate (American Health Foundation, 1987). These diseases have a low incidence in industrialized countries like Japan. This fact has lead to inquiries about dietary traditions and a key difference was noted, namely a fat intake of about 40% calories in the West and about 15% in traditional Japan. In the last 20 years the Japanese particularly in urban areas, have adopted more of a Western dietary habit resulting in increases in these cancers and also in atherosclerosis and heart disease. This is excellent evidence that dietary fat level is a major risk factor (Tajima et al., 1985; Wynder and Hiyama, 1987). This will be discussed in terms of a promotional mechanism. The question arises as to the genotoxic elements in these types of cancer, and this subject will be covered next.

Genotoxic Carcinogens

The group of Sugimura (in Finot et al., 1990) discovered that some of the most powerful mutagens were formed during charcoal broiling of meat or fish. The mutagens were much more potent than what might be accounted for through the formation of polycyclic aromatic hydrocarbons. It was found that these mutagens were formed during any type of cooking involving browning. They were new types of heterocyclic aromatic amines. Groups in Japan, the United States, and Europe provided important information on the structure of the chemicals formed, of their specific mutagenicity and of their carcinogenicity in animal models (Hatch et al., 1988). Jaegerstad (Skog et al., in Finot et al., 1990) first pointed to the need of creatinine for the formation of these mutagens. It is now clear that creatinine provides components to form the essential 2-amino-3-methyl-imidazole part of the molecule. The remainder of the ring system appears to involve of reactive aldehydes from precursor amino acids and sugars (Weisburger and Jones, in Finot et al., 1990). We have been able to demonstrate that the formation of the heterocyclic amines can be appreciably inhibited by mixing soy protein with meat prior to frying, by the addition of antioxidants, and especially of proline and tryptophan in small amounts. The latter combination effectively traps the reactive intermediate aldehydes, so that they cannot interact with creatinine. Thus practical means exist to lower the formation of important types of carcinogens in the human food chain.

Role of Dietary Fat

In addition to the observation of a distinct incidence of specific cancer types as a function of geographic area of residence, there have been detailed studies on nutrition and cancer in animal models. The first association between dietary fat and breast cancer in mice was discovered some 50 years ago, but neglected until Carroll (American Health Foundation, 1987) performed detailed research on the type and amount of fat in a model of breast cancer discovered by Huggins. At the American Health Foundation (1987), Reddy found that specific fats particularly those rich in linoleic acid, had a powerful enhancing effect in colon carcinogenesis when fed in the diet at 40% of calories, mimicking the human intake in the Western world, compared to 10% of the diet reflecting a low-risk population. These effects could be demonstrated in several distinct models involving different carcinogens for the colon. Fats involving W-3 fatty acids have a somewhat protective effect. In several models for breast cancer, including two chemically induced systems and one utilizing radiation, a Western fat level gave more mammary tumors than a low fat level (American Health Foundation, 1987). Not every fat has an identical effect. Olive oil did not much promote cancer, perhaps mimicking the situation in Southern Italy where breast cancer is lower than in Northern Italy or the United States, albeit higher than in Japan. The lower rate with olive oil has been thought to be due to the relative deficiency of linoleic acid. Nontheless, the breast cancer rate in humans is lower in the population of the Mediterranean area where olive oil is the main supply of fat (Iacono, *in* American Health Foundation, 1987; LaVecchia et al., 1988).

Mechanisms Involving Fat

Several groups, especially that of Reddy et al. (1989; also in American Health Foundation, 1987), have found a parallel in animals and and in humans between dietary fat intake and intestinal and fecal bile acid amounts. In further studies it was shown that bile acids have an enhancing or promoting effect in intestinal carcinogenesis. Fat and fiber interact and optimal amounts can be devised to lower risk (Sinkeldam et al., 1990). Bile acids and fatty acids at high levels appear to be cytotoxic to the intestinal epithelium (Bird et al., 1986) (Table 2).

The mechanism whereby fat translates to higher risk in the endocrine sensitive target organs, particularly the mammary gland, is not yet fully documented. However, it may involve estrogen and specific estradiol metabolites, in turn controlling endocrine balances, including hormones from the thyroid and pituitary glands. There are no reliable models for prostate cancer, and therefore the nutritional mechanisms in that target have not been studied in as much detail as the mammary gland. However, the findings of a low incidence of prostate cancer in Japan, and other areas with a low fat dietary tradition, suggest extension of the fat hypothesis to that organ.

The fat hypothesis has been criticized by the demonstration that animals fed a high fat diet under restricted conditions did not develop breast or colon cancer (Pariza, 1988). The relevant mechanism may involve an effect on the endocrine system. Also, an important additional parameter is the fact noted by Clayson et al. (1989), that dietary restriction lowers the rate of cell cycling in many organs. Thus, cancer development is clearly inhibited for those reasons. Some observations during the war years in Europe relative to a lower rate of fatal heart attacks and perhaps of the nutritionally-linked cancers, may permit transfer to human setting of these concepts.

Table 2. Mechanism of organ-site specific carcinogenesis

Disease	Risk Factors	Mechanism	Protective elements	Mechanism
Endocrine-related cancers; prostate, breast	Total dietary fat (saturated + w-6 polyunsatured lipids)	Complex multi-effector elements: hormonal balances, membrane & intracellular effectors; genotoxic carcinogens: heterocyclic arylamines (HAA) in cooked foods (also for colon, pancreas)	Monounsaturated oil (olive); W-3 poly-unsaturated oils; medium chain triclycerides; cereal fiber and pectin	Neutral action on hormone metabolism; protective effect in hormone metabolism; caloric equivalent to carbohydrate; affects entero-hepatic cycling of hormones
Endometrial and ovarian cancer	Same as above, and excessive body weight	Fat cells generate estrogen; high levels estrogen may yield reactive metabolite		
Colon Cancer; -proximal	?	?	?	?
-distal	Same as endocrine-related cancers	Biosynthesis of cholesterol, thence bile acids, and colon cancer promotion, including higher cell cycling; HAA as genotoxic carcinogens	Cereal fiber vegetables, pectins Calcium salts medium-chain tryglycerides	Increases stool bulk, dilutes promoters bind bile & fatty acids lower intestinal cell cycling equivalent to carbohydrates
Rectal Cancer	Alcoholic beverages, especially beer	Increases cell cycling in rectum?	?	?
Pancreas cancer	Same as endocrine, etc.; Cigarette smoking	Genotoxic carcinogens: tobacco-specific nitrosamines, or HAA; fat as enhancer through unknown mechanisms	?	?

However, these findings are not likely to be applicable in general with free and actually tempting availability of diverse foods. Therefore, other means of prevention of important types of human cancer need to be employed. Based on current knowledge, this would involve a dietary tradition with a lower amount of fat calories. In North America, the marketing of low fat variants of many traditional foods facilitates greatly the public acceptance of a decreased fat nutritional regimen (Engle et al., 1990; Shapira et al., 1990).

Mechanisms of Dietary Fiber

Fiber is a term denoting a specific component in a variety of foods, particularly fruits, vegetables, and cereals (Trowell et al., 1985; American Health Foundation, 1987). Broadly speaking, from a physiological and chemical point of view, two types of fibers are distinguished. One class is the insoluble fiber, composed mainly of cellulose and lignin, that are poorly hydrolyzed by bacterial enzymes in the intestinal tract. These components are hygroscopic, absorb fluids, and therefore increase stool bulk. The second class is soluble because bacterial enzymes can partially hydrolyze this type of fiber, leading to the formation of gums. These, in turn, carry through the intestinal tract and eliminate compounds like bile and fatty acids, and certain minerals. An adequate intake of cereal bran fibers, composed mainly of the insoluble fibers, is effective in reducing the risk of colon cancer. The best evidence for this phenomenon stems from the observation of a low colon cancer rate in populations in Finland (American Health Foundation, 1987). These people have one of the highest coronary heart disease rates in the world because of a traditional high intake of foods containing saturated fats. In contrast, the colon cancer rate is the lowest. This is accounted for by the usual intake of appreciable amounts of bran cereal fiber. In turn, this leads to a considerable stool bulk, of the order of 200-250 grams per day, compared to about 80 grams in control populations, with a high risk of colon cancer such as in New York. Animal experiments in models for colon cancer have demonstrated an inhibiting effect by moderate amounts of bran fiber (American Health Foundation, 1987). Larger amounts that appear to have an irritating effect, leading to increased DNA synthesis in the colon, increase carcinogenesis through this mechanism. Fiber also lowers the risk of breast cancer by virtue of an effect on the enterohepatic circulation of hormones (Rose, 1990' Van't Veer et al., 1990). Based on the findings in Finland, a suitable amount of bran cereal fiber is such as to produce a daily stool of about 200 grams in adults.

Chemoprevention through Vegetables and Fruits

The intake of green yellow vegetables in particular, and fruits and vegetables in general, have been found uniformly beneficial not only in lowering the risk of nutritionally-linked cancers but even those caused by tobacco use (Wattenberg, in Joossens et al., 1985). The relevant mechanisms are complex. It is probably true that individuals consuming appreciable amounts of fruits and vegetables also take in less total fat, and thus account for a lower risk from many types of cancer, and in fact also have less heart diseases and hypertension. These foods also provide vitamins and minerals. Along these lines, the Recommended Daily Allowance (RDA) represents the daily amount of vitamins, designed to avoid deficiency diseases. In the Western world, these are extremely rare, because most people have access to foods providing the RDA. The important future question is to develop information on the optimal amounts of each micronutrient, singly and in combination, in order to avoid and inhibit chronic diseases, especially the diverse type of cancer and coronary heart disease (Weisburger, 1991). Beta-carotene and vitamin A have displayed a moderating effect for several types of cancer. Vitamin C has a benefificial effect in relation to the diseases seen in the Far East, particularly cancer of the esophagus and cancer of the stomach. Vitamin E is an essential cofactor, together with selenium salts, as regards specific detoxification enzymes. Riboflavin and iron salts have been shown to be beneficial in preventing the Plummer-Vinson syndrome. Vitamin B6 may be effective in lowering the toxicity and carcinogenicity of hydrazine derivatives (Weisburger, 1991). A number of micronutrients are currently undergoing clinical trials as chemopreventive agents (Greenwald et al., 1990).

Vegetables and fruits include also other components that have been shown to have inhibiting action in cancer of several types through diverse, specific mechanisms (Wattenberg, in Joossens et al., 1985; Meyskens, 1989; Michnovicz and Bradlow, 1990). Clearly, therefore, it is important to recommend a wider, more regular use of vegetables and fruits for health promotion activities.

CONCLUSIONS

It has been established world-wide that locally prevailing nutritional traditions are associated with specific chronic disease risks. For example, the tradition of consuming salted, pickled foods in the Orient is associated with the occurrence of hypertension and stroke, and cancer of the esophagus, stomach, and perhaps liver. In countries like the United States, where the tradition was abondoned because of improved means of preservation and storage of foods through refrigeration, diseases like hypertension and cancer of the stomach have declined sharply. In Japan, likewise, a change along these lines is occurring, with consequent lower incidence rates of hypertension, stroke, and stomach cancer. Clearly, in areas of the world where such foods are still used, effective action needs to be taken to replace these habits.

In the Western world, but mostly in the United States, there have been changes in dietary customs as relates to amount and type of fat consumed. Influenced by the key findings that the traditional high intake of saturated fats from meals and dairy products were associated with atheroscleroids and coronary heart disease, the public has progressively lowered the intake of such foods, aided by the availability and indeed advertising of lower fat foods or products with a higher polyunsaturated-saturated ratio. This alteration in food habits, and perhaps the fact that fewer men are addicted smokers, has resulted in a major success in preventive medicine, a 25% reduction in mortality in coronary heart disease in the last 10 years (Committee on Diet and Health, 1989; Breslow, 1990). In addition, the awareness that moderate exercise is beneficial may have played a role as well. Table 3 defines current mechanisms underlying specific types of cancer in North America.

Changes in tradition need to be reinforced as regards the prevention of important types of cancer, specifically postmenopausal breast, distal colon, prostate, ovary, endometrium, and pancreas cancer. The heart-diseases oriented reduction of fat from 40 to 30% of calories is not likely to be adequate for a reduction of these amjor cancer types. Indeed, studies in animal models as well as observations from human populations indicate that 30% of fat, irrespective of P/S ratio, will have the same effect as 40% of fat calories (Cohen et al., 1986; Reddy in American Health Foundation, 1987). There appears to be a sharp decline in risk when the fat level is 20% of calories. Foods and recipes are available in the United States, but perhaps not yet elsewhere in the Western world, for the adoption of a new dietary tradition involving foods providing mostly complex carbohydrates, 65-70% of calories, protein, 10-15% of calories, and fat, less than 25% of calories. A recent trial

Table 3. Analysis of Causes of Human Cancers in North America

Type	% of total
Lifestyle cancers	
Tobacco-related	32
Lung, pancreas, bladder, kidney	
Nitrate-nitrite (salted, pickled foods), low vitamin C, mycotoxins - stomach, liver	5
High fat, low fiber, broiled or fried foods-colon, pancreas, breast, prostate, ovary, endometrium	31
Alcohol-rectum; mycotoxin, hepatitis B - liver	3.8
Sunlight - skin	1.7
Transplacental chemicals - brain	2.2
Occupational cancers - various organs	1
Cryptogenic cancers (virus?)	
Lymphomas, leukemias, sarcomas, cervix	11
Multifactorial	
Tobacco and alcohol - oral cavity, esophagus	2.2
Tobacco and asbestos, tobacco and mining, tobacco and uranium and radium - lung, respiratory tract	2.3
Iatrogenic cancers	
Radiation, drugs - diverse organs	1
Unspecified - diverse organs	7

Table 4. Health Promotion and Chronic Disease Prevention

	ACTION	LOWER RISK
1.	Control smoking - less harmful cigarette	Coronary heart disease; cancers of the lung, kidney, bladder, pancreas
2.	Lower total fat intake (aim for 20% of calories)	Coronary heart disease; cancers of the colon, breast, prostate, ovary, endometrium
3.	Increase fiber, Ca^{2+}	Cancer of the colon, breast, constipation, diverticulosis, appendicitis
4.	Lower intake of fried foods	Cancers of colon, breast, pancreas (?)
5.	Have regular moderate exercise	Coronary heart disease, cancer of the colon, breast
6.	Maintain proper weight, avoid obesity	Coronary heart disease, cancer of the endometrium, kidneys
7.	Moderate or no use of alcoholic beverages	Liver cirrhosis, cardiomyopathy; male impotence; head and neck cancers (in smokers)
8.	Lower salt Na^+ intake; Balance K^+ + Ca^{2+}/Na^+ ratio	Hypertension, stroke, cardiovascular disease
9.	Avoid pickled, smoked, salted foods	Cancer of the stomach, esophagus, nasopharynx, liver(?); hypertension, stroke
10.	Practice sexual hygiene	Cancer of the cervix, penis

has shown that people readily adopt this type of lifestyle, provided dietitians and recipes give clear instructions as to the composition of food (Prentice et al., 1988). Clearly, this cancer prevention dietary tradition will also lead to a considerable lower risk for atherosclerosis and heart disease. Such a dietary regimen may also bear on adult onset diabetes. Also, as noted, avoidance of highly salted foods provides the basis for a reduced risk of hypertension.

The preceding elements all bear on reducing cancer risk by limiting the phenomenon of promotion of various target sites. Promotion is not only highly dose-dependent, but also reversible. Therefore, an adoption of lower fat, high vegetables, fiber and fruit diet is deemed essential, not only for the prevention of most important types of cancer but also key to the avoidance of recurrences in patients where clinical management has satisfactorily treated a primary cancer, like in the breast or colon.

In the context of this Symposium, it can be proposed that the genotoxic carcinogens for a number of the cancers discussed stem from the mode of cooking, frying or broiling, of meat or fish. Modalities have been developed to lower appreciably the formation of these carcinogens during cooking through simple, available modalities.

Therefore, advances through research in nutritional carcinogenesis have dissected a complex problem into segments now understood as to underlying mechanisms. In turn, this comprehension has provided the basis for rational means of health promotion and prevention of important types of cancer, and other chronic diseases (Table 4).

ACKNOWLEDGEMENTS

Mrs. Julie Howard provided expert editorial support. Research in my laboratory is funded by USPHS-NIH grants, CA-17613, CA-42381, and CA-45720 from the National Cancer Institute.

REFERENCES

American Health Foundation. (1987). Workshop on New Developments on Dietary Fat and Fiber in Carcinogenesis (optimal types and amounts of fat or fiber). Prev. Med. 16, 449-595.

Bartsch, H., Ohshima, H., Shuker, D. E. G., Pignatelli, B., and Calmels, S. (1990). Exposure of humans to endogenous N-nitroso compounds: implications in cancer etiology. Mutation Res. 238, 255-267.

Berenblum, I. (1985). Challenging problems in cocarcinogenesis. Cancer Res. 45, 1917-1921.

Bird, R. P., Schneider, R., Stamp, D., and Bruce, W. R. (1986). Effect of dietary calcium and cholic acid on the proliferative indices of murine colonic epithelium. Carcinogenesis, 7, 1657-1661.

Breslow, L. (1990). The future of public health: Prospects in the United States for the 1990's. Ann. Rev. Public Health, 11, 1-28.

Bridges, B. A., Butterworth, B. E., and Weinstein, I. B. (eds.). (1982). Banbury Report 13. Cold Spring Harbor Laboratory. Indicators of genotoxic exposure.

Clayson, D. B., Nera, E. A., and Lock, E. (1989). The potential for the use of cell proliferation studies in carcinogen risk assessment. Regul.Pharmacol. Toxicol. 9, 284-295.

Cohen, L. A., Choi, K., Weisburger, J. H., and Rose, D. P. (1986). Effect of varying proportions of dietary fat on the development of N-nitrosomethylurea-induced rat mammary tumors. Anticancer Res. 6, 215-218.

Committee on Diet and Health, Food and Nutrition Board. (1989). Implications for Reducing Chronic Disease Risk. National Academy Press, Washington, D. C.

Conney, A. H. (1986). Induction of microsomal cytochrome P-450 enzymes: The first Bernard B. Brodie lecture at Pennsylvania State University. Life Sci. 39, 2493-2518.

Engle, A., Herber, J. R., and Reddy, B. S. (1990). Relationship between food consumption and dietary intake among healthy volunteers and implications for meeting dietary goals. J. Amer. Dietetic Assoc. 90, 47-48.

Finot, P. A., Aeschbacher, H. U., Hurrell, R. F., and Liardon, R. (eds.). (1990). The Maillard Reaction in Food Processing, Human Nutrition, and Physiology. Birkhauser, Basel, Switzerland.

Garro, A. J., and Lieber, C. S. (1990). Alcohol and cancer. Annu. Rev. Pharmacol. Toxicol. 30, 219-249.

Greenwald, P., Nixon, D. W., Malone, W. F., Kelloff, G. J., Stern, H. R., and Witkin, K. M. (1990). Concepts in cancer chemoprevention. Cancer Res. 65, 1483-1490.

Hartman, P. E., and Shankel, D. M. (1990). Antimutagens and anticarcinogens: A survey of putative interceptor molecules. Env. Mol. Mutagen. 15, 145-182.

Hatch, F. T., Knize, M. G., Healy, S. K., Slezak, T., and Felton, J. S. (1988). Cooked-food mutagen reference list and index. Env. Mol. Mutagen. 12 (Supplement 14), 1-85.

Hayashi, Y., Nagao, M., Sugimura, T., Takayama, S., Tomatis, L., Wattenberg, L. W., and Wogan, G. N. (eds.). (1986). Diet and Nutrition and Cancer. Japan Scientific Societies Press and VNU Science Press BV, Tokyo and Utrecht, 3-345.

Hossain, M. Z., Wilkens, L. R., Mehta, R. P., Loewenstein, W., and Bertram, J. S. (1989). Enhancement of gap junctional communication by retinoid correlates with their ability to inhibit neoplastic transformation. Carcinogenesis, 10, 1743-1748.

Howson, C. P., Hiyama, T., and Wynder, E. L. (1986). The decline in gastric cancer: Epidemiology of an unplanned triumph. Epidemiol. Rev. 8, 1-27.

Joossens, J. V., Hill, M. F., and Geboers, J. (1985). Diet and Human Carcinogenesis. Excerpta Medica, Amsterdam, Holland.

Kumar, R., Sukumar, S., and Barbacid, M. (1990). Activation of ras oncogenes preceding the onset of neoplasia. Science 248, 1101-1104.

LaVeccia, C., Harris, R. E., and Wynder, E. L. (1988). Comparative epidemiology of cancer between the United States and Italy. Cancer Res. 48, 7285-7293.

McGinnis, J. M. (1988-89). National priorities in disease prevention. Issues Science Technol. 5, 46-92.

Meyskens, F. L. (1989). Proceedings of the Third International Conference on the Prevention of Human Cancer: Chemoprevention. Prev. Med. 18, 551-757.

Michnovich, J. J., and Bradlow, H. L. (1990). Induction of estradiol metabolism by dietary indole-3-carbinol in humans. J. Natl. Cancer Inst. 82, 947-949.

Newmark, H. L., Lipkin, M., and Maheshwari, N. (1990). Colonic hyperplasia and hyperproliferation induced by a nutritional stress diet with four components of Western-style diet. J. Natl. Cancer Inst. 82, 491-496.

Pariza, M. W. (1988). Dietary fat and cancer risk: Evidence and research needs. Annu. Rev. Nutr. 8, 167-183.

Pariza, M. W., Ha, Y. L., Benjamin, H., Sword, J. T., Gruter, A., Chin, S. F., Storkson. J.. Faith, N., and Albright, K. (1991). Foramtion and action of anticarcinogenic fatty acids. This volume.

Prentice, R. L., Kakar, F., Hursting, S., Sheppard, L., Klein, R., and Kushi, L.H. (1988). Aspects of the rationale for the Women's Health Trial. J. Natl. Cancer Inst. 80, 802-814.

Reddy, B., Engle, A., Katsifis, S., Simi, B., Bartram, H-P., Perrino, P., and Mahan, C.)1989). Biochemical epidemiology of colon cancer: Effect of types of dietary fiber on fecal mutagens, acid, and neutral sterols in healthy subjects. Cancer Res. 49, 4629-4635.

Rose, D. P. (1990). Dietary fiber and breast cancer. Nutr. Cancer, 13, 1-8.

Schapira, D. V., Kumar, N. B., Lyman, G. H., and McMillan, S. C. (1990). The value of current nutrition information. Prev. Med. 19, 45-53.

Simopoulos, A. P. (1987). Diet and health: Scientific concepts and principles. Am. J. Clin. Nutr. 45, 1027-1414.

Sinkeldam, E. J., Kuper, C. F., Bosland, M. C., Hollanders, V. M. H., and Vedder, D. M. (1990). Interactive effects of dietary wheat bran and lard on N-methyl-N'-nitro-N-nitrosoguanidine-induced colon carcinogenesis in rats. Cancer Res. 50, 1092-1096.

Tajima, K., Hirose, K., Nakagawa, N., Juroshishi, T., and Tominaga, S. (1985). Urban-rural difference in the trend of colo-rectal mortality with special reference to the subsites of colon cancer in Japan. Gann, 76, 717-728.

Trowell, H., Burkitt, D., and Heaton, K. (eds.). (1985). Dietary Fibre, Fibre-Depleted Foods and Disease. Academic Press, London, England.

Van't Veer, P., Kolb, C., Verhoef, P., Kok, F. J., Schouten, E. J., Hermus, R. J. J., and Stermans, F. (1990). Dietary fiber, beta-carotene and breast cancer: Results from a case-control study. Int. J. Cancer, 45, 825-828.

Wakabayashi, K., Nagao, M., and Sugimura, T. (1989). Mutagens and carcinogens produced by the reaction of environmental aromatic compounds with nitrite. Cancer Surveys, 8, 385.

Weinberg, R. A. (1989). Oncogens, antioncogens, and the molecular bases of multistep carcinogenesis. Cancer Res. 9, 3713-3721.

Weisburger, J. H. (1990). On the mechanisms of carcinogenesis and statistical data management in tobacco smoking cessation. Epidemiology, 1, 314-317.

Weisburger, J. H. (1991). Nutritional approach to cancer prevention with emphasis on vitamins, antioxidants and carotenoids. Am. J. Clin. Nutr. 53, 226s-237s.

Williams, G. M. (1989). Methods for evaluating chemical genotoxicity. Annu. Rev. Pharmacol. Toxicol. 29, 189-211.

Williams, G. M., and Weisburger, J. H. (1991). Chemical carcinogens. In Casarett and Doull's Toxicology, the Basic Science of Poisons, 4th ed., J. Doull, C. D. Klaassen, and M. O. Amdur (eds.). McMillan, New York.

Wynder, E. L., and Hiyama, T. (1987). Comparative epidemiology of cancer in the United States and Japan: preventive implications. Gann Monogr. Cancer Res. 33, 181-191.

13

MODIFICATION OF CARCINOGEN METABOLISM BY INDOLYLIC AUTOLYSIS PRODUCTS OF *BRASSICA OLERACEAE*

[1]Christopher A. Bradfield and [2]Leonard F. Bjeldanes

[1]Department of Pharmacology and Toxicology, Northwestern University Medical School, Chicago, IL. [2]Department of Nutritional Sciences, University of California, Berkeley, CA

ABSTRACT

Cruciferous plant foods contain large quantities of secondary plant metabolites that have been shown to inhibit chemically induced carcinogenesis in animals. One mechanism by which these chemicals may inhibit carcinogenesis is through the induction of enzymes, such as cytochrome $P_1$450-dependent monooxygenases, glutathione S-transferases (GST) or epoxide hydrolases (EH), which metabolize carcinogens to more polar and excretable forms. Cruciferous vegetables of the *Brassica* genus (*e.g.* Brussels sprouts, cauliflower, broccoli) contain μg/g levels of an indolylmethyl glucosinolate commonly known as glucobrassicin. Upon disruption of the plant material, as in food preparation or chewing, a thioglucosidase-mediated autolytic process ensues generating indole-3-carbinol (I3C), glucose, and thiocyanate ion. At acid pH comparable to that found in the stomach, I3C forms a wide variety of condensation products ranging from linear and cyclic dimers, trimers and tetramers to extended heterocyclic compounds such as indolocarbazoles. Experiments reviewed here indicate that these indole-condensation products are the compounds responsible for some of the alterations in carcinogen metabolism observed in animals fed either I3C or any of several *Brassica* plant foods.

ANTICARCINOGENIC PROPERTIES OF *BRASSICA* VEGETABLES

The National Research Council, Committee of Diet, Nutrition, and Cancer has recommended increased consumption of vegetables of the *Brassica* genus as a measure to decrease the incidence of human cancer (National Research Council, 1982). This recommendation is based on epidemiological evidence (Graham, 1983) and results from animal experimentation (Stoewsand *et al.*, 1978; Wattenberg, 1983) that suggest that these vegetables possess cancer-inhibiting properties. The committee's review of the scientific literature led to the suggestion that the inhibitory effects of these vegetables may be related to the presence of a number of nutritive and nonnutritive constituents known to inhibit chemically induced carcinogenesis in experimental animals.

Stoewsand *et al.* (1978) published the first study of an anticarcinogenic effect of *Brassica* vegetables. In this study, rats that were exposed to the hepatocarcinogen aflatoxin B_1 and fed on purified diets supplemented with 25% cauliflower lived considerably longer and had smaller

Nutritional and Toxicological Consequences of Food Processing
Edited by M. Friedman, Plenum Press, New York, 1991

tumors than did the unsupplemented controls. Results of further experiments showed that diets high in cabbage had similar anticarcinogenic properties (Boyd *et al.*, 1982). The anticarcinogenic properties of these vegetables were also shown in a study by Wattenberg (1983) in which rats fed on a diet supplemented with 10% cabbage or cauliflower had fewer mammary tumors induced by dimethylbenzanthracene than did the unsupplemented controls.

Wattenberg (1983) has suggested that tumorigenesis is inhibited by carcinogen-metabolizing systems induced by compounds in *Brassica* plants. This hypothesis proposes that minor dietary constituents function as anticarcinogenic substances by virtue of their ability to enhance the activities of xenobiotic-metabolizing enzymes that shunt the metabolism of precarcinogenic substrates through detoxification pathways rather than through pathways leading to genotoxic species capable of initiating neoplasia. Additionally, agents that induce xenobiotic-metabolizing enzymes could also elicit anticarcinogenic effects *via* increases in presystemic metabolism of a carcinogenic substrate, thereby decreasing the dose of carcinogen reaching target tissues (Wattenberg, 1970).

Supporting this hypothesis is the observation that many inducers of xenobiotic-metabolizing systems have an inhibitory effect on a variety of chemically induced neoplasias. Early reports of this relationship include the demonstrations that administration of 1,2,5,6-dibenzofluorene inhibited 3-methylcholanthrene-induced tumors (Lacassagne *et al.*, 1945; Riegel *et al.*, 1951) and that 3-methylcholanthrene inhibited 3'-methyl-4-dimethylaminoazobenzene-induced tumors (Richardson *et al.*, 1951; Conney *et al.*, 1956). Since these early reports, results from a number of studies have supported the concept that induction of xenobiotic-metabolizing enzymes can lead to decreased tumor yields resulting from exposure to carcinogenic agents. For example, phenobarbitone inhibits aflatoxin-induced hepatocarcinogenesis in rats (McLean and Marshall, 1971); DDT inhibits dimethylbenzanthracene-induced mammary tumors and leukemia in rats (Silinskas and Okey, 1975); 3-methylcholanthrene inhibits dimethylbenzathracene-induced mammary tumors in rats (Wheatly, 1968); and ß-naphthoflavone inhibits benzo[a]pyrene-induced skin tumors in mice (Wattenberg and Leong, 1970).

Although it is known that many of the same xenobiotic-metabolizing systems involved in the detoxification of carcinogenic compounds can also be involved in their bioactivation (Gelboin, 1980), the relationship between inducing agents and carcinogenic outcome is not well understood. Monooxygenases, EHs, and GSTs are also involved in the bioactivation of certain chemical carcinogens *in vitro* (Rannug *et al.*, 1978; Schmassmann and Oesch, 1978; Wood *et al.*, 1976). Additionally, many inducers of xenobiotic-metabolizing systems, such as phenobarbital and 2,3,7,8-tetrachlorodibenzo-p-dioxin (TCDD), are promoters of carcinogenesis (Pitot, 1982; Pitot *et al.*, 1980). Thus, induction of xenobiotic-metabolizing enzymes may explain decreased tumor yield, but does not necessarily indicate that exposure to inducing agents is a general prescription for cancer prophylaxis.

DIETARY EFFECTS ON XENOBIOTIC METABOLISM

Early reports by Brown *et al.* (1954) and Wattenberg (1970) began to characterize the modification of xenobiotic metabolism by dietary constituents, including the description of increases in hepatic methyl-4-dimethylaminoazobenzene demethylase activity and extrahepatic aryl hydrocarbon hydroxylase (AHH) activity in rodents fed on commercial chow formulations. The presence of significant quantities of inducing agents in *Brassica* vegetables was first demonstrated by Wattenberg (1972) who noted the ability of a number of *Brassica* vegetables to induce extrahepatic AHH activity. Later, Sparnins *et al.* (1982) noted the effects of these vegetables on GST activity in rodents. Babish and Stoewsand (1975) and Stoewsand *et al.* (1978) described increases in hepatic aminopyrine N-demethylase N-methylaniline, N-demethylase, and p-nitroanisole O-demethylase in rats fed on cabbage or cauliflower. More recent reports have expanded upon these results by demonstrating that consumption of broccoli, cabbage, or Brussels sprouts leads to the induction of hepatic and intestinal microsomal and cytosolic EH and quinone reductase activities in rodents (Aspry and Bjeldanes, 1983; Hendrich and Bjeldanes, 1983; Hendrich and Bjeldanes, 1986; Bradfield *et al.*, 1985; Salbe and Bjeldanes, 1985; Salbe and Bjeldanes, 1986).

Dietary modification of xenobiotic metabolism has also been shown to alter the biological fate of therapeutic agents in rats and humans. Pantuck *et al.* (1976) showed that rats fed on cabbage or Brussels sprouts had increases in the oxidative metabolism of phenacetin and hexobarbital. Similar effects were observed on the oxidative metabolism of the analgesics phenacetin and aminopyrine in human subjects fed these vegetables at 500 g/d (Pantuck *et al.*, 1979). In another human study, these dietary treatments increased the clearance rate of acetaminophen in male subjects. The increase in clearance rate appeared to be the result of an increased capacity to form acetaminophen glucuronides, rendering this drug more readily excretable (Pantuck *et al.*, 1984).

IDENTIFICATION OF INDUCERS OF XENOBIOTIC-METABOLIZING ENZYMES FROM *BRASSICA OLERACEA*

The knowledge that cruciferous vegetables can inhibit chemically induced carcinogenesis and induce a variety of xenobiotic-metabolizing enzymes has led to attempts to isolate the constituents responsible for these properties. The relative ease of quantitating effects on enzyme activity as opposed to measuring inhibition of neoplasia, as well as the correlation between potency as inducing agents and inhibition of chemical carcinogenesis, has led to efforts to isolate compounds that modify xenobiotic-metabolizing enzymes. Fenwick *et al.* (1983) have provided the groundwork to isolating these compounds by establishing the biochemical levels and identities of a series of secondary plant metabolites, known as glucosinolates.

Glucosinolate levels in some cultivars of the genus *Brassica* are reported to be as high as 180 mg/g (Anand, 1974). Levels of glucosinolates are known to be dependent on soil nitrates, sulfates, conditioners, irrigation, growing season, cultivar, and crop spacing (Heaney and Fenwick, 1981; Heaney *et al.*, 1983; Miller *et al.*, 1983; Bible *et al.*, 1980). The generalized structure for glucosinolates is shown in Figure 1. All glucosinolates contain ß-D-thioglucose and sulfate moieties (Ettlinger and

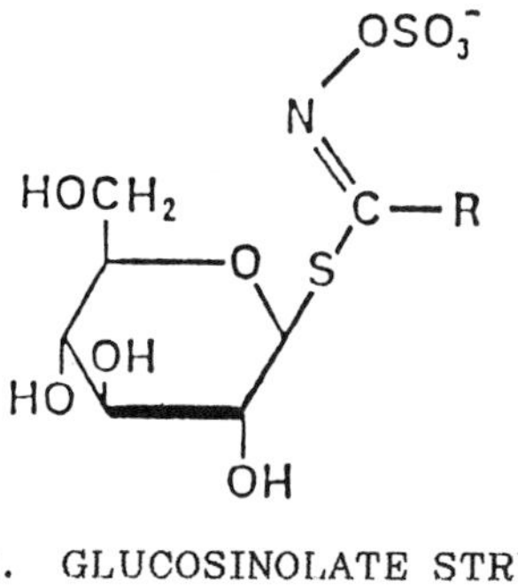

FIGURE 1. GLUCOSINOLATE STRUCTURE

Kjaer, 1968). The structure of the R-group is derived biosynthetically from amino acids (Kjaer and Olesen-Larsen, 1973). To date, over 70 unique glucosinolates have been characterized and identified in plants throughout the order of Capparales (Kjaer and Olesen-Larsen, 1976). This order contains the family *Cruciferae*, which in turn includes a number of commonly consumed plants, *e.g.* cole crops (Brussels sprouts, broccoli, cauliflower, kale, cabbage, Kohl-rabi *Brassica oleracea*), condiments (white mustard; *Brassica hirta*), radish (*Raphanus sativus*), papaya (*Carica papaya*) and forages (rapeseed; *Brassica napus*) (Fenwick *et al.*, 1983).

In addition to high levels of glucosinolates, plants of the *Cruciferae* contain high levels of thioglucosidase (EC 3.2.3.1) commonly referred to as myrosinase (Pihakaski and Pihakaski, 1978). Thioglucosidase appears as a group of isozymes which have a broad substrate specificity for all glucosinolates (Bjorkman and Lonnerdal, 1973). The thioglucosidase is localized in dilated cisternae of the rough endoplasmic reticulum. Although the exact cellular localization of glucosinolates has not been determined, this substrate does appear to reside in a compartment separate from the thioglucosidase (Pihakaski and Iversen, 1976). When the cell's integrity is disrupted, as in chewing or food preparation, enzyme and substrate come together. The generalized autolysis reaction generates glucose, sulfate, and aglucones (Ettlinger and Lundeen, 1957). The nature of the aglucone generated is dependent on the structure of the R-group, the pH at which the hydrolysis is carried out, and the presence of enzymatic modifiers such as ascorbic acid (Virtanen, 1965); Ettlinger *et al.*, 1961).

The role these thioglucosides play in plant physiology is unclear, although reports suggest a number of allelopathic effects, including insecticidal activity towards herbivorous predators (Blau *et al.*, 1978), inhibitory effects on growth and germination of competitive grasses (Kutacek, 1964; Leblova-Svobodova and Kostir, 1962), as well as functioning as an inactive storage form of plant growth hormones (Skytt Anderson and Muir, 1966). Biological effects of glucosinolates in animals include the goitrogenic activity of the autolytic products 5-vinyl-thiooxazolidinethione (goitrin) and thiocyanate ion (VanEtten and Wolff, 1973) and the hepatotoxicity and nephrotoxicity of autolytically generated epithiobutanes (Gould *et al.*, 1980).

INDOLYLIC METABOLITES, INHIBITION OF NEOPLASIA AND INDUCTION OF DRUG-METABOLIZING ENZYMES

Using the induction of AHH activity as a bioassay, I3C, indole-3-acetonitrile (IAN), indole-3-carbaldehyde (I3CHO) and 3,3'-diindolylmethane (I33') were isolated from an active fraction of an extract of *Brassica oleracea* (Loub *et al.*, 1975). These indoles are products of the thioglucosidase-mediated autolysis of indolylmethyl glucosinolate (also known as glucobrassicin (Virtanen, 1965) (Figure 2). Testing of these purified compounds demonstrated that I3C, IAN and I33' induced monooxygenases and/or GSTs (Sparnins *et al.*, 1982; Loub *et al.*, 1975). Additionally, I3C and I33' inhibited dimethylbenzanthracene-induced mammary tumors, and all three inhibited benzo[a]pyrene-initiated forestomach neoplasia in mice (Wattenberg and Loub, 1978). Although extensive comparisons were not performed, these early studies indicated that I3C was the isomer with the greatest biological potency as an inducer of monooxygenase activity and an inhibitor of neoplasia.

In our early studies of the pharmacology and chemistry of these compounds, the high sensitivity of I3C to acidic media became obvious. IAN and I33' were much less sensitive than was I3C. Subsequently, we found that I3C's chemical reactivity lies at the heart of its biological reactivity.

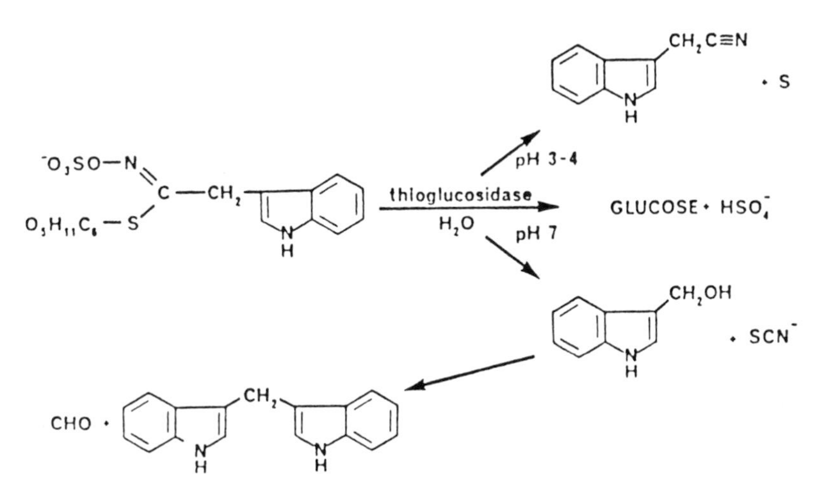

FIGURE 2. AUTOLYSIS OF GLUCOBRASSICIN

I3C and related indoles induce a variety of xenobiotic-metabolizing enzymes, most notably, cytochrome P_1-450-dependent monooxygenases, such as AHH and ethoxyresorufin O-deethylase (Loub *et al.*, 1975; Shertzer, 1982; Bradfield and Bjeldanes, 1987). These isozymes are regulated by a soluble ligand-responsive transcription factor, the Ah receptor. This receptor acts by binding ligand, interacting with genomic elements and increasing the transcriptional rates of the genes that encode these enzymes (*e.g.* the cytochrome $P_1$450 gene, Whitlock, 1987). Agonists of the Ah receptor include a variety of planar aromatic hydrocarbons, such as benzo[a]pyrene, ß-napthoflavone and 3-methylcholanthrene, as well as halogenated dibenzo-p-dioxins, biphenyls and azobenzenes. The structure-activity relationships for agonists of this receptor suggest that for nonhalogenated agonists, there is a requirement for extended planarity, and at least three aromatic rings for agonist binding (Piskorska-Pliszczynsk *et al.*, 1986; Poland, personal communication). I3C does not meet this last criterion and on a structure-activity basis appears to be an unlikely agonist of the Ah receptor. In fact, Gillner *et al.* (1985) and Poland (personal communication) have shown that I3C did not bind to the Ah receptor isolated from the rat or mouse.

The fact that I3C does not possess the structural features of an Ah receptor ligand, yet initiates biological responses characteristic of a receptor agonist suggested to us that I3C was altered *in vivo* ("bioactivated") to a form capable of receptor binding. A number of recent studies now implicate gastric acidity as the catalyst for this bioactivation and provide some clues as to the structure of the generated agonists. Evidence to suggest a role of gastric acidity in the bioactivation of I3C include:

1) results from structure-activity studies with simple indoles that describe a correlation between the instability of indoles in acidic solution and their potency as inducers of monooxygenase activity (Bradfield and Bjeldanes, 1987);

2) the observation that I3C is biologically active when administered orally, yet inactive when administered intraperitoneally (a route that allows the indole to bypass the acidity of the stomach) (Shertzer, 1982; Bradfield and Bjeldanes, 1987);

3) the fact that products generated by exposure to I3C to acidic solution are biologically active by either the intraperitoneal or oral routes;

4) the demonstration that I3C and an equivalent mass of its acid generation products have equivalent biological potency when administered orally (Bradfield and Bjeldanes, 1987).

Initial fractionation studies with I3C acid products suggested that a number of different condensation reactions occur and that many of the products possess biological activity (Bradfield and Bjeldanes, 1987). Recently, structures of some of these products have been determined. Among the products generated by treatment of I3C with acidic solution is a series of linear and cyclic methyleneindole trimers and tetramers, as well as indolocarbazoles (Figure 3). Preliminary results indicate that all these products have demonstrable binding affinity for the Ah receptor and elicit biological responses indicative of classical receptor agonists (Bradfield and Bjeldanes, manuscript in preparation). The susceptibility of substituted 3-hydroxy-methylindole derivatives to acid catalyzed elimination reactions is entirely consistent with the reactivity expected for vinylogous carbinolamines of this type. Thus, participation of the nonbonding electrons on nitrogen facilitates elimination of substituents on the α-carbon (for carbinolamines) or the δ-carbon (for vinylogous carbinolamines). The initial product of elimination of water from I3C is presumably 3-methyleneindolenine which by self condensation can produce the isolated dimeric and oligomeric products (Figure 3).

The classic studies by Virtanen (1965) on the autolysis of glucobrassicin indicated that at neutral pH (such as chewing or food preparation), I3C is the major product and can condense to I33' or react with ascorbic acid to generate ascorbigen. Our studies confirm the generation of I3C as the major autolysis product both in macerated plant material and in an *in vitro* system composed of thioglucosidase and purified glucobrassicin. Our experiments indicate that autolysis can occur even in plant material that has been boiled for 5 minutes (Bradfield and Bjeldanes, 1987B). These results suggest that I3C is likely to be found in cooked, processed, as well as in uncooked *Brassica* vegetables.

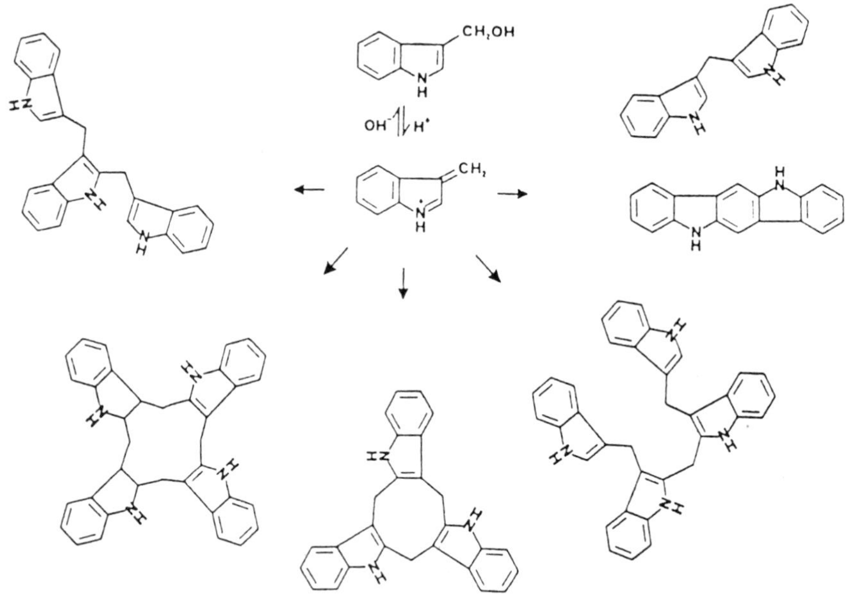

FIGURE 3. ACID CONDENSATION PRODUCTS OF I3C

The instability of I3C has important implications in assessing its dietary relevance. Our data suggest that the half-life of I3C in macerated plant material is approximately 12 hours at room temperature (Bradfield and Bjeldanes, 1987B) and that levels of I3C are substantially reduced by oxidative processes during the drying of plant material (Bradfield and Bjeldanes, 1987B, 1987C). Therefore, experiments that have utilized dried plant material or extended autolysis times may have generated levels of I3C, IAN, I3CHO, and I33' substantially lower than those levels generated during human consumption of fresh or cooked vegetables.

The National Research Council Report suggests that IAN, and not I3C, is the major autolysis product generated in *Brassica* vegetables (National Research Council, 1982). We believe that suggestion is incorrect for the following reasons:

1) HPLC analysis of the levels of I3C and IAN in freshly macerated plant material indicates that I3C is the major autolysis product and IAN is a minor one);

2) IAN has little biological potency as an inducer of monooxygenase activity (Loub *et al.*, 1975; Bradfield and Bjeldanes, 1987; Shertzer, 1982) and is the weakest indolic inhibitor of neoplasia (Wattenberg and Loub, 1978);

3) the actions of I3C, but not IAN, (Bradfield and Bjeldanes, 1987) are mediated through acid condensation products that are agonists of the Ah receptor; and

4) many agonists of the Ah receptor (*e.g.*, chlorinated dibenzo-p-dioxins and dibenzofurans) are among the most carcinogenic, teratogenic, and acutely toxic compounds known (Poland and Knutson, 1980).

The relationship between agonist activity and toxic effects suggests that consumption of large quantities of *Brassica* vegetables or their indolic constituents should be avoided.

CONCLUSION

The generation of Ah receptor agonists *via* the condensation of I3C in the acidic contents of the stomach appears to explain the mechanism by which I3C elicits induction of many xenobiotic-metabolizing enzymes. At present, it is not clear whether these condensation products also account for the anticarcinogenic activity of 3-substituted indoles and of *Brassica* vegetables. If, in fact, I3C acid condensation products bind to the Ah receptor as do highly toxic compounds, such as TCDD, we should thoroughly understand the toxicology of 3-substituted indoles before recommending them as desirable components of a diet to reduce the incidence of human cancer.

REFERENCES

Anand I. J. (1974). Mustard oil glucosides of the Indian *Brassicae*. Plant Biochem. J. 1, 26.

Aspry K. E., and Bjeldanes L. F. (1983). Effects of dietary broccoli and butylated hydroxyanisole on liver-mediated metabolism of benzo[a]pyrene. Food Chem. Toxicol. 21, 133.

Babish J. G. and Stoewsand G. S. (1975). Hepatic microsomal enzyme induction in rats fed varietal cauliflower leaves. J. Nutr. 105, 1592.

Bible B. B., Ju H. Y., Chong C. (1980). Influence of cultivar, season, irrigation and date of planting on thiocyanate ion content in cabbages. J. Am. Soc. Hort. Sci. **105**, 88.

Bjorkman R. and Lonnerdal B. (1973). Studies on myrosinases. III. Enzymatic properties of myrosinases from *Sinapis alba* and *Brassica napus* seeds. Biochem. Biophys. Acta **327**, 121.

Blau P. A., Feeny P., Contardo L. and Robson D. S. (1978). Allylglucosinolate and herbivorous caterpillars: A contrast in toxicity and tolerance. Science **200**, 1296.

Boyd J. N., Babish J. G. and Stoewsand G. S. (1982). Modification by beet and cabbage diets of aflatoxin B_1--induced rat plasma alpha-foetoprotein elevation, hepatic tumorigenesis, and mutagenicity of urine. Food Chem. Toxicol. **20**, 47.

Bradfield, C.A. and Bjeldanes, L. F. (1987). Structure-activity relationships of dietary indoles: A proposed mechanism of action as modifiers of xenobiotic metabolism. J. Toxicol. Environ. Health **21**, 311.

Bradfield, C. A. and Bjeldanes, L. F. (1987B). High performance liquid chromatographic analysis of anticarcinogenic indoles in *Brassica oleracea*. J. Agric. Food Chem. **35**, 46.

Bradfield, C. A. and Bjeldanes, L. F. (1987C). Dietary modification of xenobiotic metabolism: The contribution of indolylic compounds present in *Brassica oleracea*. J. Agric. Food Chem. **35**, 896.

Brown R. R., Miller J. A. and Miller E. C. (1954). The metabolism of methylated aminoazo dyes. IV. Dietary factors enhancing demethylation in vitro. J. Biol. Chem. **209**, 211.

Conney A. H., Miller E. C. and Miller A. J. (1956). The metabolism of methylated aminoazo dyes. V. Evidence for induction of enzyme synthesis in the rat by 3-methylcholanthrene. Cancer Res. **16**, 450.

Ettlinger M. G. and Lundeen A. J. (1957). First synthesis of a mustard oil glucoside: The enzymatic Lossen rearrangement. J. Am. Chem. Soc. **79**, 1764.

Ettlinger M. G., Kjaer A. (1968). Sulphur compounds in plants. In Recent Advances in Phytochemistry, Vol. 1, Mabry T. J., Alston R. E. and Runeckles V. C., Eds., Appleton-Century-Crofts, New York.

Ettlinger M. G., Dateo G. P., Harrison B. W., Mabry T. J. and Thompson C. P. (1961). Vitamin C as a coenzyme: The hydrolysis of mustard oil glucosides. Proc. Natl. Acad. Sci. USA **12**, 1875.

Fenwick G. R., Heaney R. K. and Mullin W. J. (1983). Glucosinolates and their breakdown products in food and food plants. CRC Crit. Rev. Food Sci. Nutr. 18, 123.

Gelboin H. V. (1980) Benzo[a]pyrene metabolism, activation and carcinogenesis: Role of regulation of mixed function oxidases and related enzymes. Physiol. Rev. **60**, 1107.

Gillner M., Bergman J., Cambillau C., Fernstrom B. and Gustafsson J-A. (1985). Interactions of indoles with specific binding sites for 2,3,7,8-tetrachlorodibenzo-p-dioxin in rat liver. Mol. Pharmacol. **28**. 357.

Gould D. H., Gumbmann M. R. and Daxenbichler M. E. (1980). Pathological changes in rats fed the crambe meal-glucosinolate hydrolytic products, 2S-1-cyano-2-hydroxy-3,4-epithiobutanes (erythro and threo) for 90 days. Food Cosmet. Toxicol. 18, 619.

Graham S. (1983). Results of case-control studies of diet and cancer in Buffalo, New York. Cancer Res. 43, 2409s.

Heaney R. K. and Fenwick G. R. (1981). Factors affecting the glucosinolate content of some *Brassicae* species. J. Sci. Food Agric. 32, 844.

Heaney R. K., Spinks E. A. and Fenwick G. R. (1983). The glucosinolate content of Brussels sprouts: Factors affecting their relative abundance. Z Pflanzenzuchtg. 91, 219.

Hendrich S. and Bjeldanes L. F. (1983). Effects of dietary cabbage, Brussels sprouts, *Illicium verum, Schizandra chinensis* and alfalfa on the benzo[a]pyrene metabolic system in mouse liver. Food Chem. Toxicol. 21, 479.

Hendrich S. and Bjeldanes L. F. (1986). Effects of dietary *Schizandra chinensis*, Brussels sprouts, and *Illicium verum* extracts on carcinogen metabolism in livers of male and female mice. Food Chem. Toxicol. 24, 903.

Kjaer A. and Olesen Larsen P. (1976). Nonprotein amino acids, cyanogenic glycosides and glucosinolates. In Biosynthesis, Vol. 5, Geissman T. A., Ed., The Chemical Society, London.

Kjaer A. and Olesen Larsen P. (1973). Nonprotein amino acids, cyanogenic glycosides and glucosinolates. In Biosynthesis, Vol. 2, Geissman T. A., Ed., The Chemical Society, London.

Kutacek M. (1964). Glucobrassicin, a potential inhibitor of unusual type affecting the germination and growth of plants: Mechanism of its action. Biol. Plant (Prague) 6, 88.

Lacassagne A., Buu-Hoi and Rudali G. (1945). Inhibition of the carcinogenic action produced by a weakly active hydrocarbon on a highly active carcinogenic hydrocarbon. Br. J. Exp. Pathol. 26, 5.

Leblova-Svobodova S. and Kostir J. (1962). Action of isothiocyanates on germinating plants. Experientia 18, 554.

Lewis J. J. (1950). Cabbage extracts and insulin-like activity. Br. J. Pharmacol. 5, 21.

Loub W. D., Wattenberg L. W. and Davis D. W. (1975). Aryl hydrocarbon hydroxylase induction in rat tissue by naturally occurring indoles of cruciferous plants. J. Natl. Cancer Inst. 54, 985.

McLean A. E. M. and Marshall A. (1971). Reduced carcinogenic effects of aflatoxin in rats given phenobarbitone. Br. J. Exp. Pathol. 52, 322.

Miller K. W., Boyd J. N., Babish J. G., Lisk D. J. and Stoewsand G. S. (1983). Alteration of glucosinolate content, pattern and mutagenicity of cabbage (*Brassica oleracea*) grown on municipal sewage sludge-amended soil. J. Food Safety 5, 131.

National Research Council (1982). Inhibitors of carcinogenesis. In Diet, Nutrition, and Cancer, p. 358, National Academy Press, Washington.

Pantuck E. J., Hsiao K. C., Loub W. D., Wattenberg L. W., Kuntzman R. and

Conney A. H. (1976). Stimulatory effect of vegetables on intestinal drug metabolism in the rat. J. Pharmacol. Exp. Ther. 35, 278.

Pantuck E. J., Pantuck C. B., Garland W. A., Min B. H., Wattenberg L. W., Anderson K. E., Kappas A. and Conney A. H. (1970). Stimulatory effect of Brussels sprouts and cabbage on human drug metabolism. Clin. Pharmacol. Ther. 25, 161.

Pantuck E. J., Pantuck C. B., Anderson K. E., Wattenberg L. W., Conney A. H. and Kappas A. (1984). Effect of Brussels sprouts and cabbage on drug conjugation. Clin. Pharmacol. Ther. 35, 161.

Pihakaski K. and Pihakaski S. (1978). Myrosinase in *Brassicaceae (Cruciferae)*. J. Exp. Bot. 29, 335.

Pihakaski K. and Iversen T. H. (1976). Myrosinase in *Brassicaceae*. I. Localization in cell fractions of roots of *Sinapis alba* L. J. Exp. Bot. 27, 242.

Pitot H. C., Goldsworthy T., Campbell H. A. and Poland A. (1980). Quantitative evaluation of the promotion by 2,3,7,8-tetrachlorodibenzo-p-dioxin of hepatocarcinogenesis from diethylnitrosamine. Cancer Res. 40, 3616.

Pitot H. C. (1982). The natural history of neoplastic development: The role of experimental models to human cancer. Cancer 315, 1206.

Poland A. and Knutson J. (1982). 2,3,7,8-Tetrachlorodibenzo-p-dioxin and related halogenated aromatic hydrocarbons: Examination of the mechanism of toxicity. Ann. Rev. Pharmacol. Toxicol. 22, 517.

Rannug U., Sundvall A. and Ramel C. (1978). The mutagenic effect of 1,2-dichloroethane on *Salmonella typhimurium*. I. Activation through conjugation with glutathione *in vitro*. Chemico-Biol. Interactions 20, 1.

Richardson H. L., Stier A. R. and Borsos-Nachtnebel E. (1952). Liver tumor inhibition and adrenal histologic responses in rats to which 3'-methyl-4-dimethylaminoazobenzene and 20-methylcholanthrene were simultaneously administered. Cancer Res. 12, 356.

Riegel B., Wartman W. B., Hill W. T., Reeb B. B., Shubik P. and Stanger D. W. (1951). Delay of methylcholanthrene skin carcinogenesis in mice by 1,2,5,6-dibenzoflourene. Cancer Res. 11, 301.

Salbe A. D. and Bjeldanes L. F. (1985). The effects of dietary Brussels sprouts and *Schizandra chinensis* on the xenobiotic-metabolizing enzymes of the rat small intestine. J. Food Chem. Toxicol. 23, 57.

Salbe A. D. and Bjeldanes L. F. (1986). Dietary influences on rat hepatic and intestinal DT-diaphorase activity. J. Food Chem. Toxicol. 24, 851.

Schmassmann H. and Oesch F. (1978). Trans-stilbene oxide: A selective inducer of rat liver epoxide hydratase. Mol. Pharmacol. 14, 834.

Shertzer H. G. (1982). Indole-3-carbinol and indole-3-acetonitrile influence on hepatic microsomal metabolism. Toxicol. Appl. Pharmacol. 64, 353.

Silinskas K. C. and Okey A. B. (1975). Protection by 1,1,1-trichloro-2,2-bis(p-chlorophenyl)ethane (DDT) against mammary tumors and leukemia during prolonged feeding of 7,12-dimethylbenz[a]anthracene to female rats. J. Natl. Cancer Inst. 55, 653.

Skytt Andersen A. and Muir R. M. (1966). Auxin activity of glucobrassicin. Physiol. Plant. 19, 1038.

Sparnins V. L., Venegas P. L. and Wattenberg L. W. (1982). Glutathione S-transferase activity: Enhancement by compounds inhibiting chemical carcinogenesis and by dietary constituents. J. Natl. Cancer Inst. 68, 493.

Stoewsand G. S., Babish J. B. and Wimberly H. C. (1978). Inhibition of hepatic toxicities from polybrominated biphenyls and aflatoxin B_1 in rats fed cauliflower. J. Environ. Pathol. Toxicol. 2, 399.

VanEtten C. H. and Wolff I. A. (1973). Natural sulfur compounds. In Toxicants Occurring Naturally in Foods, p. 210, National Academy of Sciences, Washington.

Virtanen A. I. (1965). Studies on organic sulphur compounds and other labile substances in plants. Phytochemistry 4, 207.

Wattenberg L. W., Leong J. L. and Strand P. J. (1972). Benzpyrene hydroxylase activity in the gastrointestinal tract. Cancer Res. 22, 1120.

Wattenberg L. W. (1970). The role of portal of entry in inhibition of tumorigenesis. Prog. Exp. Tumor Res. 14, 89.

Wattenberg L. W., Leong, J. L. (1970). Inhibition of the carcinogenic action of benzo[a]pyrene by flavones. Cancer Res. 3, 3022.

Wattenberg L. W. (1972). Enzymatic reactions and carcinogenesis. In Environment and Cancer, p. 241, Cumley R. D., Ed. Williams and Wilkins, Baltimore.

Wattenberg L. W. and Loub, W.D. (1978). Inhibition of polycyclic aromatic hydrocarbon-induced neoplasia by naturally occurring indoles. Cancer Res. 38, 1410.

Wattenberg L. W. (1983). Inhibition of neoplasia by minor dietary constituents. Cancer Res. 43, 2448s.

Wheatley D. N. (1968). Enhancement and inhibition of the induction by 7,12-dimethylbenz(a)anthracene of mammary tumors in female Sprague-Dawley rats. Br. J. Cancer 22, 787.

Whitlock J. P. (1987). The regulation of gene expression by 2,3,7,8-tetrachlorodibenzo-p-dioxin. Pharmacol. Rev. 39, 147.

Wood A. W., Levin L., Lu A. Y. H., Yagi H., Hernandez O., Jerina D. M. and Conney A. H. (1976). Metabolism of benzo[a]pyrene and benzo[a]pyrene derivatives to mutagenic products by highly purified hepatic microsomal enzymes. J. Biol. Chem. 251, 882.

14

DIETARY MODULATION OF THE GLUTATHIONE DETOXIFICATION PATHWAY AND THE POTENTIAL FOR ALTERED XENOBIOTIC METABOLISM

T.K. Smith

Department of Nutritional Sciences, University of Guelph
Guelph, Ontario, Canada, N1G 2W1

ABSTRACT

This review summarizes the literature regarding nutritional regulation of the pathways of glutathione synthesis and subsequent conjugation of xenobiotic compounds. The glutathione detoxification pathway includes the enzymes of the gamma-glutamyl cycle as well as sulfur conjugation reactions. This promotes bodily excretion of xenobiotics as well as normal metabolites. Regulation of intracellular glutathione concentrations is maintained largely through changes in the activity of gamma-glutamylcysteine synthetase. Availability of glutathione for detoxification purposes can be limited by the supply of intracellular cysteine to serve as a precursor for glutathione synthesis through the gamma-glutamyl cycle. Dietary methionine, cysteine and cysteine prodrugs have been examined for their potential to maximize glutathione availability for detoxification purposes. Some xenobiotic challenges have been reported to deplete hepatic glutathione reserves and toxicity correlates with the degree of depletion. Other foreign compounds, however, have been observed to increase cellular glutathione concentrations beyond normal levels despite regulation of the synthetic pathway. Such effects will be reviewed.

REGULATION OF GLUTATHIONE SYNTHESIS

The regulation of glutathione synthesis, transport and metabolism have been reviewed by Meister (1984, 1988). Glutathione is a ubiquitous tripeptide (L-gamma-glutamyl-L-cysteinyl-glycine) that performs a variety of essential cellular functions. It serves as an intracellular reductant that protects the cell from free radicals, reactive oxygen species such as peroxides and toxic compounds of both endogenous and exogenous origin. Glutathione is relatively resistant to the action of cellular peptidases but can serve as a storage and transport form of cysteine. It can undergo reversible oxidation to form glutathione disulfide in reactions catalyzed by glutathione peroxidase (uses peroxides as substrates) and glutathione reductase. Glutathione is conjugated to a variety of substrates forming S-substituted compounds through the action of glutathione S-transferase. These are usually processed to cysteine derivatives which may be acetylated. Synthesis and breakdown of glutathione (as well as some

Nutritional and Toxicological Consequences of Food Processing
Edited by M. Friedman, Plenum Press, New York, 1991

degree of amino acid transport) is accomplished through the reactions of the gamma-glutamyl cycle.

The first reaction of the gamma-glutamyl cycle is the condensation of glutamic acid and cysteine to form gamma-glutamylcysteine. This is catalyzed by gamma-glutamylcysteine synthetase. Glutathione synthetase then catalyzes the condensation of this reaction product with glycine to form glutathione. Glutathione can serve to promote amino acid transport across membranes by reacting with a substrate amino acid exterior to the cell to form gamma-glutamylamino acid and cysteinylglycine. This reaction is catalyzed by gamma-glutamyltranspeptidase which is a membrane-bound enzyme and this results in the translocation of the substrate amino acid into the cell. Cysteinylglycine is hydrolyzed by non-specific peptidases to yield cysteine and glycine while gamma-glutamylamino acid is hydrolyzed to a free amino acid and the cyclic intermediate 5-oxoproline in a reaction catalyzed by gamma-glutamylcyclotransferase. In the final reaction of the cycle, 5-oxoprolinase subsequently catalyzes the conversion of 5-oxoproline to glutamic acid. The regulation of glutathione synthesis by the gamma-glutamyl cycle is thought to be at the level of gamma-glutamylcysteine synthetase. An accumulation of glutathione is thought to inhibit the activity of this enzyme and reduce glutathione synthesis.

Not all tissues are known to contain the enzymes of the gamma-glutamyl cycle and glutathione can be removed by tissues from the blood to augment intracellular synthesis. Blood glutathione concentrations are normally much lower than those found in tissues while administration of inhibitors of glutathione synthesis lowers blood glutathione concentrations. Cellular uptake of glutathione across a concentration gradient appears to require the membrane-bound gamma-glutamyl cycle enzyme gamma-glutamyltranspeptidase. Inhibition of gamma-glutamyltranspeptidase increases blood glutathione concentration and can lead to glutathionuria. Utilization of blood glutathione is most effective in those cells with high activities of gamma-glutamyl transpeptidase. This results in the liver being the major organ of glutathione synthesis with kidney being the major organ recipient.

MAINTAINING TISSUE GLUTATHIONE CONCENTRATIONS

Tissue glutathione concentrations can be depleted during xenobiotic challenges following conjugation reactions involving compounds that form reactive oxygen intermediates, free radicals or cause other cellular damage. Williamson et al. (1982) administered acetaminophen to mice depleting hepatic glutathione. When hepatic glutathione concentrations were maintained during the acetaminophen challenge by increasing precursor amino acid concentrations, toxicity was thereby reduced. Cysteine is the amino acid limiting the synthesis of glutathione and the utility of dietary supplements of cysteine and analogues in maintenance of cellular glutathione concentrations has been explored.

Boebel and Baker (1983) examined the effect of deficient, adequate and excess dietary levels of cysteine on blood and liver concentrations of glutathione in the chick. The crystalline amino acid diet used was adequate in methionine but devoid of cysteine. Hepatic glutathione concentrations increased with increasing dietary supplements of sulfur amino acids until the dietary requirement for maximal growth rate was reached. Further supplementation did not increase hepatic glutathione concentrations beyond control levels. This would presumably be due to feedback inhibition of gamma-glutamylcysteine synthetase by glutathione.

A similar study with rats was conducted by Cho et al. (1984) who noted that dietary supplements of cystine above requirements for maximal growth did not increase tissue glutathione concentrations beyond control levels although deficient diets resulted in glutathione depletion in liver, muscle, spleen, heart and thymus.

The effectiveness of dietary cysteine as an agent to increase tissue cysteine concentrations for glutathione synthesis is poor compared to some other compounds. Cysteine can be oxidized or used for protein synthesis prior to uptake by the cell in which glutathione will be synthesized. The use of L-2-oxothiazolidine-4-carboxylate as an effective intracellular cysteine delivery system has been described by Williamson et al. (1982). This compound can serve as a substrate for 5-oxoprolinase, the gamma-glutamyl cycle enzyme that catalyzes the metabolism of 5-oxoproline to glutamic acid. This same enzyme catabolizes L-2-oxothiazolidine-4-carboxylate to yield cysteine. Such treatments have been shown to be effective against acetaminophen (Williamson et al., 1982), bromobenzene (Goyal and Brodeur, 1987) and other xenobiotic compounds that conjugate glutathione to form mercapturic acids. Other intracellular cysteine delivery systems that have been described include gamma-glutamylcysteine and N-acetyl-L-cysteine (Wong et al., 1986). Although it has been possible to demonstrate hepatic glutathione concentrations higher than control levels following cysteine supplementation, this requires that animals be fasted (Williamson et al., 1982) or be prefed sulfur amino acid deficient diets before testing (Bauman et al., 1988a,b; Chung et al., 1990).

ELEVATING TISSUE GLUTATHIONE CONCENTRATIONS

Some xenobiotic compounds have been shown to increase hepatic glutathione concentrations beyond control levels despite the regulation of tissue glutathione concentrations through feedback inhibition of gamma-glutamylcysteine synthetase. The consumption of cruciferous vegetables such as cabbage has been reported to alter the activities of hepatic and intestinal xenobiotic-metabolizing enzymes (Whitty and Bjeldanes, 1987). Rats were fed diets containing 25% freeze-dried cabbage for 21 days and this resulted in increased liver weight and a 2.1 and 2.3 fold increase in hepatic and intestinal glutathione-S-transferase activity respectively. Hepatic glutathione-S-transferase activity was also increased in coturnix quail fed cabbage and this effect was correlated with the glucosinolate content of the diet (Stoewsand et al., 1986). Glucosinolate hydrolysis products such as allyl isothiocyanate and goitrin have also been shown to have this effect (Bogaards et al., 1990). Glutathione has been further linked to glucosinolate metabolism in that it has been shown to be destroyed by allyl isothiocyanate (Kawakishi and Kaneko, 1985) while glutathione and other thiol compounds have been shown to promote nitrile compound formation during glucosinolate hydrolysis (Uda et al., 1986). The feeding of 1-cyano-2-hydroxy-3-butane, a glucosinolate hydrolysis product formed from progoitrin, has been shown to increase glutathione concentrations beyond control levels as well as to elevate hepatic and pancreatic activities of glutathione-S-transferase (Wallig and Jeffrey, 1990). Cabbage feeding produces similar effects (Stohs et al., 1986) and this has been shown to be caused by the presence of dithiolthiones which are found in this vegetable (Ansher et al., 1984).

A particularly interesting example of the ability of some xenobiotic compounds to increase intracellular glutathione concentrations beyond control levels is that described by Jaeschke and Wendel (1986). These authors tested various doses of the phenolic antioxidants butylated hydroxytoluene (BHT) and butylated hydroxyanisole (BHA) over time (14 days) and examined both tissue glutathione concentrations and glutathione-S-transferase activities. Oral doses of BHA increased tissue glutathione concentrations above controls and this effect was most obvious in liver. Kidney was not affected presumably indicating a lack of ability of liver and other tissues to release glutathione into the blood for uptake by the kidney. Liver glutathione concentrations were initially depleted by both BHT and BHA before rising above control levels for BHA but not for BHT. The authors indicated that the initial depletion of glutathione could be due to the use of glutathione for mercapturic acid synthesis however only BHT and not BHA forms such metabolites. The authors further suggest that glutathione must be released by liver triggering increased synthesis.

SUMMARY

It is clear from the above review that protection against the toxicity of certain xenobiotic compounds results from the maintenance of optimal tissue concentrations of glutathione for detoxification through conjugation reactions. Intracellular cysteine concentrations must be maintained to ensure repletion of glutathione lost in conjugation reactions and this can be provided for through dietary supplements of sulfur amino acids or, more effectively, cysteine prodrugs such as L-2-oxothiazolidine-4-carboxylate. Regulation of glutathione synthesis through the feedback inhibition of gamma-glutamylcysteine synthetase, however, prevents accumulation of glutathione beyond control levels which might afford additional protection against xenobiotic compounds. Administration of a wide variety of naturally-occurring and synthetic compounds, however, can actually result in the accumulation of hepatic glutathione beyond control levels even when glutathione is required for metabolism of these compounds through conjugation reactions. It is not clear why normal regulation of intracellular glutathione concentrations is deranged under such conditions. It is also not clear what effect dietary supplements of sulfur amino acids or cysteine prodrugs might have on the toxicity of such compounds.

REFERENCES

Ansher, S.S., Dolan, P., and Bueding, E. (1986). Biochemical effects of dithiolthiones. Fd. Chem. Toxic., 25, 581-587.

Bauman, P.F., Smith, T.K., and Bray, T.M. (1988a). The effect of dietary protein and sulfur amino acids on hepatic glutathione concentration and glutathione-dependent enzyme activities in the rat. Can. J. Physiol. Pharmacol., 66, 1048-1052.

Bauman, P.F., Smith, T.K., and Bray, T.M. (1988b). Effect of dietary protein deficiency and L-2-oxothiazolidine-4-carboxylate on the diurnal rhythm of hepatic glutathione in the rat. J. Nutr., 118: 1048-1054.

Boebel, K.P. and Baker, D.H. (1983). Blood and liver concentrations of glutathione, and plasma concentrations of sulfur-containing amino acids in chicks fed deficient, adequate, or excess levels of dietary cysteine. Proc. Soc. Exper. Biol. Med., 172, 498-501.

Bogaards, J.J. P., van Ommen, B., Falke, H.E., Willems, M.I., and van Bladeren, P.J. (1990). Glutathione S-transferase subunit induction patterns of brussels sprouts, allyl isothiocyanate and goitrin in rat liver and small intestinal mucosa: a new approach for the identification of inducing xenobiotics. Fd. Chem. Toxic., 28, 81-88.

Cho, E.S., Johnson, N., and Snider, B.C.F. (1984). Tissue glutathione as a cyst(e)ine reservoir during cystine depletion in growing rats. J. Nutr., 114, 1853-1862.

Chung, T.K., Funk, M.A., and Baker, D.H. (1990). L-2-oxothiazolidine-4-carboxylate as a cysteine precursor: Efficacy for growth and hepatic glutathione synthesis in chicks and rats. J. Nutr., 120, 158-165.

Goyal, R. and Brodeur, J. (1987). Effect of a cysteine prodrug (L-2-oxothiazolidine-4-carboxylic acid) on the metabolism and toxicity of bromobenzene: a repeated exposure study. J. Toxicol. Environ. Health, 21, 325-340.

Jaeschke, H. and Wendel, A. (1986). Manipulation of mouse organ glutathione contents II: Time and dose-dependent induction of the glutathione conjugation system by phenolic antioxidants. Toxicology, 39, 59-70.

Kawakishi, S. and Kaneko, T. (1985). Interaction of oxidized glutathione with allyl isothiocyanate. Phytochemistry, 24, 715-718.

Meister, A. (1984). New aspects of glutathione biochemistry and transport - selective alteration of glutathione metabolism. Nutr. Rev., 42, 397-410.

Meister, A. (1988). Glutathione metabolism and its selective modification. J. Biol. Chem., 263, 17205-17208.

Stoewsand, G.S., Anderson, J.L., and Lisk, D. (1986). Changes in liver glutathione S-transferase activities in coturnix quail fed municipal sludge-grown cabbage with reduced levels of glucosinolates. Proc. Soc. Exper. Biol. Med., 182, 95-99.

Stohs, S.J., Lawson, T.A., Anderson, L., and Bueding, E. (1986). Effects of oltipraz, BHA, ADT and cabbage on glutathione metabolism, DNA damage and lipid peroxidation in old mice. Mech. Ageing. Dev., 37, 137-145.

Uda, Y., Kurata, T., and Arakawa, N. (1986). Effects of thiol compounds on the formation of nitriles from glucosinolates. Agric. Biol. Chem., 50, 2741-2746.

Wallig, M.A. and Jeffrey, E.H. (1990). Enhancement of pancreatic and hepatic glutathione levels in rats during cyanohydroxybutene intoxication. Fundam. Appl. Toxicol., 14, 144-159.

Whitty, J.P. and Bjeldanes, L.F. (1987). The effects of dietary cabbage on xenobiotic metabolizing enzymes and the binding of aflatoxin B_1 to hepatic DNA in rats. Fd. Chem. Toxic., 25, 581-587.

Williamson, J.M., Boettcher, B., and Meister, A. (1982). Intracellular cysteine delivery system that protects against toxicity by promoting glutathione synthesis. Proc. Natl. Acad. Sci. U.S.A., 79, 6246-6249.

Wong, B.K., Chan, H.C., and Corcoran, G.B. (1986). Selective effects of N-acetylcysteine stereoisomers on hepatic glutathione and plasma sulfate in mice. Toxicol. Appl. Pharmacol., 86, 421-429.

15

PREVENTION OF ADVERSE EFFECTS OF FOOD BROWNING

Mendel Friedman

USDA, ARS, Western Regional Research Center
800 Buchanan Street
Albany, CA 947100

ABSTRACT

Amino-carbonyl interactions of food constituents encompass those changes commonly termed browning reactions. Such reactions are responsible for deleterious post-harvest changes during processing and storage and may adversely affect the appearance, organoleptic properties, nutritional quality, and safety of a wide spectrum of foods. A growing area of concern is nutritional carcinogenesis, in which nutritionally linked cancer has been associated with amino-carbonyl reaction products.

Specific practical and theoretical approaches to prevent adverse effects of food browning include: (1) modification and removal of primary reactants and endproducts in the browning reaction; (2) prevention of deleterious browning reactions through the use of antioxidants; (3) blocking of _in vivo_ toxicant formation from browning products by means of dietary modulation; (4) accurate estimation of low levels of browning products in whole foods and their removal through antibody complexation; and (5) stimulation of inactivation _in vivo_ toxicants from browning products by use of amino acids and sulfur-rich proteins.

INTRODUCTION

Reactions of amino acids, peptides, and proteins with sugars (nonenzymatic browning) and quinones (enzymatic browning) cause deterioration of food during storage and commercial or domestic processing. The loss of nutritional quality is attributed to the destruction of essential amino acids and a decrease in digestibility. The production of antinutritiitonal and toxic compounds may further reduce the nutritional value and possibly the safety of foods.

Although extensive efforts have been made to elucidate the chemistry of both desirable and undesirable compositional changes during browning, parallel studies on the nutritional and toxicological consequences of browning are limited. Reported studies in this area include (a) influence of damage to essential amino acids, especially lysine, methionine, and tryptophan, on nutritional quality; (b) attempts to restore nutritional quality by fortifying browning products with

Nutritional and Toxicological Consequences of Food Processing
Edited by M. Friedman, Plenum Press, New York, 1991

essential amino acids; (c) nutritional damage as a function of processing conditions; (d) biological utilization of characterized browning products; and (e) formation of food toxicants including kidney-damaging compounds, growth inhibitors, mutagenic (DNA-damaging), clastogenic (chromosome-damaging), and carcinogenic compounds.

Until recently, sulfites have been used to inhibit enzymatic and nonenzymatic browning reactions. Sulfite substitutes are needed since sulfites are reported to induce asthmatic crises in 4 to 8% of exposed asthmatics, and their use is being discontinued (Fan and Book, 1987; Giffon et al., 1989; Wedzicha, 1987). Our studies show that certain sulfur amino acids such as N-acetylcysteine and the natural peptide glutathione, are nearly as effective as sulfite in preventing browning of a wide variety of foods including heated protein-sugar mixtures, apples, potatoes, and fruit juices. These sulfur amino acids merit further study to assess their potential for preventing long-term food browning under practical storage and processing conditions. Studies are also needed to find out whether these sulfur amino acids can also prevent the formation of heterocyclic amines (IQ compounds), potent carcinogens formed during food processing.

This review largely covers our own studies on the nature of Maillard and non-Maillard browning reactions and possible approaches to prevent them. It complements more extensive recent overviews of this subject (Finot et al., 1990; Phillips and Finley, 1989; Somogyi and Muller, 1989; Walker and Quattrucci, 1989; Friedman, 1973, 1974, 1975, 1977, 1978, 1982, 1984a, 1986, 1989).

MUTAGENS AND CARCINOGENS FORMED DURING BROWNING

Formation of mutagens and carcinogens during browning, a major human health concern, will be examined in some detail (Ames, 1983, 1986, 1989; Ames et al., 1987; Archer, 1988, Ayatase et al., 1983; Barnes et al., 1983; Bjeldanes and Chew, 1979; Bjeldanes et al., 1982; Bosin et al., 1986; Cuzzoni et al., 1989; Degawa et al., 1987; Friedman and Cuq, 1988; Friedman and Henika, 1990; Friedman et al., 1987; 1989, 1990; Gazzini et al., 1987; Hatach and Felton, 1986; Hatch et al., 1988; Kinae, 1986; Krone et al., 1986; Lovelette et al., 1987; MacGregor et al., 1989; Miller, 1989; Niemand, 1983; Pariza et al., 1979; Powrie et al., 1986; Reuterswad et al., 1987; Rhee et al., 1987; Rogers and Shibamoto, 1982; Spingarn et al., 1980, 1983; Stich and Rosin, 1985; Sugimura, 1985; Sugimura et al., 1990; Van der Hoeven, 1982; Vuolo and Schuessler, 1985; Wakabayashi, 1986; Wang et al., 1982; Weisburger, 1987).

Risk Assessment

The potential for formation of mutagens and carcinogens in foods during processing is a major area of concern for human health and safety, as documented in the following review on risk assessment.

Human carcinogenic hazard associated with dietary mutagens and carcinogens has been estimated using an index called the Human Exposure/Rodent Potency Ratio (HERP) (Ames et al., 1987). This is the ratio of the chronic human exposure level to the lifetime dose rate in rodents which results in 50% of the animals being tumor-free, both human and animal dosages being expressed per unit of body weight.

This method was applied to nine heterocyclic amines isolated from cooked foods and heated amino acids and found to be potent carcinogens in rodents (Sugimura, 1986). Based on estimated daily human intake of 100 micrograms of heterocyclic amines, HERP was calculated at 0.02%, a value which indicates that these browning products are a potentially significant dietary risk to man (Weisburger, 1987).

The hazard associated with the consumption of browned foods in general is several hundred times greater than that associated with inhaling severely polluted air. In grilled chicken, for example, the dose of ingested carcinogenic nitro-pyrenes formed during cooking will be much higher than can be expected from air pollution (Ames et al., 1987; Kinouchi et al., 1986).

Bjeldanes et al. (1982) estimated the daily mutagen intake from cooked protein-rich foods in the average American diet. They based their calculations on the observed mutagenic activities of major food items listed in surveys of the average daily protein intake. The items included ground beef, beef steak, eggs, pork, ham, roast beef, chicken, milk, bread, and fish. The total mutagenic activity (assayed with S. typhimurium TA1538) ingested from cooked proteins in an average American diet is equivalent to 5,600 revertants per day. People who consume their food well-done or overdone will consume a greater quantity of mutagens. These mutagens are formed during normal cooking, baking, and broiling conditions used by the average household.

In a review on mutagen formation during commercial processing of foods, Krone et al. (1986) report that in foods subjected to commercial baking, canning, dehydration, and related thermal treatments, levels of bacterial mutations are 3 to 17 times greater than spontaneous rates. Mutagen formation was related to heating time and to processing temperature, which ranged from about 100 to 200°C.

The survey suggests that mutagen formation during commercial food processing is largely assocated with formation of browning products and that perhaps mutagens produced in different foods, e.g., bread, fish, fruits, and meat are not chemically identical. Inhibitors of browning, such as sodium sulfite, were found to reduce mutagen formation.

With respect to trends now evident in the food industry, Krone et al., (1986) predict that retort pouches will be used increasingly because of certain economic advantages. They, however, note that pouch retorting of the one item examined to date (beef stew) resulted in significantly higher levels of mutagenic activity than a similar product in a standard metal can. The problem may worsen, since the Food and Drug Administration has recently increased the maximum retort temperature from 121 to 135°C.

Tables 1-6 summarize relative activities associated with browning-derived mutagens and carcinogens of a number of different food types. It should be noted that the potential intake of mutagens and carcinogens from food sources listed in the tables is quite high relative to mutagens known to be a significant human risk. For example, the estimated daily intake of aflatoxins in the U.S. diet (0.017 μg) and in areas of Thailand in which human liver damage and cancer due to aflatoxin were reported (3.7 μg) (Carlborg, 1979) are equivalent to 68 and 15,000 revertants per day, respectively. The mutagenic activity in the smoke of one cigarette ranges from about 1000 to 10,000 revertants depending on tar content (Kier et al., 1974).

Mutagen Formation During Baking

Heat-induced nonenzymatic browning occurs in most foods. Such browning reactions often lead to the formation of antinutritional and toxic compounds (Fujimaki et al., 1986). Mutagen formation in baked products such as cookies and bread has been reported by Pariza et al. (1979), Spingarn et al. (1979), and Van der Hoeven et al. (1982). Mutagenic products in cooked protein-rich foods are formed by several mechanisms, including carbohydrate caramelization, protein pyrolysis, amino acid/creatinine reactions, and amino-carbonyl (Maillard) reactions, in which free amino groups condense with reducing sugars to produce brown melanoidins, furans, carbolines, and a variety of other heterocyclic amines (Powrie et al., 1986; Sugimura et al., 1990; Friedman and Cuq, 1988; Hatch et al., 1988; Miller, 1988; Cuzzoni et al., 1989; Friedman et al., 1990b,c; Ziderman and Friedman, 1985; Ziderman et al., 1987, 1989). Amino-acid-reducing sugar mixtures have been widely used to study mutagen formation. Less frequently, nonenzymatic browning of some foods also may occur by chemical reaction of acidic and neutral amino acid residues with non-reducing sugar or even with polysaccharide carbohydrates. These precursors are in fact much more abundant in foods than the reactants required for classical Maillard reactions. Such an alternative source of food browning would involve chemical mechanisms quite different from the amino-carbonyl reaction, and correspondingly different mutagens may be formed.

Friedman et al. (1990c) heated gluten, carbohydrates, and gluten-carbohydrates blends in a simulation of low moisture baking. The baked materials were then assayed by the Ames Salmonella his-reversion test in order to evaluate the formation of mutagenic browning products. An aqueous acetonitrile extract of heated gluten was highly mutagenic when assayed with Salmonella typhimurium strain TA98 with metabolic activation (Figures 1-4). Weak mutagenicity was also observed with strains TA100 and TA102. Gluten heated in a vacuum oven yielded less extract than did the air-baked protein, but its mutagenic activity (revertants/mg extract) was similar. Baked D-glucose, maltose, lactose, sucrose, wheat starch, potato amylose, cellulose, microcrystalline hydrocellulose, sodium ascorbate and L-ascorbic acid, sodium (carboxymethyl)cellulose, or (hydroxy- propyl)methylcellulose were moderately mutagenic in strain TA98 with microsomal (S9) activation and were weakly mutagenic without microsomal activation in strains TA100, TA102, and TA1537. Heated blends of gluten with 20% of these carbohydrates were also mutagenic, but the total activity recovered did not exceed levels of the individual ingredients baked separately. Maillard-type melanoidins prepared from L-lysine and D-fructose were very weakly active with strain TA98 but were mutagenic without S9 activation in strains TA100, TA102, and TA2637.

Nonclastogenicity of Browning Products

Heated sugar/amino acid reaction mixtures known to contain products clastogenic and/or mutagenic to cells *in vivo* were evaluated for clastogenic activity in mice using the erythrocyte micronucleus assay (MacGregor et al., 1989). Heated fructose/lysine reaction mixtures were also evaluated in the Salmonella his-reversion assay and the Chinese hamster ovary cell (CHO) chromosomal aberration assay to confirm and extend previous *in vitro* observations (Table 7). Significant mutagenicity of fructose/lysine mixtures was observed in Salmonella strains TA100, TA2637, TA98, and TA102, with greater activity in mixtures heated at pH 10 than at pH 7. S9 decreased the activity in strains TA100, TA2637, and TA98, but increased the activity in strain TA102. Both pH 7 and pH 10 reaction mixtures of the fructose/lysine browning reaction were highly clastogenic (chromosome-damaging) in CHO cells. Heated mixture of fructose and lysine, and of glucose or ribose with lysine, histidine, tryptophan, or cysteine, did not increase the frequency of micronucleated erythrocytes in mice when administered by the oral route. This indicates the absence of chromosomal aberrations

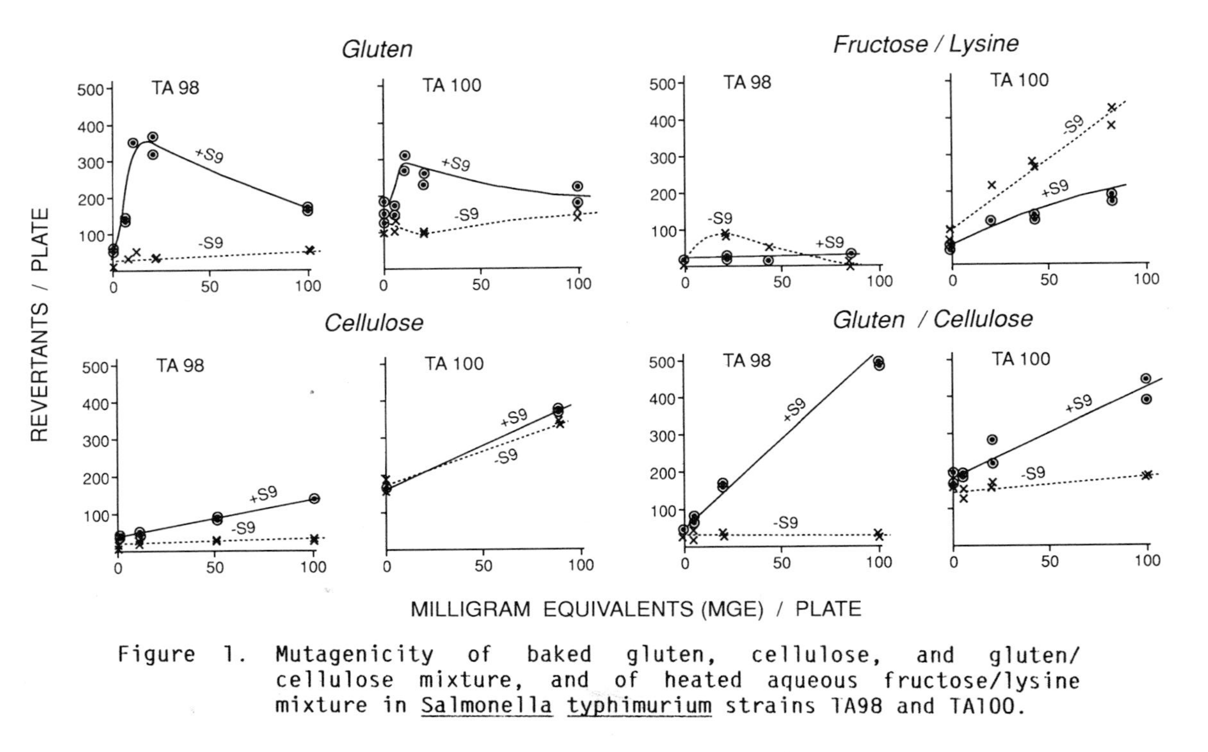

Figure 1. Mutagenicity of baked gluten, cellulose, and gluten/cellulose mixture, and of heated aqueous fructose/lysine mixture in Salmonella typhimurium strains TA98 and TA100.

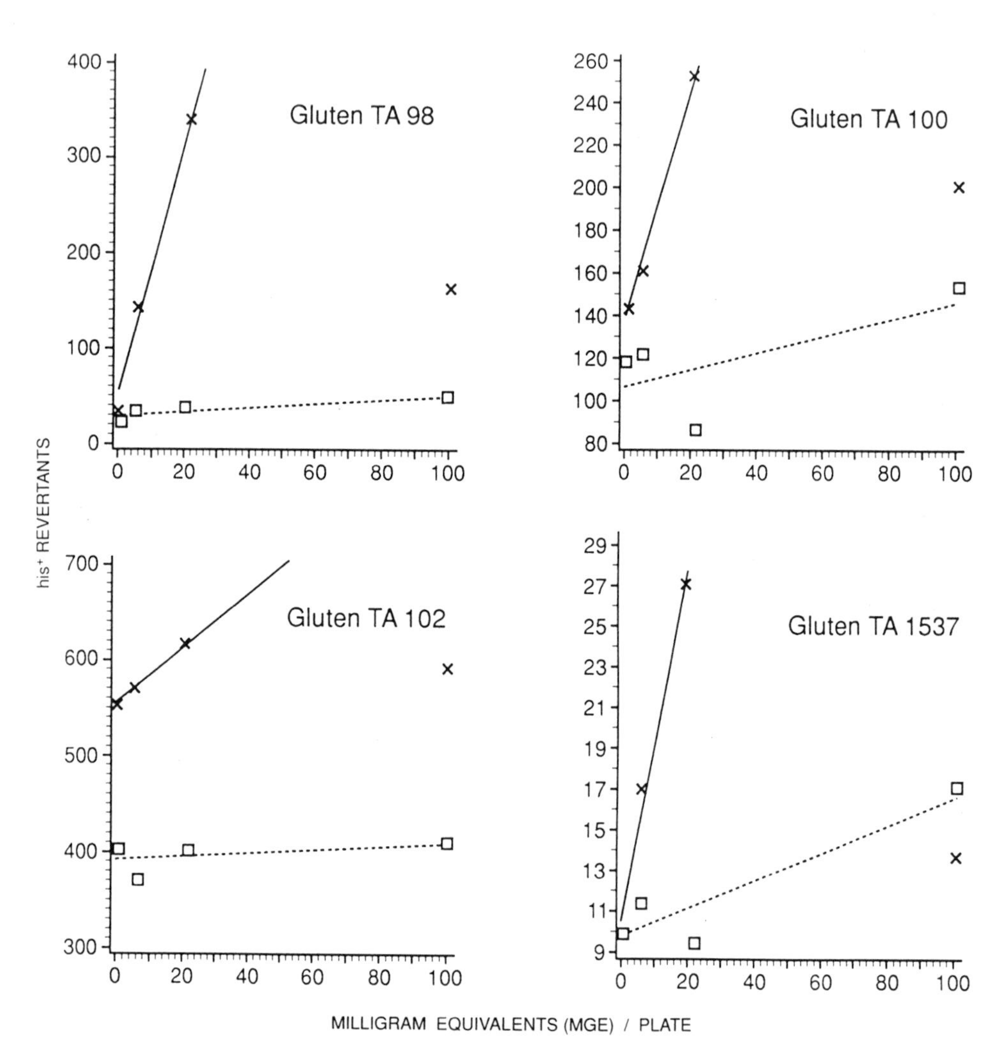

Figure 2. Mutagenicity of heated gluten extracts in four strains of *Salmonella typhimurium*: (X), +S9; (□), -S9.

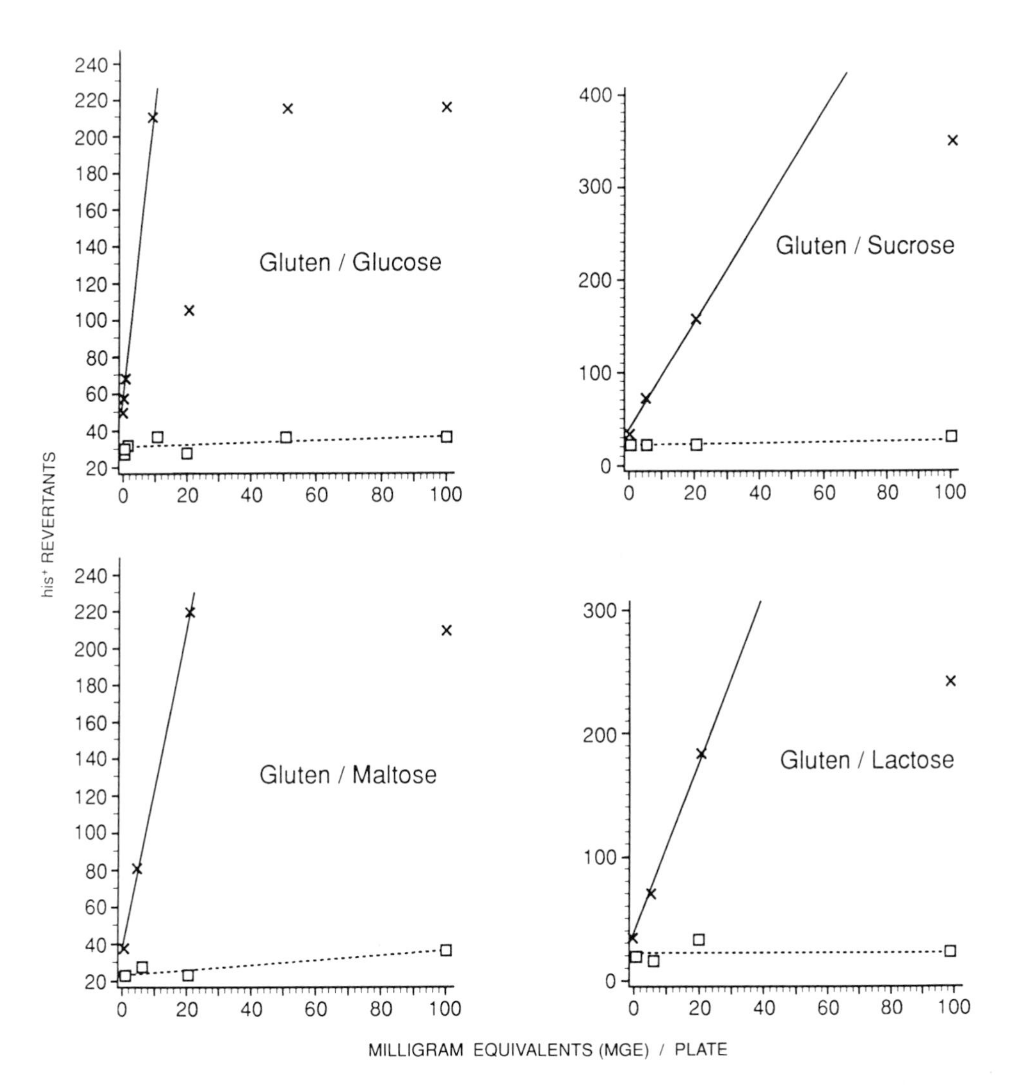

Figure 3. Mutagenicity of heated gluten/glucose, gluten/sucrose, gluten/maltose, and gluten/lactose extracts in <u>Salmonella typhimurium</u> strain TA98: (X), +S(; (□), -S9.

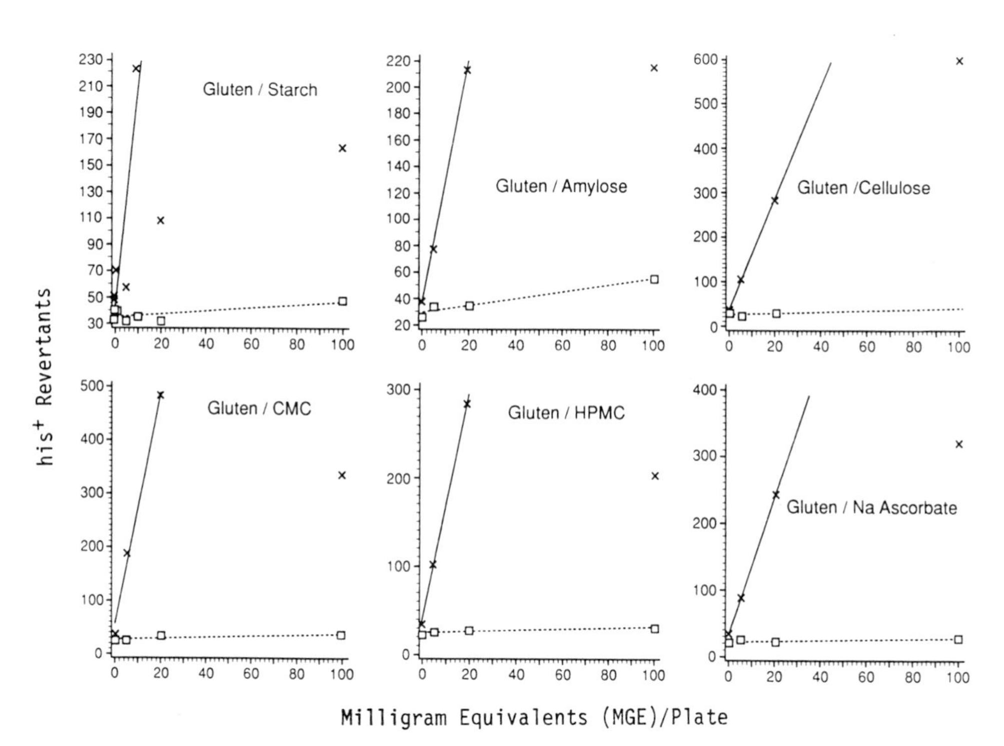

Figure 4. Mutagenicity of heated gluten/starch, gluten/amylose, gluten/cellulose, gluten/carboxymethylcellulose (CMC), gluten/hydroxypropylmthylcellulose (HPMC), and gluten/sodium ascorbate extracts in Salmonella typhimurium strain TA98: (X), +S9; (□), -S9.

Table 1. Mutagenicities and Carcinogenicities of Heterocyclic Amines Isolated from Cooked Foods

Heterocyclic	Mutagenicity (TA98)	Mouse carcinogenicity
IQ	433,00	Liver, forestomach, lung
MeIQ	661,000	Liver, forestomach
MeIQx	145,000	Liver
Trp-P-1	39,000	Liver
Trp-P-2	104,000	Liver
Glu-P-1	49,000	Liver, blood vessel
Glu-P-2	1,900	Liver, blood vessel
Aα-C	300	Liver, blood vessel
MeA-α-C	200	Liver, blood vessel
Aflatoxin B_1 for comparison	6,000	Liver

(Adapted from Sugimura, 1986).

Table 2. Mutagen Formation in Meat and Fowl with Various Cooking Procedures

Meat	Mode of Cooking	Revertant/100g
Beefsteak	Broiled 11 min/side at 300°C	10,000
Ground beef	Fried at 300°C	30,000
Ham	Fried at 200°C	5,400
	Fried at 242°C	22,000
Pork chop	Fried at 245°C	12,000
	Fried at 280°C	23,000
Sausage	Fried at 200°C	9,000
	Fried at 245°C	18,000
Chicken	Deep fried 12 min at 101°C	1,000
(white meat)	Broiled 17 min/side at 274°C	16,000
Lamb chop	Pan-broiled 5 min/side at 210°C	14,000

(Adapted from Barnes et al., 1983).

Table 3. Mutagen Formation in a Variety of Foods

Food	Sample	Cooking Proceddure	Cooking time(min)	Revertants per sample
White bread	1 slice	Broiling	6	205
Pumpernickel bread	1 slice	Broiling	12	945
Biscuit	1 each	Baking	20	735
Pancake	1 each	Frying	4	2,500
Potato	1 sm. sl.	Frying	30	200
Beef	1 patty	Frying	14	21,700

(Barnes et al., 1983).

Table 4. The Clastogenic Effect of Caramelized Sugars

Caramelized Sugar	Concentration (mg/ml)	% Metaphase plates with chromosome aberrations[a]	Average number of	
			breaks/cell	exchanges/cell
Sucrose	7.5	23.2	0.19	0.29
Glucose	15.0	21.5	0.16	0.58
Mannose	15.0	20.9	0.76	2.00
Arabinose	15.0	21.9	0.55	1.90
Maltose	15.0	30.0	0.90	3.04
Fructose	15.0	66.6	2.16	6.60

(Stich et al., 1981).

Table 5. Clastogenic Activity of Commercially Available Dried Fruits

Sample Exchanges	Concentration of aqueous extract (mg/ml)	Chromosome aberrations	
		% Metaphase plates	Breaks/cell
Raisins	60	46.9	0.59
Prunes	90	31.8	0.23
Dates	125	21.9	0.17
Bananas	30	33.3	0.35
Apricots	125	41.0	0.23
Control		0.5	0.01

(Stich et al., 1981).

Table 6. Clastogenic Activities Resulting from an Enzymatic Browning Reaction

Sample	Concentration (μL/mL)	% Metaphase	Chromosome aberrations: Average number of breaks	Chromosome aberrations: Average number of exchanges
Freshly prepared apple juice	500	1.8	0.02	0.00
1-Hour-old apple juice	500	40.0	00.01	0.81
Canned apple juice	162	42.7	0.05	1.06
Control	-	0.0	0.00	0.00

(Stich et al., 1981).

in erythrocyte precursor cells, and indicates that the genotoxic components of the browned mixtures are not absorbed and distributed to bone marrow cells in amounts sufficient to induce micronuclei when given orally or are metabolized to an inactive form. Because sugar/amino acid browning reactions occur commonly in heated foods, it is important to evaluate further the *in vivo* genotoxicity of browning products in cell populations other than bone marrow.

Carcinogenicity of Browning-Derived Food Ingredients

In view of the extremely high mutagenic activity of the heterocyclic amines derived from browning reactions towards both bacteria (Ames test) and mammalian cell lines, it is important to know if these short-term tests of genotoxicity are good predictors of carcinogenicity. According to Sugimura et al. (1990) and Weisburger (1987), this is indeed the case. They found that the mutagenic heterocyclic amines tested induce multiple tumors in rats and mice. Target organs include breast, colon, and pancreas - major types of human cancer throughout the world.

Sugimura (1985) estimates, as stated earlier, that the average person consumes about 100 micrograms of heterocyclic amines per day. This is a significant amount in view of these compounds' reported extremely high mutagenic activities and the possibility that the effects in animals and humans could be cumulative. The presence of tumor promoters in the diet may further enhance the risk of consuming heterocyclic amines.

These considerations suggest an urgent need, repeatedly emphasized by active workers in the field (Sugimura, 1985; Hatch and Felton, 1986; Weisburger, 1987), to develop new approaches and strategies to prevent the formation during food processing of heterocyclic amines and other browning products.

Metabolism of Heterocyclic Amines

The following is a brief summary of our present knowledge of the metabolism of heterocyclic amines (Vuolo and Schuessler, 1986; Degawa et al., 1987):

a. Structure-activity studies revealed that the primary amino group and an unsubstituted 3-position are important for genotoxicity.

b. The ultimate mutagens and carcinogens are formed only after activation by the cytochrome P-448/450 enzyme system. Degawa et al. (1987) report that the extent of activation of the heterocyclic amines Trp-P-1 and Trp-P-2 derived from tryptophan by cytochrome P-450 in mice and rats depends on species, sex, and organ type.

c. The heterocyclic amines appear to be metabolized through oxidation and acetylation of the primary amino group.

d. After activation, the heterocyclic amines bind to the 8-position of guanine in DNA to form covalent adducts.

e. Whole-body autoradiography of Trp-P-2 in mice revealed that the compound is distributed in the brain, endocrine system, kidney, liver, and lymphatic system.

Vuolo and Schuessler (1985) state that studies on the metabolism and toxicology of the heterocyclic amines are important in determining their potential danger to humans.

Chemistry of Mutagen and Carcinogen Formation Related to Food Browning

Amino-carbonyl reactions in foods, also called Maillard browning reactions after Louis-Camille Maillard, who discovered them in 1911, occur widely in all foods subjected to heat processing and storage. Transformations in which amino acids, peptides, proteins, carbohydrates, oxidized fatty acids and occasionally aldehydes, ketones, and quinones participate fall under the general category of such nonenzymatic browning reactions in which intermediates or products have brown coloration. Vitamin C is a special case. It forms brown products resulting from degradation as well as from interaction with amino compounds.

The chemical nature of the transformations of the initial stages of the Maillard reactions is shown in Figure 5 (Friedman, 1982). The first step is the formation of a Schiff's base (aldimine) between the amino and carbonyl group. This is followed by the rearrangement of the Schiff's base to an Amadori compound (1-amino-1-deoxy-2-ketose). The Amadori compounds then react further by several pathways, including enolization, dehydration, aldol condensations, and Strecker degradation, to form a large number of compounds. Some of these compounds contribute to the color and flavor of baked and fried foods.

In order to design rational approaches to preventing mutagen and carcinogen formation during food processing, it is of paramount importance to establish the detailed chemical pathways leading to their formation. This will make it possible to devise conditions to block these transformations at intermediate stages.

The mechanism of formation of pyrrole and pyridine derivatives and precursors for the production of mutagenic compounds in the Maillard reaction is shown in Figures 6 and 7. The precursor then undergoes further dehydration and cyclization to form the observed pyrrole and pyridine derivatives. The heterocyclic pyridines and pyrazines then undergo further transformations with participation of aldehydes and creatinine, to produce imidazoquinolines and imidazoquinoxalines, heterocyclic amines that are among the most potent mutagens known.

Another class of heterocyclic amines, the so-called carbolines, are formed when free or protein-bound tryptophan is exposed to heat under food processing conditions (Friedman and Cuq, 1988; Figure 8). These compounds have been found in commercial foods such as beef extracts and

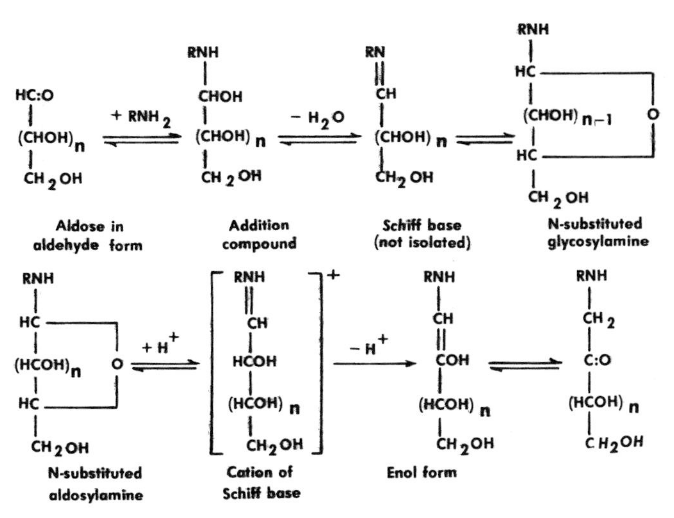

Figure 5. Initial stages of the classical Maillard reaction (Friedman, 1982).

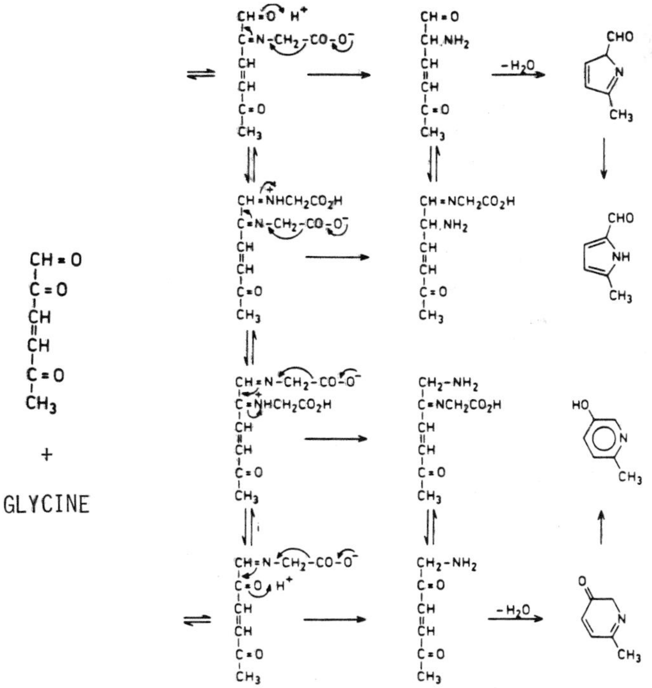

Figure 6. Formation of pyrrole and pyridine derivatives in the Maillard browning reaction (Adapted from Nyhammar, 1986).

$C_6H_{12}O_6$ + $RCH(NH_3^{\oplus})CO_2^{\ominus}$

hexose amino acid

creatine

Strecker degradation

Pyridine (Z = CH)

Pyrazine (Z = N)

aldehyde

creatinine

	X	Y	Z	R
IQ:	H	H	CH	H
MeIQ:	H	H	CH	Me
MeIQx:	H	Me	N	H
7,8-DiMeIQx:	Me	Me	N	H
4,8-DiMeIQx:	H	Me	N	Me

Figure 7. Formation of carcinogenic imidazoquinoline and imidazoquinaxoline browning product (Adapted from Nyhammer, 1986).

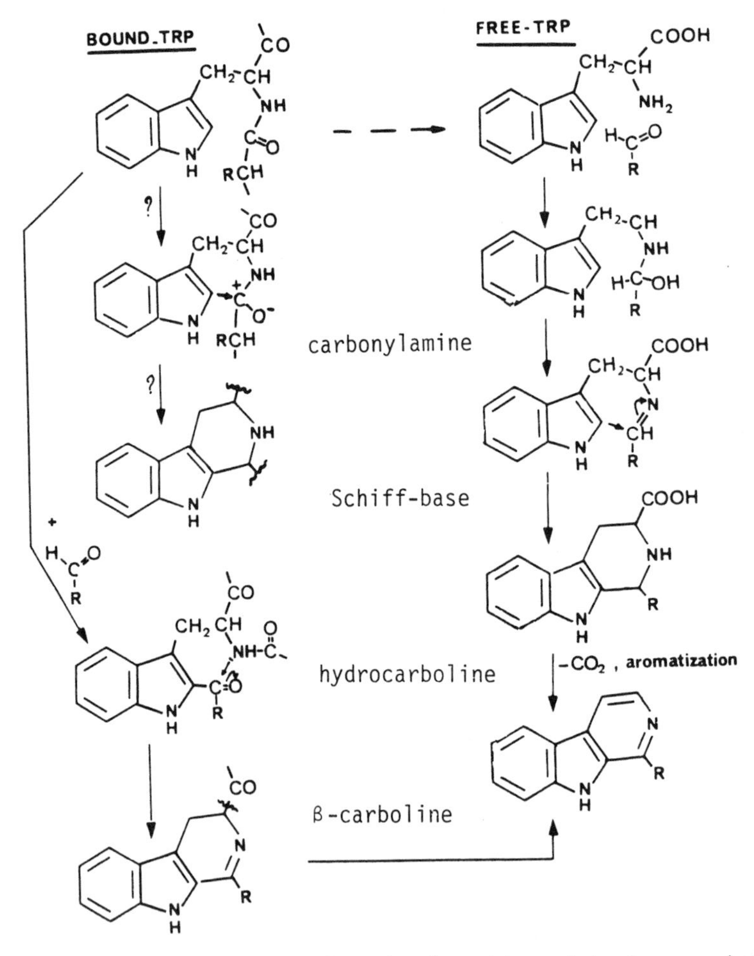

Figure 8. Possible pathways for the formation of toxic β-carbolines during heating of free and protein-bound tryptophan (Friedman and Cuq, 1988).

Table 7. Effect of Heated Fructose/Lysine Mixtures on the Incidence of Chromosomal Aberations in Chinese Hamster Ovary Cells.

	DOSE (mM)	NORMAL CELLS	ABNORMAL CELLS	CHROMATID DELETIONS	CHROMOSOME DELETIONS	CHROMATID EXCHANGES	CHROMOSOME EXCHANGES	PULVERIZED CELLS
CONTROL	0	94	6	3	1	0	2	0
pH 7	3.12	96	4	1	2	1	0	0
	6.25	94	6	3	1	2	0	0
	12.50	85	15	4	3	8	0	0
	25.00	59	41	30	5	34	4	3
pH 10	1.60[a]	64	36	10	10	37	2	0
	3.12[a]	97	3	1	4	0	0	0
	6.25[a]	NO METAPHASE						
	12.50[a]	98	2	1	0	1	0	0

[a] Severe cytotoxicity.

(MacGregor et al., 1989).

fried hamburger (Bjeldanes et al., 1982; Hatch et al., 1986), heated milk (Rogers and Shibamato, 1982), and beer and wine (Bosin et al., 1986). Among these compounds, two γ-carbolines (Trp-P-1, Trp-P-2) and two α-carbolines (A-α-C, Me-A-α-C) show significant mutagenic activity in Salmonella typhimurium tester strains after metabolic activation. Although not mutagenic themselves, the β-carbolines (harman and norharman) enhance the activity of the α- or γ-mutagenic carbolines and become potent mutagens after exposure to nitrites (Kinae, 1986). As mentioned earlier, bread crust and other baked products, dried fruit, and processed fruit juices also contain significant concentrations of processing-induced mutagens.

Our studies showed that the loss of nutritional quality of casein and casein-carbohydrate mixtures heated under crust-baking condition can be related to decreased nitrogen digestibility as opposed to actual destruction of essential amino acids. We also showed that such heat treatments of wheat gluten produce mutagenic substances (Friedman et al., 1990b).

Further studies are needed to explain the extent and nature of (a) the heat-induced destruction of essential amino acids and (b) formation of undigestible browned and crosslinked products. These changes impair intestinal absorption and nutritional quality in general. The possible presence of toxic compounds formed under these conditions might also modulate nutritional quality. Thus, such studies should differentiate antinutritional and toxicological interrelationships and develop means for preventing or minimizing the formation of deleterious compounds in foods.

Mutagens in Fruit Juices

The following information, summarized from a review (Handwerk and Coleman, 1988), offers a brief overview of our current knowledge of browning in citrus products:

1. Browning of citrus products has been a major problem throughout the history of the citrus processing industry.

2. Browning induces deterioration in color, flavor, and taste during processing of citrus fruit juices and concentrates, and during long-term storage of dehydrated products.

3. The food ingredients responsible for browning include: (a) free amino acids; (b) carbohydrates such as glucose, fructose, and sucrose; and (c) vitamin C.

4. Characterized browning products include: (a) furfuraldehyde derivatives; (b) pyrrole derivatives; (c) brown polymers; (d) aldehydes and ketones, and (e) cyclopentenones.

5. Maillard-type nonenzymatic browning plays a prominent and diverse role in that it produces beneficial flavors as well as objectionable color and taste changes.

6. Vitamin C browning is a special case, leading to the destruction and loss of this important vitamin and to the production of furfurals as a result of degradative transformations. Vitamin C and its degradation products also participate with amino acids in Maillard browning.

The authors conclude that better and more practical means need to be developed to reduce deleterious changes during browning of citrus products.

Nonenzymatic browning has generally been recognized to occur in all fruit juices tested, including apple (Toribio and Lozano, 1984), grapefruit (Lee and Nagy, 1988), lemon (Robertson and Samanego, 1986), and pear (Beveridge and Harrison, 1984). Such browning damages the appearance, quality, and safety of the juices. With respect to safety, a recent report by Ekasari et al. (1986, 1989) reported that fresh orange juice was mutagen-free, but when heated for about 2 min at 93°C (the usual pasteurization conditions), the juice produced significant amounts of mutagens as measured by the Ames test. Mutagen formation was related to dose and to heating time. Commercial juice, also contained significant mutagen levels.

Stich et al. (1981) found that although freshly prepared apple juice had only a minor effect on chromosomes of Chinese hamster ovary cells, stored and canned apple juice induced strong damage to cells. They ascribe this effect to the formation of browning products during storage and canning. Mutagen formation could indeed be due to browning products (Powrie et al., 1986). It could also be induced by aldehydes and ketones produced from carbohydrates and vitamin C (Bjeldanes and Chew, 1979).

Friedman et al. (1990a) investigated mutagenic responses of freshly squeezed orange juice, fresh orange juice heated for 0.5-30 min at 100°C, and several commercial orange juices in four bacterial strains of Salmonella typhimurium (TA98, TA100, TA102, TA2637) with and without microsomal activation. The response without activation was similar in all cases, ranging 2-3 times the background values. The experiments with microsomal activation produced barely visible pinpoint colonies that do not appear to be normal revertants (Tables 8-11). Our results do not duplicate previously reported high mutagenic activity of fresh orange juice after heating or of commercial varieties (Ekasari et al., 1986, 1989). In preliminary studies, no increase in the number of revertants was apparent after freshly prepared apple, grape, and pear juices were heated (Table 12).

These observations suggest that research to improve both the quality and safety of the juices is needed. Studies are needed to define the cause-effect relationship of mutagen formation, to assess the risk of consuming such browning products, and to devise improved processing conditions to minimize or prevent nonenzymatic browning.

Browning in Dried Fruits

During drying, storage, and irradiation of fruits a number of chemical changes affect their appearance, nutritional quality, and safety (Krone et al., 1986; Niemand, 1983). The chemical changes that take place during drying and storage of fruits include (a) enzymatic browning; (b) nonenzymatic Maillard browning; and (c) caramelization of sugars.

Since nonenzymatic browning is implicated in the formation of mutagens in heated food, Stich et al., (1981) studied the genotoxicity of a variety of dried fruits. They discovered that water extracts of dried fruits including raisins, prunes, table dates, bananas, figs, and apricots induced significant chromosome damage in Chinese hamster ovary cells. They suggest that dried fruits represent a widely consumed food product with strong genotoxic activity, as evidenced by the following findings: One chromosome exchange per diploid chromosome complement (22 chromosomes) was induced by 2 golden seedless raisins, 2 California black mission figs, 0.23 of a pitted table date, 0.06 of a dried prune, and 0.01 of a dried banana. The exact causes and possible approaches to the prevention of the chromosome-damaging genotoxic risk of dried fruit are not known (Niemand, 1983).

Kidney Damage by Browning Products

Browning products produce kidney damage which closely resembles the nephrocytomegaly induced by lysinoalanine, and is similar to precarcinogenic changes observed with other compounds (Gould and MacGregor, 1977; Friedman et al., 1984). The consumption of browning products by humans is several orders of magnitude greater than of lysinoalanine. Kawamura and Hayashi (1987) reported that extracts from human kidneys are less effective in metabolizing lysinoalanine than are corresponding extracts from kidneys of pigs, cattle, rats, mice, rabbits, chickens, and quail. This implies that lysinoalanine may be more toxic for humans than for the other species.

The mechanism of lysinoalanine-induced kidney damage is unknown. Recent studies demonstrated a strong affinity of lysinoalanine for copper and other metal ions. These findings and computer simulation studies of possible effects of lysinoalanine *in vivo* suggest that kidney damage is probably related to the perturbation by lysinoalanine of copper transport, storage, and utilization (Pearce and Friedman, 1988; Friedman and Pearce, 1989).

Von Wagenheim et al. (1984) examined kidney histopathology of rats fed casein heated with glucose for four days at 65°C. Enlarged epithelial cells and nuclei (giant cells) were observed after two weeks of feeding. The average size of the nuclei increased with feeding time, with significant differences being observed after 6, 8, or 10 weeks of feeding. The authors state that the observed kidney damage is similar to that reported for lysinoalanine. The etiology of the effect is presently unknown, but it seems to be due to the formation of early or advanced stage browning products.

In a subsequent study, Finot and Furniss (1986) confirmed and extended the above findings about browning-product-induced nephrotoxicity. They report that feeding either lysinoalanine or casein heated with glucose at 37°C for 3 or 15 days (storage conditions) induced nephrocytomegaly and was associated with urinary loss of zinc. Increased urinary loss of copper was also reported. Kidney levels of zinc and copper were higher than controls in the group fed heated casein-glucose.

These observations suggest that alteration of the mineral content of the diet could minimize browning-induced kidney damage (Friedman and Pearce, 1989; Pearce and Friedman, 1990).

Growth Inhibitors Derived from Ascorbate Browning

Sodium ascorbate heated with wheat gluten and other proteins under conditions of crust baking strongly inhibits the growth of mice when added to an otherwise nutritionally adequate diet (Figures 9, 10; Ziderman and Friedman, 1985; Friedman et al., 1987, 1989; Ziderman et al., 1989). Sodium ascorbate is widely used in many food applications; for example, the vitamin is added to flour before baking to improve bread dough characteristics and to bacon to prevent nitrosamine formation.

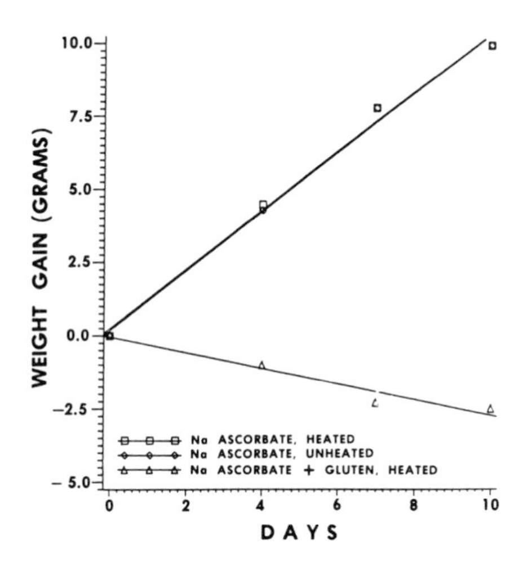

Figure 9. Weight gain in mice fed unheated and heated sodium ascorbate compared to that of mice fed sodium ascorbate heated in the presence of wheat gluten (Friedman et al., 1987).

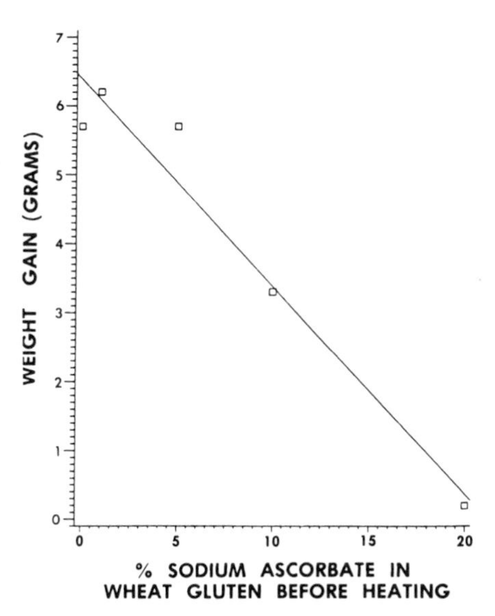

Figure 10. The relationship of weight gain in mice to the concentration of sodium ascorbate in gluten during heating (Friedman et al., 1987).

Table 8. Quantitative Plate Test of Fresh Orange Juice Heated for Various Time Periods in Salmonella typhimurium TA100 Strain. (Duplicate Values Are Revertants per Plate (1 mL of Juice)

Test Compound	TA100
Orange juice, unheated	185, 213
Orange juice, heated, 0.5 min	249
Orange juice, heated, 2 min	237, 276
Orange juice, heated, 3 min	272, 273
Orange juice, heated, 5 min	258, 283
Orange juice, heated, 30 min	272, 304
H_2O control	118, 132
Positive control	
(NQNO, 0.1 µg/plate)	574, 596

(Friedman et al., 1990b).

Table 9. Quantitative Plate Tests of Fresh Unheated and Heated Orange Juices in Salmonella typhimurim Strain TA100 and Histidine and Lysine Content of the Juices[a] (Duplicate Values are Revertants per Plate)

Test Compound	TA100	His (µg/mL)	Lys (µg/mL)
Orange juice, unheated	227,253	6.6	26.8
Orange juice, heated, 0.5 min	239, 270	4.0	19.4
Orange juice, heated, 2 min	224, 247	3.8	22.6
Orange juice, heated, 4 min	218, 258	3.4	19.8
H_2O control	99, 128		
Positive control			
(NQNO, 0.1 µg/plate)	544		

[a] Freeze-dried samples were resuspended to 5X concentration of fresh juice. Each test was done with 0.2 mL of the suspension. There was no change in pH of the orange juices following heat treatments. (Friedman et al., 1990b).

Table 10. Quantitative Plate Tests of Commercial Orange Juices in *Salmonella typhimurium* Strains TA98, TA100, TA102, abd TA2637 without Microsomal Activation. (Duplicate Values are Revertants per Plate; 1 ml of Juice)

Test compound	TA98	TA100	TA102	TA2637
Brand A ready to drink orange juice	96, 187	362, 365	754, 764	126, 141
Brand B ready to drink orange juice	102, 104	266, 268	735, 837	75, 77
Brand C orange juice from concentrate	118, 126	215, 215	678	81, 150
Brand D orange juice from concentrate	104, 119	219	661, 854	112, 132
Brand E orange juice from concentrate	140	243, 256	779, 902	60, 114
H_2O control	35, 45	148, 161	413, 448	56, 63
Positive controls (all with microsomal activation)	>aflatoxin 0.4 µg: 1200, 1050	>aflatoxin, 0.4 µg: 1200, 1200	>danthron, 45 µg: 1500, 1500	>emodin, 30 µg: 399, 458

(Friedman et al., 1990b)

Table 11. Quantitative Plate Tests of Orange Juices in Salmonella typhimurium Strains TA98, TA100, and TA2637 Without Microsomal Activation. (Duplicate Values are Revertants per Plate; 1 ml of Juice)

Test Compound	TA98	TA100	TA2637
Orange juice, freshly squeezed unheated	77,105[a]	310,324	79, 80
Orange juice, freshly squeezed heated (2 min at 100°C)	61, 78	305,313	93, 87
Orange juice, commercial from carton (brand C)	64, 72	331,380	67, 72
H_2O controls	22, 25	167,173	52, 59
NQNO, 0.1 µg/plate	159,186	636,749	198,203

(Friedman et al., 1990b).

Table 12. Quantitative Plate Tests of Fresh Apple, Grape, and Pear Juices in Salmonella typhimurium Strain TA100 Without Microsomal Activation. (Duplicate Values are Revertants per Plate; 1 ml of Juice)

	TA100	% H_2O control
Apple juice, freshly squeezed, unheated[a]	212, 236	118
Apple juice, freshly squeezed, heated[b]	220, 240	121
Grape juice, freshly squeezed, unheated[c]	364, 418	206
Grape juice, freshly squeezed, heated[b]	372, 377	197
Pear juice, freshly squeezed, unheated[d]	262, 266[e]	152
Pear juice, freshly squeezed, heated[b]	210, 216[e]	122
Pear juice, freshly squeezed, heated[b]	200, 230[e]	124
H_2O control	161, 219	100
H_2O control	177, 171[e]	
Positive control	458	
(NQNO, 0.1 µg/plate)	478, 437[e]	

[a] Prepared from Washington State Golden Delicious apples.
[b] 2 min at 100°C.
[c] Prepared from Thompson white seedless grapes.
[d] Prepared fro d'Anjou pears.
[e] Separate experiment.
(Friedman et al., 1990b).

In a related study, mice were fed for 14 days a nutritionally adequate casein diet supplemented at the expense of starch-dextrose with a 5% series of amino acids previously heated in the dry state with sodium ascorbate (Oste and Friedman, 1990). Growth inhibition by the heated mixtures ranged from none for arginine to significant for tryptophan (Figure 11). Additional studies revealed that (a) L-tryptophan-ascorbate significantly decreased weight gain when heated with an oven temperature of 200 or 215°C but not when heated at 180°C; (b) growth inhibition was less with N-acetyl-L-tryptophan/sodium ascorbate than with the tryptophan mixture; and (c) heated tryptophan or sodium ascorbate alone or tryptophan heated with glucose or ascorbic adid did not affect weight gain. These results complement the studies with heated food protein-sodium ascorbate and ascorbic acid mixtures and suggest that tryptophan interacts with sodium ascorbate, forming growth inhibitors.

Elucidation of the nature and potency of the antinutritional material is needed (1) to reveal the extent to which concern for food safety is warranted and (2) to develop food processing conditions to prevent formation of toxic material. This work will lead to a better understanding of the fate of vitamin C in processed foods (Ekasari et al., 1986, 1989; Friedman et al., 1987; Hayashi, 1985; Handwerk and Coleman, 1988; Ishii, 1986; Kincal and Giray, 1987; Krone et al., 1986; Robertson and Samanego, 1986; Stich et al., 1981, 1982; Toribio and Lozano, 1984; Wolfrom et al., 1974; Yamaguchi and Nakagawa, 1983; Ziderman and Friedman, 1985, 1989) and to the prediction and rapid discovery of related adverse nutrient interactions induced by food processing.

The safety of the red pigment derived from vitamin C browning (Hayashi, 1985; Kurata et al., 1986) also needs to be evaluated.

POSSIBLE APPROACHES TO PREVENT ADVERSE EFFECTS OF NONENZYMATIC BROWING

Conditions could be devised to minimize or prevent the formation of browning products during food processing or to prevent their deleterious effects in animals (Adams, 1987; Alldrick et al., 1986; Ames, 1983; Bradfield and Bjeldanes, 1987; De Flora, 1989; Deshpande et al., 1984; Friedman, 1984a, b, 1986; Friedman and Finot, 1991; Friedman and Gumbmann, 1990; Friedman and Molnar-Perl, 1990; Friedman and Smith, 1984; Friedman et al., 1982; Friedman et al., 1987; Friedman et al., 1989c; Molnar-Perl and Friedman, 1990a,b; Oste et al., 1990; Ramel et al., 1986; Rhee et al., 1987; Stich and Rosin, 1984; Troll, 1986; Wall et al., 1988; Wallace and Friedman, 1985; Wattenberg, 1983; Wilpart, 1985; Yen and Lai, 1987). This research should focus on browning reactions in bread crust and in fruits and vegetables.

The following overlapping approaches could, in principle, be used to reduce the toxic potential of our food supply and prevent or minimize food browning and the consequent antinutritional and toxicological manifestations. These represent the most promising current experimental avenues to the solution of these problems. Experience with any specific approach derived from preliminary experiments will dictate whether it should be pursued in depth.

Thermal Inactivation of Toxic Alkaloids

Commercial grain shipments may contain non-grain components, including the seeds from plants that coexist with the crop to be harvested. Some of these seeds may contain highly toxic components and may not be readily separable during the normal cleaning process. Such

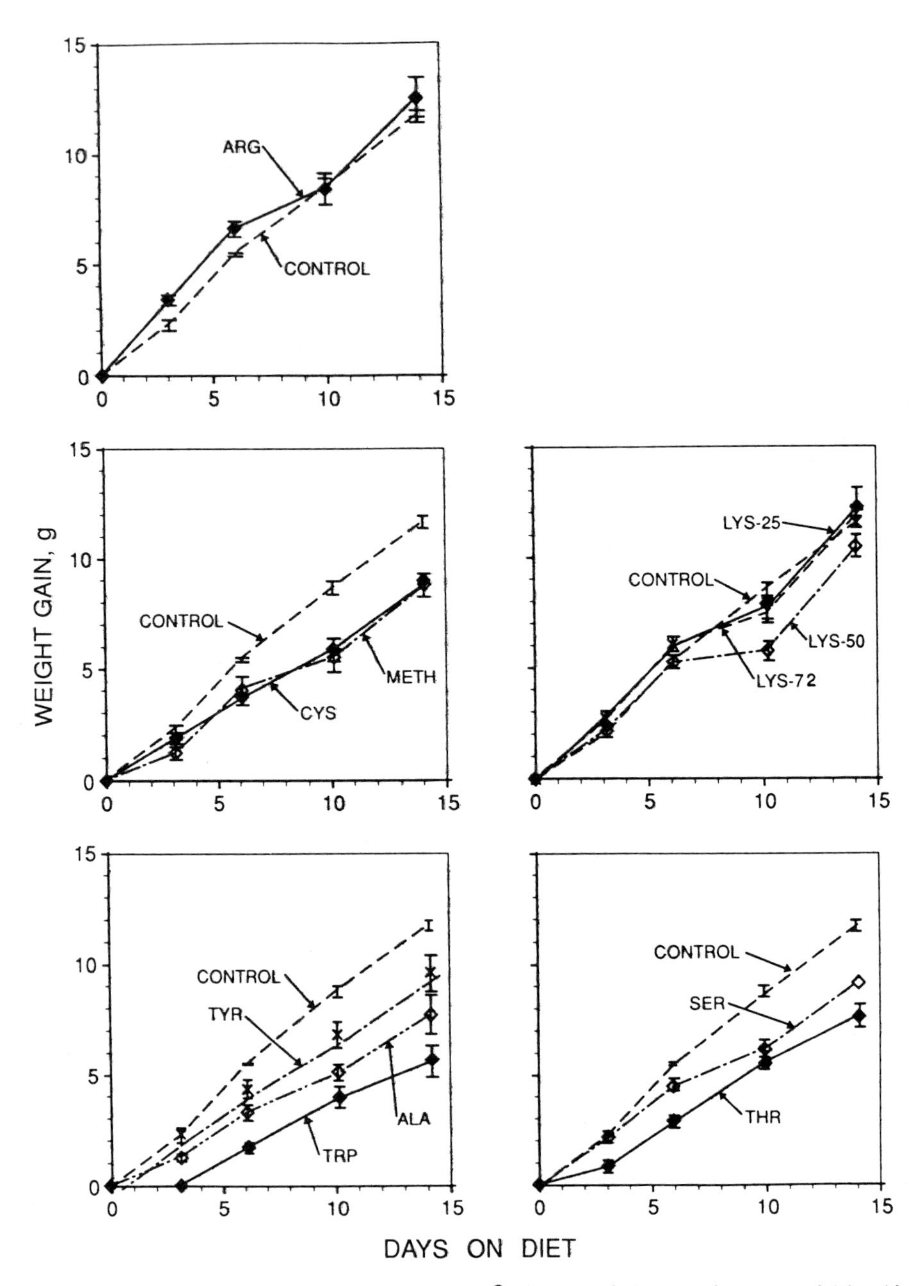

Figure 11. Effect of heated (215°C/72 min) amino acid/sodium ascorbate mixtures (5% in diet) on weight gains in mice fed a standard casein diet. Lys-25, Lys-50, and Lys-72 designate lysine/ascorbate mixtures heated for 25, 50, and 75 min, respectively (Oste and Friedman, 1990).

seeds have been documented to present serious problems, including illness and death of live-stock and concern over food safety for humans. Since most grain is not consumed as such but is subjected to food processing such as baking, it is of practical interest to find out whether, for example, lysergic acid derivatives and ergot alkaloids in morning glory seeds, atropine and scopolamine in Jimson weed seeds, and anthraquinones in sicklepod seeds, survive processing. Our studies show that autoclaving and conventional and microwave baking cause various degrees of thermal destruction of ergot alkaloids in a typical flour mix at ordinary temperatures. These findings imply that if these toxic weed seeds should enter our grain supply, their toxicity will be reduced after baking. They also imply that grain standards need to take into consideration the fate of the toxic principles during storage and processing (Crawford and Friedman, 1990; Crawford et al., 1990; Friedman and Dao, 1990; Friedman and Levin, 1989; Dugan et al., 1989; Friedman et al., 1989a; Scott and Lawrence, 1982).

Modification and Removal of Reactants and Endproducts in the Browning Reaction

a. Oxidation of the free amino acids by the enzyme L-amino oxidase. Since oxidation occurs naturally during the normal metabolism of amino acids (Friedman, 1989), the enzymatic oxidation holds promise of being both safe and inexpensive. A special effort is needed to oxidize or otherwise modify creatine, a key participant in the formation of carcinogenic heterocyclic amine browning products. If this approach is successful, the enzyme could be produced inexpensively by recombinant DNA techniques.

b. Acylation of amino groups. Modification of amino groups to stop them from participating in browning reactions. For example, treatment of foods with the enzyme transglutaminase will transform lysine amino groups to amide groups. The former initiate browning whereas the latter do not (Friedman, 1978; Friedman and Finot, 1991).

c. Adsorption of the amino acids or browning products by synthetic materials and natural compounds. Adsorbents include natural fibers (Barnes et al., 1983), cyclodextrins derived from carbohydrates (Pszczola, 1988), and naturally occurring porphyrin-like compounds such as chlorophylls, heme, and protoporphyrin. The heterocyclic amines show a strong affinity for phytalocyanines (blue cotton) presumably forming charge-transfer complexes. Phythalocyanines cannot be used in food applications. Expectations are that related naturally occurring compounds such as the porphyrins or chlorophyll immobilized on columns are expected to also bind the amines and thus provide a safe scientific basis for their inactivation and removal. The same holds true for the cyclodextrins and natural fibers which participate in strong hydrogen-bonding interactions with food ingredients.

Prevention of Browning Reactions Through the Use of Antioxidants

Our studies of browning in simulated bread crust (Ziderman and Friedman, 1985) revealed that oxygen seems to be required for nonenzymatic browning. We showed that baking in a vacuum oven caused less damage. These results imply that antioxidants should suppress browning in foods. A number of oxidants which are on the GRAS (generally accepted as safe) meed to be evaluated for their potential to minimize nonenzymatic browning during baking and storage (Ramel et al., 1986).

Blocking of in vivo Toxicant Formation by Means of Dietary Modulation

The lower incidence of cancer among groups whose diet is rich in vegetables, and high incidence of cancer among groups whose diet is deficient in vegetables, have been ascribed to the protective effects of dietary ingredients with antioxidant properties (Stich and Rosin, 1984), and the preventation of absorption of deleterious constituents by dietary fiber, tannins, and other natural polymers. These dietary ingredients include vitamins A, C, and E, beta-carotenes, nonflavonoids and flavanoid phenolic compounds, indole compounds, and natural fibers, nondigestible polysaccharides, tannins, etc. (Deshpande et al., 1984). All of these have the ability to act as antioxidants both in vitro and in vivo. The effectiveness of these compounds in trapping intermediates in the conversion of browning products to toxic substances, both ionic and free-radical types, have been little studied. Some of these food ingredients may have the ability to minimize nonenzymatic browning reactions by trapping free radical (one-electron) or other reactive intermediates in the browning process. These ingredients also may prevent activation of the procarcinogens to active forms in vivo (Whitty and Bjeldanes, 1987; Wall et al., 1988; Yamaguchi, 1982). The cited dietary ingredients should be evaluated for their ability to (a) prevent formation of browning products in vitro and (b) reduce the toxicity of browned foods in vivo (MacGregor et al., 1989; De Flora, 1989).

Analysis and Removal of Browning Products Through Antibody Complexation

According to Hatch and Felton (1986) improved methods of analysis are required so that research on mutagens and carcinogens formed during cooking can be advanced from exploration to intensive surveys of the diet to assess human risk. Since some of these compounds are formed at relatively low levels, the currently used chromatographic and mass spectrophotometric techniques are time-consuming, expensive, and inadequate.

A promising approach to detect and measure food toxicants is the use of monoclonal antibodies (Brandon et al., 1988, 1989, 1990; Friedman et al., 1989c). In addition to accurately measuring low levels of food toxicants in complex food mixtures, such antibodies may form antibody-antigen complexes with the toxicants, thereby preventing them from acting as carcinogens. If this approach is successful, the antibodies could be produced by recombinant DNA techniques (Oste et al., 1990).

Inactivation by Use of Sulfur-Rich Compounds

Sulfur-containing amino acids such as cyteine, N-acetylcysteine, and the tripeptide glutathione play key roles in the biotransformation of xenobiotics by actively participating in their detoxification. These sulfur-containing compounds also inhibit the action of mutagens, carcinogens, and other toxic compounds by direct interaction. These antioxidant and antitoxic effects are due to a multiplicity of mechanisms including the ability to act as (a) reducing agents; (b) scavengers of reactive oxygen (free-radical species); (c) strong nucleophiles which can trap electrophilic compounds and intermediates; (d) precursors for intracellular reduced glutathione; and (e) inducers of cellular detoxification. For example, we have shown that cysteine and related thiols inactivated the mutagenicity of aflatoxin in vitro (Friedman et al., 1982). Other examples include (a) the striking demonstration by De Flora (1988) that co-administration of N-acetyl-cysteine dramatically decreased urethane-induced tumor formation in mice; (b) the report by Troll (1986) that the sulfur-rich protein called the Bowman-Birk inhibitor suppressed nitrosamine-induced carcinogenicity in the digestive tract of rats; (c) the reported reduction in mutagen formation in fried beef by cottonseed flour (Rhee et al., 1987) and by soy protein concentrate (Wang et al., 1982); and (d) the observed synergistic effects of SH-containing compounds on heat-inactivation of trypsin inhibitors and lectins (Friedman and Gumbmann, 1986; Friedman et al., 1990a; Wallace and Friedman, 1985).

For these reasons, fruitful results are expected from evaluation of the effectiveness of sulfur amino acids and sulfur-rich proteins (purothionin from wheat, solubilized keratins, etc) to (a) prevent the formation of toxic browning products by trapping intermediates; and (b) reduce the toxicity of browning products in animals by preventing activation of such compounds to biologically active forms.

These expectations were realized as evidenced by the following observations on the prevention of both enzymatic and nonenzymatic by sulfur amino acids (Friedman and Molnar, 1990; Molnar and Friedman, 1990 a, b).

To demonstrate whether SH-containing sulfur amino acids minimize nonenzymtic browning, β-alanine, N-α-acetyl-L-lysine, glycyl-glycine, and a mixture of amino acids were each heated with glucose in the absence and presence of the following potential inhibitors: N-acetyl-L-cysteine, L-cysteine, reduced glutathione, sodium bisulfite, and urea. Inhibition was measured as a function of temperature, time of heating, and concentration of reactants. The extent of browning was estimated by absorbance measurements at 420 nm (Table 13). The minimum concentrations for optimum inhibition, in moles of inhibitor per mole of D-glucose, were as follows: sodium bisulfite, 0.02; L-cysteine, 0.05; N-acetyl-L-cysteine, 0.2; reduced glutathione, 0.2; urea, 8. An "Index of Prevention" (IP) was used to calculate the inhibition at the optimum mole ratio range. The calculated values were about 90% in all cases.

In the pathway toward Maillard browning, it has been postulated that an aldehyde group of a reducing sugar interacts with an amino group to form a hemiketal adduct, which then dehydrates rapidly to form an aldimine (Schiff's base). After an Amadori rearrangement, the latter is further transformed to dark browning products (equation 1). In the presence of a thiol, the aldehyde can competively interact with one or two SH groups to form a thiohemiketal or thioketal, thus blocking Maillard browning (equations 2 and 3)

$$\underset{\text{Carbohydrate}}{\text{R-CHO}} + \underset{\text{Amine}}{\text{R'-NH}_2} \xrightarrow{H^+} \underset{\substack{\text{Hemiketal}\\\text{(carbinolamine)}}}{\text{R-CH(OH)-NH}_2\text{-R'}} \xrightarrow{-H_2O} \underset{\substack{\text{Aldimine}\\\text{(Schiff's base)}}}{\text{RCH=NH-R'}} \quad \text{Dark Products} \qquad (1)$$

$$\text{R-CHO} + \underset{\text{Thiol}}{\text{R'-SH}} \rightarrow \underset{\substack{\text{Thiohemiketal}\\\text{(Carbinol sulfide)}}}{\text{R-CH(OH)-S-R'}} \qquad (2)$$

$$\text{R-CH(OH)-S-R'} + \text{R'-SH} \xrightarrow{-H_2O} \underset{\text{Thioketal}}{\text{R-CH-(SR')}_2} \qquad (3)$$

Table 13. Effectiveness of Inhibitors on Browning Reactions (I-IV) of D-glucose with β-Alanine (I), N-acetyl-L-lysine (II), Mixture of Amino Acids (III), and Glycylglycine (IV).

Reaction	Inhibitors	Index of prevention (IP)[a] (%)
I	N-acetyl-L-cysteine	70
II	N-acetyl-L-cysteine	83
III	N-acetyl-L-cysteine	91
IV	N-acetyl-L-cysteine	89
I	L-cysteine	79
II	Glutathione	83
I	Sodium bisulfite	79
II	Sodium bisulfite	96
III	Sodium bisulfite	74
IV	Sodium bisulfite	91
I.	Urea	91
II	Urea	88
III	Urea	95
IV	Urea	89

[a] Index of Prevention (IP) = 100 - (molar absorptivity value of the amine compound + glucose + inhibitor) x 100/(molar absorptivity value of the amine compound + glucose). (Adapted from Friedman and Molnar, 1990).

tested include ascorbic acid, a commercial formulation called "Sporix", sodium sulfite, N-acetyl-L-cysteine, L-cysteine, and reduced glutathione. For comparison, related studies were also carried out with several protein-containing foods such as casein, barley flour, soy flour, and nonfat dry milk, and the commercial infant formula "Isomil". The results revealed that under certain conditions SH-containing N-acetyl-L-cysteine and the tripeptide reduced glutathione may be as effective as sodium sulfite in preventing both enzymatic and nonenzymatic browning. These sulfur amino acids merit further study to

Relative reactivities of aldehyde or ketone groups with SH or NH_2 groups in structurally different environments will dictate the extent of inhibition of Maillard browning by various thiols (Friedman et al., 1966).

Reflectance measurements were used to compare the relative effectiveness of a series of compounds in inhibition browning in freshly prepared and commercial fruit juices including apple, grape, grapefruit, orange, and pineapple juices (Tables 14-16). The potential inhibitors assess their potential for preventing long-term food browning under practical storage and processing conditions.

In a related study designed to develop sulfite alternatives, Russet Burbank potatoes, Washington golden delicious apples, and Washington red delicious apples were subjected to enzymatic browning in air and evacuated plastic pouches in the absence and presence of the following potential browning inhibitors: L-cysteine, N-acetyl-L-cysteine, reduced glutathione, sodium bisulfite, sodium sulfhydrate, and sodium hydrosulfite (Tables 17-20). Studies on the effects of concentration of inhibitors, storage conditions and pH revealed that N-acetyl-L-cysteine and reduced glutathione were nearly as effective as sodium sulfite in preventing browning of both apples and potatoes. In contrast, a previously proposed mixed solution of salicylic and ascorbic acids and potassium sorbate was effective only for short periods. These results suggest that N-acetyl-L-cysteine and reduced glutathione are promising alternatives to sulfite in preventing browning in fruits and vegetables.

CONCLUSIONS

Studies should emphasize the prevention of browning and the consequent antinutritional and toxicological manifestations of browning products in whole foods as consumed. Many of the safety concerns cited, especially those of genotoxic potential, are based on *in vitro* data which may not always be relevant to *in vivo* effects following the consumption of whole food products containing the browning-derived constituents. The presence of other dietary constituents in the food and the process of digestion and metabolism can be expected in some cases to decrease or increase the adverse manifestations of browning products.

The proposed approaches to the prevention of adverse consequences of food browning need to be coordinated with analytical, chemical, and appropriate animal studies to identify which food ingredients have the greatest antinutritional or toxicological potential. In cases where only *in vitro* effects have been reported, for example from clastogenic browning products formed from sugar-amino acid interactions in dried fruits and fruit juices, appropriate *in vivo* genotoxicity assays should be used to assess the potential for DNA and chromosome damage (MacGregor et al., 1989). Such information will lead to the development of a ranking scale of relative toxicity of browning products in specific foods. This, in turn, will permit our efforts to be directed to minimize or prevent the formation of the most deleterious food ingredients.

ACKNOWLEDGMENTS

I take great pleasure in thanking my colleagues whose names are listed on the cited references for excellent scientific collaboration.

Table 14. Prevention of Browning in Fresh Apple Juice with Various Inhibitors

Inhibitor	Treatment time	Inhibition (%)					
Ascorbic acid (mM):		0.62	1.55	3.10	4.65	6.20	12.40
	1-2 min	24	82	94	100	96	105
	1 hr	0	57	77	92	94	105
	2 hr	0	26	38	44	71	98
	6 hr	-2	20	13	23	40	69
	24 hr	-8	20	8	21	24	32
Sporix (10^{-3}%):		4.55	11.36	22.72	34.13	45.50	91.00
	1-2 min	26	29	29	35	21	24
	1 hr	11	9	22	34	14	13
	2 hr	20	8	19	25	13	11
	6 hr	11	6	12	22	14	13
	24 hr	9	9	-1	1	1	1
Sodium bisulfite (mM):		0.227	0.568	1.136	1.704	2.27	4.54
	1-2 min	66	63	106	106	92	103
	1 hr	30	38	103	107	94	96
	2 hr	21	25	105	106	96	95
	6 hr	20	16	101	109	104	106
	24 hr	28	6	104	102	105	93
L-cysteine (mM):		0.568	1.136	1.704	2.27	4.54	9.08
	1-2 min	103	101	101	78	99	114
	1 hr	96	96	98	88	103	108
	2 hr	100	99	96	95	107	119
	6 hr	94	101	97	104	106	121
	24 hr	83	89	86	97	97	106
Glutathione (mM): (reduced)		0.568	1.136	1.704	2.27	4.54	9.08
	1-2 min	103	98	98	107	96	99
	1 hr	94	108	93	114	89	91
	2 hr	94	109	96	110	92	102
	6 hr	94	116	99	114	95	103
	24 hr	74	102	96	110	99	93
N-acetyl-L-cysteine (mM):		0.568	1.136	1.704	2.27	4.54	9.08
	1-2 min	92	101	100	96	114	97
	1 hr	87	98	92	99	112	102
	2 hr	83	102	95	98	108	108
	6 hr	65	97	100	90	107	99
	24 hr	43	100	93	96	107	103

(Molnar-Perl and Friedman, 1990 a).

Table 15. Prevention of Browning in Fresh Pear Juice with Various Inhibitors

Inhibitor	Treatment time	Inhibition (%)					
Ascorbic acid (mM):		0.62	1.55	3.10	4.65	6.20	12.40
	1-2 min	78	91	82	79	84	73
	1 hr	24	60	78	81	87	81
	2 hr	5	36	32	54	81	80
	4 hr	0	35	49	32	38	61
	24 hr	40	40	28	24	29	48
Sporix (10^{-3}%):		4.55	11.36	22.72	34.13	45.50	91.00
	1-2 min	50	45	71	54	68	65
	1 hr	37	23	59	31	43	52
	2 hr	10	9	36	15	25	20
	4 hr	10	10	36	8	24	18
	24 hr	17	19	46	24	32	45
Sodium bisulfite (mM):		0.227	0.568	1.136	1.704	2.27	4.54
	1-2 min	62	66	74	104	100	99
	1 hr	35	55	71	106	101	97
	2 hr	15	55	76	105	98	100
	4 hr	7	73	84	115	100	105
	24 hr	45	58	86	120	106	116
L-cysteine (mM):		0.568	1.136	1.704	2.27	4.54	9.08
	1-2 min	107	100	100	93	99	99
	1 hr	96	102	101	90	106	112
	2 hr	84	98	103	95	106	113
	4 hr	68	94	97	94	103	107
	24 hr	56	70	82	99	117	116
Glutathione (mM): (reduced)		0.568	1.136	1.704	2.27	4.54	9.08
	1-2 min	117	101	106	98	96	97
	1 hr	119	101	106	98	97	96
	2 hr	122	115	112	105	100	101
	4 hr	124	106	113	104	102	102
	24 hr	124	123	122	110	111	106
N-acetyl-L-cysteine (mM):		0.568	1.136	1.704	2.27	4.54	9.08
	1-2 min	103	105	106	106	107	111
	1 hr	48	105	106	103	104	111
	2 hr	28	103	105	102	107	111
	4 hr	22	105	108	103	110	115
	24 hr	39	106	117	110	112	117

(Molnar-Perl and Friedman, 1990 a).

Table 16. Prevention of Browning in Commercial Fruit Juices by Sodium Bilsulfite and N-acetyl-L-cysteine

Protein Source	Sodium bisulfite (mM)					N-acetyl-L-cysteine					
	1.0	2.0	4.0	8.0	16.0	2.5	1.0	25	50	100	200
	Inhibition (%)										
Grape	10	25	69	72	100	6	18	35	79	100	99
Apple (cider)	42	50	59	93	100	1	36	52	87	100	89
Apple (juice)	10	40	49	100	100	5	40	55	92	93	100
Pineapple	11	39	68	102	105	10	35	54	76	100	102
Grapefruit	18	25	39	63	100	32	61	75	87	93	96
Orange	13	49	69	92	108	3	69	66	72	107	100

(Molnar-Perl and Friedman, 1990 a).

Table 17. Prevention of Browning in Protein-Containing Food by Sodium Bisulfite and N-acetyl-L-cysteine

Protein	Sodium bisulfite (mM)					N-acetyl-L-cysteine (mM)				
	2.5	25	50	100	200	25	6.2	125	250	500
	Inhibition (%)									
Casein	4	12	44	82	100	0	25	42	101	101
Barley flour	3	43	61	98	95	36	42	79	96	104
Soy flour	10	27	80	98	102	19	38	84	99	101
Non fat dry milk	3	23	44	94	104	19	43	78	98	101
Isomil	7	29	72	88	100	7	43	65	93	109

(Molnar-Perl and Friedman, 1990 b).

Table 18. Prevention of Enzymatic Browning of Cut Surfaces of White Russet Burbank Potato by Treatment with Various Inhibitors[a]

			Treatment bath concentration: mM			
		Time	5	10	25	50
Inhibitor	Treatment	(hr)	Inhibition (%)[a]			
Sodium bisulfite	A	4	93	105	102	101
		7	97	102	101	99
	B	2	94	89	103	98
		5	73	85	96	103
Sodium sulfhydrate	A	4	97	98	99	100
		7	77	99	95	98
	B	2	97	99	98	98
		5	100	99	95	98
Sodium hydrosulfite	A	4	96	97	102	93
		7	100	104	103	102
	B	2	100	99	98	96
		7	97	99	101	95
L-Cysteine	A	4	96	97	102	96
		7	93	95	96	98
	B	2	83	82	92	96
		5	79	80	95	100
N-acetyl-L-cysteine	A	4	95	97	102	101
		7	88	98	96	106
	B	2	80	92	99	99
		5	78	96	99	99
Glutathione (reduced)	A	4	100	93	97	100
		7	95	85	95	101
	B	2	83	87	96	96
		5	70	90	95	98
Sodium salicylate	A	4	70	76	71	76
		7	72	84	92	89
	B	2	33	39	39	25
		5	30	44	33	-27

[a] Inhibition % = (Δ control - ΔL treatment) X 100/ΔL control.

ΔL values are differences in L values between 2 min and specified time. The values measured after slicing from the basis of comparison for all conditions tested.

(Molnar-Perl and Friedman, 1990 b).

Table 19. Prevention of Enzymatic Browning of Cut Surfaces of Washington Red Delicious Apple by Treatment with Various Inhibitors

Inhibitor	Treatment	Time (hr)	Treatment bath concentration: mM 5	10	25	50
			Inhibition (%)			
Sodium bisulfite	A	6	106	101	102	102
		24	99	98	102	102
	B	2	93	72	100	98
		5	91	66	110	98
Sodium sulfhydrate	A	6	101	100	94	100
		24	99	102	101	100
	B	2	83	78	75	77
		5	75	74	72	70
Sodium hydrosulfite	A	6	101	103	109	102
		24	100	102	106	102
	B	2	86	88	94	92
		25	90	81	86	72
L-Cysteine	A	6	102	103	95	103
		24	98	98	95	99
	B	2	86	90	86	70
		5	71	96	81	74
N-acetyl-L-cysteine	A	6	85	102	102	103
		24	82	103	97	100
	B	2	97	97	95	100
		5	105	93	100	102
Glutathione (reduced)	A	6	102	98	98	100
		24	101	98	98	100
	B	2	96	104	95	92
		5	84	100	93	93
Sodium salicylate	A	6	89	87	93	92
		24	89	87	85	92
	B	2	-81	5	-31	8
		5	-90	4	-46	0

(Molnar-Perl and Friedman, 1990 b).

Table 20. Prevention of Enzymatic Browning of Cut Surfaces of Washington Golden Delicious Apple by Treatment with Various Inhibitors

Inhibitor	Treatment	Time (hr)	Treatment bath concentration: mM 5	10	25	50
			Inhibition (%)			
Sodium bisulfite	A	6	100	95	116	116
		24	95	95	93	104
	B	2	97	83	97	105
		5	109	80	91	100
Sodium sulfhydrate	A	6	85	89	105	105
		24	92	92	123	103
	B	2	120	94	91	91
		5	109	106	97	91
Sodium hydrosulfite	A	6	103	103	103	108
		24	100	97	99	93
	B	2	100	106	94	94
		5	97	106	103	100
L-Cysteine	A	6	94	95	94	94
		24	100	95	99	94
	B	2	57	91	97	117
		5	0	83	100	80
N-acetyl-L-cysteine	A	6	98	111	102	111
		24	99	103	105	120
	B	2	14	41	86	100
		5	0	43	100	106
Glutathione (reduced)	A	6	111	100	106	98
		24	107	93	107	96
	B	2	82	91	100	89
		5	94	93	96	103
Sodium salicylate	A	6	34	50	85	55
		24	29	33	64	52
	B	2	-345	-342	-360	-377
		5	-328	-414	-397	385

(Molnar-Perl and Friedman, 1990 b).

RFERENCES

Adams, J.D., Jr., Heins, M.C. and Yost, G.S. 3-Methylindole inhibits lipidd peroxidation. Biochem. Biophys. Res. Commun., 149, 73-78. 1987.

Alldrick, A.J., Flynn, J. and Rowland, I.R. Effect of plant-derived flavonoids and polyphenolic acids on the activity of mutagens from cooked food. Mutation Res., 163, 225-232. 1986.

Ames, B.N. Dietary carcinogens and anticarcinogens: oxygen radical and degenerative diseases. Science, 221, 1256. 1983.

Ames, B.N. Carcinogens and anticarcinogens. In "Antimutagenesis and Anticarcinogenesis Mechanisms", D.M. Shankel, Ed., Plenum, New York, p. 7. 1986.

Ames, B.N. Mutagenesis and carcinogenesis: endogenous and exogenous factors. Environ. Mol. Mutagen. 14, 66-77. 1989.

Ames, B.N., Magaw, R. and Gold, L.S. Ranking possible carcinogenic hazards. Science, 236, 271-280. 1987.

Archer, V.E. Cooking methods, carcinogens, and diet-cancer studies. Nutrition and Cancer, 11, 75-79. 1988.

Ayatase, J.O., Eka, O.U., and Ifon, E.T. Chemical evaluation of the effect of roasting on the nutritive value of maize. Food Chemistry, 12, 135-147. 1983.

Barnes,W.S., Maiello, J. and Weisburger, H.H. In vitro binding of the food mutagen 2-amino-3-methylimidazo-(4,5-f) quinoline to dietary fiber. J. National Cancer Institute, 70, 757-760. 1983.

Beverdige, T. and Harrison, J.E. Nonenzymatic browning in pear juice concentrate at elevated temperatures. J. Food Sci., 49, 1335-1340. 1984.

Bjeldanes, L.F. and Chew, H. Mutagenicity of 1,2-dicarbonyl compounds: maltol, kojic acid, diacetyl, and related substances. Mut. Res., 67, 367-371. 1979.

Bjeldanes, L.F., Morris, M.M., Felton, J.S., Healty, S., Stuermer, D., Berry, P., Timourian, H., and Hatch, F.T. Mutagens from the cooking of food. II. Protein-rich foods of the American diet. Food Chem. Toxicol., 20, 357-363. 1982.

Bosin, T.R., Krogh, S. and Mzis, D. Identification and quantification of 1,2,3-4-tetrahydro-beta-carboline-3-carboxylic acid and 1-methyl-1,1,3,4-tetrahydro=beta-carboline-3-carboxylic acid in beer and wine. J. Agric. Food Chem., 34, 843-847. 1986.

Bradfield, C.A. and Bjeldanes, L. Dietary modification of xenobiotic metabolism. Contribution of indolylic compounds present in Brassica oleracea. J. Agric. Food Chem. 35, 896-900. 1987.

Brandon, D.L., Hague, S. and Friedman, M. Interaction of monoclonal antibodies with soybean trypsin inhibitors. J. Agric. Food Chem., 34, 195-200. 1987.

Brandon, D. L., Bates, A. H. and Friedman, M. Enzyme-linked immunoassay of soybean Kunitz trypsin inhibitor using monoclonal antibodies. J. Fd. Sci. 53, 97-101. 1988.

Brandon, D. L., Bates, A. H. and Friedman, M. Antigenicity of soybean protease inhibitors. In "Protease Inhibitors as Potential Cancer Chemopreventive Agents", W. Troll and R. A. Kennedy, Eds., Plenum:New York, 1991.

Carlborg, F. W. Cancer, mathematical models and aflatoxin. Food Cosmetic Toxicology, 17, 159-166. 1979.

Crawford, L. and Friedman, M. The effects of low levels of dietary toxic weed seeds on realative size of rat liver and levels and function of cytochrome P-450. Toxicol. Letters, 54, 175-181. 1990.

Crawford, L., McDonald, G. M. and Friedman, M. Composition of sicklepod (Cassia obtusifolia) toxic weed seeds. J. Agric. Food Chem. 38, 2169-2175. 1990.

Cuzzoni, M. T., Stoppini, G. and Gazzani, G. Effect of storage on mutagenicity, absorbance and content of furfural in the ribose-lysine Mailard system. Food Chem. Toxicol., 27, 209-213. 1989.

Degawa, M., Hishinuma, T., Yoshida, H. and Hashimoto, Y. Species and sex organ differences in induction of a cytochrome P-450 isozyme responsible for carcinogen activation effects of dietary hepatocarcinogenic tryptophan pyrolysate components in mice and rats. Carcinogenesis,8,1913-1918. 1987.

Deshpande, S. S. Sathe, S.K. and Salunkhe, D. K. Chemistry and safety of plant polyphenols. In "Nutritional and Toxicological Aspects of Food Safety", M. Friedman, Ed., Plenum Press, New York, pp. 457-495. 1984.

Ekasari, I., Jongen, W. M. F. and Philnik, W. Use of bacterial mutagenicity assay as a rapid method for the detection of early stage of Maillard reactions in orange juices. Food Chem.,21, 125-13. 1986.

Dugan, G. M., Gumbmann, M. R. and Friedman, M. Toxicological evaluation of Jimson weed (Datura stramonium) toxic weed seeds. Food Chem. Toxicol., 27, 501-510. 1989.

Fan, A. M. and Book, S. A. Sulfite hypersensitivity: a review of current issues. J. Apllied Nutr., 39, 71-78. 1989.

Finot, P. A. and Furniss, D. E. Nephrocytomegaly in rats induced by Maillard reaction products: the involement of metal ions. In "Amino-Carbonyl Reactions in Food and Biological Systems", M. Fujimaki, M. Namiki, and H. Kato, Eds., Elsevier, Tokyo, pp. 493-502. 1986.

Finot, P. A., Aeschbacher, H. U., Hurrell, R. F. and Liardon, R. "The Maillard Reaction in Food Processing, Human Nutrition, and Physiology", Birkhauser, Basel, Switzerland. 1990.

Finot, P. A., Mottu, F., Bujard, E. and Mauron, J. N-substituted lysine as a source of lysine in nutrition. In "Nutritional Improvement of Food and Feed Proteins", M. Friedman, M., Ed., Plenum Press, New York, pp. 549-569. 1978.

Friedman, M. "The Chemistry and Biochemistry of the Sulfhydryl Group in Amino Acids, Peptides, and Proteins", Pergamon Press, Oxford, England, 1973.

Friedman, M. (Editor). "Protein-Metal Interactions" - Adv. Exp. Med.Biol. 40, Plenum Press, New York. 1974.

Friedman, M. (Editor). "Protein Nutritional Quality of Foods and Feeds", Marcel Dekker, New York. 2 volumes. 1975.

Friedman, M. (Editor). "Protein Crosslinking: Biochemical and Molecular Aspects" - Adv. Exp. Med. Biol. 86A, Plenum Press, New York. 1977.

Friedman, M. (Editor). "Protein Crosslinking: Nutritional and Medical Consequences", Adv. Exp. Med. Biol. 86B, Plenum Press, New York. 1977.

Friedman, M. (Editor). "Nutritional Improvement of Food and Feed Prote - ins", Adv. Exp. Med. Biol. 105, Plenum Press, New York. 1978a.

Friedman, M. Inhibition of lysinoalanine synthesis by protein acylation. Adv. Exp. Med. Biol., 105, 613-648. 1978b.

Friedman, M. Chemically reactive and unreactive lysine as an index of browning. Diabetes, 31, 5-14. 1982.

Friedman, M. Effect of lysine modification on chemical, physical, nutri - tive, and functional properties of proteins. In "Food Proteins", J.R. Whitaker and S. Tannenbaum, Ed., AVI, Westport CT., pp. 446-463. 1977.

Friedman, M. Crosslinking amino acids - stereochemistry and nomenclature Adv. Exp. Med. Biol. 86B, 1-27. 1977.

Friedman, M. (Editor). "Nutritional and Toxicological Aspects of Food Safety", Adv. Exp. Med. Biol. 177, Plenum Press, New York. 1984a.

Friedman, M. Sulfhydryl groups and food safety. Adv. Exp. Med. Biol., 177, 31-64. 1984.

Friedman, M. (Editor). "Nutritional and Toxicological Significance of Enzyme Inhibitors in Foods", Adv. Exp. Med. Biol. 199, Plenum Press, New York. 1986.

Friedman, M. and Gumbmann, M. R. Nutritional improvement of legume prot- eins through disulfide interchange. Adv. Exp. Med. Biol.,199,357-389. 1986.

Friedman, M. (Editor). "Absorption and Utilization of Amino Acids", CRC Press, Boca Raton, Florida, 3 volumes. 1989.

Friedman, M. and Cuq., J. L. Chemistry, analysis, nutrition, and toxi- cology of tryptophan in food. A review. J. Agric. Food Chem., 36, 1079-1083. 1988.

Friedman, M. and Dao, L. Effect of heating on the ergot alkaloid and chlorogenic acid contetents of morning glory seeds. J. Agric. Food Chem. 38, 805-808. 1990.

Friedman, M. and Finot, P. A. Improvement in the nutritional quality of bread. 1991. This Volume.

Friedman, M. and Henika, P. Mutagenicity of toxic weed seeds. J. Agric. Food Chem. 39, No 3, 1991.

Friedman, M. and Molnar-Perl, I. Inhibition of food browning by sulfur amino acids. Part I. Heated amino acid-glucose systems. J. Agric. Food Chem., 38, 1642-1647.

Friedman, M. and Pearce, K.N. Copper (II) and cobalt (II) affinities of LL- and LD-lysinoalanine diastereoisomers: implication for food safety and nutrition. J. Agric. Food Chem., 37, 123-127. 1989.

Friedman, M. and Smith, G.A. Inactivation of quercetin mutagenicity. Fd. Chem. Toxicol., 22, 535-539. 1984.

Friedman, M., Wehr,C.M., Schade, J.E. and MacGregor, J.T. Inactivation of aflatoxin B_1 mutagenicity by thiols. Fd. Chem Toxicol., 20, 887-892. 1982.

Friedman, M., Gumbmann, M.R. and Masters, P.M. Protein-alkali reactions -chemistry, toxicology, and nutritional consequences. In "Nutritional Improvement of Food and Feed Proteins", M. Friedman, Ed., Plenum Press, New York, pp. 367-412. 1984.

Friedman, M., Gumbmann, M.R. and Dao, L. Ergot alkaloid and chlorogenic acid content of morning glory (Ipomoea spp.) toxic weed seeds. J. Agric. Food Chem., 37, 708-712. 1989.

Friedman, M., Gumbmann, M.R. and Ziderman, I.I. Nutritional value and safety in mice of proteins and their admixtures with carbohydrates and vitamin C after heating. J. Nutrition, 117, 508-518. 1987.

Friedman, M., Gumbmann, M.R. and Ziderman, I.I. Nutritional and toxicological consequences of browning during simulated crust-baking. In "Protein Quality and the Effects of Processing", R.D. Phillips and J. W.Finley, Eds., Marcel Dekker, New York, pp. 189-219. 1989.

Friedman, M., Gumbmann, M.R., Brandon, D.L. and Bates, A.H. Inactivation and analysis of soybean inhibitors of digestive enzymes. In "Food Proteins", J.E. Kinsella and W.G. Soucie, Eds., American Oil Chemists Society, Champaign, Illinois, pp. 297-328. 1989.

Friedman, M., Brandon, D.L., Bates, A.H. and Hymowitz, T. Comparison of a commercial soybean cultivar and an isoline lacking the Kunitz trypsin inhibitor: composition, nutritional value, and effects of heating. J. Agric. Food Chem., 39, 327-335. 1991.

Friedman, M., Rayburn, J.R. and Bantle, J.A. Developmental toxicology of potato alkaloids in the frog embryo teratogenesis assay-Xenopus (FETAX). Submitted for publication. 1991

Friedman, M., Wilson, R.E. and Ziderman, I.I. Effects of heating on the mutagenicity of fruit juices in the Ames test. J. Agric. Food Chem., 38 740-743. 1990.

Friedman, M., Wilson, R.E. and Ziderman, I.I. Mutagen formation in heated wheat gluten, carbohydrates, and gluten/carbohydrate blends. J. Agric. Food Chem., 38, 1019-1028. 1990.

Gazzini, G., Vagnarelli, P., Cuzzoni, M.T. and Mazza, P.G. Mutagenic activity of the Maillard reaction products of ribose with different amino acids. J. Food Sci., 52, 757-760. 1987.

Giffon,E., Vervloet, D. and Charpin, J. Suspicion sur les sulfites. Rev. Nal. Resp., 6, 303-310. 1989.

Gould, D. H. and MacGregor, J. T. Biological effects of alkali-treated protein and lysinoalanine. An overview. In "Protein Crosslinking: Nutritional Medical consequences." Friedman, M., Ed., Plenum Press, New York, pp. 29-48. 1977.

Gumbmann, M. R., Friedman, M. and Smith, G.A. The nutritional value and digestibilities of heat-damaged casein and casein-carbohydrate mixtures. Nutrition Reports International, 28, 355-367. 1983.

Handwerk, R. G. and Coleman, R. L Approaches to the citrus browning problem. A review. J. Agric. Food Chem., 36, 231-236. 1988.

Hatch, F. T. and Felton, J. S. Toxicoligic strategy for mutagens formed in foods during cooking: status and needs. In "Genetic Toxicology of the Diet," I. Knudsen, Ed., Alan R. Liss, New York, pp. 109-131. 1986.

Hatch, F.T., Knize, M.G., Healy, S.K., Slezak, T. and Felton, J.S. Cooked food mutagen reference list and index. Environ. Mol. Mutagen, 12, 1-85. 1988.

Hayashi, T. Red pigment formation by reaction of oxidized ascorbid acid and protein. Agric. Biol. Chem., 49, 3139-3144. 1985. See also, "Amino- Carbonyl Reactions in Food and Biological Systems," M. Fujimaki, Ed., Elsevier, N. Y., p.105. 1986.

Ishii, K. Characterization of brown pigment formed by the reaction of dehydroascorbic acid with adenine. In "Amino-Carbonyl Reactions in Food and Biological Systems," M. Fujimaki, Ed., Elservier, Amsterdam, pp. 207-213. 1986.

Kier, L.D., Yamasaki, E. and Ames, B.N. Detection of mutagenic activity in cigarette smoke condensate. Proc. Nat. Acad. Sci. USA, 71, 4159-4163. 1974.

Kinae, N. Isolation of beta-carboline derivatives from Maillard reaction mixtures that are mutagenic after nitrite treatment. In "Amino-Carbonyl Reactions in Food and Biological Systems," M. Fujimaki, Ed, Elsevier, Amsterdam, 343-352. 1986.

Kincal, N. S. and Giray, C. Kinetics of ascorbic acid degradation during potato blanching. Int. J. Food Sci. and Technol., 22, 249-254. 1987.

Kinouchi, T., Tsutsui, H. and Ohnishi, Y. Detection of 1-nitropyrene in yakatori (grilled chicken). Mutation Res., 171, 105-113. 1986.

Krone, C. A., Yeh, S. M. J. and Iwaka, W. T. Mutagen formation during commmercial processing of foods. Environmental Health Perspectives, 67, 75-88. 1986.

Kurata, T., Imai, T. and Arakawa, N. The structure of the red pigment (derived from vitamin C). In "Amino-carbonyl Reactions in Food and Biological Systems." M. Fujimaki, Ed., Elsevier, 1986, pp. 67-75.

Lee, H.S. and Nagy, S. Quality changes and nonenzymatic browning intermediates in grapefruit juice during storage. J. Food Science, 53, 168-172. 1988.

Lee, C. M., Lee, T. C. and Chichester, C. O. Kinetics of the production biologically active Maillard browning products in apricots and glucose-L-tryptophan. J. Agric. Food Chem., 27, 478-482. 1979.

Lovelette, C., Barnes, W. S., Weisburger, J. H. and Williams, G. M. Improved synthesis of the mutagen 2-amino-3, 7, 8-trimethyl-3H-imidazo (4,5-f) quinoxaline and acivity in a mammalian DNA repair system. J. Agric. Food Chem., 35, 912-915. 1987.

McGregor, J. T., Ducker, J. D., Ziderman, I. I., Wehr, C. M., Wilson, R. E. and Friedman, M. Nonclastogenicity in mouse bone marrow of fructose -lysine and other sugar-amino acid browning products with in vitro genotoxicity. Food Chem. Toxicol., 27, 715-721. 1989.

Miller, A. J. Thermally induced mutagens in protein foods. In "Protein Quality and Effects of Processing", R. D. Phillips and J. W. Finley, Eds., Marcel Dekker, New York, pp. 145-188. 1989.

Molnar-Perl, I. and Friedman, M. Inhibition of food browning by sulfur amino acids. Part 2. Fruit juices and protein-containing foods. J. Agric. Food Chem., 38, 1648-1651. 1990.

Molnar-Perl, I. and Friedman, M. Inhibition of food browning by sulfur amino acids. Part 3. Apples and potatoes. J. Agric. Food Chem., 38, 1652-1656. 1990.

Niemand, J. G. A study of the mutagenicity of irradiated sugar solutions. Implication for the radiation preservation of subtropical fruits. J. Agric. Food Chem., 31, 1016-1020. 1983.

Oste, R.E. and Friedman,M. The nutritional value and safety of heated amino acid-sodium ascorbate mixtures. J. Agric. Food Chem., 38, 1687 - 1690. 1990.

Oste, R. E., Brandon, D. L., Bates, A. H. and Friedman, M. Effect of the Maillard browning reactions of the Kunitz soybean trypsin inhibitor on its interaction with monoclonal antibodies. J. Agric. Food Chem., 38, 258-261. 1990.

Pariza, M. W., Ashoor, S. H. and Chu, F. S. Mutagens in heat -processed meat, baked and cereal products. Food Cosmet. Toxicol., 17, 429-430, 1979.

Pearce, K. N. and Friedman, M. The binding of copper (II) and other metal ions by lysinoalanine and related comounds and its significance for food safety. J. Agric. Food Chem., 36, 707-717. 1988.

Pearce, K. N., Karahalios, D. and Friedman, M. A ninhydrin assay for proteolysis during cheese ripening. J. Food Science, 53, 432-438. 1988.

Phillips, R. D. and Finley, J. W. (Editors). "Protein Quality and the Effects of Processing", Marcel Dekker, New York. 1989.

Powrie, W. D., Wu, C. and Mulund, W. P. Browning reaction systems as sources of mutagens and carcinogens. Env. Health Perspectives, 57, 47-54, 1986.

Pszczola, D. E. Production and potential food applications of cyclodex - trins. Food Technol., 96-100. 1988.

Ramel, C., Alekperov, U. K., Ames, B. N., Kada, T., Wattenberg, L.W. Inhibitors of mutagenesis and their relevance to carcinogenesis. Mutation Res., 168, 47. 1986.

Reuterswad, A. L. K. Skogg, and M. Jagestad, Mutagenicity of pan-fried bovine tissue in relation to their content of creatine, creatinine, monosaccharides, and free amino acids. Food Chem. Toxcol., 25, 755-762. 1987.

Rhee, K. S., Donnelly, K. C. and Ziprin, V. A. Reduction of mutagen formation in fried ground beef by glandless cottonseed flour. J. Food Protection, 50, 753-755. 1987.

Robertson, G. L. and Samaniego, C. M. L. Effect of initial dissolved oxygen levels on the degradation of ascorbic acid and the browning of lemon juice during storage. J. Food Sci., 51, 184-187. 1986.

Rogers, A.M. and Shibamoto, T. Mutagenicity of the products obtained from heated milk systems. Food Chem. Toxicol., 20, 259-263. 1982.

Scott, M. and Lawrence, A.A.G. Losses of ergot alkaloids during making of bread and pancakes. J. Agric. Food Chem., 30, 445-450. 1982.

Shibamoto, T. Occurrence of mutagenic products in browning model systems. Food Technology, 59-62. 1982.

Somogyi, J.C. and Muller, H.R. (Editors). "Nutritional Impact of Food Processing", Karger, Farmington, CT. 1989.

Spingarn, N. E., Slocum, L. A., and Weisburger, J. H. Formation of mutagens in cooked foods. II. Foods with high starch content. Cancer Letters, 9, 7-12. 1980.

Spingarn, W. E., Garvie-Gould, C. T. and Slocum, L. A. Formation of mutagens in sugar-amino acid model systems. J. Agric. food Chem., 31, 301-304. 1983.

Stich, H. F. and Rosin, M. P. Naturally occurring phenolics as antimutagenic and anticarcinogenic agents. In "Nutritional and Toxicological Aspects of Food Safety," M. Friedman, Ed., Plenum, New York, pp. 1-30. 1984.

Stich, H. F. and Rosin, M. P. Towards a more comprehensive evaluation of a genotoxic hazard in man. Mutation Res., 150, 43-50. 1985.

Stich, H. F., Stich, W., Rosin, M. P. and Powrie, W. D. Clastogenic activity of caramel and caramelized sugars. Mutation Res., 91, 129-136. 1981.

Stich, H. F., Rosin, M. P., Wu, C. H. and Powrie, W. D. Clastogenic activity of dried fruit. Cancer Letters, 12, 1-8. 1982.

Sugimura, T. Carcinogenicity of mutagenic heterocyclic amines formed during the cooking process. Mutation Res., 150, 33-41. 1984.

Sugimura, S. Studies on environmental chemical carcinogenesis in Japan. Science, 233, 312-318. 1986.

Sugimura, T., Takayama, S., Ohgaki, H., Wakabayashi, K. and Nagao, M. Mutagens and carcinogens formed by cooking meat and fish: heterocyclic amines. In "The Maillard Reaction in Food Processing, Human Nutrition and Physiology", P.A. Finot, H.U. Aeschbacher, R.F. Hurrell and R. Liardon, Eds., Birkhauser, Basel, Switzerland, pp. 323-334. 1990.

Toribio, J. L. and Lozano, J. F. Nonenzymatic browning in apple juice concentrate during storage. J. Food Sci. 49, 899-892. 1984.

Trammell, D. J., Dalsis, D. E. and Malone, C. T. Effect of oxygen on taste, ascorbic acid loss, and browning of pasteurized, single-strength orange juice. J. Food Sci., 51, 1021-1023. 1986.

Troll, W. Protease inhibitors: their role as modifiers of the carcinogenic process. In "Nutritional and Toxicological Significance of Enzyme Inhibitors in Foods", M. Friedman, Ed., Plenum Press, New York. Adv. Exp. Med. Biol., 199, 153-166. 1986.

Van der Hoeven, J. Ć. M. Mutagens in food products of plant origin. In "Mutagens in Our Environment," I. Knudsen, Ed., Alan R. Liss, Inc., pp. 327-338. 1982.

Von Wagenheim, B., Hanichen, T. and Ebersdobler, H. H. Histopathological investigation of the rat kidneys after feeding heat damaged proteins. Z. Ernahrungswissenschaft, 23, 219-229. 1984. (German).

Vuolo, L. L. and Schuessler, G. J. Review: putative mutagens and carcinogens in foods. Env. Mutagenesis, 8, 577-597. 1986.

Wakabayashi, K. Quantification of mutagenic and carcinogenic heterocyclic amines in cooked foods. In "Amino-Carbonyl Reactions in Food and Biological Systems," M. Fujimaki, Ed., Elsvier, Amsterdam, pp. 363-371. 1986.

Wall, M.E., Taylor, H., Perera, P. and Wani, M.C. Indoles in edible members of the Cruciferae. J. Natural Products, 51, 129-135. 1988.

Wallace, J. M. and Friedman, M. Inactivation of hemagglutinin in lima bean flour by N-acetylcysteine, pH, and heat. Nutr. Repts. Int., 32, 743-748. 1985.

Wang, Y. Y., Vuolo, L. L., Spingarn, L. E. and Weisburger, J. H. Formation of mutagens in cooked foods. V. The mutagen reducing effect of soy protein concentrates and antioxidants during frying of beef. Cancer Letters, 16, 179-189. 1982.

Wattenberg, L. W. Anticarcinogenic effects of several minor dietary components. In "Environmental Aspects of Cancer. The Role of Macro and Micro Components of Foods." E. L. Wynder, Ed., Food & Nutrition Press, Westport, CT, pp. 157-165. 1982.

Wedzicha, B.L. Review: chemistry of sulfur dioxide in vegetable dehydration. Int. J. Food Sc. Technol., 22, 433-450. 1987.

Weisburger, J. H. On the mechanisms relevant to nutritional carcinogenesis. Preventive Medicine, 16, 586-591. 1987.

Whitty, J. P. and Bjeldanes, L. F. The effects of dietary cabbage on xenobiotic-metabolizing enzymes and the binding of aflatoxin B_1 to hepatic DNA in rats. Food Chem. Toxicol., 25, 581-587. 1987.

Wilpart, M. Desmutagenic effects of N-acetylcysteine on direct and indirect mutagens. Mutation Res., 142, 169. 1985.

Wolfrom, M. L., Kashimura, N. and Horton, D. Factors affecting the Maillard browning reactions between sugars and amino acids. Studies on the nonenzymatic browning of dehydrated orange juice. J. Agric. Food Chem. 22, 796-800. 1974.

Yamaguchi, T. Reduction of induced mutability with biologically active quinones through inhibition of metabolic activation. Agric.Biol. Chem. 46, 2373-2375. 1982.

Yamaguchi, T. and Nakagawa, K. Mutagenicity and formation of oxygen radicals by trioses and glyoxal derivatives. Agric. Biol. Chem., 47, 2461-2465. 1983.

Yen, G. C. and Lai, Y. H. Influence of antioxidants on Maillard browning reaction in a casein-glucose model system. J. Food Sci., 52, 115-116, 1987.

Ziderman, I. I. and Friedman, M. Thermal and compositional changes of dry wheat gluten carbohydrate mixtures during simulated crust-baking. J. Agric. Food Chem., 33, 1096-1102. 1985.

Ziderman, I. I., Gregorski, K. S. and Friedman, M. Thermal analysis of protein-carbohydrate mixtures in oxygen. Thermochimica Acta, 114, 109-114, 1987.

Ziderman, I. I., Gregorski, K. S., Lopez, S. and Friedman, M. Thermal interaction of ascorbic acid and sodium ascorbate with proteins in relation to nonenzymatic browning and Maillard reactions. J. Agric. Food. Chem., 37, 1480-1486. 1989.

Additional References

Jiang, Z. and Ooraikul, B. Reduction of nonenzymatic browning in potato chips and french fried potatoes with glucose oxidase. J. Food Proces. and Preserv. 13, 175-186. 1989.

Oszmianski, J. and Lee, C. Y. Inhibition of polyphenol oxidase activity and browning by honey. J. Agric. Food Chem., 38, 1892-1895. 1990.

16

INHIBITION OF BROWNING BY SULFITES

B.L. Wedzicha, I. Bellion and S.J. Goddard

Procter Department of Food Science
University of Leeds, Leeds, LS2 9JT, Great Britain

ABSTRACT

The present state of understanding of the mechanisms by which sulfites inhibit browning reactions in food is reviewed. The difficulties of specifying the composition of sulfur(IV) oxospecies in sulfited foods arise from the existence of labile equilibria between SO_2, HSO_3^-, SO_3^{2-} and $S_2O_5^{2-}$, whose position depends on concentration, ionic strength and the presence of non-electrolytes. A proportion of the additive is also found in a reversibly bound form. The main reason why sulfites are able to inhibit a wide range of browning reactions is the nucleophilic reactivity of sulfite ion.

The mechanism of reactions between sulfite species and intermediates in the model Maillard browning reaction, glucose + glycine, are considered in depth and are supported by kinetic data. A most interesting feature is the fact that sulfites seem to catalyse the reactions they are added to control. Reaction products include 3,4-dideoxy-4-sulfohexosulose which is formed initially and polymeric substances arise from the reaction of sulfite species with melanoidins. It is found that melanoidins from glucose + glycine react with sulfite to such an extent that one sulfur atom is incorporated for every two glucose molecules used to make up the polymer.

The mechanisms of inhibition of ascorbic acid browning, enzymic browning and lipid browning are reviewed briefly. The known toxicological consequences of the formation of reaction products when sulfites are used for the control of Maillard browning give little cause for concern. Little is known of the implications of the formation of reaction products during the inhibition of other forms of browning. Consideration of the requirements for alternatives to sulfites is given.

INTRODUCTION

It is generally regarded that sulfite species (SO_2, HSO_3^-, SO_3^{2-}) are unique in their action as inhibitors of browning in food. The mechanisms of the various known browning reactions (enzymic, Maillard, ascorbic acid, lipid) are very different from one another, and it is tantalising to consider the features of these reactions and of the additive which cause it to be so versatile. A review on this subject is

Nutritional and Toxicological Consequences of Food Processing
Edited by M. Friedman, Plenum Press, New York, 1991

timely because there is considerable interest in possible replacements for sulfites (Friedman and Molnar-Perl, 1990; Molnar-Perl and Friedman, 1990a, 1990b) in applications where those suffering from asthma may be exposed to significant levels of gaseous sulfur dioxide (Giffon _et al._, 1989). A detailed understanding of the mode of action is essential to progress towards alternatives but we will show here that a study of the reactivity of sulfites in model browning systems also provides a new route to detailed information about the mechanisms of the browning reactions themselves.

When used in the context of a food additive, the term _sulfite_ or _sulfur dioxide_ is somewhat imprecise as it refers to a mixture of oxospecies of sulfur in oxidation state +4. The actual species present depend on many variables including the pH, ionic environment, water activity, presence of non-electrolytes and concentration of the medium in which they are dissolved. In some instances the term includes that additive which is reversibly bound to food components and no longer represents sulfur(IV) species; such bound additive is, however, released during the standard analytical procedures used to enforce the level of the additive in food. In order to simplify the specification of the amount of additive present in a food without the need to assess its detailed ionic distribution, it is normal to refer to the weight of sulfur dioxide equivalent to all the forms present. Many procedures for the analysis of the additive do, indeed, involve conversion of all sulfur(IV) species to gaseous SO_2. In order that the terminology be unambiguous here, the term S(IV) will be used to denote a mixture of oxospecies of sulfur in oxidation state +4 when it is not desired, or even possible, to specify the actual forms of the additive present. The terms sulfur dioxide, sulfite etc. will only be used when it is desired to refer to specific molecules or ions.

We begin this review with a critical examination of the nature of sulfur(IV) species present in solution as a function of pH, ionic environment, concentration and water activity.

SULFUR(IV) OXOSPECIES IN AQUEOUS SYSTEMS

Sulfites are salts of sulfurous acid, the existence of which has been questioned from experimental evidence (Davis and Chatterjee, 1975) and on thermodynamic grounds (Guthrie, 1979). It is unlikely that significant concentrations of H_2SO_3 exist in solution; dissolved sulfur dioxide is best regarded as $SO_2.H_2O$ in which any interaction between SO_2 and H_2O is very weak and undetectable spectroscopically. The dibasic acid ionizes according to:

$$SO_2.H_2O \rightleftharpoons HSO_3^- + H^+$$

$$HSO_3^- \rightleftharpoons SO_3^{2-} + H^+$$

with pK_a values of 1.89 and 7.18 (25 °C, zero ionic strength) for the first and second ionizations respectively (Huss and Eckert, 1977; Smith and Martell, 1976).

The fact that ionization of the acids changes the ionic charge and the number of ions present implies that the equilibria should be affected by ionic strength. The effects of 3 salts (NaCl, $NaNO_3$ and Na_2SO_4) on the pK_a of HSO_3^- are illustrated in _Figure 1_ (Wedzicha and Goddard, 1988; Wedzicha and Goddard, 1991). The data are consistent with predictions of the effect of ionic strength on activity coefficients of the ions in question, using the extended Debye-Huckel formula (Robinson and Stokes, 1965). The differences between the 3 salts tried are accounted for by choice of parameters in the Debye-Huckel equation and are not necessarily the result of specific salt effects. Whilst some ion pairing between Na^+ and SO_3^{2-} has been identified (Wedzicha and Goddard, 1991), the significance of this is not certain. The effect of ionic strength on the ionization of $SO_2.H_2O$ follows the same trend as in the case of HSO_3^- and is similarly explained in terms of general salt effects. A weak association between SO_2 and chloride ions is also evident.

The many non-electrolytes present in foods, including humectants, are likely to affect the solvation of ionic species, particularly at high non-electrolyte concentration. The effects of polyethylene glycol (PEG-400), glycerol, ethanol and sucrose on the pK_a of $SO_2.H_2O$ and HSO_3^- are illustrated in Figures 2 and 3 respectively (Wedzicha and Goddard, 1991). These effects are seen to be substantial and opposite in direction to the effects of moderate concentrations of salts. Unlike the effect of salt, it is not possible to quantitatively reconcile the effect of non-electrolyte with any basic or empirical theory, but it is possible that the greater tendency for charged species to be solvated is the reason why non-electrolytes tend to increase the bias for less highly charged molecules. Thus, SO_2 is favoured in the SO_2/HSO_3^- equilibrium and HSO_3^- in HSO_3^-/SO_3^{2-}. If one is to extrapolate these findings to concentrated foods, e.g. dehydrated or partially dehydrated fruits and vegetables, the view (Wedzicha, 1986) that these systems are at c. pH 5.5 and hence the predominating form of S(IV) is HSO_3^-, is very naive and possibly far from the truth. The effect of ions is to increase the concentration of sulfite ion at low to moderate ionic strength and perhaps to reduce it at very high ionic strength. On the other hand the effect of non-electrolytes would be to reduce the tendency for formation of sulfite ion at the pH of these foods. It is interesting that sucrose has little or no effect on the equilibria in question over the range of concentrations shown in Figures 2 and 3, but it has not yet proved possible to measure pK_a values in systems at the limit of solubility as found in dehydrated foods.

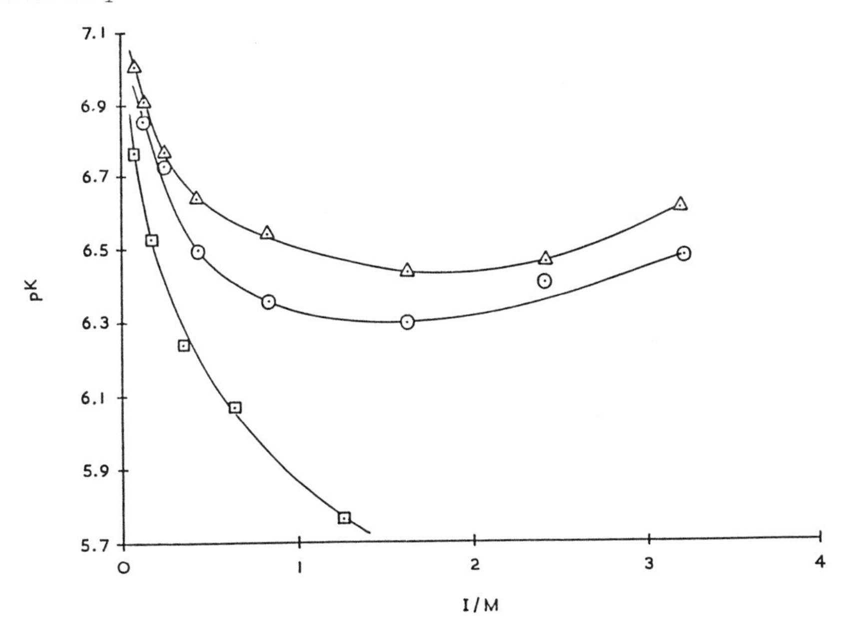

Figure 1. Effect of ionic strength, I, on the pK of a 50 mM solution of HSO_3^- at 30 °C. Ionic strength adjusted using ⊙ NaCl, △ $NaNO_3$, ⊡ Na_2SO_4.
Reproduced from Wedzicha and Goddard (1991).

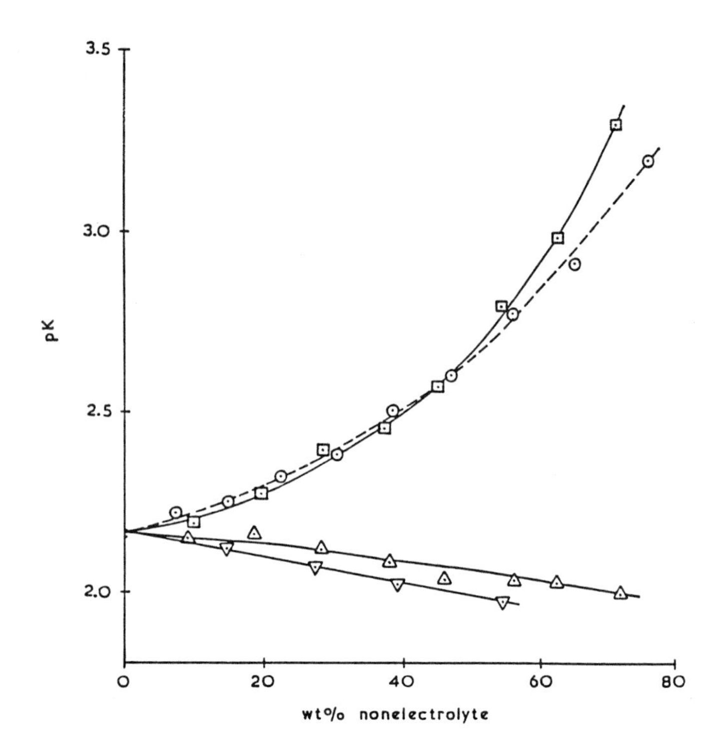

Figure 2. Effect of concentration of non-electrolyte on the pK of a 45 mM solution of $SO_2.H_2O$ at 30 ^oC. ⊙ Ethanol; ▲ Glycerol; ⊡ PEG-400; ▽ Sucrose.
Reproduced from Wedzicha and Goddard (1991).

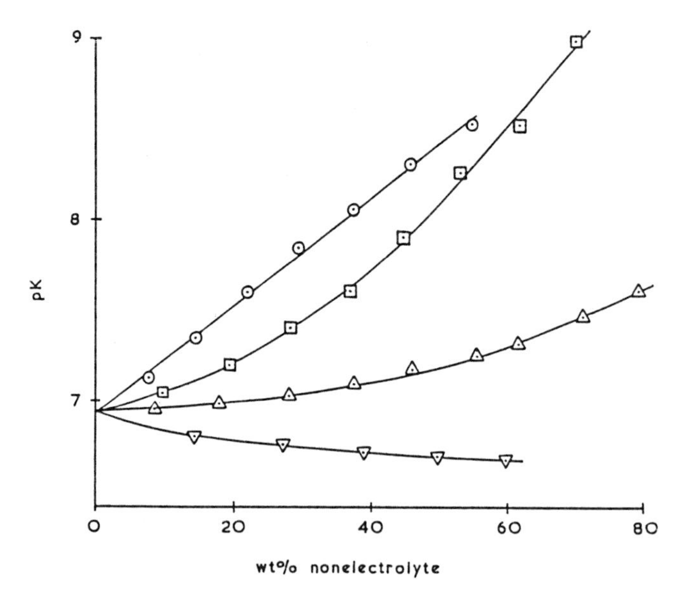

Figure 3. Effect of concentration of non-electrolyte on the pK of a 50 mM solution of $NaHSO_3$ at 30 ^oC. ⊙ Ethanol; ▲ Glycerol; ⊡ PEG-400; ▽ Sucrose.
Reproduced from Wedzicha and Goddard (1991).

The likely predominance of HSO_3^- in concentrated foods renders it important to examine the formation of metabisulfite ($S_2O_5^{2-}$) ion, according to:

$$2\ HSO_3^- \rightleftharpoons S_2O_5^{2-} + H_2O$$

In dilute solution, the equilibrium is well over to the left (equilibrium constant =0.033 l mol^{-1}) (Connick et al., 1982) but the degree of association of HSO_3^- increases with concentration for two reasons. First, the law of mass action predicts that the position of equilibrium be concentration-dependent. Secondly, the value of equilibrium constant is dependent on ionic strength, the variation being described approximately by,

$$K = 10^{(-1.398 + 0.35\sqrt{I})}$$

for $0.15\ M<I<4\ M$. . Additionally, there is the possibility that, in foods with low water activity, the concentration of water becomes limiting in the reaction, pushing the equilibrium further to the right. Solutions of metal bisulfites crystallize as the corresponding metabisulfites confirming this tendency for metabisulfite formation at high concentration and suggests, perhaps, also a lower solubility of the metabisulfite salt than the bisulfite salt.

Thus, in a dehydrated food, we expect S(IV) to exist mainly as metabisulfite ion in equilibrium with HSO_3^- and SO_3^{2-}. Much of the heterolytic reactivity of S(IV) in foods is due to the nucleophilicity of the sulfite ion (Wedzicha, 1984a). It has been shown that metabisulfite ion shows no such reactivity but it is possible that it might act as a particularly good general acid-base catalyst for other reactions (Ivanov and Lavrent'ev, 1967; Slae and Shapiro, 1978).

An assessment of the nucleophilic reactivity of the predominant ion, HSO_3^-, presents problems. This species can be envisaged as existing in one of two isomeric forms, i.e.,

H
|
S⁻ (O, O, O)   S⁻ (O, O, O — H)

Evidence for the existence of the S-H bonded ion in solution and the solid state is available from Raman spectroscopic data (Meyer et al., 1979) and X-ray crystallographic studies (Johansson et al., 1980). However, additional peaks in the Raman spectrum have been noted and assigned to the O-H bonded isomer (Connick et al., 1982) and good evidence confirming the existence of this species has been obtained from ^{17}O NMR studies of the ion in labelled water (Horner & Connick, 1986). Indeed, it is suggested that the equilibrium constant for the isomerization of the two ions, given as the concentration of O-H bonded isomer divided by the concentration of S-H bonded isomer is in the region of 5 at 25 ^oC. Molecular orbital calculations of the electron distributions in these ions originally disfavoured the formation of an O-H bonded species (Meyer et al., 1977) but refined calculations suggest that the two species are of comparable energy (Strömberg et al., 1983). If the bisulfite ion is indeed protonated on oxygen, it should still show nucleophilic reactivity through a lone pair of electrons on the sulfur atom. We (Wedzicha and Goddard, 1991) have found no evidence that bisulfite ion is a nucleophile towards malachite green, chosen as a good substrate for nucleophilic attack. It is, of course, possible that the single charge on HSO_3^- causes the ion to be less easily attracted to areas of partial positive charge on a molecule, making it seem a much less effective nucleophile than SO_3^{2-}.

Extrapolating once more to concentrated foods, we suggest that the formation of metabisulfite ion reduces the amount of SO_3^{2-} available and, hence the nucleophilic reactivity of the additive.

The situation in less concentrated systems should be less complicated. One must, nevertheless, exercize caution even in dilute systems such as wines because an ethanol content of some 12 % v/v is sufficient to alter the pK_a of $SO_2.H_2O$ by 0.3 units, thereby causing the concentration of $SO_2.H_2O$ in the wine at, say, pH 3.5 to be double that expected from the normally accepted pK_a value.

FREE AND BOUND SULFITES

It is conventional to consider S(IV) added to foods as existing in 3 forms (Wedzicha, 1984a):

(a) Free S(IV), that is, additive which is present as gaseous or aqueous SO_2, HSO_3^-, SO_3^{2-} or $S_2O_5^{2-}$;
(b) reversibly bound S(IV), that is, the additive which is in reversible combination, as hydroxysulfonate adducts, with carbonylic constituents of foods. Such bound S(IV) may be released during distillation analysis of the additive or by raising the pH of the system;
(c) irreversibly bound S(IV), which represents that additive which can no longer be measured as free or reversibly bound.

INHIBITION OF MAILLARD BROWNING

Preliminary observations. By far the most work on the inhibition of Maillard browning by S(IV) has been on the model system: glucose-glycine. It is generally recognised that the early stages of the browning reaction involve formation of the aldosylamine, glucosylglycine, followed by Amadori rearrangement to monofructose glycine (MFG). This, in turn, may react with a second molecule of glucose and undergo a second Amadori rearrangement to give difructose glycine (DFG). Both MFG and DFG decompose spontaneously to form osuloses, i.e. 3-deoxyhexosulose (DH) and 3,4-dideoxyhexosulos-3-ene (DDH), the latter resulting from dehydration of the former. Further dehydration of DDH gives rise to hydroxymethylfurfural (HMF) but this is much less reactive towards browning than its precursors and HMF is not regarded as an important intermediate in browning (Reynolds, 1963, 1965; McWeeny *et al.*, 1974).

The characteristics of the inhibition of Maillard browning by S(IV) are (i) a concentration-dependent induction period in color formation (Song and Chichester, 1967), (ii) a gradual increase in the stability of products formed as a result of reversible binding of S(IV) to carbonylic constituents of model systems and (iii) gradual decrease in the amount of S(IV) (free + reversibly bound) which may be recovered from model reaction mixtures (McWeeny *et al.*, 1969).

Irreversible binding of S(IV). The irreversible binding of S(IV) is accompanied by the formation of 3,4-dideoxy-4-sulfohexosulose (DSH) (Knowles, 1971) as a result of nucleophilic addition of sulfite ion to DDH, as follows:

$$\begin{array}{ccccc}
\text{CHO} & & \text{CHO} & & \text{CHO} \\
| & & | & & | \\
\text{C=O} & & \text{C-O}^- & & \text{C=O} \\
| & & \| & & | \\
\text{CH} & & \text{CH} & \xrightarrow{H^+} & \text{CH}_2 \\
\| & & | & & | \\
\text{CH} \;\; :SO_3^{2-} & \longrightarrow & \text{CHSO}_3^- & & \text{CHSO}_3^- \\
| & & | & & | \\
\text{CHOH} & & \text{CHOH} & & \text{CHOH} \\
| & & | & & | \\
\text{CH}_2\text{OH} & & \text{CH}_2\text{OH} & & \text{CH}_2\text{OH} \\
 & & & & \\
\text{DDH} & & & & \text{DSH}
\end{array}$$

This sulfonate is much less reactive towards browning than DH or DDH, presumably because it is unable to form the α,ß-unsaturated carbonyl compound which is the most reactive known intermediate in browning. DSH is, however, not completely unreactive in model Maillard systems as illustrated in Figure 4 (Wedzicha and Garner, 1991). This shows the formation of DSH and loss of S(IV) as a function of time, using the specific detection of radio-labelled S(IV) and DSH. The formation of DSH is seen to mirror the loss of S(IV), but the former appears to undergo a slow reaction observed when S(IV) has run out. The reactivity of DSH may

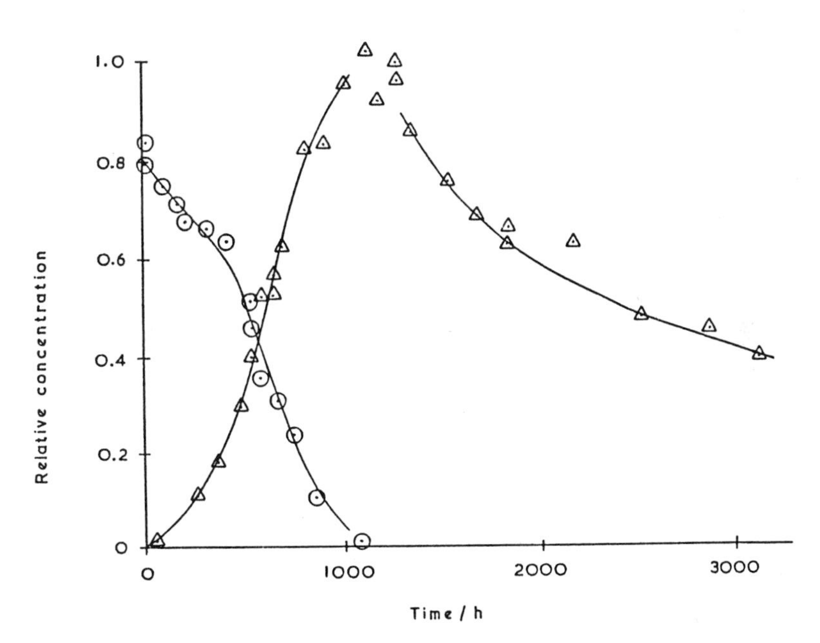

Figure 4. Time-dependent relative concentration of S(IV) and relative ^{35}S-activity due to DSH when a mixture of glucose (1 M), glycine (0.5 M) and ^{35}S(IV) (0.05 M), initial pH 5.5, is allowed to react at 55°C. The concentration of DSH is proportional to its ^{35}S-activity. ⊙ S(IV); △ ^{35}S-DSH. Reproduced from Wedzicha and Garner (1991).

be more critically appraised by adding ^{35}S-labelled DSH to a mixture of glucose, glycine and S(IV) at zero time and observing the distribution of radio-labelled products. The progress of this reaction is illustrated in Figure 5 where analyses were carried out soon after the start of the reaction (72 h), when all the S(IV) had run out (1540 h) and after the mixture had undergone considerable browning (3530 h). The only discreet

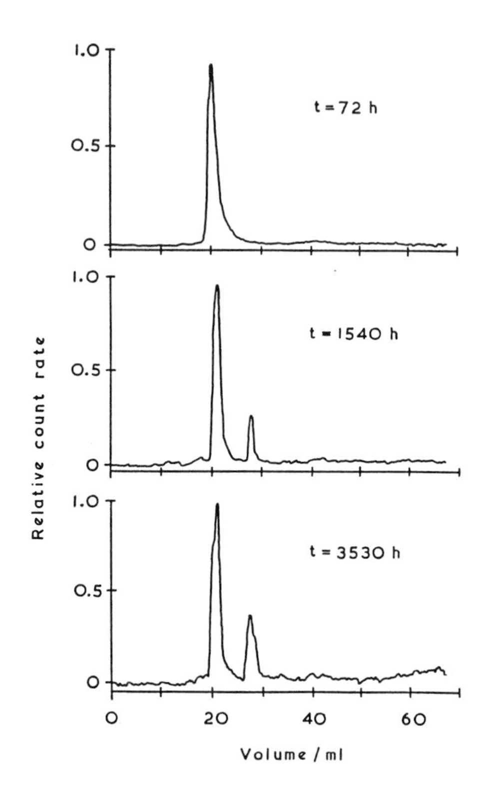

Figure 5. Chromatograms of ^{35}S-labelled reaction products formed when a mixture of glucose (1 M), glycine (0.5 M) and S(IV) (0.05 M) is spiked with ^{35}S-labelled DSH at time zero. Initial pH 5.5, 55°C. Chromatograms shown are for reaction mixtures at times 72 h, 1540 h and 3530 h.
Reproduced from Wedzicha and Garner (1991).

new product formed has a retention volume at c. 28 ml and is the metasaccharinic acid obtained by hydration of DSH. At long times the sulfonate is converted to products which could not be resolved by hplc and are suspected as being polymeric in nature. Thus, it is possible that DSH is incorporated into high molecular weight polymers (melanoidins) formed as a result of browning.

It has been verified unequivocally (Wedzicha and Garner, 1991), by spiking model reaction mixtures with ^{14}C-labelled DH, that all the S(IV) which is bound irreversibly undergoes reaction with intermediates derived from DH. Comparison of the distribution of ^{14}C-labelled products with that of ^{35}S-labelled products when mixtures of glucose, glycine and $^{35}SO_3^{2-}$ are allowed to react shows no additional ^{35}S-labelled products. The primary reaction product is also seen to be DSH.

Kinetic model. Typical concentration-time data for the reversible and irreversible binding of S(IV) in mixtures of glucose, glycine and S(IV) are illustrated in Figure 6 (McWeeny *et al.*, 1969). Initially (time zero), a substantial amount of the S(IV) is present as glucose hydroxysulfonate and the extent of this reversible binding seems to be unaffected by the presence of glycine despite the fact that glycine forms a Schiff's base with the reducing sugar. The equilibrium constant for the formation of glucose hydroxysulfonate is close to unity suggesting only a weak interaction compared with simple aldehydes and ketones whose hydroxysulfonates have formation constants of the order 10^5-10^6 l mol^{-1} (Wedzicha, 1984a). In this and similar S(IV)-inhibited model Maillard reactions, the concentration of glucose generally far exceeds that of S(IV) and the reversible binding of glucose does little to reduce its effective concentration in the Maillard reaction. Unless the concentration of S(IV) is high it is unlikely, therefore, that the formation of hydroxysulfonates of reducing sugars is the method by which browning is inhibited.

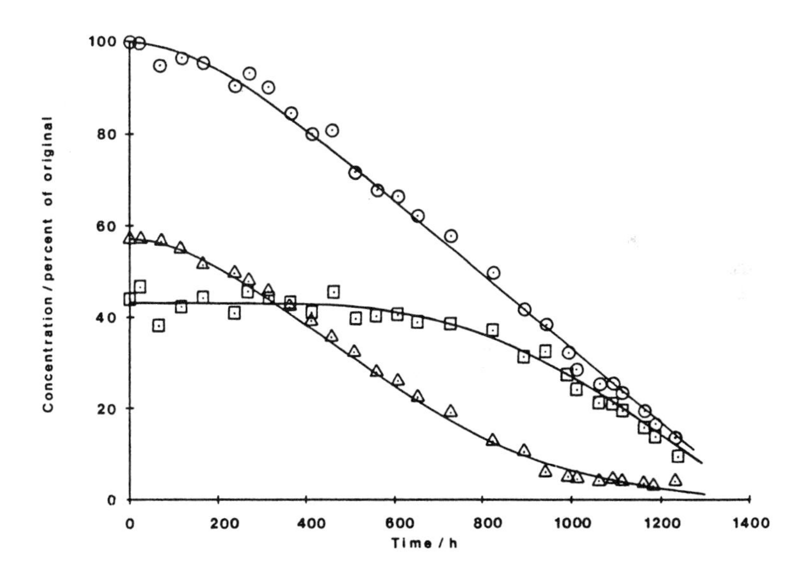

Figure 6. Concentration-time data for S(IV) in the sulfite-inhibited Maillard reaction of glucose and glycine after McWeeny *et al.* (1969). Reaction conditions: [glucose] = 1.0 M; [glycine] = 0.5 M; [S(IV)] = 0.039 M; pH 5.5; 55 °C. ⊙ Total S(IV); ▲ free S(IV); ⊡ reversibly bound S(IV).

The concentration of reversibly bound S(IV) remains relatively constant as the concentration of total (free + reversibly bound) S(IV) begins to fall. Thus, the stability of hydroxysulfonate adducts appears to increase and it is argued that carbonyl compounds which bind S(IV) more strongly than does glucose are being formed. Obvious candidates for such compounds are the 3-deoxyosuloses and early suggestions regarding their binding power considered that the α-dicarbonyl moiety might cause the formation of particularly stable dihydroxysulfonates (Knowles, 1971; Wedzicha and Imeson, 1977). This is now considered unlikely (Wedzicha and Smith, 1987) for steric reasons but the possible involvement of 3-deoxyosuloses in reversible binding of S(IV) has to be critically assessed. This may be done by considering the kinetics of the S(IV)-inhibited Maillard reaction in more detail.

The simplest mechanism which gives rise to the concentration-time behaviour for total S(IV) shown in Figure 6 is a two-stage consecutive process. The major part of the concentration-time profile is linear and, therefore, consistent with a reaction of zero order with respect to S(IV). As an initial hypothesis, the following simple scheme may be proposed (Wedzicha, 1984b):

$$\text{Glycine + glucose} \xrightarrow{\text{slow}} \text{I1} \xrightarrow{\text{slow}} \text{I2} \xrightarrow[\text{S(IV)}]{\text{fast}} \text{Product}$$

The early, "induction" phase of the reaction (up to 200 h in Figure 6) is the build-up of the concentration of intermediate I1 and if the conversion of this to intermediate I2 is non-zero order, the concentration of I1 will eventually reach a steady state (rate of formation = rate of loss). The zero order behaviour with respect to S(IV) is taken account of by the fast conversion of I2 to product which is expected to be DSH. The conversion of glucose to DSH involves 1 mole of glucose and 1 mole of S(IV) and if no other reaction is taking place then provided that the concentration of glucose is much greater than the concentration of S(IV), one can regard the glucose concentration as approximately constant and unchanged even at the stage when most of the S(IV) has undergone reaction. If glycine concentration is also relatively unaffected by this reaction, the rate of formation of I1 should be constant during the course of a kinetic run. Thus, the rate of formation of I2 and of product would be constant once steady state conditions are established. This is seen to be the case and the kinetic model can now be investigated further.

The effect of glucose and glycine on the constant rate phase of the loss of S(IV), is found to be of first order with respect to each reactant, consistent with one molecule of each being involved in the rate determining step for the formation of I1. It is interesting and very relevant to compare this result with the statement (Labuza et al., 1977) that the reaction which produces precursors of browning is of first order with respect to each of reducing sugar and amine. Separate experiments (Wedzicha and Kaban, 1986) to investigate the kinetics of the reaction of isolated DH with S(IV) show that the reaction is of first order with respect to DH and zero order with respect to S(IV) and the rate constant is of similar value to that for the conversion of I1 to I2 measured from data such as those shown in Figure 6. We therefore conclude that I1 is DH but the exact nature of I2 is speculative. It might, for example, be DDH and the rate determining step in the conversion of DH to DSH could be the dehydration of DH. It has proved too difficult to measure the rate of conversion of DH to DDH or the conversion of DDH to DSH in isolation.

It is interesting that the conversion of DH to DSH is catalysed by glycine but more exciting is the fact that the constant rate for loss of S(IV) in the glucose-glycine-S(IV) reaction depends on initial S(IV)-concentration as illustrated in Figure 7 (Wedzicha and Vakalis, 1988). On this basis, it is proposed that the formation of I1 involves two parallel reactions one independent of S(IV) and the other catalysed by S(IV). The latter reaction is of first order with respect to S(IV) and the presence of a non-zero order process throws into doubt the original suggestion that the overall binding of S(IV) is of zero order with respect to S(IV). However, when the rate equations for the contributing reactions are integrated numerically, the rate of binding of S(IV) is still found to be constant over the greater part of the reaction, and consistent with the observed kinetics.

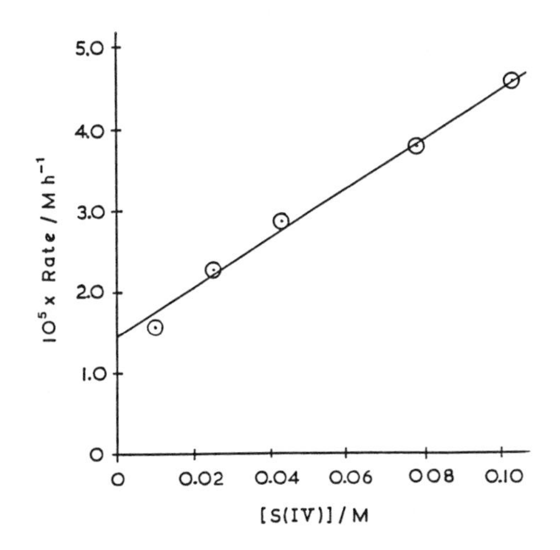

Figure 7. Effect of initial [S(IV)] on the constant rate of loss of total [S(IV)] in the sulfite-inhibited Maillard reaction of glucose and glycine. Reaction conditions: [glucose] = 1.0 M; [glycine] = 0.5 M; [S(IV)] = 0.01-0.10 M; pH 5.5; 55°C. Reproduced from Wedzicha and Vakalis (1988).

The nature of the catalytic step is not known but it is tempting to suggest that it is the result of general acid-base catalysis. It is known that weak acids and their salts affect the rate of the Maillard reaction (Reynolds 1959) and it has been reported that sulfur(IV) species can act as effective acid-base catalysts in some reactions (Ivanov and Lavrent'ev, 1967; Slae and Shapiro, 1978). Indeed, sulfite ion is somewhat better as a catalyst than acetate or phosphate species which are often regarded as moderately good catalysts. Metabisulfite ion is an even better general acid-base catalyst than sulfite ion and the possible presence of this species in concentrated foods, and particularly those at low water activity, could be of some significance to the causes and inhibition of Maillard browning in S(IV)-treated dehydrated fruit and vegetable products. It is tempting to suggest that this might be one of the reasons why it is necessary to treat these foods with high levels of the additive for successful dehydration but some recent evidence indicates that metabisulfite ion formation may not be a dominant factor in practice. Thus, in the dehydration of cabbage, carrot and potato, the yield of C-sulfonate is proportional to the amount of S(IV) added (Wedzicha, 1987). Catalysis of the formation of DSH by metabisulfite ion would have been reflected in a non-linear relationship between yield and S(IV)-concentration, tending towards second order behaviour with respect to S(IV) if the catalytic reaction were the dominant one. It is relevant to emphasise that S(IV) actually catalyses the reaction it is added to control and the success of its control of browning rests with the "mopping up" of reactive intermediates in color formation, in a fast, zero order, reaction. Thus, S(IV) is effective even as the concentration of free S(IV) becomes small but the efficiency of S(IV) should be greater at lower concentrations where the extent of catalytic action is reduced.

The amount of S(IV) bound reversibly during the kinetic experiment shown in Figure 6 is completely accounted for by the formation of the hydroxysulfonates of glucose and of DSH (Wedzicha, 1984b). The equilibrium constant for the formation of DSH hydroxysulfonate, obtained by measurements using pure samples of DSH (Wedzicha *et al*., 1985), is identical to that which may be inferred from the kinetic data. It is interesting that there is no evidence from kinetic data of the formation of a hydroxysulfonate of DH in the Maillard system despite it having the same functional grouping as DSH. On the other hand, isolated pure DH appears to bind S(IV) as strongly as does DSH (Wedzicha and Kaban, 1986). Some circumstantial evidence also points to there being no DH-hydroxysulfonate formed. It is found, for example, that the rate of browning in glucose-glycine-S(IV) mixtures, once the S(IV) has run out, is similar to that had no S(IV) been added in the first place. Had DH-hydroxysulfonate been formed, this would have accumulated as an unreactive by-product, only to be released once the free S(IV)-concentration became small. This release of DH would cause its concentration to far exceed any which may be found in the uncatalysed, uninhibited Maillard reaction, and would give rise to a higher rate of browning at the end of the S(IV)-inhibited process.

Any reason why isolated DH might react differently from DH formed as part of the Maillard reaction in some respects and yet show the same overall rate constant for its conversion to DSH is intriguing. It is speculated that the Maillard intermediate might be formed as a 2-enol- rather than 2-keto-compound, and undergo conversion to DDH as such. On the other hand, the preparations of DH which give rise to samples of excellent purity involve the decomposition of a bis(benzoyl)hydrazone of DH (Madson and Feather, 1981), which presumably gives rise to the α-dicarbonyl isomer. It is not unreasonable to suggest that the enol form might bind S(IV) less strongly, but it should also be appreciated that there will exist several isomeric forms of DH and the actual distribution of these in the various situations considered here has not yet been elucidated.

It is, of course, possible that weak reversible DH-S(IV) interactions exist but are not detectable kinetically. The position of equilibrium for the formation of hydroxysulfonate, given by

$$K = \frac{[\text{hydroxysulfonate}]}{[C{=}O]\ [HSO_3^-]}$$

is concentration-dependent and it is conceivable that even in the relatively concentrated model systems used in the experiments reported here, the concentration of DH and S(IV) is insufficiently great to cause a significant formation of such unstable hydroxysulfonate. On the other hand, the concentrations of glucose and S(IV) in, say, dehydrated fruits and vegetables probably exceed their normal solubilities in water. Kinetic experiments have not yet been carried out using model systems under such conditions.

It is reasonable criticism to say that a relatively straight forward mechanism is being proposed for the inhibition of a reaction which is perhaps one of the most complicated reactions identified in foods. The ideas presented here are defensible because the kinetic approach involves identification of rate determining steps and always gives a simplified view of the mechanism. The consecutive reaction scheme proposed above has also been rigorously tested by spiking glucose-glycine-S(IV) reaction mixtures with ^{14}C-labelled DH and ^{35}S-labelled S(IV) (Wedzicha & Garner, 1991). The addition of ^{14}C-DH when the reaction has reached its steady state allows one to estimate the turnover of DH, its steady state concentration and kinetics of reaction with S(IV) in the Maillard system, by direct measurement instead of inference from the kinetics of the overall binding of S(IV). The turnover of DH is in keeping with the proposed mechanism and for a reaction mixture containing initially the concentrations of glucose, glycine and S(IV) used to obtain the data illustrated in Figure 6, the measured DH concentration is 6.0 mM compared with 6.0 mM obtained from the kinetic data alone and 5.8 mM obtained after fitting, to our model, kinetic data published over 20 years ago (McWeeny *et al.*, 1969).

Reaction with melanoidins. A well known observation is that S(IV) is able to partially bleach the color formed during Maillard browning reactions (McWeeny *et al.*, 1974). In the case of glucose-glycine browning, the pigments which have molecular weights in excess of *c.* 10,000 daltons are composed approximately of 5 molecules of glucose per 4 molecules of glycine, incorporated into the polymer with the elimination of 3 molecules of water per molecule of glucose and 0.17 molecules of CO_2 per molecule of glycine (Wedzicha and Kaputo, 1987). Interestingly, this composition is relatively insensitive to the composition of the mixture of glucose and glycine used to form the melanoidins (Wedzicha and Kaputo, unpublished). When such melanoidins are allowed to react with S(IV) for >39 days (pH 5.5, 40 ^{o}C) up to 1 atom of sulfur is incorporated for every 2 molecules of glucose used to form the polymer (Wedzicha and Kaputo, 1987). This binding is the same as the frequency of C=C bonds in the melanoidin; the chromophore is not extensively conjugated with respect to olefinic C=C. It is likely that the double bond which is reactive towards S(IV) is part of one type of chromophore, the bleaching of the melanoidin by S(IV) being only partial. Evidence suggests that the binding of S(IV) is irreversible and a sulfonated polymer should result.

Application to study of Maillard browning mechanism. The elucidation of the mechanism of binding of S(IV) in the glucose-glycine-S(IV) reaction opens a new possibility of following the mechanism of the Maillard reaction. Of the many reported investigations of Maillard browning a high proportion fall into one of two categories: either they involve the measurement of color formation or attempts have been made to isolate reaction products and intermediates. The former approach uses a technique which cannot be easily related to the progress of specific reactions because color is due to a complex mixture of products. The latter approach does not necessarily lead to the study of important reaction intermediates or products and much emphasis could be placed on substances which are easily identifiable or have distinct chromatographic properties. On the other hand the approach advocated in this work is to restrict the observed reaction to a small number of definable steps and interrupting the sequence of events in browning by removing an intermediate whose importance is unambiguous. We have yet to demonstrate by our approach whether or not the formation of DH and intermediate I2 is rate determining in the formation of color. However, it is evident that our approach can provide fundamental data on the sequence of events which give rise to color-forming intermediates by separating the S(IV)-dependent and S(IV)-independent processes during the analysis of kinetic data. We are currently using this technique to investigate the effects of pH and water activity on the reactions in question and extending the approach to fructose-amino acid systems.

INHIBITION OF ASCORBIC ACID BROWNING

Anaerobic browning. In the absence of oxygen ascorbic acid decomposes spontaneously, losing carbon dioxide and water, to form 3-deoxypentosulose (DP) which may subsequently dehydrate to furfural (Kurata and Sakurai, 1967a). This decomposition is accompanied by browning, which is made more intense by the presence of amino acids. Colour formation is retarded by the addition of S(IV) and the characteristics of the inhibition of browning resemble those of the Maillard reaction (Wedzicha and McWeeny, 1974a). During the course of inhibition S(IV) binds irreversibly to intermediates; the kinetics of this binding in the ascorbic acid-glycine-S(IV) reaction are of zero order with respect to S(IV) and glycine, and of first order with respect to ascorbic acid (Wedzicha and Davies, unpublished). It is not surprising, therefore, that color formation is only delayed by S(IV), and the "induction" period for color formation depends on S(IV) concentration. The main difference between ascorbic acid and Maillard browning is, therefore, that the latter depends on the presence of the amino acid, whereas the decomposition of ascorbic acid is spontaneous. Also, there is no "induction" phase in the kinetics of irreversible binding of S(IV); in systems containing [ascorbic acid]>>[S(IV)] the rate of loss of S(IV) is constant from time zero.

It is established that irreversible binding is the reaction of $H^+SO_3^{2-}$ with DP, probably a result of nucleophilic attack by SO_3^{2-} on 3,4-dideoxypentosulos-3-ene as illustrated for the six carbon analog, DDH, above. The product of this reaction, 3,4-dideoxy-4-sulfopentosulose, has been characterized and identified in model ascorbic acid browning systems (Wedzicha and McWeeny, 1974a, Wedzicha and Imeson, 1977) as well as in dehydrated cabbage (Wedzicha and McWeeny, 1974b). Partial bleaching of melanoidins formed as a result of ascorbic acid browning is similar to that found using melanoidins from glucose and glycine (Wedzicha and McWeeny, 1974a), but the nature of these polymers and the extent of their reaction with S(IV) has not yet been established.

Oxidative browning In the presence of air ascorbic acid undergoes oxidation to dehydroascorbic acid, by way of the ascorbyl radical (Mushran and Agrawal, 1977). Dehydroascorbic acid browns on its own much more readily than does ascorbic acid, the pathway involving opening of the lactone ring to 2,3-diketogulonic acid followed by degradation (Kurata and Sakurai, 1967b). The mechanism of color formation is not known but it is likely to involve the condensation of α-dicarbonyl compounds formed from 2,3-diketogulonic acid with each other or with amino compounds if these are also present.

The oxidative browning of ascorbic acid is inhibited by S(IV) though the mechanism of its inhibition is not known. It is possible that sulfite ion might reduce the partially oxidized ascorbyl radical (Kalus and Filby, 1977) but the browning of dehydroascorbic acid, once formed, is likely to be inhibited by formation of the hydroxysulfonate adduct. The latter is quite stable with a formation constant of 1750 $l\ mol^{-1}$ (Wisser *et al.*, 1970).

INHIBITION OF ENZYMIC BROWNING

Enzymic browning is the term used to describe the formation of high molecular weight pigments by the polymerisation of *ortho*-quinones formed by the enzymic oxidation of *ortho*-diphenols by polyphenol oxidase. The mechanism of the inhibition of enzymic browning by S(IV) is by reaction of quinone intermediates with S(IV) rather than the inhibition of the enzyme itself (Embs and Markakis, 1965). In the case of the simplest possible substrate, catechol, inhibition of browning involves nucleophilic attack by sulfite ion in position 4 of the *ortho*-quinone as follows:

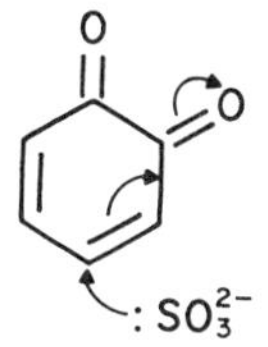

to give 4-sulfocatechol after subsequent addition of a hydrogen ion (Wedzicha *et al.*, 1987). The quinone has, therefore, been reduced in this reaction. It is interesting to find that 4-sulfocatechol is unreactive towards polyphenol oxidase and the reaction of catechol with sulfite ion under such oxidising conditions does not, therefore, proceed beyond the monosulfonate.

Natural substrates for enzymic browning are substituted catechols, often with the substituent in position 4. The reaction of 3,4-dihydroxyphenylalanine with sulfite ion in the presence of mushroom tyrosinase probably gives rise to the 6-sulfonate (Wedzicha and Churchill, unpublished) which is unreactive towards the enzyme. Thus, substrates for enzymic browning may be converted into products which are unreactive and the effect of S(IV) in preventing enzymic browning is permanent if all the substrate is converted to sulfonate. The stability of the products means that it should be possible to detect their presence in fresh foods which have been treated with S(IV) but which no longer contain the additive (e.g. S(IV) had undergone reaction to sulfate or to organic products).

INHIBITION OF LIPID BROWNING

Lipid browning is the least well studied of the browning reactions and is believed to be the polymerization or reaction of carbonylic oxidation products of unsaturated lipids with amino compounds. It is envisaged that S(IV) inhibits these browning reactions either by forming hydroxysulfonates with these carbonylic intermediates or by preventing their formation by reducing hydroperoxide intermediates in lipid oxidation (Davies, 1961). It is possible, also, that hydroperoxides themselves interact with amines to form colored products (Shimasaki *et al.*, 1977).

In contrast to its role as an antioxidant described above, S(IV) is capable of causing the oxidation of unsaturated organic compounds (Kaplan *et al.*, 1975; Inouye *et al.*, 1980; Lizada and Yang, 1981; Wedzicha and Lamikanra, 1983) including unsaturated fats, fatty acids and carotenoids. Whilst such oxidation should theoretically occur in S(IV)-treated foods, its significance is not established.

TOXICOLOGICAL IMPLICATIONS

Based on experiments using rats there does not appear to be any toxicological hazard associated with the formation of DSH in foods (Walker *et al.*, 1983a, 1983b) The product is metabolically inert and significantly less toxic than S(IV). The other known reaction products arising from reaction between S(IV) and Maillard browning intermediates, hydroxysulfonate adducts, are thought to be relatively stable under acid stomach conditions but are decomposed in the alkaline environment of the gut releasing S(IV). The hazard associated with this form of S(IV) is considered to be no greater than with free S(IV) (Gibson and Strong, 1976). It may be significant that S(IV) also inhibits bovine trypsin (Wedzicha and Luck, 1987) by cleaving disulfide bonds in a simple non-competitive mechanism ($K_i \approx 1$ mM) (Wedzicha and Abdel Karim, unpublished). The reversibility of this enzyme inhibition is not certain but it is likely that S(IV) also interferes with the process of trypsin activation. The possibility that S(IV) interferes with trypsin and other enzymes *in vivo* should now be explored.

There are no toxicological data on other compounds known to be formed as a result of inhibition of browning reactions.

REQUIREMENTS FOR ALTERNATIVES TO SULFITES

It has been shown in this review that the important property of S(IV) which leads to its versatility as an inhibitor of browning reactions is the nucleophilic reactivity of sulfite ion. In principle, inhibition of browning could be achieved with any powerful nucleophile and thiols appear to be the most promising choice (Friedman and Molnar-Perl, 1990; Molnar-Perl and Friedman, 1990a, 1990b). Earlier work (Song and Chichester, 1967) suggests that the characteristics of the inhibition of Maillard browning by a wide range of thiols are similar to those of inhibition by S(IV) but the mechanism of the reaction between DH and simple thiols is unexpectedly different. For example the stoichiometry of the DH-mercaptoethanol reaction is 1:2 (Wedzicha and Edwards, 1991). Thiols form relatively unstable adducts (hemimercaptals) with carbonyl compounds; the rendering of carbonyl groups inactive by forming such adducts is likely to be less effective when thiols are used in place of S(IV) but their action may well be more efficient than S(IV) at high concentration when perhaps less catalysis of the Maillard reaction might be observed. The possible use of naturally occurring thiols, or their derivatives, is attractive but toxicological studies on any reaction products formed are nevertheless essential.

ACKNOWLEDGEMENT

We acknowledge generous support from the Agricultural and Food Research Council for a Research Assistantship to S.J.G. and a Research Studentship to I.B.

REFERENCES

Connick, R.E., Tam, T.M. & Von Deuster, E. (1982). Equilibrium constant for the dimerization of bisulfite ion to form $S_2O_5^{2-}$. *Inorg. Chem.*, 21, 103-7.

Davies, A.G. (1961). *Organic Peroxides*, Butterworths, London.

Davis, A.R. and Chatterjee, R.M. (1975). A vibrational-spectroscopic study of the SO_2-H_2O system. *J. Solution Chem.*, 4, 399-412.

Embs, R.J. and Markakis, P. (1965). The mechanism of sulfite inhibition of browning caused by polyphenol oxidase. J. Food Sci., 30, 753-758.

Friedman, M. and Molnar-Perl, I. (1990). Inhibition of browning by sulfur amino acids. 1. Heated amino acid-glucose systems. J. Agric. Food Chem., 38, 1642-1647.

Gibson, W.B. and Strong, F.M. (1976). Metabolism and elimination of α-hydroxyethanesulfonate by rats. Food Cosmet. Toxicol., 14, 41-43.

Giffon, E., Vervloet, D. and Charpin, (1989). J. Suspicion sur les sulfites. Rev. Mal. Respir., 6, 303-310.

Guthrie, P.J. (1979). Tautomeric equilibria and pK_a values for 'sulfurous acid' in aqueous solution. A thermodynamic analysis. Can. J. Chem., 57, 454-457.

Horner, D.A. and Connick, R.E. (1986). Equilibrium quotient for the isomerization of bisulfite ion from HSO_3^- to SO_3H^-. Inorg. Chem., 25, 2414-2417.

Huss Jr, A. & Eckert, C.A. (1977). Equilibria and ion activities in aqueous sulfur dioxide solutions. J. Phys. Chem., 81, 2268-70.

Inouye, B., Morita, K., Ishida, T and Ogata, M. (1980). Cooperative effect of sulfite and vanadium compounds on lipid peroxidation. Toxicol. Appl. Pharmacol., 53, 101-107.

Ivanov, M.A. and Lavrent'ev, S.P. (1967). Mutarotation of glucose. Izv. Vyssh. Ucheb. Zaved., Les. Zh., 10, 138-141, through Chemical Abstracts, 68, 96959x.

Johansson, L.G., Linqvist, O. and Vannerberg, N.G. (1980). The structure of caesium hydrogen sulfite. Acta Cryst.,B36, 2523-2526.

Kalus, W.H. and Filby, W.G. (1977). The effect of additives on the free radical formation in aqueous solutions of ascorbic acid. Int. J. Vit. Nutr. Res., 47, 258-264

Kaplan, D., McJilton, C. and Luchtel, D. (1975). Bisulfite induced lipid oxidation. Arch. Environ. Health., 30, 507-509.

Knowles, M.E. (1971). Inhibition of non-enzymic browning by sulfite: identification of sulfonated products. Chem. Ind. (London), 910-911.

Kurata, T. and Sakurai, Y. (1967a). Degradation of L-ascorbic acid and mechanism of non-enzymic browning reaction. Part II. Non-oxidative degradation of L-ascorbic acid including the formation of 3-deoxy-L-pentosone. Agric. Biol. Chem., 31, 170-176.

Kurata, T. and Sakurai, Y. (1967b). Degradation of L-ascorbic acid and mechanism of non-enzymic browning. Part III. Oxidative degradation of L-ascorbic acid (degradation of dehydro-L-ascorbic acid). Agric. Biol. Chem., 31, 177-184.

Labuza, T.P., Warren, R.M. and Warmbier, H.C. (1977). The physical aspects with respect to water and non-enzymatic browning. Adv. Exp. Biol., 86B, 379-418.

Lizada, M.C.C. and Yang, S.F. (1981). Sulfite induced lipid oxidation. Lipids, 16, 189-194.

Madson, M.A. & Feather, M.S. (1981). An improved preparation of 3-deoxy-D-erythro-hexos-2-ulose via the bis(benzoylhydrazone) and some related constitutional studies. Carbohydr. Res., 94, 183-91.

McWeeny, D.J., Biltcliffe, D.O., Powell, R.C.T. and Spark, A.A. (1969). The Maillard reaction and its inhibition by sulfite. J. Food Sci., 34, 641-643.

McWeeny, D.J., Knowles, M.E. and Hearne, J.F. (1974). The chemistry of non-enzymic browning in foods and its control by sulfites. J. Sci. Food Agric., 25, 735-746.

Meyer, B., Peter, L. and Shaskey-Rosenlund, C. (1979). Raman spectra of isotopic bisulfite and disulfite ions in alkali salts and aqueous solutions. Spectrochim. Acta, 35A, 345-354.

Meyer, B., Peter, L. and Spitzer, K. (1977). Trends in charge distributions in sulfanes, sulfanesulfonic acids, sulfanedisulfonic acids and sulfurous acid. Inorg. Chem., 16, 27-33.

Molnar-Perl, I. and Friedman, M. (1990a). Inhibition of browning by sulfur amino acids. 2. Fruit juices and protein-containing foods. J. Agric. Food Chem., 38, 1648-1651.

Molnar-Perl, I. and Friedman, M. (1990b). Inhibition of browning by sulfur amino acids, 3. Apples and potatoes. J. Agric. Food Chem., 38, 1652-1656.

Mushran, S.P. and Agrawal, M.C. (1977). Mechanistic studies on the oxidation of ascorbic acid. J. Sci. Ind. Res., 36, 274-283.

Reynolds, T.M. (1959). Chemistry of non-enzymic browning. III. Effect of bisulfite, phosphate and malate on the reaction of glycine and glucose. Aust. J. Chem., 12, 265-274.

Reynolds, T.M. (1963). Chemistry of non-enzymic browning. I. The reaction between aldoses and amines. Adv. Food res., 12, 1-52.

Reynolds, T.M. (1965). Chemistry of non-enzymic browning. II. Adv. Food Res., 14, 167-283.

Robinson, R.A. & Stokes, R.H. (1965). Electrolyte Solutions. (2nd edn (revised)), Butterworths, London.

Shimasaki, H., Privet, O.S. and Hara, I. (1977). Studies on the fluorescent products of lipid oxidation in aqueous emulsion with glycine on the surface of silica gel. J. Amer. Oil Chem. Soc., 54, 119-123.

Slae, S. and Shapiro, R. (1978). Kinetics and mechanism of the deamination of 1-methyl-5,6-dihydrocytosine. J. Org. Chem., 43, 1721-1726.

Smith, R.M. and Martell, A.E. (1976). Critical Stability Constants. Vol. 4. Inorganic Complexes. Plenum Press, New York.

Song, P.S. and Chichester, C.O. (1967). Kinetic behaviour and mechanism of inhibition in the Maillard reaction. III. Kinetic behaviour of the inhibition in the reaction between D-glucose and glycine. J. Food Sci., 32, 98-106.

Strömberg, A., Gropen, O., Wahlgren, U. and Lindqvist, O. (1983). Ab initio calculations on the sulfite ion, SO_3^{2-}, and hydrogen sulfite ion, HSO_3^- or SO_2OH^-. Inorg. Chem., 22, 1129-1133.

Walker, R., Mendoza-Garcia, M.A., Ioannides, C. and Quattrucci, E. (1983a). Acute toxicity of 3-deoxy-4-sulfohexosulose in rats and mice, and in vitro mutagenicity in the Ames test. Food Cosmet. Toxicol., 21, 299-303.

Walker, R., Mendoza-Garcia, M.A., Quattrucci, E. and Zerilli, M. (1983b). Metabolism of 3-deoxy-4-sulfohexosulose, a reaction product of sulfite in foods, by rat and mouse. Food Cosmet. Toxicol., 21, 291-297.

Wedzicha, B.L. (1984a). Chemistry of Sulphur Dioxide in Foods, Elsevier Applied Science Publishers, London.

Wedzicha, B.L. (1984b). A kinetic model for the sulfite-inhibited Maillard reaction. Food Chem., 14, 173-87.

Wedzicha, B.L. (1986). Interactions involving sulfur dioxide in foods. In: "Interactions of Food Components", G.G. Birch and M. Lindley, eds, Elsevier Applied Science Publishers, London, pp 99-116.

Wedzicha, B.L. (1987). Chemistry of sulfur dioxide in vegetable dehydration. Int. J. Food Sci. Technol., 22, 433-50.

Wedzicha, B.L. and Edwards, A.S. (1991). Kinetics of the reaction of 3-deoxyhexosulose with thiols. Food Chem., in press.

Wedzicha, B.L. and Garner, D.N. (1991). The formation and reactivity of osuloses in the sulfite-inhibited Maillard reaction of glucose and glycine. Food Chem., in press.

Wedzicha, B.L. & Goddard, S.J. (1988). The dissociation constant of hydrogen sulfite ion at high ionic strength. Food Chem., 30, 67-71.

Wedzicha, B.L. and Goddard, S.J. (1991). The state of sulfur dioxide at high concentration and low water activity. Food Chem., in press.

Wedzicha, B.L., Goddard, S.J. and Garner, D.N. (1987). Enzymic browning of sulfocatechol. Int. J. Food Sci. Technol., 22, 653-657.

Wedzicha B.L. and Imeson, A.P. (1977). The yield of 3-deoxy-4-sulfopentosulose in the sulfite-inhibited browning of ascorbic acid. J. Sci. Food Agric., 28, 669-672.

Wedzicha, B.L. & Kaban, J. (1986). Kinetics of the reaction between 3-deoxyhexosulose and sulfur(IV) oxospecies in the presence of glycine. Food Chem., 22, 209-23.

Wedzicha, B.L., Kaban, J. and Aldous, D.G. (1985). The dissociation constant of the hydroxysulfonate of 3,4-dideoxy-4-sulfohexosulose in the sulfite-inhibited Maillard reaction of glucose and glycine. Food Chem., 17, 125-129.

Wedzicha, B.L. and Kaputo, M.T. (1987). Reaction of melanoidins with sulfur dioxide: stoichiometry of the reaction. Int. J. Food Sci. Technol., 22, 643-651.

Wedzicha, B.L. and Lamikanra, O. (1983). Sulfite-mediated destruction of ß-carotene: the partial characterisation of reaction products. Food Chem., 10, 275-283.

Wedzicha, B.L. and Luck, S. (1987). Effect of sulfur dioxide on the activity of trypsin. Food Chem., 26, 237-243.

Wedzicha, B.L. and McWeeny, D.J. (1974a). Non-enzymic browning reactions of ascorbic acid and their inhibition. The production of 3-deoxy-4-sulfopentosulose in mixtures of ascorbic acid, glycine and bisulfite ion. J. Sci. Food Agric., 25, 577-587.

Wedzicha, B.L. and McWeeny, D.J. (1974b). Non-enzymic browning reactions of ascorbic acid and their inhibition. The identification of 3-deoxy-4-sulfopentosulose in dehydrated, sulfited cabbage after storage. J. Sci. Food Agric., 25, 589-592.

Wedzicha, B.L. and Smith, G. (1987). Geometry of 1,2-dihydroxy-1,2-disulfonates. Food Chem., 25, 165-174.

Wedzicha, B.L. & Vakalis, N. (1988). Kinetics of the sulfite-inhibited Maillard reaction: The effect of sulfite ion. Food Chem., 27, 259-71.

Wisser, K., Völter, I and Heimann, W. (1970). About the reaction of sulfurous acid during oxidation. II. Binding of sulfurous acid with oxidation products of ascorbic acid. Z. Lebensm. Untersuch.-Forsch., 42, 180-185.

17

HEPATOTOXICITY CAUSED BY DIETARY SECONDARY PRODUCTS ORIGINATING FROM LIPID PEROXIDATION

Kazuki KANAZAWA

Department of agricultural chemistry
Kobe University
Nada-ku, Kobe 657, JAPAN

ABSTRACT

Hepatic dysfunction caused by oxidative stress when secondary peroxidation products were administered orally was investigated in rat. In serum at 24 hr after the administration of secondary products, the contents of lipid peroxides reached a maximum, the level of tocopherol reached a minimum, and the transaminase activities were elevated. In the liver, the lipid peroxide contents were kept high between 6 and 24 hr and tocopherol level was kept low between 15 and 48 hr after the does. Therefore, the hepatic oxidative stress was most severe around 15 hr after the dose. Dysfunction in the liver having oxidative stress was then made clear. One was a disturbance in synthetic system of glucose 6-phosphate. The decreases in activities of phosphoglucomutase and glucokinase reduced a level of glucose 6-phosphate, which suppressed the supply of NADPH in pentose cycle, while the NADPH was consuming well for detoxification of endogenous lipid peroxides. Another was specific inactivations of mitochondrial succinate dehydrogenase and aldehyde dehydrogenase. A third was the depletion of CoASH, which induced the decreases in activities of citrate cycle and lipogenesis. The other was a formation of lipofuscin. Even after the liver was recovering from the oxidative stress, the liver was getting hypertrophy and lipofuscin was accumulating. To make the cause of hepatic dysfunction clear, it was examined whether the incorporated secondary products in the liver could directly attack the enzymes or not. A reasonable amount of secondary products present in the liver was estimated, and then the amount of secondary products was added in hepatic subcellular organelles *in vitro*. It was found that mitochondrial NAD-dependent aldehyde dehydrogenase, glucokinase, and CoASH were directly attacked and inactivated by the incorporated secondary products in the liver. Thus, a part of dietary secondary products was incorporated into liver, and was not detoxified, but injured the enzymes and CoASH. Then it resulted in lipofuscin formation.

INTRODUCTION

Peroxidation products of lipids are toxic, but occur in our daily food. The intake of peroxidation products is believed to give a living body oxidative stress. Hydroperoxides, which are the first products at the autoxidation, may be most toxic. Dietary hydroperoxides have acute toxicity to digestive tract such as diarrhea and hemorrhage (Kanazawa et al., 1988). However, hydroperoxides are almost not incorporated into body, because most hydroperoxides is decomposed in the digestive tract mainly to their hydroxy forms (Holman and Greenberg, 1958; Glavind and Tryding, 1960; Bergan and Draper, 1970). Even if hydroperoxides are absorbed into body, they are easily reduced by glutathione peroxidase (Flohé, 1982). It is, therefore, considered that hydroperoxides exhibit the toxicity to the digestive tract, but the toxicity do not extend into the body.

Secondary peroxidation products are major products in the peroxidation of lipids, for example, they occupy 35% in the products when linoleic acid was autoxidized for 7 days, while hydroperoxides do 18% (Kanazawa et al., 1983). Secondary products are formed by the further peroxidation of hydroperoxides (Kanazawa et al., 1973), and consists of high molecular-weight polymers and decomposed products. The decomposed products are composed mainly of low molecular-weight aldehydes. In the case of peroxidation products of linoleic acid, hexanal and 9-oxononanoic acid are the major components (Terao et al., 1975). Polymers are difficult to absorb into the body (Yoshida and Alexander, 1983), but low molecular-weight aldehydes are easily incorporated into liver when secondary products were administered orally, because we have detected hexanal and 9-oxononanoate in the hepatic mitochondria or microsomes after the oral administration (Kanazawa and Natake, 1986a).

Aldehydes have been considered to be toxic (Schauenstein, 1967). Toxic aldehydes containing in the peroxidation products are not simple aldehydes, but are complex aldehydes such as carboxy semialdehydes like 9-oxononanoic acid (Minamoto et al., 1985 and 1988) and hydroxyalkenals like 4-hydroxynonenal (Benedetti et al., 1980; Cadenas et al., 1983), since it has been suggested that the complex aldehydes are difficult to detoxify by aldehyde dehydrogenase in the liver (Mitchell and Petersen, 1989). The oxidative stress induced by dietary peroxidation products should be due to these aldehydes.

The oxidative stress gives the liver a dysfunction. It has been reported that the oral dose of secondary products disturbed the hepatic carbohydrate and lipid metabolisms (Ashida et al., 1987a, 1987b and 1988). The aldehyde components incorporated into liver may be not detoxified and may attack some hepatic enzymes. It is interesting, therefore, to make clear that the incorporated secondary products could directly inactivate the hepatic enzymes.

Two questions must be solved: Which hepatic enzymes are inactivated after the oral dose of secondary products, and how much amount of secondary products are accumulated in the hepatocyte after the dose? In this study, secondary products were obtained from autoxidized linoleic acid and administered orally to rats, and then the oxidative stress was measured in the rat body. Since oxidative stress in the liver was most severe at 15 hr after the dose, changes in the enzyme activities and the related metabolite levels were determined 15 hr after the dose. Then, it was made clear that four hepatic enzymes were inactivated and one coenzyme was decomposed. On the other hand, a reasonable amount of secondary products present in the liver was estimated after the oral dose of secondary products. This amount of secondary products was added to the hepatic subcellular organelles *in vitro*, and then changes in the activities of above four enzymes and in the level of coenzyme were observed.

EXPERIMENTAL SECTION

Secondary peroxidation products. Linoleic acid was purchased from Tokyo Kasei Kogyo Co., Ltd., and autoxidized at 37°C for 7 days. The secondary product fraction was obtained from the autoxidized linoleic acid by silica gel column and thin layer chromatographies, and was analyzed by gas chromatography-mass spectrometry and Sephadex LH-20 gel filtration-chromatography (Kanazawa et al., 1983). The secondary product fraction consisted of 36% mixture of polymers, 26% epoxyhydroperoxides or endoperoxides (identified on gas chromatography (Neff et al., 1983)), 4.8% 9-oxononanoic acid, 3.7% hexanal, 2.5% nonanedioic acid, 2.4% short chain carboxylic acids, 0.75% 8-oxooctanoic acid, 0.34% 12-oxododecadienoic acid, and other smaller unidentified compounds. Peaks of hydroxy alkenals were observed on the gas-chromatogram (Oarada et al., 1986) and their amounts were estimated roughly to be less than 1%.

Animals and diet. Male Wistar rats, 5 weeks old and each weighing about 110 g (KY, SPF: Japan SLC), were housed in room in which the light cycle (a light and dark of 12 hr each), temperature, and humidity were controlled. An 1 week period was allowed for acclimatization before the animals were used. The diet was prepared daily and its peroxide value was maintained at less than 0.5 meq/kg. The detailed composition of the diets has been described previously (Kanazawa et al., 1986b). Briefly, it consisted of 30% sucrose, 25% casein, 24% corn starch, 15% soybean oil, 4% McCollum's salt mixture, 1% cellulose powder, and 1% vitamin mixture. Their foods were withheld for 4 hr before the treatments. Fifty six rats were intragastrically given 400 mg/rat of secondary products using a tuberculin syringe equipped with a stomach tube. Eight rats each was sacrificed at random to measure the serum and hepatic conditions 0, 6, 15, 24, 48, 72, and 96 hr after the dose. The other 56 rats were given 400 mg/rat saline solution and also treated as control.

Preparations of serum and a liver homogenate. Blood was collected by cardiac puncture, and serum was obtained after centrifugation at 3,000 rpm for 5 min. The rat liver was perfused for 10 sec with a saline solution using a cannula. Then, the liver was removed, washed with cold 1.15% KCl solution, blotted, weighed and homogenized with 10 volumes of 1.15% KCl solution. The serum and liver homogenate were submitted to the following 6 kinds of analyses.

Measurement of the serum transaminase activities. The activities of glutamic oxaloacetic transaminase and glutamic pyruvic transaminase in serum were determined spectrophotometrically by measuring the disappearance of NADH (Rej and Hørder, 1983).

Determinations of tocopherol level and lipid peroxide content. The serum and hepatic tocopherol levels were determined fluorescence-spectrophotometrically (Taylor et al., 1976). To observe the serum and hepatic lipid peroxidation, thiobarbituric acid (TBA) (Masugi and Nakamura, 1977) and hemoglobin-methylene blue (HMB) (Kanazawa et al., 1985a) tests were carried out. Lipid peroxides can be specifically detected with the HMB-test and aldehydes originating from the decomposition of lipid peroxides can be done with the TBA-test (Kanazawa et al., 1989a).

Analysis of hepatic fluorescent pigment. The liver homogenate was centrifuged at 105,000 x *g* for 60 min, and 0.1 ml of the supernatant (cytosol fraction) was mixed with 3 ml of saline solution. Fluorescent intensity of the solution was measured with a Hitachi Fluorescence Spectrophotometer F-3010 using an excited wavelength at 365 nm and an emission wavelength at 435 nm, when the intensity of 0.1 μg/ml quinine sulfate was taken 100. The fluorescent intensity is believed to be in proportion to the content of lipofuscin (Shimasaki et al., 1980).

Statistic analysis. When *F*-test for homogeneity of variance showed that variances were heterogeneous, the Student's *t* test was employed to determined the statistical significance, and a 0.05 probability level was chosen.

Treatment of liver to determine hepatic metabolite levels and enzyme activities. The other two groups of rats (8 rats each) were given 400 mg/rat of secondary products or 400 mg/rat of saline solution as control, and were sacrificed 15 hr after the doses. The liver was perfused for 10 sec with a saline solution, and then treated immediately according to the freezed-clamped method (Williamson et al., 1967). The liver was frozen with lead blocks precooled in liquid nitrogen. The frozen liver was pulverized in a mortar with the frequent addition of liquid nitrogen. The liver powder was subjected to the following analyses within 24 h. The protein concentration was estimated using the phenol reagent (Lowry et al., 1951).

Levels of metabolites. A part of the liver powder was homogenized with 10 volumes of a 1.15% KCl solution. The levels of reduced form of glutathione (GSH) and its oxidized form (GSSG) were quantified by fluorometry (Hissin and Hilf, 1976). The another part of liver powder (3 g) was transferred to a tube containing 12 ml of cold 9% (w/v) perchloric acid, and gently homogenized. The homogenate was centrifuged at 30,000 x *g* for 10 min at 4°C. The supernatant was adjusted with 40% KOH to a pH of around 6, allowed to stand for 30 min, and centrifuged to remove the $KClO_4$ precipitate. Florisil (0.1 g/ml) was added to the supernatant and shaken vigorously. After removing the florisil by centrifugation, clear supernatant was used for spectrophotometric analyses of metabolites: NADP and NAD (Klingenberg, 1985); glucose (Kunst et al., 1984); glucose 1-phosphate and glucose 6-phosphate (Michal, 1984); CoASH (Michal and Bergmeyer, 1985); and acetyl-CoA (Decker, 1985). Acid-insoluble long-chain acyl-CoA present in the precipitate in the acid-extract solution was converted to CoASH (Garland, 1985), and then quantified. The other 1 g of liver powder was submitted to alkaline-extraction with 10 ml of 0.5N KOH in ethanol for assaying of NADPH and NADH levels (Klingenberg, 1985). The metabolite levels per whole liver were estimated.

Preparation of subcellular fractions. The liver powder (1.5 g) was homogenized with 10 volumes of a 1.15% KCl solution. The homogenate was centrifuged at 700 x *g* for 10 min and the supernatant was further centrifuged at 5,000 x *g* for 20 min. The pellet was suspended in the original volume of the KCl solution and centrifuged again at the same *g*. The pellet was suspended in 150 mM KCl solution and used as a mitochondrial fraction. The supernatant from 5,000 x *g* was recentrifuged at 24,000 x *g* for 10 min and the supernatant was further centrifuged at 57,000 x *g* for 60 min. The pellet was referred to as microsomal fraction and suspended in 150 mM KCl solution. The supernatant of 57,000 x *g* was recentrifuged at 105,000 x *g* for 60 min and the supernatant was used as a cytosol fraction.

Determination of hepatic enzyme activities. The activity of glutathione peroxidase was determined by the method of Little et al. (1970) using the whole liver homogenate, when pure linoleic acid hydroperoxides was used as its substrate. The activities of glutathione reductase was also measured by the methods of Zanetti (1979). The activities aldehyde dehydrogenases in the mitochondrial and cytosolic fractions were determined by the rate of formation of NADH or NADPH spectrophotometrically (Black, 1955). The activities of citrate cycle enzymes, isocitrate dehydrogenase (Bergmeyer et. al., 1983) and succinate dehydrogenase (Veeger et al., 1969), were spectrophotometrically determined using the mitochondrial fraction. The activities of NADPH-supplementary enzymes, glucose 6-phosphate dehydrogenase (Glock and McLean, 1953), phosphogluconate dehydrogenase (Glock and McLean, 1953), and malate dehydrogenase (E.C.1.1.1.40) (Ochoa, 1955), were measured using the cytosol fraction. The activities of phosphoglucomutase (Bergmeyer et al., 1983), glucokinase (Bergmeyer et al., 1983), and phosphofructokinase (Massey and Deal, 1975) were also measured using the cytosol fraction. The activities of microsomal glucose-6-phosphatase were determined by the measurement of the production of inorganic phosphate (Swanson, 1955). The activities of lipolytic and lipogenic enzymes were determined as follows: Carnitine palmitoyltransferase activity in the mitochondrial fraction was assayed according to (Markwell et al. 1973). Acetyl-CoA carboxylase activity was measured by the ^{14}C-labeled sodium bicarbonate fixation method using the cytosol fraction (Nakanishi and Numa, 1970).

Measurement of de novo synthesis of CoA derivatives. A part of each cytosolic fraction from the 4 rats dosed with secondary products was stood in 80 mM of tris-HCl buffer (pH 7.5) with 5 mM pantothenic acid, 10 mM ATP, 18 mM cystein, and 10 mM $MgCl_2$. The mixture was incubated at 25^oC for 10 min and the incubation was stopped by an addition of cold perchloric acid (finally 9%), since CoA derivatives were produced linearly with time until 10 min-incubation. The synthesized amounts of CoASH, acetyl-CoA, and long-chain acyl-CoA (nmol/min/mg protein) were measured enzymatically with the method as described above. The cytosol fraction of saline-dosed rats were also treated as control.

Determination of direct effects of secondary products on the hepatic enzymes in in vitro system. The hepatic subcellular fractions were prepared from untreated rats. Secondary products in ethanol solution were added to the fractions, and then incubated at 37^oC. The enzyme activities were monitored with the same procedures as described above. The control incubation was carried out with the same volume of ethanol. Remaining activities of the respective enzymes were determined after 60 min-incubation, because the activity of most enzymes was decreased linearly with time in the incubation mixtures.

RESULTS AND DISCUSSION

Oxidative stress induced by dietary secondary products

Dietary secondary products stimulates the endogenous lipid peroxidation (Kanazawa et al., 1986b), which could give the animals oxidative stress. Then, the oxidative stress was measured after the dose of secondary products (*Figure 1 and 2*). The content of lipid peroxides was estimated with HMB-test and the amount of aldehydes originated from the decomposition of lipid peroxides was determined with the TBA-test. Since the living body consumes tocopherols to protect the endogenous lipid peroxidation, the tocopherol level reflects the oxidative stress. Hepatic injury can be evaluated by measurements of the increasing in serum transaminase activities. *Figure 1* shows the changes in these values in serum after the dose of secondary products. Both HMB and TBA values were increased and reached a maximum around 24 hr after the dose. The inverse change in serum tocopherol level and the proportional changes in serum transaminase activities were also observed. *Figure 2* shows the changes in these values in the liver. The HMB and TBA values were reached a maximum between 6 and 24 hr after the dose. The HMB value was then decreased and returned to normal 72 hr after the dose, but the TBA

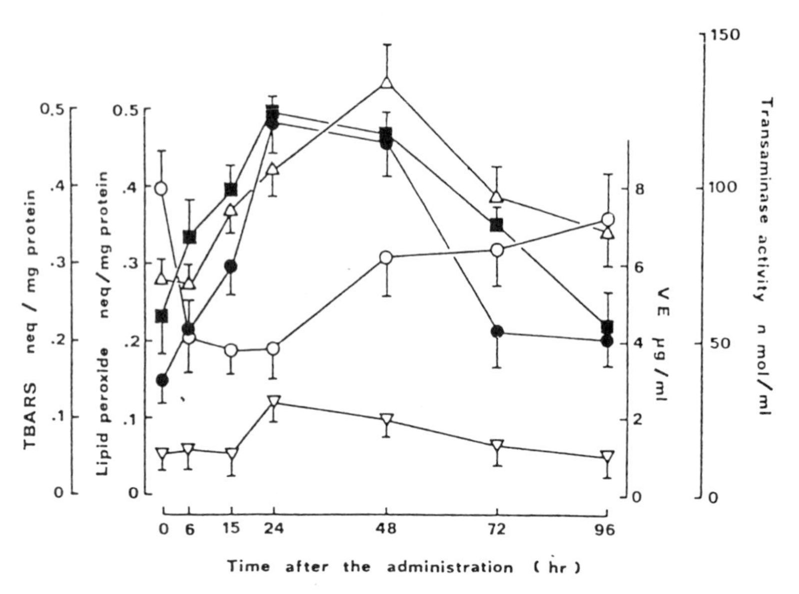

Figure 1. Changes in the content of lipid peroxide (●), content of thiobarbituric acid reactive substances (■), level of tocopherol (vitamin E) (○), activity of glutamate oxaloacetate transaminase (△), and activity of glutamate pyruvate transaminase (▽), in serum after the administration of secondary products.

value was kept high even after the decrease in HMB value. The hepatic tocopherol level was very low between 15 and 48 hr after the does. Thus, the incorporated secondary products induced the hepatic lipid peroxidation, and the lipid peroxides (HMB-reactive substances) was decomposed to aldehydes (TBA-reactive) with time in the liver. The accumulation of them and depletion of tocopherol caused the injure of hepatic membrane, which resulted in leakage of transaminases into serum. It was concluded that the oxidative stress in liver was most severe between 15 and 24 hr after the dose of secondary products.

When the hepatic oxidative stress was recovering, hypertrophy of the liver was observed and fluorescent pigment (lipofuscin) was producing (*Figure 2*). Moreover, lipofuscin was accumulating in the liver even after the recovery from hypertrophy, and did not disappear more few days later.

Hepatic dysfunction by dietary secondary products

Activities of hepatic enzymes and levels of their related-metabolites were measured 15 hr after the dose of secondary products (*Table 1 and 2*). The endogenous lipid peroxides is detoxified by glutathione peroxidase using GSH (Sure, 1924). Its activity in the secondary products-dosed group was increased three-folds comparing to that in control group (*Table 1*). The activity of its coupling enzyme, glutathione reductase, which produced GSH from GSSG, remained unchanged. Both contents of GSH and GSSG remained also unchanged (*Table 2*). These results indicate that the

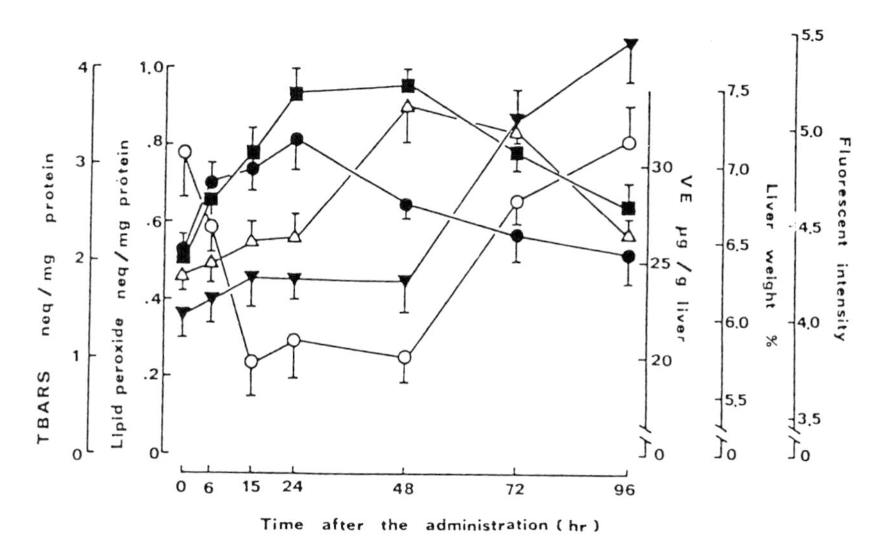

Figure 2. Changes in the content of lipid peroxide (●), content of thiobarbituric acid reactive substances (■), level of tocopherol (vitamin E) (○), amount of fluorescent pigment (▼), and weight (△), in/of the liver after the administration of secondary products.

hepatic protecting system worked sufficiently for the endogenous lipid peroxidation induced by dietary secondary products.

Aldehydes originating from lipid peroxides should be detoxified by the hepatic aldehyde dehydrogenases. However, the activity of mitochondrial NAD-dependent aldehyde dehydrogenase was decreased to half, but the decreases in the other aldehyde dehydrogenase activities were less (*Table 1*). This enzyme was a major aldehyde dehydrogenase in the hepatocyte of rat, because its activity occupies around 60% in total activity of hepatic aldehyde dehydrogenases. Though NAD-dependent aldehyde dehydrogenase should consume NADH, the hepatic NAD(H) levels remained unchanged after the dose (*Table 2*). It has been made clear that aldehydes were incorporated into the liver when secondary products were administered orally (Kanazawa et al., 1986a). Aldehydes containing in secondary products are not only simple aldehydes but also complex aldehydes such as carboxy semialdehydes and hydroxyalkenals (see the EXPERIMENTAL SECTION). It has been, moreover, reported that the complex aldehydes were difficult to be detoxified by aldehyde dehydrogenase (Mitchell and Petersen, 1989), and that 9-oxononanoate (Minamoto et al., 1985 and 1988) and 4-hydroxy nonenal (Benedetti et al., 1980; Cadenas et al., 1983) were most toxic components originating from lipid peroxidation. It is considered that such complex aldehydes rather inactivated mitochondrial NAD-dependent aldehyde dehydrogenase than were detoxified by the enzyme in the liver.

Glutathione reductase consumes NADPH as a reductant to produce GSH.

Table 1. Changes in the hepatic enzyme activities

		Hepatic enzyme activity (%)[a]
Glutathione peroxidase		285
Glutathione reductase		115
Mitochondrial aldehyde dehydrogenase	NAD-dependent	48.9
	NADP-dependent	80.5
Cytosolic aldehyde dehydrogenase	NAD-dependent	84.5
	NADP-dependent	75.1
Glucose-6-phosphate dehydrogenase		80.9
Phosphogluconate dehydrogenase		84.9
Cytosolic malate dehydrogenase (NADP-dependent)		89.8
Phosphofructokinase		61.6
Glucose-6-phosphatase		88.7
Phosphoglucomutase		74.3
Glucokinase		73.0
Isocitrate dehydrogenase		72.9
Succinate dehydrogenase		66.3
Carnitine palmitoyltransferase		141
Acetyl-CoA carboxylase		39.2

[a]Means of the respective enzyme activities in the secondary products-dosed rats (8 rats) were compared to those in the control group (saline-dosed group) which was taken 100.

Table 2. Changes in the metabolite levels

	Level (%)[a]
GSH	90.9
GSSG	72.7
NADH	110
NAD	87.3
NADPH	54.1
NADP	128
Glucose 6-phosphate	51.3
Glucose 1-phosphate	141
Glucose	87.2
CoASH	18.3
Acetyl-CoA	62.3
Long-chain acyl-CoA	93.1

[a]Means of the respective metabolite levels in the secondary products-dosed rats (8 rats) were compared to those in the control group (saline-dosed rats) which was taken 100.

The NADPH level was reduced to half after the dose of secondary products, but NADP level remained unchanged. NADPH is supplied mainly with pentose cycle and the other is done with malic cycle in rat (Flatt and Ball, 1964; Katz et al., 1966). The activities of glucose-6-phosphate dehydrogenase and phosphogluconate dehydrogenase which are the rate limiting enzymes in citrate cycle, were slightly decreased, but the activity of malic enzyme (cytosolic malate dehydrogenase) remained unchanged (*Table 1*). The activity of phosphofructokinase in glycolysis was also decreased. Both pentose cycle and glycolysis need glucose 6-phosphate as the substrate. The decreases in these activities may be complicated with the level of glucose 6-phosphate, which is synthesized *via* glucose 1-phosphate from glycogen and also from glucose. Then, the levels of these metabolites were measured (*Table 2*). The glucose 6-phosphate level was very low and glucose 1-phosphate was accumulated after the dose of secondary products. The productions of glucose 6-phosphate from glucose 1-phosphate and glucose were catalyzed by phosphoglucomutase and glucokinase, respectively, whose activities were decreased (*Table 1*). Therefore, these results indicate that depletion of glucose 6-phosphate caused by the reduction in activities of the synthetic systems disturbed pentose cycle and glycolysis, and then resulted in the depletion of NADPH.

Citrate cycle activity was also changed by dietary secondary products. The activities of succinate dehydrogenase and isocitrate dehydrogenase (the rate limiting enzyme) were decreased (*Table 1*). On the other hand, when secondary products administered daily for 3 days, the activity of succinate dehydrogenase was reduced much more, but the rate limiting enzyme was not (Kanazawa et al., 1989b). It is suggested that the inactivation of succinate dehydrogenase was due to a different cause from the decrease in citrate cycle activity, and was specifically effected by the incorporated aldehydes.

Lipid metabolism was much affected by dietary secondary products (*Table 1*). The activity of carnitine palmitoyltransferase, one of the rate limiting enzymes of lipolysis, was increased, and the activity of acetyl-CoA carboxylase, of lipogenesis, was decreased. It has been believed that the lipolytic system is stimulated to detoxify the incorporated secondary products, because when labeled secondary products were administered orally, isotope was incorporated in hepatic lipids and labeled carbon dioxide was respired well (Kanazawa et al., 1986b). On the other hand, acetyl-CoA (the substrate of lipogenesis) became low level but long-chain acyl-CoA (the substrate of lipogenesis) did not, and CoASH was depleted (*Table 2*). This shows that the reduction in lipogenic activity was due to the decrease in supply of acetyl-CoA which was caused by the depletion of CoASH. Acetyl-CoA is also the substrate for citrate cycle. The reduction in citrate cycle activity is considered also to relate with the low level of acetyl-CoA.

From these results, the hepatic dysfunction caused by dietary secondary products are summarized as shown in *Figure 3*. One is a disturbance in synthetic system of glucose 6-phosphate. The decreases in activities of phosphoglucomutase and glucokinase reduced the level of glucose 6-phosphate, and which suppressed the supply of NADPH with pentose cycle, though NADPH was consuming well by the detoxifying system of endogenous lipid peroxidation. Another is specific inactivations of mitochondrial succinate dehydrogenase and NAD-dependent aldehyde dehydrogenase. A third is the depletion of CoASH, which induced the decreases in citrate cycle and lipogenesis activities. The other is a formation of lipofuscin. Even after the liver was recovering from the dysfunctions, lipofuscin was accumulating little by little, which may induce the liver senescence.

Inactivations of hepatic enzymes and CoASH directly caused by the incorporated aldehydes

It must be answered how the inactivations of above four enzymes (phosphoglucomutase, glucokinase, succinate dehydrogenase, and NAD-dependent aldehyde dehydrogenase) and depletion of CoASH were caused by the incorporated secondary products in the liver. A simple answer is that the incorporated aldehydes directly attack the enzymes and CoASH.

Phosphoglucomutase **Glucokinase**

1) Reduction in Glucose 6-phosphate level

Pentose cycle → Decrease in NADPH level

2) Inactivation of mitochondrial Succinate dehydrogenase and Aldehyde dehydrogenase

3) Depletion of CoASH

Citrate cycle **Acetyl-CoA carboxylase**

4) Lipofuscin was accumulating → Senescence

Figure 3. Summary of the hepatic dysfunction caused by dietary secondary products.

There are many data on that the products of lipid peroxidation easily inactivated enzymes and destroyed the amino acid residues (Roubal and Tappel, 1966; Matoba et al., 1984). Especially, secondary products injured the basic and sulfur containing amino acids (Kanazawa et al., 1975). These data, however, had been obtained from the effects of very large amount of peroxidation products on the enzymes in *in vitro*, for example, in the report by Kanazawa et al. (1975), 800 μg secondary products/mg protein was used. It can not be imagined that such a high concentration of secondary products be in the living cell, and besides it can be suggested that secondary products react with the other biological components besides the amino acids (Kanazawa et. al., 1987) and are removed before attacking the enzymes.

A reasonable amount of secondary products present in the liver was estimated by the oral dose of secondary products (*Table 3*). Uniformly labeled products were used and the distribution of isotope in hepatocyte was determined at 12 hr after the dose, because the incorporation of isotope and the accumulation of aldehydes (TBA-reactive substances) in the liver reached maximums at that time (Kanazawa et. al., 1985b). The incorporated amounts nearly remained constant being independent of the dosed amounts, and were 1.7 μg/mg protein in mitochondrial, 2.9 μg in microsomal, and 3.7 μg in cytosolic fractions. After considering a possibility for temporary accumulation of the larger amounts, the five times-amounts were added to the respective subcellular fractions of hepatocyte, and then their effects on the enzyme activities were observed.

Table 3. Estimation of the incorporated amounts of secondary products into subcellular fractions of hepatocyte

Subcellular fraction	Incorporated amount (μg/mg protein)[a]
Whole liver	3.3
Mitochondria	1.7
Microsomes	2.9
Cytosol	3.7

[a]The distribution of ^{14}C in the liver was determined at 12 hr after the oral administration of [^{14}C-U] secondary products (115-570 mg/rat). The data were means with 20 rats.

Table 4. The remaining enzyme activities and amount of CoASH after the 60 min-incubation with secondary products *in vitro*

	Remaining activity or amount(%)[a]
Phosphoglucomutase	94
Glucose-6-phosphate dehydrogenase	43
Phosphogluconate dehydrogenase	79
Glucokinase	61
Glucose-6-phosphatase	70
Phosphofructokinase	76
Isocitrate dehydrogenase	96
Succinate dehydrogenase	95
Carnitine palmitoyltransferase	89
Mitochondrial	
NAD-aldehyde dehydrogenase	40
Cytosolic	
NAD-aldehyde dehydrogenase	78
NAD-alcohol dehydrogenase	86
CoASH[b]	70

[a]The hepatic cytosolic, microsomal and mitochondrial fractions were incubated for 60 min with 20 μg, 15 μg and 10 μg/mg protein of methyl-esterified secondary products, respectively, and the enzyme activities were determined. As the control, these fractions were incubated with the same volumes of ethanol, and the enzyme activities were taken 100.

[b]CoASH was incubated with secondary products, and the remaining amount was calculated with the same method as the enzyme activity.

Table 4 shows the remaining activities of enzymes in the subcellular organelles after the 60 min-incubation with the small amounts of secondary products. The enzymes inactivated markedly were glucose-6-phosphate dehydrogenase, glucokinase, glucose-6-phosphatase, and mitochondrial NAD-dependent aldehyde dehydrogenase. The other enzymes were almost not inactivated by the addition of such small amount of secondary products. Glucose-6-phosphate dehydrogenase is considered to be inhibited by fatty acids (Muto and Gibson, 1970), and its remaining activity after 60 min-incubation with methyl linoleate was 58% (data not shown). Secondary products contain a few % of short-chain fatty acids (see the EXPERIMENTAL SECTION). The decrease in glucose-6-phosphate dehydrogenase activity may be due to the fatty acids containing in secondary products. The reduction in glucose-6-phosphatase activity was not observed by the oral dose of secondary products (*Table 1*), but was done here. Glucose-6-phosphatase is a microsomal membrane-dependent enzyme and the activity is sensitive to a membrane disturbance. The membrane disturbance is caused by the peroxidation products in *in vitro* (Wills, 1971), while it could be easily repaired in *in vivo* (Kanazawa et al., 1986b).

The enzymes inactivated by dietary secondary products were phosphoglucomutase, succinate dehydrogenase, glucokinase, and mitochondrial NAD-dependent aldehyde dehydrogenase (*Figure 3*), and in *Table 4* the inactivations of only two of them were observed. The incorporated aldehydes in liver directly attacked and inactivated glucokinase and aldehyde dehydrogenase, but did not phosphoglucomutase and succinate dehydrogenase.

The amount of CoASH was also decreased by 30% after the 60 min-incubation with secondary products (*Table 4*). Esterbauer et al. (1975 and 1976) reported that SH-compounds such as CoASH easily react with aldehydes originating from lipid peroxides. CoASH may also receive the direct attack of aldehydes in the liver. There is, however, one unsolved problem that the synthetic pathway of CoASH could be affected by dietary secondary products. Then, the synthetic activities of CoA derivatives in the hepatic cytosol of rats dosed with secondary products were compared with those in the control rats (*Table 5*). The synthetic activities were elevated after the oral dose of secondary products, and especially, the activity for CoASH was done to 2-times. Therefore, it is considered that the depletion of hepatic CoASH by dietary secondary products was caused by direct attack of the incorporated aldehydes.

Table 5. Changes in synthetic activities of CoA derivatives after the oral administration of secondary products

Synthetic activity (%)[a] of		
CoASH	Acetyl-CoA	Long-chain acyl-CoA
235	116	128

[a]Synthetic activities of CoA derivatives (nmol/min/mg protein) in hepatic cytosol fractions of rats administered orally with secondary products was compared with those in control rats which were taken 100.

CONCLUSIONS

This study demonstrates that a part of dietary secondary products was incorporated into liver and not detoxified, and aldehydes containing in the incorporated products directly injured mitochondrial NAD-dependent aldehyde dehydrogenase, glucokinase and CoASH. The attacks of aldehydes to these biological components resulted in the formation of lipofuscin pigment, which was accumulating little by little in the liver.

Mechanism for the inactivations of succinate dehydrogenase and phosphoglucomutase by the incorporated aldehydes is not clear. There are two ideas; One is that the incorporated aldehydes injured a biomembrane and then membrane-dependent succinate dehydrogenase was inactivated. The other is that a synthetic system of enzymes was disturbed by the aldehydes. I have negative evidences for the first idea. The changes in composition of fatty acids in the mitochondrial membrane were not detected after the oral dose of secondary products (Ashida et al., 1988), and it is believed that even if the phospholipid membrane was injured it was easily repaired *in vivo* (Terao et al., 1984). Moreover, the activity of a microsomal membrane-enzyme, glucose-6-phosphate dehydrogenase, remained unchanged after the dose of secondary products (*Table 1*). Thus, the first idea can be denied. Then, a study for the second idea is in progress using primary cultured hepatocytes.

ACKNOWLEDGEMENTS

I thank Dr. Hitoshi Ashida for his kind technical assistance.

REFERENCES

Ashida, H., Kanazawa, K., Minamoto, S., Danno, G., and Natake, M. (1987a). Effect of orally administered secondary autoxidation products of linoleic acid on carbohydrate metabolism in rat liver. *Arch. Biochem. Biophys.*, **259**, 114-123.

Ashida, H., Kanazawa, K., and Natake, M. (1987b). Decrease of the NADPH level in rat liver on oral administration of secondary autoxidation products of linoleic acid. *Agric. Biol. Chem.*, **51**, 2951-2957.

Ashida, H., Kanazawa, K., and Natake, M. (1988). Comparison of the effects of orally administered linoleic acid, and its hydroperoxides and secondary autoxidation products on hepatic lipid metabolism in rats. *Agric. Biol. Chem.*, **52**, 2007-2014.

Benedetti, A., Comporti, M., and Esterbauer, H. (1980). Identification of 4-hydroxynonenal as a cytotoxic product originating from the peroxidation of liver microsomal lipids. *Biochim. Biophys. Acta*, **620**, 281-296.

Bergan, J.G. and Draper, H.H. (1970). Absorption and metabolism of 1-^{14}C-methyl linoleate hydroperoxide. *Lipids*, **5**, 976-982.

Bergmeyer, H.U., Grassl, M., and Walter, H.-E. (1983). Hexokinase; Isocitrate dehydrogenase; and Phosphoglucomutase. *In*: "Methods of Enzymatic Analysis", H.U. Bergmeyer, J. Bergmeyer, and M. Grassl, eds., VCH Verlagsgesellschaft, Weinheim, Vol. II, pp. 222-223; pp. 230-231; and pp. 277-278, respectively.

Black, S. (1955). Potassium-activated yeast aldehyde dehydrogenase. *Methods Enzymol.*, **I**, 508-511.

Cadenas, E., Müller, A., Brigelius, R., Esterbauer, H., and Sies, H. (1983). Effects of 4-hydroxynonenal on isolated hepatocytes. *Biochem. J.*, **214**, 479-487.

Decker, K. (1985). Acetyl Coenzyme A. *In*: "Methods of Enzymatic Analysis", H.U. Bergmeyer, J. Bergmeyer, and M. Grassl, eds., VCH Verlagsgesellschaft, Weinheim, Vol. VII, pp. 186-193.

Esterbauer, H., Zollner, H., and Scholz, N. (1975). Reaction of glutathione with conjugated carbonyls. *Z. Naturforsch.*, **30**, 466-473.

Esterbauer, H., Ertl, A., and Scholz, N. (1976). The reaction of cysteine with α,β-unsaturated aldehydes. *Tetrahedron*, **32**, 285-289.

Flatt, J.P. and Ball, E.G. (1964). Studies on the metabolism of adipose tissue. *J. Biol. Chem.*, **239**, 675-685.

Flohé, L. (1982). Role of GSH peroxidase in lipid peroxide metabolism. *In*: "Lipid Peroxides in Biology and Medicine", K. Yagi ed., Academic Press, New York, pp. 149-159.

Garland, P.B. (1985). Coenzyme A Derivatives of Long-chain Fatty Acids. *In*: "Methods of Enzymatic Analysis", H.U. Bergmeyer, J. Bergmeyer, and M. Grassl, eds., VCH Verlagsgesellschaft, Weinheim, Vol. VII, pp. 206-211.

Glavind, J. and Tryding, N. (1960). On the digestion and absorption of lipoperoxides. *Acta Physiol. Scand.*, **49**, 97-102.

Glock, G.E. and McLean, P. (1953). Further studies on the properties and assay of glucose 6-phosphate dehydrogenase and 6-phosphogluconate dehydrogenase of rat liver. *Biochem. J.*, **55**, 400-408.

Hissin, P. J. and Hilf, R. (1976). A fluorometric method for determination of oxidized and reduced glutathione in tissues. *Anal. Biochem.*, **74**, 214-226.

Holman, R.T. and Greenberg, S.I. (1958). A note on the toxicities of methyl oleate peroxide and ethyl linoleate peroxide. *J. Am. Oil Chem. Soc.*, **35**, 707.

Kanazawa, K., Mori, T., and Matsushita, S. (1973). Oxygen absorption at the process of the degradation of linoleic acid hydroperoxides. *J. Nutr. Sic. Vitaminol.*, **19**, 263-275.

Kanazawa, K., Danno, G., and Natake, M. (1975). Lysozyme damage caused by secondary degradation products during the autoxidation process of linoleic acids. *J. Nutr. Sci. Vitaminol.*, **21**, 373-382.

Kanazawa, K., Danno, G., and Natake, M. (1983). Some analytical observations of autoxidation products of linoleic acid and their thiobarbituric acid reactive substances. *Agric. Biol. Chem.*, **47**, 2035-2043.

Kanazawa, K., Minamoto, S., Ashida, H., Yamada, K., Danno, G., and Natake, M. (1985a). Determination of lipid peroxide contents in rat liver by a new coloration test. *Agric. Biol. Chem.*, **49**, 2799-2801.

Kanazawa, K., Kanazawa, E., and Natake, M. (1985b). Uptake of secondary autoxidation products of linoleic acid by the rat. *Lipids*, **20**, 412-419.

Kanazawa, K. and Natake, M. (1986a). Identification of 9-oxononanoic acid and hexanal in liver of rat orally administered with secondary autoxidation products of linoleic acid. *Agric. Biol. Chem.*, **50**, 115-120.

Kanazawa, K., Ashida, H., Minamoto, S., and Natake, M. (1986b). The effect of orally administered secondary autoxidation products of linoleic acid on the activity of detoxifying enzymes in the rat liver. *Biochim. Biophys. Acta*, **879**, 36-43.

Kanazawa, K., Ashida, H., and Natake, M. (1987). Autoxidizing process interaction of linoleic acid with casein. *J. Food Sci.*, **52**, 475-478.

Kanazawa, K., Ashida, H., Minamoto, H., Danno, G., and Natake, M. (1988). The effects of orally administered linoleic acid and its autoxidation products on intestinal mucosa in rat. *J. Nutr. Sci. Vitaminol.*, **34**, 363-373.

Kanazawa, K., Inoue, N., Ashida, H., Mizuno, M., Natake, M. (1989a). What do thiobarbituric acid and hemoglobin-methylene blue tests evaluate in the endogenous lipid peroxidation of rat liver? *J. Clin. Biochem. Nutr.*, **7**, 69-79.

Kanazawa, K., Ashida, H., Inoue, N., Natake, M. (1989b). Succinate dehydrogenase and synthetic pathways of glucose 6-phosphate are also the markers of the toxicity of orally administered secondary autoxidation products of linoleic acid in rat liver. *J. Nutr. Sci. Vitaminol.*, **35**, 25-37.

Katz, J., Landau, B.R., and Bartsch, G.E. (1966). The pentose cycle, triose phosphate isomerization, and lipogenesis in rat adipose tissue. *J. Biol. Chem.*, **241**, 727-740.

Klingenberg, M. (1985). Nicotinamide-adenine dinucleotides and dinucleotide phosphates (NAD, NADP, NADH, NADPH). *In*: "Methods of Enzymatic Analysis", H.U. Bergmeyer, J. Bergmeyer, and M. Grassl, eds., VCH Verlagsgesellschaft, Weinheim, Vol. VII, pp. 251-271.

Kunst, A., Draeger, B., and Ziegenhorn, J. (1984). D-Glucose. *In*: "Methods of Enzymatic Analysis", H.U. Bergmeyer, J. Bergmeyer, and M. Grassl, eds., VCH Verlagsgesellschaft, Weinheim, Vol. VI, pp. 163-172.

Little, C., Olinescu, R., Reid, K.G., and O'Brien, P.J. (1970). Properties and regulation of glutathione peroxidase. *J. Biol. Chem.*, **245**, 3632-3636.

Lowry, O. H., Rosebrough, N. J., Farr, A. L., and Randall, R. J. (1951). Protein measurement with the folin phenol reagent. *J. Biol. Chem.*, **193**, 265-275.

Markwell, M. A. K., McGroarty, E. J., Bieber, L. L., and Tolbert, N. E.

(1973). The subcellular distribution of carnitine acyltransferases in mammalian liver and kidney. *J. Biol. Chem.*, **248**, 3426-3432.

Massey, T. and Deal, Jr., W.C. (1975). Phosphofructokinases from porcine liver and kidney and from other mammalian tissues. *Methods Enzymol.*, **XLII**, 99-110.

Masugi, F. and Nakamura, T. (1977). Measurement of thiobarbituric acid value in liver homogenate solubilized with sodium dodecylsulphate and variation of the values affected by vitamin E and drugs. *Vitamin*, **51**, 21-29.

Matoba, T., Yonezawa, D., Nair, B.M., and Kito, M. (1984). Damage of amino acid residues of proteins after reaction with oxidizing lipids: estimation by proteolytic enzymes. *J. Food Sci.*, **49**, 1082-1084.

Michal, G. (1984). D-Glucose 1-Phosphate; and D-Glucose 6-Phosphate and D-Fructose 6-Phosphate. *In*: "Methods of Enzymatic Analysis", H.U. Bergmeyer, J. Bergmeyer, and M. Grassl, eds., VCH Verlagsgesel-lschaft, Weinheim, Vol. VI, pp. 185-198.

Michal, G. and Bergmeyer, H.U. (1985). Coenzyme A and Derivatives. *In*: "Methods of Enzymatic Analysis", H.U. Bergmeyer, J. Bergmeyer, and M. Grassl, eds., VCH Verlagsgesellschaft, Weinheim, Vol. VII, pp. 156-165.

Minamoto, S., Kanazawa, K., Ashida, H., Danno, G., and Natake, M. (1985). The induction of lipid peroxidation in rat liver by oral intake of 9-oxononanoic acid contained in autoxidized linoleic acid. *Agric. Biol. Chem.*, **49**, 2747-2751.

Minamoto, S., Kanazawa, K., Ashida, H., and Natake, M. (1988). Effect of orally administered 9-oxononanoic acid on lipogenesis in rat liver. *Biochim. Biophys. Acta*, **958**, 199-204.

Mitchell, D.Y. and Petersen, D.R. (1989). Oxidation of aldehydic products of lipid peroxidation by rat liver microsomal aldehyde dehydrogenase. *Arch. Biochem. Biophys.*, **269**, 11-17.

Muto, Y. and Gibson, D.H. (1970). Selective dampening of lipogenic enzymes of liver by exogenous polyunsaturated fatty acids. *Biochem. Biophys. Res. Commun.*, **38**, 9-15.

Nakanishi, S. and Numa, S. (1970). Purification of rat liver acetyl coenzyme A carboxylase and immunochemical studies on its synthesis and degradation. *Eur. J. Biochem.*, **16**, 161-173.

Neff, W.E., Frankel, E.N., Selke, E., and Weisleder, D. (1983). Photosensitized oxidation of methyl linoleate monohydroperoxides: Hydroperoxy cyclic peroxides, dihydroperoxides, keto esters and volatile thermal decomposition products. *Lipids*, **18**, 868-876.

Oarada, M., Miyazawa, T., and Kaneda, T. (1986). Distribution of ^{14}C after oral administration of [U-^{14}C]labeled methyl linoleate hydroperoxides and their secondary oxidation products in rats. *Lipids*, **21**, 150-154.

Ochoa, S. (1955). "Malic" · Enzyme. *Methods Enzymol.*, **I**, 739-753.

Rej, R. and Hørder, M. (1983). Aspartate aminotransferase.

In: "Methods of Enzymatic Analysis", H.U. Bergmeyer, J. Bergmeyer, and M. Grassl, eds., VCH Verlagsgesellschaft, Weinheim, Vol. III, pp. 416-424; and Alanine aminotransferase., pp. 444-450.

Roubal, W.T. and Tappel, A.L. (1966). Damage to proteins, enzymes, and amino acids by peroxidizing lipids. *Arch. Biochem. Biophys.*, **113**, 5-8.

Schauenstein, E. (1967). Autoxidation of polyunsaturated esters in water: chemical structure and biological activity of the products. *J. Lipid Res.*, **8**, 417-428.

Shimasaki, H., Ueta, M., and Privett, O.S. (1980). Isolation and analysis of age-related fluorescent substances in rat testes. *Lipids*, **15**, 236-241.

Sure, B. (1924). Dietary requirements for reproduction. *J. Biol. Chem.*, **58**, 681-704.

Swanson, M. A. (1955). Glucose-6-phosphatase from liver. *Methods Enzymol.*, **II**, 541-543.

Taylor, S. L., Lamden, M. P., and Tappel, A. L. (1976). Sensitive fluorometric method for tissue tocopherol analysis. *Lipids*, **11**, 530-538.

Terao, J., Ogawa, T., and Matsushita, S. (1975). Degradation process of autoxidized methyl linoleate. *Agric. Biol. Chem.*, **39**, 397-402.

Terao, J., Asano., Matsushita, S. (1984). High-performance liquid chromatographic determination of phospholipid peroxidation products of rat liver after carbon tetrachloride administration. *Arch. Biochem. Biophys.*, **235**, 326-333.

Veeger, C., DerVartanian, D.V., and Zeylemaker, W.P. (1969). Succinate dehydrogenase. *Methods Enzymol.*, **XIII**, 81-90.

Williamson, D. H., Lund, P., and Krebs, H. A. (1967). The redox state of free nicotinamide-adenine dinucleotide in the cytoplasm and mitochondria of rat liver. *Biochem. J.*, **103**, 514-527.

Wills, E.D. (1971). Effects of lipid peroxidation on membrane-bound enzymes of the endoplasmic reticulum. *Biochem. J.*, **123**, 983-991.

Yoshida, H. and Alexander, J.C. (1983). Enzymatic hydrolysis in vitro of thermally oxidized sunflower oil. *Lipids*, **18**, 611-616.

Zanetti, G. (1979). Rabbit liver glutathione reductase. Purification and properties. *Arch. Biochem. Biophys.*, **198**, 241-246.

18

DIETARY N-3 POLYUNSATURATED FATTY ACIDS OF FISH OILS, AUTOXIDATION EX VIVO AND PEROXIDATION IN VIVO: IMPLICATIONS

John E. Kinsella

Institute of Food Science
Cornell University
Ithaca, NY 14853

INTRODUCTION

Because of their apparent efficacy in reducing the occurrence of a number of pathophysiologies related to heart disease, thrombosis, inflammation and immune functions the omega (w), or n-3 polyunsaturated fatty acids (n-3 PUFA) of seafoods and fish oils are receiving much attention (Kinsella et al., 1990a,b,c). Increased consumption of n-3 PUFA is being promoted and products containing n-3 PUFA are being developed. The n-3 PUFA being highly unsaturated have a tendency to undergo rapid autoxidation and cause off-flavors (Hsieh and Kinsella, 1989). Hence, there is some concern about the potential deleterious effects resulting from peroxidation of n-3 PUFA, especially _in vivo_.

Autooxidation of n-3 Polyunsaturated Fatty Acids (n-3 PUFA) The polyunsaturated fatty acids, particularly the eicosapentaenoic and docosahexaenoic acids of fish oils and seafoods can undergo rapid autoxidation ex vivo (Ke et al. 1977; Chan, 1987; Frankel, 1987a,b; Hsieh and Kinsella, 1989). The mechanisms which are responsible for the initiation of autoxidation, the reaction mechanisms and products of autoxidation have recently been reviewed (Kanner et al., 1987; Frankel, 1987; Hsieh and Kinsella, 1989). Generally, the autoxidation of polyunsaturated fatty acids increases with extent of unsaturation. For example, in model systems, the rate of oxidation of linolenic acid is 2.03 x $10^{-2}M^{1/2} \cdot sec^{1/2}$ while the corresponding rate for DHA is five-fold more rapid (Cosgrove et al., 1987).

It is generally recognized that the n-3 PUFA of fish oils can rapidly oxidize during storage at ambient and even refrigeration/frozen storage temperatures (Hsieh and Kinsella, 1989). Fritche and Johnston, (1988) demonstrated that menhaden oil when added to animal feeds underwent rapid autoxidation within 1 to 2 days during storage at room temperature with peroxide values increasing from 10 to around 200 milliequivalents (meq) per kilogram (Kg) of oil. There was a concurrent decrease in both EPA and DHA to around 90% of initial values, i.e. a 10% loss. Storage at -20°C in sealed polyethylene bags markedly decreased rates of oxidation as indicated by the peroxide values which only increased from 10 to 17 meq per Kg within one week. Samples stored at the -20°C but opened every other day and held at 23°C for up to 30 minutes showed a six fold increase in PV after one week. The inclusion of the antioxidant TBHQ at 0.02% by weight of lipids significantly reduced the rate of autoxidation at ambient temperatures and appeared to be an effective antioxidant in feed applications. The use of tocopherols especially in acetate form as an antioxidant in dietary oils is not very effective, ex vivo. However, tocopherol should be included in products intended for feed and/or food use to reduce peroxidation in vivo.

Nutritional and Toxicological Consequences of Food Processing
Edited by M. Friedman, Plenum Press, New York, 1991

There is concern about the possible adverse effects of peroxidized polyunsaturated fatty acid when ingested in foods because of the potential deleterious effects of acylhydroperoxides and polymerized fatty acids. The potential toxicity of oxidized fish oil was studied by Matuso, (1961), who reported that impaired growth rate and mortality of oxidized fish oils was correlated with the actual peroxide content. However, in these studies extremely oxidized oils were used at high levels which would be very improbable for human consumption. Rasheed et al. (1963) fed peroxidized menhaden oil (peroxide value (PV) 2,125 and 310 mg peroxides/100g respectively) at a level of 10% in the diet for 21 days. They observed lower body weight gain as the level of peroxides in the diet were increased. However, no statistically significant differences were found for any of the oils, except those containing over 300 milligrams peroxides per 100 grams of peroxidized oil.

Many of the studies on peroxidized oils are suspect because of the inordinate high amounts of peroxidized oils that have been fed, and in many cases, these diets contained inadequate amounts of vitamin E or other antioxidant components. Studies are needed to quantify the absorption of acyl hydroperoxides and to study their metabolic effects using modern analytical techniques and monitoring metabolic indices which are sensitive to acyl hydroperoxides. Fortunately, for most human applications, slightly oxidized oils, particularly fish oils, have markedly undesirable odors and flavors which causes immediate rejection. Nevertheless, because of the concern about oxidized products, fish oils intended for food uses must be rigorously refined, processed and stored under conditions that essentially eliminate autoxidative deterioration (Kinsella, 1987). Thus, the use of antioxidants, BHA, BHT and TBHQ, storage under nitrogen or carbon dioxide and storage at a cold/refrigerated temperatures is necessary. Overall, the general consensus indicates that fish oil becomes unpalatable at very low levels of oxidation and hence, the consumption of significant amounts of oxidized fish oil per se or foods containing fish oil is unlikely.

Metabolism of Fish Oils: In vivo Oxidation and Peroxidation The lipids of fish oil are efficiently digested by lipolytic enzymes in the intestinal tract and absorbed in a manner similar to other unsaturated fatty acids (Nelson and Ackman, 1988). The absorbed polyunsaturated fatty acids are facilely incorporated into chylomicron particles, transported to the liver and peripheral tissues where they are rapidly incorporated into lipoprotein particles, membrane lipids, and depending upon intake levels, stored in adipose tissue (Kinsella et al., 1990b).

Some dietary n-3 PUFA are oxidized in liver and probably other tissues as a source of energy. Because of the limited oxidation of long chain n-3 PUFA-CoA esters by mitochondria, the consumption of fish oil induces hepatic peroxisomal enzymes which facilely oxidize these fatty acids. Thus, in rat liver peroxisomal acyl-CoA oxidase increased over 250% following two weeks ingestion of fish oil and this was associated with an increase in catalase activity (150%) and a significant reduction of plasma triglycerides. Thus, fish oils stimulates peroxisomal beta-oxidation activity (Yamazaki, et al., 1987; Hagve and Christophersen, 1986), and the consumption of n-3 PUFA increases ketogenesis via the peroxisomal beta-oxidation system (Flatmark et al., 1988).

One of the most significant effects of dietary n-3 PUFA is the resultant increase in concentrations of n-3 PUFA (both EPA and DHA) in membrane phospholipids, particularly phosphatidyethanolamine and phosphatidylcholine the major membrane phospholipid components. These changes are usually associated with a decrease in long chain n-6 PUFA, particularly arachidonic acid (AA). These alterations may induce changes in membrane fluidity, microviscosity thereby affecting membrane bound enzymes and receptors and reducing eicosanoid synthesis (Lands, 1986,1987; Kinsella 1987; Kinsella et al., 1989; Chandra, 1989; Kinsella et al., 1990a,b).

However, in addition, the increased concentrations of n-3 PUFA may render such membranes more susceptible to peroxidative changes in vivo. Thus, in the classical literature, there is evidence that the consumption of fish is associated with yellow fat disease commonly observed in monogastric animals consuming fish oils. Yellow fat pigment may be frequently referred to as lipofuscin or ceroid material, which represents the polymerized, peroxidized polyunsaturated fatty acid/protein complexes that accumulate in tissue and generally represents advanced stages of peroxidation. (Jubb et al., 1985). The reactions responsible for the initiation of lipid peroxidation in vivo, the mechanisms of peroxidation and potential toxic effects have been reviewed (Horton and Fairhurst, 1987, Kanner et al., 1987, Marx, 1987).

Ruiter, et al. (1978) has discussed the peroxidation of mackerel oil in young swine as related to yellow fat disease. They attributed this to the ingestion of large amounts of n-3 PUFA without an adequate concomitant intake of vitamin E (Hartog et al., 1987). It is probable that the development of yellow fat disease mostly reflects chronic ingestion of dietary n-3 PUFA over long periods with inadequate amounts of supplemental vitamin E and other dietary antioxidants, e.g. selenium, vitamin C (Meydani et al., 1987, 1988).

Peroxidation of Dietary n-3 Fatty Acids: Peroxidation of lipids in cellular membranes has been implicated in several pathological conditions, including liver injury by toxins, depletion of glutathione, lung damage upon exposure to active oxygen, increased red blood cell permeability and fragility, and several facets of oxygen toxicity. Horton and Fairhurst, (1987) recently summarized the mechanisms of peroxidation and the various protective strategies available to cells to minimize these effects. Determination of lipid oxidation is an important criterion in assessing the effects of dietary factors on the rates and extent of peroxidation. Several methods have been used to quantify peroxidation of PUFA based on the measurement of products of oxidation, post-facto and usually ex vivo. These methods include the quantification of malondialdehyde (MDA) by thiobarbituric acid (TBA); detection of conjugated dienes by ultraviolet absorption, fluorescence analysis of lipid peroxidation products, (lipofuscins); detection of chemiluminescence, measurement of oxygen uptake, determination of lipid hydroperoxides in tissue, measurement of carbonyl compounds by dinitrophenylhydrazine; adduct formation and measurement of ethane and pentane formation (Pompella et al., 1987; Frankel et al., 1989).

Many of these can be facilely used ex vivo but the quantification of in situ peroxidation in vivo still remains a challenge. The quantification of ethane or pentane exhalation has been used as an index of in vivo peroxidation, however because of possible microbial generation of these gases, this method can be inaccurate. Duthie et al. (1987) has observed that ethane and pentane production was apparently not directly reflective of in vivo peroxidation. They suggested that other sources of hydrocarbons may mask their production from peroxidation of PUFA. Nevertheless, Hietanen et al., (1987) monitored ethane exhalation from rats consuming increasing concentrations of PUFA. There was an increase in exhalation of ethane as the PUFA intake was increased from 2 to 12.5% with little further change up to 25% intake of PUFA. Rats consuming PUFA exhaled considerably more ethane than those consuming saturated fatty acids.

Malondialdehyde (MDA), a major product of lipid peroxidation of PUFA, is a normal component of rat and human urine and the concentration of excreted MDA may provide a useful indicator of endogenous lipid peroxidation. Dhanakoti and Draper, (1987) demonstrated that the composition of dietary fat affected MDA excretion and was significantly higher in animals consuming n-3 PUFA (cod-liver oil) compared to animals ingesting corn oil (Table 1). The administration of adrenal corticotrophic hormones and epinephrine increased excretion of MDA suggesting that following lipolysis free fatty acids were oxidized more rapidly. MDA excretion was also markedly increased during fasting. The inclusion of antioxidants (30 IU and 300 IU DL-alpha tocopherol per kg diet) significantly decreased MDA excretion indicating in vivo reduction of PUFA peroxidation. These workers observed that the addition of 300 IU DL-alpha tocopherol acetate plus 0.5% BHT plus 0.5% vitamin C per kilogram of diet was very effective in reducing in vivo peroxidation i.e. by 37%. Incidentally, the quantity of MDA excreted represents only a small proportion of total endogenous production because much of the MDA is extensively degraded in vivo prior to excretion. Most urinary MDA is excreted as n-acetyl ϵ-propenyl-lysine (Dhanakoti and Draper, 1987).

Table 1. Effect of Dietary Fatty Acids and Antioxidants on Malondialdehyde (MDA) Excretion

Amount of Antioxidant included	HCO	10% CO Fed	10% CO Fasted	10% CO + 5% CLO Fed	10% CO + 5% CLO Fasted
			(μg MDA/24 hr)		
Low	2.7	2.1	4.1	6.4	5.6
High	2.6	2.1	2.6	6.1	3.8

Low = 30 IU Vit E/Kg; High = 300 IU DL α-tocol acetate + 0.5% BHT + 0.5% Vit C/Kg diet. MDA μg MDA excreted/24 hr. (Dhanakoti and Draper, 1987).

Humans consuming 30 mls of cod liver oil (CLO) per day for 14 days (providing about 5.0 grams of n-3 PUFA per day) excreted significantly greater amounts of MDA, i.e. 34.7 versus 24.5 ug MDA per day for controls on a normal diet. Consumption of 10 grams of MaxEPA per day for 50 days, (providing approximately 3 grams of EPA plus DHA, 10 IU of DL alpha-tocopherol acetate plus 100 ppm dodecyl gallate per day) caused no significant increase in MDA excretion compared to subjects consuming a normal free choice diet (Piche et al., 1988). The increase in MDA excretion in the subjects on CLO may have reflected the increased peroxide value of the ingested oil and the lack of added antioxidants as indicated by the marked reduction in MDA when refined fish oil i.e. menhaden oil, (maxEPA) with low peroxide value and added antioxidants were consumed.

Table 2. Relative Excretion of Malondialydehyde by Humans Consuming Fish Oils

	Control	CLO (14 days)	Max EPA (50 days)
	(MDA (μg/morning volume))		(MDA μg/day)
A (n=6)	24.5 ± 3.5	34.7 ± 2.5	-
B (n=7)	31.7 ± 5.9	49.1 ± 8.0	139 ± 15

Cod Liver Oil (CLO): 30 ml/day no antioxidants; MaxEPA 10 ml/day supplying 3g n-3 PUFA 10 IU Vit E + 100 ppm dodecylgallate (Piche et al., 1988).

This study suggested that prolonged consumption of n-3 PUFA increases vitamin E requirements, and indicated that the amount of antioxidants in MaxEPA capsules, i.e. 30 IU vitamin E and 100 ppm dodecyl gallate were effective in minimizing autoxidation prior to consumption and apparently had beneficial effects in minimizing peroxidation in vivo after consumption (Piche et al., 1988). The potential toxicity of peroxides and the concomitant generation of MDA a mutagen and possible carcinogen warrants further study. However, it should be noted that dietary n-6 polyunsaturated fatty acids also generate hydroperoxides and MDA (Horton and Fairhurst, 1987). Hence, the potential danger of these compounds is questionable because populations consuming high amounts of seafoods generally show a longer life expectancy (e.g. Japan, Iceland). The increased excretion of MDA derivatives of lysine should perhaps be examined to determine if extra amounts of lysine might be needed in the subjects consuming large amounts of dietary n-3 and/or n-6 polyunsaturated fatty acids.

Peroxidation in vivo: Biochemical Effects The PUFA acylated in membrane phospholipids can readily undergo peroxidation under physiological conditions to generate oxygenated compounds including hydroperoxides, epoxides, aldehydes, ketones (Kanner et al., 1987; Frankel, 1988). The primary autoxidation products of PUFA are the hydroperoxides (which can be further metabolized to alcohols, hydroperoxy, hydroxy, or keto-fatty acids) the earliest products of lipid peroxidation in biological membranes and are involved in cellular effects induced by oxidative stress. Acyl hydroperoxides are potent oxidizing agents either directly or via free radical mechanisms (Horton and Fairhurst, 1987; Kanner et al., 1987). These can perturb membrane structure, membrane function and also affect the activities of other oxidative enzymes for example, cyclooxygenase and lipoxygenase (Lands, 1985; Lands and Kulmac, 1986; Kanner et al., 1987; Hsieh and Kinsella, 1989).

Numerous oxygenated products of polyunsaturated fatty acids, particularly arachidonic acid (AA), are synthesized in vivo in response to normal metabolic and/or physiological stimuli by highly regulated enzymes namely cyclooxygenase and lipoxygenases. Generally the products of these reactions are the biologically active eicosanoids, i.e. prostanoids, leukotrienes and lipoxins which have many functions, but are rapidly inactivated and catabolized for excretion (Lands, 1986; Curtis-Prior, 1988). Under normal circumstances, the generation of these compounds is tightly regulated and excess hydroperoxides generated by these enzymes are rapidly converted to more stable hydroxy products (Curtis-Prior, 1988; Lands, 1986; Willis, 1987). However, in conjunction with the activities of these enzymes, particularly when endogenous antioxidant capacity and/or peroxidases may be inadequate or impaired, these hydroperoxides may accumulate and via free radical mechanisms initiate extensive unregulated peroxidation (Kanner et al., 1987; Frankel, 1987; Horton and Fairhurst, 1987). Generally, however, in the presence of adequate amounts of tocopherol (vitamin E) and peroxide scavenging systems deleterious peroxidative mechanisms are minimized.

Popella et al. (1987) recently compared many of the common methods used to assess lipid peroxidation using mouse livers depleted of GSH. Thus, measurements of MDA content, diene conjugation, decrease in polyunsaturated fatty acids and the formation of carbonyl compounds were all positively correlated and they concluded that these provided a reliable method for measuring lipid peroxidation, both in vivo and in vitro. However, it should be noted that measurement of concentration of polyunsaturated fatty acids is generally not a reliable index of peroxidation when changes are less than 10%, because these fatty acids may be utilized by pathways unrelated to lipid peroxidation.

The n-3 PUFA of fish oils and seafoods are highly unsaturated, (with several reactive alpha methylene groups), which increase the propensity towards random peroxidation in conjunction with the activities of monoxygenases/and dioxygenase enzyme systems. However, in certain cases, increased levels of acyl hydroperoxides (i.e. peroxide tone) may impart beneficial effects (Lands, 1986). Thus, increased peroxide tone enhances the conversion of EPA to the antiaggregatory PGI_3 and may thereby reduce platelet aggregability and thrombotic tendencies (Leaf and Weber, 1988). The observation that AA presumably via its hydroperoxides, can enhance the antiaggregatory action of both EPA and DHA is relevant in this regard (Croset et al., 1988).

Because plasma acyl hydroperoxides may provide a sensitive index of peroxidation in vivo, Lands et al. (1987) monitored the effects of dietary fish oils on plasma hydroperoxide levels. Acyl hydroperoxides can stimulate oxygenases in vitro and thereby generate additional levels of intracellular hydroperoxides. This amplification mechanism may occur in vivo and is associated with inflammatory processes (Lands, 1986). Elevated circulating hydroperoxides may represent a chronic oxidant stress which can be conducive to atherogenesis, inflammatory reactions and if uncontrolled, can result in pathophysiological deterioration. Concentrations of hydroperoxides from 0.5 to 3 uM are in the range reported to be damaging to vascular tissue (Lands et al., 1987; Warso and Lands, 1984). Lands et al., (1987) reported that the inclusion of EPA in diets fed to rabbits reduced the levels of acyl hydroperoxides in plasma, suggesting that n-3PUFA actually decreased acylhydroperoxide concentrations and could thereby decrease both cyclooxygenase and perhaps lipoxygenase activity. Such a mechanism could be involved in the beneficial impact of dietary n-3 PUFA in relation to atherosclerosis (Kinsella et al., 1990).

Nelson et al., (1988) O'Connor et al., (1989) and Karmali (1987) have reported that dietary n-3 PUFA can reduce chemically induced carcinogenesis and also reduce tumor growth. This has been attributed to the effective reduction of prostaglandin E_2 (PGE_2) synthesis by n-3 PUFA which reflects competitive inhibition of the prostaglandin synthetase and perhaps reduction of the peroxide tone, thereby reducing the activity of cyclooxygenase. Nelson et al. (1988) showed that the induction of colorectal tumors in response to a carcinogen was reduced in rats fed fish oil compared to rats ingesting corn oil. Coincidentally, the mean levels of peroxides in plasma were 1.4 and 20.3 uM for the fish oil and corn oil treated animals respectively. However, the authors questioned the validity of these data because of the marked variability in quantifying these peroxides. Nevertheless the data are consistent with the reduction of eicosanoids in animals consuming n-3 PUFA.

There has been reports that dietary polyunsaturated fatty acids in conjunction with anti-cancer agents can generate lipid hydroperoxides which may be useful in killing tumor cells. Thus, EPA is effective in killing tumor cells in vitro (Begin et al., 1986, 1987). Begin et al., (1987) suggested that specific polyunsaturated fatty acids, at the appropriate concentrations, may have a protective role against cancer development by inducing or mediating cytotoxic reactions in malignant cells, directly or indirectly via immune cells. Das et al., (1987) reported that tumor cells incorporated less AA and EPA than normal cells in culture, however, following their incorporation the extent of free radical generation was augmented in the tumor cells which could be a basis of cytotoxic mechanisms. They indicated that pro-oxidants accelerated whereas antioxidants, (vitamin E, BHA), inhibited this effect. However, curiously, in these studies DHA was much less effective than arachidonic acid or gamma-linolenic acid which is somewhat puzzling since DHA is much more susceptible to oxidation in vitro (Cosgrove et al., 1987).

Fox and DiCorletto (1988) reported that the decreased risk of cardiovascular disease and atherosclerosis associated with fish oil may be caused by reactions associated with peroxidation of n-3PUFA which decrease the production of a substance like the endothelial cell derived growth factor. This conclusion was based on the fact that antioxidants suppress the inhibitory activity of dietary fish oil (MaxEPA). The significance of this in vitro finding to in vivo situations obviously requires clarification.

PEROXIDATION: EFFECTS ON TISSUE AND ENZYME

The peroxidation of PUFA in membranes can affect organelle, tissue and organ functions (Tappel, 1973; Yagi, 1982; Horton and Fairhurst, 1987).

Heart: Because of the pronounced oxidative metabolism occurring in cardiac tissue, a number of workers have examined the effects of lipid peroxidation on cardiac functions. Oxygen radicals are actively involved in myocardial necrosis resulting from ischemia and reperfusion and these have been reports of increased generation of free radicals and MDA during ischemia (McCord, 1985). In this regard, there is concern that the relative quantities of polyunsaturated fatty acids in cardiac membranes might affect the extent of lipid peroxidation and injury to heart during ischemia. Dietary fish oil significantly alters the fatty acid composition of heart tissues (Swanson and Kinsella, 1986; Charnock et al., 1986; Hock et al., 1987; Nalbone et al., 1988).

Lamers et al., (1988) studied the effects of dietary mackerel oil (9.1% by weight) or lard following 8 weeks ingestion on the composition of cardiac membranes, the formation of MDA and changes in glutathione GSH following ischemia and reperfusion using swine. The mackerel oil reduced both LA and AA levels in these animals by approximately 50%; this was associated with a 16-fold increase in EPA and a 5-fold increase in DHA in heart tissue after 8 weeks. During the feeding trial, the plasma levels of MDA and of peroxidized lipids were no higher in pigs receiving the mackerel oil. Experimental ischemia and reperfusion did not result in enhanced MDA production in cardiac tissue of animals receiving the fish oil and the recovery of regional functions after ischemia was similar for both groups. The authors concluded that the dietary fish oil had no adverse effects and did not increase susceptibility to free radical peroxidation (Lamers et al., 1988).

These data are not inconsistent with the reports that dietary fish oil in fact significantly reduce cardiac injury following ischemia and reperfusion and also reduce the extent of arrythmia (Leaf and Weber, 1988; Kinsella et al., 1990). These beneficial effects are apparently partly attributed to the reduce generation of thromboxane and lipoxygenase products in animals consuming n-3 PUFA (Leaf and Weber, 1988; Lamers et al., 1987). Dietary fish oil significantly alters the fatty acid composition of heart tissues (Swanson and Kinsella, 1986; Charnock et al., 1986; Hock et al., 1987; Nalbone et al., 1988). The accumulation of n-3 PUFA in rat heart following consumption of diets containing 12.5% salmon oil plus 4.5% corn oil for 8 weeks was associated with the appearance of lipofuscin materials in the ventricles (Nalbone et al., 1989). There was a significant increase in activity of heart GSH peroxidase whereas liver GSH peroxidase activity and GSH-S-transferase remained unchanged indicating that dietary n-3 PUFA apparently did not affect selenium status.

A number of workers have reviewed the occurrence of cardiac lipidosis and chronic lesions in rat hearts (Charnock et al., 1987). The accumulation of lipid droplets in heart muscle cells has been frequently observed in experimental animals, especially the rat, and may represent the accumulation of unoxidizable fatty acids. Charnock et al., (1987) reviewed different aspects of this and demonstrated that rats consuming a diet containing 12% tuna oil developed cardiac lipidosis compared to animals consuming normal chow or sunflower oil. In addition, lipofuscin pigmentation of the perirenal adipose fat of rats consuming tuna fish oil (which incidentally had a rather similar fatty acid composition to menhaden oil, being low in 22:1) was observed. However, there was also considerable lipidosis in the hearts of animals consuming safflower oil diets. A causal relationship between cardiac lipidosis and necrosis has not been demonstrated and Charnock et al. (1987) concluded that lesions observed in animals are most likely age related rather than related to a specific dietary effect. In fact Charnock et al., (1987) stated that in long term studies with rats, there is strong evidence that dietary PUFA are of great benefit to cardiac performance, significantly decrease the susceptibility of heart to experimentally induced arrhythmias and infarction and result in a general reduction of thrombogenesis. They suggest that additional dietary vitamin E may be desirable in the long term feeding of n-3 PUFA to animals.

The cumulating evidence generally indicates that dietary omega-3 fatty acids provide significant benefits in minimizing cardiac malfunctioning, arrthymia and ischemic damage (Kinsella, et al., 1990). The evidence suggests the need for more long term studies using animals species other than rats, (for example swine), to study the long term effects of ingesting n-3 PUFA in diets containing adequate amounts of vitamin E.

<u>Brain:</u> Chaudiere et al., (1987) reported that rats consuming a diet enriched in n-3 PUFA i.e. menhaden oil, caused an increase in the content of EPA and DHA in brain phospholipids. These changes were not associated with any alteration in either vitamin E concentration, glutathione peroxidase nor catalyase activities in the cerebrum and cerebellum. No increase in peroxidative damage was observed in these animals, in fact n-3 PUFA may indirectly reduce ischemic damage by attenuating eicosanoid synthesis (Kinsella et al., 1990).

Effects on Enzymes: Consumption of n-3 PUFA may affect the activity of enzymes involved in antioxidant functions and scavenging of active oxygen species or their precursors, however, the literature on these enzymes is scant and some studies report increases, whereas others report decreases in the activities of these enzymes. The consumption of herring oil enhances cytochrome P-450 activity in liver microsomes. (Mounie et al., 1986). Microsomal epoxide hydrolase, uridine diphosphate glucuronidase and plasma transaminase activities were higher in animals consuming fish oils. In these particular studies, however, comparable increases were noticed in animals consuming corn oil. Wade et al., (1986) reported that the activity of glutathione S transferase was affected by dietary lipid levels including dietary menhaden oil.

The consumption of fish oil, (MaxEPA, 20 ml per day) by 20 adult patients suffering from hyperlipidemia caused a decrease in triglycerides after 8 weeks and an increase in glutathione peroxidase activity in erythrocytes and platelets and a concomitant reduction of MDA production. There was no change in the concentration of plasma selenium nor alpha-tocopherol and the authors suggested that the possible increase in the formation of lipoperoxides was counteracted by the increased activity of the scavenger enzyme glutathione peroxidase (Olivieri et al., 1988). Adequate glutathione peroxidase activity is important for the maintenance of prostacyclin synthesis by blood vessels since hydroperoxides could inhibit the formation of prostacyclin (Yamazaki et al., 1987).

Yamazaki (1987) noted that the n-3 PUFA caused no impairment of glutathione dependent detoxification activities, i.e. glutathione peroxidase nor glutathione S transferase. This is noteworthy because clofibrate and other hypotriglyceridemic agents may cause oxidative damage by depletion of these enzymes. The lack of detrimental effects of fish oil on these activities suggest their increased safety over other hypotriglyceridemic agents in therapeutic applications (Yamazaki et al., 1987).

Erythrocytes from psoriatic patients contain elevated arachidonic acid levels and show increased generation of MDA and enhanced glutathione peroxidase activity even though alpha-tocopherol levels are within to normal range. Supplementation of psoriatic subjects with fish oil for two months increased EPA and DHA concentration in erythrocyte membranes with a subsequent marked reduction of MDA generation and a stimulation of glutathione peroxidase in both erythrocytes and platelets of these patients (Corrocher et al., 1989). The mechanism of this effect was not explained.

VITAMIN E AND DIETARY N-3 PUFA

Diets supplemented with cod liver oil have been reported to produce muscle lesions in various animals. This has been attributed to deficient vitamin E since this effect could be prevented by the administration of alpha-tocopherol (McKenzie et al., 1941). An adequate vitamin E intake is essential for stabilizing biological membranes and providing a protective effect against random peroxidation of PUFA. An increased intake of polyunsaturated fatty acids increase the requirements for vitamin E in order to provide adequate protection and also to compensate for the impaired absorption and increased utilization of vitamin E in diets containing high amounts of PUFA, particularly n-3 fatty acids (Meydani et al., 1987; Muggli, 1989).

Dietary fish oils can affect tocopherol status and Meydani et al., (1987) observed that animals consuming fish oil had a lower tissue level of alpha-tocopherol than animals fed an equivalent amount of corn or coconut oil. Thus, tocopherol levels which may be adequate in certain diets may be inadequate in diets containing n-3 PUFA and result in symptoms of muscular dystrophy. Mice fed increasing levels of vitamin E (i.e. 30, 100, 500 ppm) showed progressive increases in plasma tocopherol concentrations in animals consuming either coconut oil or corn oil, whereas there was no significant increase in those animals on fish oil. This was attributed to decreased absorption of tocopherol in the presence of n-3 PUFA or to the enhanced post-absorptive utilization of tocopherol. Mice fed very high levels of tocopherol, i.e. 500 ppm had plasma and liver tocopherol levels comparable to those fed 30 ppm in the presence of corn or coconut oil. This agrees with previous observations that concurrent administration of fish oil and tocopherol may not alleviate the signs of tocopherol deficiency. These studies indicate the need to supplement the fish oils with adequate amounts of tocopherol to ensure adequate absorption and tissue concentrations.

Nalbone et al. (1988) fed salmon oil (12.5% by weight, plus 4.5% corn oil) to rats for 2 months. There was extensive replacement of n-6 by n-3 PUFA and serum vitamin E levels decreased in rats consuming salmon oil despite the fact that these diets were supplemented with 100 mg of alpha-tocopherol per 100g of oil. This observation emphasized the need to supplement the diets with vitamin E when n-3 PUFA are included. Kockmann, et al., (1988) reported that low doses of vitamin, E which increased plasma and platelet alpha-tocopherol levels had no significant effect on arachidonic acid concentrations, cyclooxygenase nor lipoxygenase metabolism. Vitamin E had no effect on the lipoxygenation of eicosapentaenoic acid.

Shapiro et al., (1988) examined the effects of fish oil supplementation i.e. 18 capsules of maxEPA per day (providing around 6g n-3 PUFA) on plasma alpha-tocopherol levels in 10 human subjects consuming a normal diet. The alpha-tocopherol content of oil was 0.3 mg per gram of oil providing a total of 5.4 mg of vitamin E per day. There was a significant increase in plasma tocopherol from 0.81 initially versus 63 mg per deciliter after six weeks of fish oil supplementation. The fish oil was consumed as 18 grams per day in addition to a normal western type diet. Recently, Leika et al., (1989) reported that alpha-tocopherol absorption was lower in the presence of fish oil than an equivalent amount of corn oil, indicating that dietary n-3 PUFA may decrease vitamin E absorption.

Because of the apparent decreased absorption of tocopherol in the presence of n-3 PUFA, supplementation of diets may be necessary when fish oils are being ingested. In this regard, MaxEPA capsules contain approximately 1300 μg α-tocopherol per gram of oil containing 300 mg of n-3 PUFA. Consumption of MaxEPA (18 grams per day) resulted in significantly increased plasma alpha-tocopherol levels following six week ingestion by ten healthy adult males (Shapiro et al., 1989).

Leedle and Aust (1986) reported that liposomes having PUFA to vitamin E ratios around 250 nmol PUFA per nmol vitamin E tended to be resistant to peroxidation. However, is not known how relevant this is to in vivo situations where oxygen concentration and peroxidative tendency may be markedly different. Vitamin E is located in the membrane in association with unsaturated fatty acids and hence is effective in minimizing localized peroxidation. Peroxidation of tocopherol occurs during normal biological functions where peroxides are generated, e.g. during the phagocytosis or in pathological processes. Thus, a constant supply of tocopherol, or regeneration of tocopherol by vitamin C or its preservation or conservation through the preferential utilization of other biological antioxidants are all factors that need to be considered in the determining vitamin E intake in conjunction with n-3 PUFA. In this regard, vitamin C may be significant in terms of regenerating tocopherol (Burton and Ingold, 1986).

Data concerning appropriate levels of supplementation for humans are scarce. Generally the amount of vitamin E required for protection against autoxidation and peroxidation increases with the degree of unsaturation of the component fatty acids. The data from the literature indicate that for each gram of linolenic acid in the diet, a minimum additional requirement of 0.6 mg α-tocopherol is desirable. Muggli (1989) developed a formula to estimate the amount of vitamin E needed to compensate for the elevated vitamin E required upon increasing the intake ofpolyunsaturated fatty acids.

$$m_{\text{Vit E}} = 0.2 \times 10^{-3} (0.3\, m_1 + 2\, m_2 + 3\, m_3 + 4\, m_4 + 5\, m_5 + 6\, m_6)$$

where $m_{\text{Vit E}}$ = moles of d-alpha-tocopherol and m_n = moles of unsaturated fatty acid where n denotes the number of double bonds. Using this equation, the following amounts i.e. 0.89, 1.34, 2.24 and 2.24 mg of supplementary dl-α-tocopherol are required per gram of dietary linoleic, linolenic, eicosapentaenoic and docosahexaenoic acid, respectively. Supplementation of a diet with 500 - 1000 mg of EPA and 400-700 mg DHA requires the addition of 1.5 to 3.0 mg d-alpha-tocopherol or an additional 15 to 30% of current US RDA for vitamin E. Muggli, (1989) suggests that the formula can be used to estimate intake but he concluded that it may be prudent to double or triple this amount in practice.

Conclusion

Based on the clinical studies with MaxEPA preparations which contain vitamin E, the data overall strongly suggests that the intake of the dietary omega-3 fatty acids should not necessarily exert any deleterious effects via peroxidation when adequate levels of vitamin E and other complementary antioxidants are concurrently consumed.

ACKNOWLEDGEMENT: This work is a result of research sponsored by the NOAA Office of Sea Grant, U.S. Department of Commerce, under Grant #NA90AA-D-SG078 to the New York Sea Grant Institute.

REFERENCES

Begin, M.E., Ells, G. and Das, U.N. (1986) Selected fatty acids as possible intermediates for selective cytotoxic activity of anticancer agents involving oxygen radicals. Anticancer Res. 6(2), 291-5.

Begin M.E. (1987) Effects of polyunsaturated fatty acids and of their oxidation products on cell survival. Chem. Phys. Lipids. 45(2-4), 269-313.

Burton, G.W. and Ingold, K.U. (1986) Vitamin E applications of the principles of physical organic chemistry to the exploration of its structure and function. Acc. Chem. Res. 19, 194.

Chan, H.W. (1987) Autoxidation of unsaturated lipids. Academic Press, London.

Charnock, J.S., Abeywardena, M.Y. and McLellan, R.L. (1986) Comparative changes in the fatty acid composition of rat cardiac phospholipids after long-term feeding of sunflower seed oil or tuna fish oil supplemented diets. Ann. Nutr. Metab. 30:393-406.

Charnock, J.S., Turner, J. and MacIntosh, G.H. (1987) The occurrence of cardiac lipidosis and the chronic lesions in the hearts of rats following long term feeding of different lipid supplemented diets. J. Nutr. Sci. Vitaminol. 33, 75.

Chaudiere, J., Clement, M., Driss, J. and Bourre, J.M. (1987) Unaltered brain membranes after prolonged intake of highly oxidizable long-chain fatty acids of the (n-3) series. Neurosci-Lett. 82(2), 233-9.

Corrocher, R., Ferrari, S., de-Gironcoli, M., Bassi, A., Olivieri, O., Guarini, P., Stanzial, A., Barba, A.L. and Gregolini, L. (1989) Effect of *fish* oil supplementation on erythrocyte lipid pattern, malondialdehyde production and glutathione-peroxidase activity in psoriasis. Clin. Chem. Acta. 179(2), 121-31.

Cosgrove, J.P., Church, D.F. and Prior, W.A. (1987) The kinetics of the autoxidation of polyunsaturated fatty acids. Lipids. 22(5), 299-304.

Croset, M., Guichardant, M. and Lagarde, M. (1988) Different metabolic behavior of long chain n-3 PUFA in human platelets. Biochim. Biophys. Acta. 961, 262.

Curtis-Prior, P.B. (1988) Prostaglandins, Biology and Chemistry of Prostaglandins and Related Eicosanoids. Churchill Livingston, NY.

Das, U.N., Huang, Y.S., Begin, M.E., Ellis, G. and Horrobin, D.F. (1987) Uptake and distribution of cis-unsaturated fatty acids and their effect on free radical generation in normal and tumor cells in vitro. Free Radic. Biol. Med. 3, 9-14.

Dhanakoti, S.N. and Drapper, H.H. (1987) Response of urinary malondioldehyde to factors that stimulate lipid peroxidation in vivo. Lipids 22, 643.

Duthie, G., Arthur, J.R. and Mills, C.F. (1987) Tissue damage in vitamin E deficient rats is not detected by expired ethane and pentane. Free Radical Research Commun. 4, 21.

Flatmark, T., Nilsson, A., Kvannes, J., Eikhom, T.S., Fukami, M.H., Kryvi, H. and Christiansen, E.N. (1988) On the mechanism of induction of the enzyme systems for peroxisomal beta-oxidation of fatty acids in rat liver by diets rich in partially hydrogenated fish oil. Biochem. Biophys. Acta. 962(1),122-30.

Frankel, E.N., Hu, M.L. and Tappel, A.L. (1989) Rapid head space gas chromatography as a measure of lipid peroxidation in biological samples. Lipids 24, 976.

Fox, P.L. and DiCorleto, P.E. (1988) Fish oils inhibit endothelial cell production of platelet-derived growth factor-like protein. Science 241(4864), 453-6.

Frankel, E.N. (1987) Secondary products of lipid oxidation. Chem. Phys. Lipids 44, 73.

Frankel, E.N. (1987) Biological significance of secondary lipid oxidation products. Free Rad. Res. Comm. 3, 213.

Fritsche, K. and Johnston, P.V. (1988) Rapid autoxidation of fish oil in diets without added antioxidants. J. Nutr. 118:425.

Hagve, T.A. and Christophersen, B.O. (1986) Evidence for peroxisomal retroconversion of adrenic acid (22:4(n-6)) and docosahexaenoic acids (22:6(n-3)) in isolated liver cells. Biochem. Biophys. Acta. 875(2), 165-73.

Harris, W.S. (1989) Fish oils and plasma lipid and lipoprotein metabolism in humans: A critical review. J. Lipid Res. 30, 785.

Hartog, J.N., Glammers, J.M., Montfoort, A., Becker, A.E., Klompe, N., Morse, H., Tencate, F.J., Vanderwert, L., Huelsman, N.W.C., Hugenholtz, P. and Verdouw, P.D. (1987) Comparison of mackerel oil and lard fed enriched diets on plasma lipids, cardiac membrane phospholipids, cardiovascular performance and morphology in young pigs. Am. J. Clin. Nutr. 46, 258.

Hietanen, E., Ahotupa, M., Bereziat, J.C., Bussacchini, V., Camus, A.M. and Bartsch, H. (1987) Elevated lipid peroxidation in rats induced by dietary lipids and N-nitrosodimethylamine and its inhibition by indomethacin monitored via ethane exhalation. Toxicol. Pathol. 15(1), 93-6.

Hock, E.C., Holahan, M.A. and Reibel, D.K. (1987) Effect of dietary fish oil on myocardial phospholipids and myocardial ischemic damage. Am. J. Physiol. 252, H554-H560.

Horton, A.A. and Fairhurst, S. (1987) Lipid peroxidation and mechanisms of toxicity. CRC Rev. in Toxicology 18:27.

Hsieh, R. and Kinsella, J.E. (1989) Oxidation of polyunsaturated fatty acids: Mechanisms, products and inhibition with emphasis on fish. Advances Food & Nutr. Res. 33, 233.

Jubb, J.V., Kennedy, P.C., and Palmer, N. (1985) Pathology of domestic animals 3rd edition, Academic Press, Orlando, FL.

Kanner, J., German, B. and Kinsella, J.E. (1987) Initiation of lipid peroxidation in biological systems. Crit. Rev. Food Sci. and Nutr. 25, 317.

Karmali, R.A. (1987) Omega-3 fatty acids and Cancer: A review. In Polyunsaturated Fatty Acids and Eicosanoids. Lands, W.E.M., editor, Champaign, IL. AOCS p. 225.

Ke, P.J., Nash, D. and Ackman, R.G. (1977) Mackerel skin lipids as an unsaturated fat model system for determination of antioxidant potency of TBHQ and other antioxidant compounds. J. Am. Oil. Chem. Soc. 54, 417.

Kinsella, J.E. (1987) Seafoods and Fish Oils in Human Health and Disease. Marcel Dekker, Inc. Publishers, NY. p. 1-400.

Kinsella, J.E., Lokesh, B. and Stone, R. (1989) Dietary n-3 polyunsaturated fatty acids and amelioration of cardiovascular disease: Possible mechanisms. Am. J. Clin. Nutr. 46 (press).

Kinsella, J.E., Broughton, S.K. and Whelan, J.W. (1990) Dietary unsaturated fatty acids: interactions and possible needs in relation to eicosanoid synthesis. J. Nutr. Biochem. 1, 123-141.

Kinsella, J.E. and Lokesh, B. (1990) Dietary lipids, eicosanoids and the immune system. Critical Care Med. 18(2), S94-S113.

Kockmann, V., Vericel, E., Croset, M. and Lagarde, M. (1988) Vitamin E fails to alter the aggregation and oxygenated metabolism of arachidonic acid in normal human platelets. Prostaglandins, 36, 67.

Lamers, J.M., Hartog, J.M., Guarins, F., Vaona, I., Verdouw, P.D. and Koster, J.F. (1988) Lipid peroxidation in normoxic and ischaemic-reperfused hearts of fish oil and lard fat fed pigs. J. Mol. Cell. Cardiol. 20(7), 605-15.

Lamers, J.M., Hartog, J.M, Verdouw, P.D. and Hulsamm, W.C. (1987) Dietary fatty acids and myocardial function. Basic Res. Cardiol. 82 Suppl 1., 209-21.

Lands, W.E.M. (1986) Fish and Human Health. Academic Press, Orlando, FL.

Lands, W.E.M. Editor. (1987) Polyunsaturaed fatty acids and eicosanoids. Am. Oil Chem. Soc., Champaign, IL.

Lands, W.E.M. (1985) Interactions of lipid hydroperoxides with eicosanoid biosynthesis. J. Free Rad. Biol. Med. 1, 97.

Lands, W.E.M. and Kulmac, R.J. (1986) The regulation of the biosynthesis of prostaglandins and leukotrienes. Prog. Lipid Res. 25, 105.

Lands, W.E.M., Miller, J.F. and Rich, S. (1987) Influence of dietary fish oil on plasma lipids hydroperoxides. Advance PG. TX and Leukotr. Res. 17, 876.

Leaf, A. and Weber, P.C. (1988) Cardiovascular effects of n-3 fatty acids. New Eng. J. Med. 318, 549.

Leedle, R.A. and Aust, S.D. (1986) Importance of the polyunsaturated fatty acid to vitamin E ration in the resistance of rat lung microsomes to lipid peroxidation. J. Free. Radic. Biol. Med. 2(5-6), 397-403.

Leika, L.S., Remali, K.M., Bizinkausakas, P.A. and Meydani, M. (1989) Effect of fish oil in intestinal absorption of vitamin E in the rat. FASEB J. 3, #4214.

McKenzie, G.G., McKenzie, J.B. and Mccollum, E.V. (1941) J. Nutr. 5, 21, 225.

Marx, J.L. (1987) Oxygen free radicals linked to many diseases. Science 235, 529.

Matsuo, N. (1961) Studies on the toxicity of fish oil. Toxicity of Autoxidized unsaturated fatty acid ester applied to rat. Tokushima, J. Expt. Med. 8, 90.

McCord, J.M. (1985) Oxygen derived free radicals in post-ischemic tissue injury. New Engl. J. Med. 312, 159.

Meydani, S.N., Yogeesarran, G., Liu, S., Baskar, F. and Meydani, M. (1988) Fish oil and tocopherol induced changes in natural killer cell - mediated cytotoxicity and PGE2 synthesis in young and old mice. J. Nutr. 118, 1245.

Meydani, S.N., Shapiro, S.C., Meydani, M., Macaulley, J.V. and Blumberg, J.B. (1987) Effect of age and dietary fat on tocopherol status of C57BL mice. Lipids 22, 345.

Meydani, S.N., Stocking, L.M., Shapiro, A.C., Meydani, M. and Blumberg, J.B. (1988) Fish oil and tocopherol induced changes in ex-vivo synthesis of spleen and lung leukotriene B4 in mice. Ann. NY Acad. Sci. 524:395-398.

Meydani, S.N., Shapiro, S.C., Meydani, M., Macaulley, J.V. and Blumberg, J.B. (1987) Effect of age and dietary fat on tocopherol status of C57BL mice. Lipids 22, 345.

Meydani, S.N., Stocking, L.M., Shapiro, A.C., Meydani, M. and Blumberg, J.B. (1988) Fish oil and tocopherol induced changes in ex-vivo synthesis of spleen and lung leukotriene B4 in mice. Ann. NY Acad. Sci. 524:395-398.

Mounie, J.B., Faye, J., Maqdalou, H., Gondonnet, R., Reuchot, and Siest, G. (1986) Modulation of UDP glucoronosyltransferase activity in rats by dietary lipids. J. Nutr. 116, 2034.

Muggli, R. (1989) Dietary fish oils increase the requirement for vitamin E in humans. In Health Effects of Fish and Fish Oils, Chandra, S.K. editor, ARTS Press, St. Johns, Newfoundland, 1989.

Nalbone, G., Termine, E., Leonardi, J., Portugal, H., Lechene, P., Calaf, R., Lafonte, R. and Lafonte, H. (1988) Effect of dietary salmon oil feeding on rat heart lipid status. J. Nutr. 118, 809-17.

Nalbone, G., Leonardi, J., Termine, E., Portugal, H., Lechene, P., Pauli, A.M. and Lafonte, H. (1989) Effects of fish oil, corn oil and lard diets on lipid peroxidation status and glutathione peroxidase activities in rat heart. Lipids 24, 179.

Nelson, G.J. and Ackman, R. (1988) Absorption and transport of fat in mammals with emphasis on n-3 polyunsaturated fatty acids. Lipids 23, 1005.

Nelson, R.L., Tenaure, J.C., Andrianopolous, G., Souza, and Lands, W.E. (1988) A comparison of dietary fish oil and corn oil in experimental colo-rectal carcinogenesis. Nutr. Cancer 11, 215.

O'Connor, T.P., Roebuck, B.D., Peterson, F.J., Lokesh, B., Kinsella, J.E. and Campbell, T. (1989) Effect of dietary w-3 fatty acids on development of azaserien induced preneoplastic lesions in rat pancreas. J. Natl. Cancer Institute 81, 858.

Olivieri, O, Negri, M., De-Gironcoli, M., Bassi, A., Guarini, P., Stanzial, A.M., Grigolini, L., Ferrari, S. and Corrocher, R. (1988) Effects of dietary fish oil on malondialdehyde production and glutathione peroxidase activity in hyperlipidaemic patients. Scand. J. Clin. Lab. Invest. 48(7), 659-65.

Piche, L.A., Draper, H.H. and Cole, P.D. (1988) Malendehyde excretion by subjects consuming cod liver oil and a concentrate of n-3 fatty acids. Lipids 23, 370.

Pompella, A., Maellaro, E., Casine, A., Ferrali, M. Ciccoli, L. and Comporti, M. (1987) Measurement of lipid peroxidation in vivo, a comparison of different procedures. Lipids 22, 206.

Rasheed, A.A., Oldfield, J., Kaufmes, J., Sinnhuber, R.O. (1963) Nutritional value of marine oils. 1. Menhaden oil at varying oxidation levels with and without antioxidants in rat diets. J. Nutr. 79, 323.

Ruiter, A., Jung-Bloed, W., Vangent, C.M., Danse, L.H. and Metts, S.H. (1978) The influence of dietary mackerel oil on the condition of organ and on blood lipid composition in the young growing pig. Am. J. Clin. Nutr. 31, 2159.

Shapiro, A.C., Meydani, S.N., Meydani, M., Blumberg, J., Andres, S. and Dinarello, C. The effect of fish oil supplementation on human plasma alphatocopherol levels. In Health Effects of Fish and Fish Oils, R.K. Chandra editor. St. Johns Newfoundland, Canada, Press, (1989).

Shapiro, A., Meydani, S., Meydani, N. and Shaeffer, A.E. (1989) Effect of fish oil supplementation on plasma alpha-tocopherol, retinol, triglyceride and cholesterol levels. FASEB Abstr. #4204, FASEB J. 3.

Swanson, J.E. and Kinsella, J.E. (1986) Dietary n-3 polyunsaturated fatty acids: modification of rat crdiac lipids and fatty acid composition. J. Nutr. 116:514-523.

Tappel, A. (1973) Lipid peroxidation damage to cell components. Fed. Proc. 32, 1870.

Wade, A., Bellows, J. and Dharwadkar, S. (1986) Influence of dietary menhaden oil on enzymes metabolizing drugs and carcinogens. Drug. Nutr. Interactions 4, 339.

Warso, M.A. and Lands, W.E.M. (1984) Presence of lipid hydroperoxide in human plasma. J. Clin. Invest. 75, 667.

Willis, A.L. (1987) Handbook of Eicosanoids, Prostaglandins and Related Compounds. CRC Press, Boca Rato, FL.

Yamazaki, R.K., Shen, T. and Schade, G.B. (1987) A diet rich in (n-3) fatty acids increases peroxisomal beta-oxidation activity and lowers plasma triacylglycerols without inhibiting glutathione-dependent detoxication activities in the rat liver. Biochem. Biophys. Acta. 920(1), 62-7.

Yagi, K. Lipid peroxides in biology and medicine. Academic Press, NY. (1982).

19

FORMATION AND ACTION OF ANTICARCINOGENIC FATTY ACIDS

M. W. Pariza, Y. L. Ha, H. Benjamin, J. T. Sword, A. Grüter, S. F. Chin, J. Storkson, N. Faith, and K. Albright

Food Research Institute, Department of Food Microbiology and Toxicology, University of Wisconsin--Madison
1925 Willow Drive, Madison, WI 53706

ABSTRACT

Conjugated dienoic derivatives of linoleic acid (referred to by the acronym CLA) constitute a newly recognized class of anticarcinogenic fatty acids. Of the eight major CLA isomers, the cis-9, trans-11 isomer alone is incorporated into phospholipid and may to be the most biologically relevant isomer. CLA exhibits potent antioxidant activity; evidence is presented indicating that CLA acts both as an in vitro and in vivo antioxidant. The formation of CLA in foods, and its possible biological significance in cell membranes, is discussed.

INTRODUCTION

The relationship between dietary fat and cancer is complicated by many factors, in particular total caloric intake and fatty acid composition (Pariza, 1988). The first of these is general, in that any digestible fat may serve as a source of excess calories. By contrast, the effect of fatty acid composition is much more specific. Among the numerous fatty acids that comprise lipid, only linoleic acid is clearly linked to the enhancement of carcinogenesis. The mechanism of linoleic acid enhancement is not understood but is probably related, at least in part, to the fact that this fatty acid is essential.

Given the well-established enhancing effect of linoleic acid on carcinogenesis, we find it of great interest that conjugated dienoic derivatives of linoleic acid (CLA) should inhibit carcinogenesis in animal models (Ha et al., 1987; 1990). We have shown that CLA inhibits carcinogen-induced neoplasia in both mouse epidermis and forestomach. The mechanism of action of CLA appears related in part to its antioxidant properties. In addition, there is preliminary evidence indicating that CLA inhibits the induction of ornithine decarboxylase (ODC) activity by 12-0-tetradecanoylphorbol-13-acetate (TPA) (Benjamin et al., 1990), indicating that CLA may inhibit the promotion as well as the initiation stage of carcinogenesis.

FORMATION OF CLA IN FOODS

We isolated CLA from grilled ground beef extracts and established

Nutritional and Toxicological Consequences of Food Processing
Edited by M. Friedman, Plenum Press, New York, 1991

that it was an anticarcinogen (Ha et al., 1987). We then developed methods for isolating and quantifying the material from foods, and applied the methods to natural and processed cheeses (Ha et al., 1989). Virtually all of the CLA in dairy products was esterified in triglycerides. One cheese spread contained particularly high levels of CLA, leading us to conclude that processing conditions could influence the concentration of these anticarcinogenic fatty acids.

The amount of CLA in ground beef is increased by cooking (Ha et al., 1989). For example, the fat phase of uncooked ground beef contained 2050 ppm CLA, whereas after grilling the value was almost 9300 ppm. Since the cooking conditions per se were unlikely to have directly transformed linoleic acid into CLA, it is likely that factors comprising the microenvironment of the beef patty are also involved. In this regard the proteins in the beef patty matrix would seem to be particularly attractive candidates for further study.

One of the ingredients in the cheese spread containing particularly high levels of CLA is whey (Ha et al., 1989). This led us to postulate that whey protein might be a critical factor in converting linoleic acid to CLA. We have obtained evidence in support of this hypothesis. For example, in one experiment the CLA content of butterfat was doubled (from 935 ppm to 1888 ppm) following addition of whey, gassing with nitrogen, and subjecting the mixture to a boiling water bath for 15 minutes. The whey itself contained only a trace amount of CLA (23 ppm CLA), indicating that the observed increase was due to the de novo production of CLA. We are currently investigating which whey protein(s) are involved in this reaction. Dormandy and Wickens (1987) reported that CLA is produced when linoleic acid, either free or bound in triglycerides, is mixed with albumin and exposed to UV light. This mechanism was proposed to explain their finding of CLA in human blood, and its apparent modulation by oxidative stress (e.g., CLA is reported to be increased in the blood of alcoholics, and to decrease during periods of alcohol withdrawal) (Dormandy and Wickens, 1987).

CLA is a normal product of the metabolism of linoleic acid by rumen bacteria. The pathway has been studied extensively in the strict anaerobe, *Butyrivibrio fibrisolvens* (Kepler and Tove, 1967). The first step in the biohydrogenation of linoleic acid by *B. fibrisolvens* is the formation of cis-9, trans-11 CLA, a reaction catalyzed by a membrane-bound isomerase. The isomerase acts only on free linoleic acid, and requires neither nucleotide cofactors nor a hydrogen atmosphere. It is not yet known if the reaction catalyzed by this isomerase is related to reactions resulting in the conversion of linoleic acid to CLA that may occur in food, particularly where whey or other proteins may be involved.

MECHANISM OF ACTION OF CLA

Incorporation of CLA into phospholipid

Understanding the mechanism whereby CLA inhibits neoplasia is dependent on knowing the fate of CLA following ingestion. We found that CLA administered p.o. is taken up and subsequently incorporated into body lipids. All isomers appear in triglycerides whereas only the cis-9, trans-11 isomer is found in phospholipid. This observation led us to theorize that the cis-9, trans-11 isomer is the active anticarcinogenic CLA isomer (Ha et al., 1990). Interestingly, this is the same isomer that is synthesized from linoleic acid during its biohydrogenation by *B. fibrisolvens* (Kepler and Tove, 1967), as discussed above.

Antioxidant activity of CLA

We have shown that CLA is an effective antioxidant both in vitro and in vivo. Synthetically-prepared CLA was tested for its capacity to protect linoleic acid from oxidation. Dose-response studies were conducted and it was determined that CLA worked best when used at a ratio of 1:1000 (CLA:linoleic acid) (Ha et al., 1990).

In another very recent experimental series we isolated liver microsomes from mice treated with CLA in olive oil, or olive oil alone (control). Under these conditions CLA is absorbed and incorporated into phospholipid throughout the body, including that of liver cell membranes. The microsomes were then subjected to oxidative stress using a non-enzymatic iron-dependent lipid peroxidation system. The results, to be reported elsewhere, establish that microsomes from CLA-treated animals are far more resistant to oxidation than microsomes from control animals. Hence, CLA is an effective antioxidant when incorporated into membrane lipid.

In related experiments (to be reported in full elsewhere) we have studied the effect of CLA on benzo[a]pyrene (BP) metabolism in cultured mouse fibroblasts. For cells cultured in the presence of CLA (so that the cis-9, trans-11 isomer is incorporated in their membranes), BP metabolism is shifted away from activation and towards detoxification. Since much of the activation of BP in fibroblasts is mediated by peroxidative mechanisms rather than cytochrome P-450, these results provide additional support for the thesis that CLA acts as an antioxidant in vivo as well as in vitro.

A perplexing question is the mechanism of CLA antioxidant activity, since there is nothing in the structure of CLA per se to suggest that it should possess such activity. Our current hypothesis is that an oxidized derivative of CLA is the actual ultimate antioxidant form rather than CLA itself. The most likely candidate, based on in vitro and in vivo UV spectrophotometric evidence, is that a beta-hydroxy acrolein moiety is introduced across the conjugated double bond system (Ha et al., 1990). A similar compound with similar antioxidant activity has been reported in Eucalyptus leaf waxes (Osawa and Namiki, 1985). Current evidence indicates that antioxidant activity may result by the chelation of transition metals, especially iron, by the beta-hydroxy acrolein derivative of CLA (Ha et al., 1990).

Inhibition of TPA action by CLA

To further explore the anticarcinogenic action of CLA we initiated experiments aimed at elucidating the possible effects of CLA on the process of tumor promotion in the mouse forestomach (Benjamin et al., 1990). We employed the induction of ODC by the well-studied tumor promoter TPA.

It was first necessary to establish that ODC was induced in mouse forestomach by TPA. We found that treating 7 week old female ICR mice with TPA (6 μg/animal p.o.) resulted in ODC induction similar to that observed in other tissues that have been investigated. The peak activity, about 5-times control, occurred at 6 hours after TPA intubation. Pre-treatment with CLA significantly reduced the ODC response effected by TPA.

There are a number of possible explanations for these observations. We are particularly attracted to the possibility that the cis-9, trans-11 CLA isomer, incorporated into phospholipid, may directly affect the interaction of TPA with its principal receptor, protein kinase C (PK-C) (Merrill, 1989). If this were so then CLA might be a natural and possibly essential regulator of PK-C activity. Since PK-C controls superoxide generation (Merrill, 1989) CLA might serve not only as a direct antioxidant but also as an indirect antioxidant (prevention of superoxide generation via its effect on PK-C).

CONCLUSION

CLA may represent a newly recognized metabolic defense mechanism against the generation of hydroxyl radical and other active species of oxygen. The effectiveness of CLA as an anticarcinogen may in fact directly result from that normal physiological function. The relationship between dietary sources of CLA, and the synthesis of CLA in vivo, remains to be elucidated.

ACKNOWLEDGMENTS

This work was supported in part by the College of Agricultural and Life Sciences, University of Wisconsin-Madison; USPHS Training Grant 5-T32CA-08451; USPHS Training Grant 5-T32ES-07015; a grant from the Wisconsin Milk Marketing Board; and gift funds administered through the Food Research Institute, University of Wisconsin-Madison.

REFERENCES

Benjamin, H., Storkson, J. M., Albright, K., and Pariza, M. W. (1990). TPA-mediated induction of ornithine decarboxylase activity in mouse forestomach and its inhibition by conjugated dienoic derivatives of linoleic acid. FASEB J., 4, A508 (abstract # 1403).

Dormandy, T.L. and Wickens, D. G. (1987). The experimental and clinical pathology of diene conjugation. Chem. Phys. Lipids 45, 353-364.

Ha, Y. L., Grimm, N. K., and Pariza, M. W. (1987). Anticarcinogens from fried ground beef: heat altered derivatives of linoleic acid. Carcinogenesis 8, 1881-1887.

Ha, Y. L., Grimm, N. K., and Pariza, M. W. (1989). Newly recognized anticarcinogenic fatty acids: identification and quantification in natural and processed cheeses. J. Agric. Fd. Chem., 37, 75-81.

Ha, Y. L., Storkson, J., and Pariza, M. W. (1990). Inhibition of benzo[a]pyrene-induced mouse forestomach neoplasia by conjugated dienoic derivatives of linoleic acid. Cancer Research 50, 1097-1101.

Kepler, C. R. and Tove, S. B. (1967). Biohydrogenation of unsaturated fatty acids. J. Biol. Chem. 242, 5686-5692.

Osawa, T. and Namiki, M. (1985). Natural antioxidants isolated from Eucalyptus leaf waxes. J. Agric. Food Chem. 33, 777-780.

Merrill, A.H. (1989). Lipid modulators of cell function. Nutrition Rev. 47, 161-169.

Pariza, M. W. (1988). Dietary fat and cancer: evidence and research needs. Ann. Rev. Nutrit. 8, 167-183.

20

RESIDUE TRYPSIN INHIBITOR: DATA NEEDS FOR RISK ASSESSMENT

John N. Hathcock

Center for Food Safety and Applied Nutrition
Food and Drug Administration
Washington, DC 20204

ABSTRACT

Trypsin inhibitor (TI) occurs naturally in many foods from plants, notably soybean protein products. Heat treatment inactivates TI and improves nutritional quality, but residual TI activity of 5 to 20 % remains after typical commercial treatments. Chronic feeding of TI or products that contain TI can inhibit trypsin and chymotrypsin, stimulate their secretion, cause hypertrophy and hyperplasia of the pancreas, and lead to adenomas and carcinomas of the exocrine pancreas. In the rat, TI promotes pancreatic carcinogenesis initiated by azaserine. Data needed for possible risk assessment on TI would include 2-year bioassays from animals treated with TI and fed diets carefully controlled for type and amount of fat (which also promotes pancreatic carcinogenesis). The effects of TI on protein nutrition would have to be considered when identifying the maximum tolerated dose. Major reductions in human dietary TI exposure may not be feasible because of the multiple sources of TI, the substantial promotion by other factors such as fat, and the adverse effects of excessive heat on food products. For risk assessment of TI in a particular food, other promotors and the feasibility of decreasing TI intake must be considered.

TRYPSIN INHIBITORS IN FOODS

Natural protease inhibitors are a diverse group of proteins with wide distribution in plants commonly used as human foods. Within the broad category of natural protease inhibitors are trypsin inhibitors (TIs), many of which also inhibit chymotrypsin and other proteinases. TIs occur in many common food plants, including soybeans, other beans and seeds, and potatoes (Rackis and Gumbmann, 1981; Wilson, 1981). Certain nonprotein compounds of low molecular weight are effective inhibitors of trypsin activity (Muller, et al., 1988).

The feeding of heat-treated soybeans promotes the growth of rats (Liener, 1986). Because raw soybeans contain heat-labile TIs, it was assumed at first that their antinutritive effect was caused by the inhibition of protein digestion and that heat treatment would eradicate this effect. These assumptions, however, were not consistent with the subsequent observation that the addition of TI to free amino acid diets also inhibited the growth of rats.

Nutritional and Toxicological Consequences of Food Processing
Edited by M. Friedman, Plenum Press, New York, 1991

Feeding raw soybeans or TI concentrates to rats led to hypertrophy and hyperplasia of the pancreas. The increased output of pancreatic enzymes is quantitatively sufficient to explain the antinutritive effect of TI on protein nutrition and account for the growth depression caused by the fecal loss of endogenous nitrogen as digestive enzymes, which substantially drains the protein supply. The pancreatic hypertrophy and hyperplasia, however, produced neoplastic effects which were not directly related to protein nutrition.

Soybeans have two major types of TIs: the Kunitz inhibitor (about 20,000 daltons) and the extensively disulfide cross-linked Bowman-Birk inhibitor (about 8,000 daltons). Although heat treatment improves the nutritional quality of soybean products by destroying TI activity, standard heat-processing methods often leave 5 to 20% of the original TI content (Rackis and Gumbmann, 1981). Enzyme activity assays do not distinguish between the two types of TIs. Enzyme-linked immunosorbent assay (ELISA) methods are being developed (Brandon et al., 1988) for rapid quantitation of the amounts of the two types of TIs in raw and processed soy products. The importance of the proportion of the two inhibitors on possible adverse effects is uncertain.

EFFECTS ON THE PANCREAS

The use of soybean products has increased in the US diet during a period in which deaths from pancreatic cancer have also increased (Mack, 1982; Roebuck, 1987). Except for the role of smoking, the etiology of human pancreatic cancer is not known, but results from animal research on TI effects on pancreatic oncogenesis suggest a causal relationship.

Ingestion of protein by the rat increases output of trypsin and chymotrypsin, which is regulated through a feedback control mechanism, probably involving cholecystokinin (CCK) (Fushiki and Iwai, 1989). Treatment of rats with dietary soybean TI causes similar increases in enzyme secretion, and continued treatment with TI is associated with hypertrophy and hyperplasia of the exocrine (acinar) cells (Roebuck, 1987; Smith, et al., 1989).

Although feeding of TI to animals for a few weeks causes hypertrophy and hyperplasia, no overt neoplastic changes occur unless the animals are also treated with a pancreatic carcinogen, such as azaserine (Liener, 1986; Roebuck, 1987). Treatment with azaserine followed by 4 to 8 weeks of dietary TI leads to atypical acinar cell foci; with continued TI treatment, these foci grow and adenomas are observed as early as 4 months, and adenomas and adenocarcinomas may be observed after about a year. In this paradigm, TI seems to be a classic promotor of azaserine-initiated pancreatic cancer. Long-term feeding of TI (1 or 2 years) without treatment with azaserine or any other initiator may result in progression of hypertrophy and atypical acinar cell foci to adenoma and carcinoma (Melmed et al., 1976; Rackis et al., 1979; McGuiness et al., 1980).

The relationship between TI and pancreatic carcinogenesis is still obscure, partly because many studies in this area probably were inherently confounded by use of full-fat soybean flour. High levels of dietary fat promote pancreatic carcinogenesis (Roebuck et al., 1981). The low molecular weight nonpolymer Camostat [N,N-dimethylcarbamoyl-methyl-p-(p-guanidobenzoyloxy)phenylacetate] has a powerful trypsin inhibitory action (Fujii, 1977) and also produces CCK-8-sensitive pancreatic hypertrophy and hyperplasia (Muller et al., 1988) similar to that produced by soybean TI (Liddle et al., 1984). These studies imply that trypsin inhibition may enhance pancreatic pathology associated with pancreatic carcinogenesis and that the origin and identity of the TI are not important. Nevertheless, if soybean protein is a major dietary source of TI, studies of soybean TI may be important. Most evidence suggests that TI promotes pancreatic carcinogenesis, i.e., it is effective only if the animals are treated with an initiator such as azaserine (Roebuck, 1987; Hathcock, in press). Prolonged consumption of TI without treatment with a recognized initiator such as azaserine may produce nodular hyperplastic foci in the rat pancreas, but the effect is very weak compared with that observed when an initiator is also used (McGuiness et al., 1984).

The relevance of these studies to human health has been questioned because in many species TI treatment does not cause pancreatic hypertrophy, hyperplasia, and oncogenesis. Dogs are not appropriate models for the possible effects of TI on human health because they do not respond to either dietary protein or TI with increased pancreatic enzyme output. Apparently, pancreatic proteinase secretion is controlled by stimulation with free amino acids (at relatively high concentrations in the dog's natural diet) rather than by the feedback control found in the rat (Sale, et al., 1977, Fushiki and Iwai, 1989).

Species that respond to TI with increased enzyme secretion, hypertrophy and hyperplasia, perhaps enhanced oncogenesis, and related changes include rats, mice, and chickens (McGuiness et al., 1984; Roebuck, 1987; Tudor and Dayan, 1987; Gumbmann et al., 1989; Schingoethe et al., 1974; Schneeman and Gallaher, 1986) (Table 1). Unresponsive species include the calf, pig, dog, Rhesus monkey, Cebus monkey, and marmoset (Tudor and Dayan, 1987; Kakade et al., 1975; Yen et al., 1974; Patten et al., 1971; Fushiki and Iwai, 1989; Struthers et al., 1983; Ausman et al., 1985; Schneeman and Gallaher, 1986) (Table 1).

Because of the reports that primates and large species do not respond to TI, it has been assumed that human consumption of TI is not likely to result in the adverse effects seen in responsive species. A recent study with humans, however, indicated that a single dose of the soybean Bowman-Birk TI elicits increased pancreatic secretion of trypsin and chymotrypsin (Liener et al., 1988). If the temporal sequence of trypsin inhibition, CCK release, enhanced trypsin and chymotrypsin secretion, hypertrophy and hyperplasia, altered acinar cell foci, and oncogenesis observed in responsive species represents a causal cascade, then the results of this human study suggest that humans are susceptible to pancreatic oncogenesis by TI. No such conclusion can be reached, however, from present data because this sequence of effects was not studied in the experiments which indicated a lack of response of the primate pancreas to chronic TI consumption. More data are needed to resolve this question.

RISK ASSESSMENT DATA NEEDS

Risk assessment is part of the process of evaluating a potential hazard. For a substance in food, risk may be defined as the probability of toxicity at a specified dose, and hazard may be defined as the likelihood that it will produce adverse effects under the usual circumstances of exposure (Hathcock, 1976). The formal process of risk assessment involves calculation of risk and evaluation of hazard. The steps usually included are qualitative and quantitative toxicity evaluation, exposure estimation, and either calculation of risk at achievable dosages or calculation of dosage for acceptable risk. If the needed epidemiological, mechanistic, dose-response, and exposure data are available, the alternatives should be evaluated before feasibility and desirability of decreasing TI intake from a specific source are

Table 1. Trypsin Inhibitor Effects on Pancreatic Feedback Mechanisms in Several Species

Species	Substance	Effect(s)
Rat	Raw soy, soy TI (Kunitz and Bowman-Birk)	CCK release, enzyme secretion, hypertrophy, hyperplasia, oncogenesis
Rat	Camostat	Hypertrophy, hyperplasia, CCK, enzyme release
Chick	Raw soy, soy TI	Hypertrophy, hyperplasia
Rhesus monkey	Raw soy protein	No hypertrophy or hyperplasia
Cebus monkey	Raw soy protein	No hypertrophy or hyperplasia
Marmoset	Raw soy	No hypertrophy or hyperplasia
Humans	Soy (Bowman-Birk) TI	Secretion of trypsin and chymotrypsin

decided. The need for risk assessment depends on:

- epidemiological data sufficient to confirm or deny a relationship of TI to pancreatic cancer or other specific adverse effect,
- experimental studies to identify the mechanism in susceptible species and to determine whether it is present in humans,
- availability of reliable dose-response data over a sufficiently wide range to allow estimation of risk at relevant dosages,
- availability of reliable exposure data, and
- evaluation of alternatives.

Mechanistic and quantitative questions about observed effects of TI on the pancreas include:

- Is the effect species-specific?
- Is there specificity for body size, enzyme profile, hormones, or other factors?
- Are other dietary components needed for the effect?
- What is the dose-response relationship (slope, threshold, and variability of response)?

In any risk assessment for TI, the many food sources of potential promotors of pancreatic carcinogenesis must be considered. TI is found not only in soybeans and soy protein products but also in many other legumes, potatoes, and vegetables (Gumbmann et al., 1989; Ryan and Hass, 1981). If TI is a promotor of pancreatic carcinogenesis (and present data do not rule out initiation), it is not the only promotor present in foods. High dietary fat is also an effective promotor of chemically initiated pancreatic carcinogenesis (Roebuck, 1987).

A second major factor in any contemplated risk assessment for TIs is the possible benefit from their consumption. Several studies involving transformed cells in vitro and others involving chemically initiated animals have shown that TIs can have anticarcinogenic effects (Billings et al., 1988; Troll and Kennedy, 1989; Witschi and Kennedy, 1989). Soybean Bowman-Birk TI administered by intraperitoneal injection decreased frequency of lung tumors in mice initiated with 3-methylcholanthrene. The relevance of these studies to the question of safety for dietary exposure to TI will depend on the risk related to pancreatic effects as compared with the beneficial effects in other organs.

REFERENCES

Ausman, L. M., Harwood, J. P., King, N. W., Sehgal, P. K., Nicolosi, R. J., Hegsted, D. M., Liener, I. E., Donatucci, D., and Tarcza, J., 1985, The effects of long-term soy protein and milk protein feeding on the pancreas of Cebus albifrons monkeys, J. Nutr., 115:1691-1701.

Billings, P. C., St. Clair, W., Owen, A. J., and Kennedy, A. R., 1988, Potential intracellular target proteins of the anticarcinogenic Bowman Birk protease inhibitor identified by affinity chromatography, Cancer Res., 48:1798-1802.

Brandon, D. L., Bates, A. H., and Friedman, M., 1988, Enzyme-linked immunoassay of soybean Kunitz trypsin inhibitor using monoclonal antibodies, J. Food Sci., 53:102-106.

Fujii, S., 1977, Synthetic protease inhibitor, Metab. Dis., 14:1087-1092.

Fushiki, T., and Iwai, K., 1989, Two hypotheses on the feedback regulation of pancreatic enzyme secretion, FASEB J., 3:121-126.

Gumbmann, M. R., Dugan, G. M., Spangler, W. L., Baker, E. C., and Rackis, J. J., 1989, Pancreatic response in rats and mice to trypsin inhibitors from soy and potato after short- and long-term dietary exposure, J. Nutr., 119:1598-1609.

Hathcock, J. N., 1976, Nutrition: Toxicology and pharmacology, Nutr. Rev., 34:65-70.

Hathcock, J. N., 1990, Nutritional toxicology: Basic principles and actual problems, Food Addit. Contam., in press.

Kakade, M. L., Thompson, R. D., Engelstad, W. E., Behrens, G. C., Yoder, and R. D., Crane, F.M., 1975, Failure of soybean trypsin inhibitor to exert deleterious effects in calves, J. Dairy Sci., 59:1484-1489.

Liddle, R. A., Goldfine, I. D., and Williams, J. A., 1984, Bioassay of plasma cholecystokinin in rats: Effects of food, trypsin inhibitors and alcohol, Gastroenterology, 87:542-549.

Liener, I. E., 1986, Trypsin inhibitors: Concern for human nutrition or not? J. Nutr., 116:920-923.

Liener, I. E., Goodale, R. L., Desmukh, A., Satterberg, T. L., Ward, G., DiPietro, C. M., Bankey, P. E., and Borner, J. W., 1988, Effect of a trypsin inhibitor from soybeans (Bowman-Birk) on the secretory activity of the human pancreas, Gastroenterology, 94:419-427.

Mack, T. M., 1982, Pancreatic cancer, in: "Cancer Epidemiology and Prevention," D. Schottenfelf and J. Fraumen, eds., W.B. Saunders, Philadelphia.

McGuiness, E. E., Morgan, G. H., Levison, D. A., Frape, D. L., Hopwood, D., and Wormsley, K. G., 1980, The effects of long-term feeding of soya flour on the rat pancreas, Scand. J. Gastroenterol., 15:497-502.

McGuiness, E. E., Morgan, H. G., and Wormsley, K. G., 1984, Effects of soybean flour on the pancreas of rats, Environ. Health Perspect., 56:205-212.

Melmed, R. N., El-Asser, A. A., and Holt, S. J., 1976, Hypertrophy and hyperplasia in the neonatal rat exocrine pancreas induced by orally-administered soy bean trypsin inhibitor, Biochim. Biophys. Acta, 421:280-288.

Muller, M. K., Goebell, H., Alfen, R., Ehlers, J., Jager, M., and Plumpe, H., 1988, Effects of Camostat, a synthetic protease inhibitor, on endocrine and exocrine pancreas of the rat, J. Nutr., 118:645-650.

Patten, J. R., Richards, E. A., and Pope, H., 1971, The effect of raw soybean on the pancreas of adult dogs, Proc. Soc. Exp. Biol. Med., 137, 59-63.

Rackis, J. J., and Gumbmann, M. R., 1981, Protease inhibitors: physiological properties and nutritional significance, in: "Antinutrient and Natural Toxicants in Foods," R. L. Ory, ed., Food and Nutrition Press, Inc., Westport.

Rackis, J. J., McGee, J. E., Gumbmann, M. R., and Booth, A. N., 1979, Effects of soy proteins containing trypsin inhibitors in long-term feeding studies in rats, J. Assoc. Off. Anal. Chem., 56:162-168.

Roebuck, B. D., 1987, Trypsin inhibitors: Potential concern for humans? J. Nutr., 117:398-400.

Roebuck, B. D., Yager, J. D., and Longnecker, D. S., 1981, Effects of dietary fats and soybean protein on azaserine-induced pancreatic carcinogenesis and plasma cholecystokinin in the rat, Cancer Res., 41:888-893.

Ryan, C. A., and Hass, G. M., 1981, Structural, evolutionary and nutritional properties of proteinase inhibitors from potatoes, in: "Antinutrients and Natural Toxicants in Foods," R. L. Ory, ed., Food and Nutrition Press, Inc., Westport.

Sale, J. K., Goldberg, D. M., Fawcett, A. N., and Wormsley, K. G., 1977, Chronic and acute studies indicating absence of exocrine pancreatic feedback inhibition in dogs, Digestion, 15:540-555.

Schingoethe, D. J., Tidemann, L. J., and Uckert, J. R., 1974, Studies in mice on the isolation and characterization of growth inhibitors from soybeans, J. Nutr., 104:1304-1312.

Schneeman, B. O., and Gallaher, D., 1986, Pancreatic response to dietary trypsin inhibitor: variations among species, in: "Nutritional and Toxicological Significance of Enzyme Inhibitors in Foods," M. Friedman, ed., Plenum Press, New York.

Smith, J. C., Wilson, F. D., Allen, P. V., and Berry, D. L., 1989, Hypertrophy and hyperplasia of the rat pancreas produced by short-term dietary administration of soya-derived protein and soybean trypsin inhibitor, J. Appl. Toxicol., 9:175-179.

Struthers, B. J., MacDonald, J. R., Dahlgren, R. R., and Hopkins, D. T., 1983, Effects on the monkey, pig and rat pancreas of soy products with varying levels of trypsin inhibitor and comparison with the administration of cholecystokinin, J. Nutr., 113:86-97.

Troll, W., and Kennedy, A. R., 1989, Protease inhibitors as cancer chemopreventive agents, Cancer Res., 49:499-502.

Tudor, R. J., and Dayan, A. D., 1987, Comparative subacute effects of dietary raw soya flour on the pancreas of three species, the marmoset, mouse and rat, Food Chem. Toxicol., 25:739-745.

Wilson, K. A., 1981, The structure, function, and evaluation of legume proteinase inhibitors, in: "Antinutrients and Natural Toxicants in Foods," R. L. Ory, ed., Food and Nutrition Press, Inc., Westport.

Witschi, H., and Kennedy, A. R., 1989, Modulation on lung tumor development in mice with the soybean-derived Bowman-Birk protease inhibitor. Carcinogenesis, 10:2275-2277.

Yen, J. T., Hymowitz T., and Jensen, A. H., 1974, Effects of soybeans of different trypsin inhibitor activities on performance of growing swine, J. Anim. Sci., 38:304.

21

STUDIES OF FOOD ALLERGENS: SOYBEAN AND EGG PROTEINS

R. Djurtoft, H. S. Pedersen, B. Aabin, and V. Barkholt

Center for Food and Process Biotechnology
Department of Biochemistry and Nutrition
The Technical University of Denmark
DK-2800 Lyngby, DENMARK

INTRODUCTION

Food allergens are food components giving rise to a hypersensitivity reaction, food allergy. Food allergy is defined as adverse reactions to food based on immunological mechanisms. The present study is concerned with immediate type allergy in adults mediated by immunoglobulins of the IgE subclass.

Food allergy is often difficult to diagnose, and at present the recommended treatment is total avoidance of the identified components. Many food items are, however, difficult to avoid because they are common ingredients in industrially prepared food. On the other side, the industrial processing will give ample opportunity to modify food components into less offensive products provided that suitable methods are at hand. Our aim is to describe procedures that will minimize possible allergenicity.

In spite of much interest in food allergy we still need (a) reliable and easy methods for diagnosis; (b) methods for the modification of foods to reduce adverse reactions in people who are already allergic; and (c) evaluation of the possibility to prevent the induction of allergy towards a given food item. This last goal, especially, implies a much more profound understanding of the whole phenomena than we do have at present (Reinhardt and Schmidt, 1988).

The IgE antibodies are the tools which are available to us for investiagations of allergies. Whereas production of IgE towards food is a disadvantageous reaction, the production of IgG antibodies at least towards some dietary antigens is a normal reaction (Husby, 1988). By comparing IgE and IgG, we wish to determine whether one of the two types of antibodies, IgE or IgG, react preferentially with a degraded form of the allergen/antigen. We also would like to characterize IgE binding epitopes.

Antibodies raised in laboratory animals are easily obtained. It would be convenient if they could be used to evaluate allergenicity. It is often assumed that this is the case. However, there is no reason to take for granted that the IgG antibodies raised by subcutaneous injections in animals represent the IgE antibodies raised by oral administration in man. We have, therefore, included rabbit IgG in our studies.

We have focused our interest on two food items, soybean protein and egg white protein, both of which have already been the subject of several investigations in the field of food allergy. Soy protein is now a commercial product sold under different names and often industrially pretreated to suit special purposes. Egg proteins are not only consumed in whole eggs but they are also common ingredients in prepared foods, and spray dried egg fractions are sold as commercial products. Egg is one of the commonly mentioned food allergens, and the egg white proteins are well suited as model proteins for investigations because they are easily obtained, well characterized, and can be expected to represent a broad range of proteins, as they are very different with respect to size, carbohydrate content, and susceptibility to proteolysis (Burley and Vadhera, 1989).

SOY PROTEINS

Soy protein allergy is frequently referred to as a problem among children fed a soy-based infant formula. The soybeans contain more than 30 antigenic proteins as demonstrated by crossed imunoelectrophoresis using rabbit antisera. Some of these seem to be more allergenic than others. The 2S fraction has occasionally been reported as being allergenic, and in one patient a specific reaction towards the Kunitz trypsin inhibitor was observed Moroz and Yang, 1980). Shibasaki (1980) found that the 11S, 7S, and 2S fractions were reactive by the radioallergosorbent test (RAST) and RAST in inhibition techniques with sera from soy-sensitive individuals. The 2S fraction seems to be the most allegenic in these studies. The experiments of Burks et al. (1988) suggest, however, that there are no specific major allergens in the soybean.

The 11S protein glycinin is quantitatively the most important protein the soybean and accounts for approximately 35% of the total protein by weight. It is classified as a storage protein as it has no known biological functions. The physical and chemical properties of glycinin have been closely studied by Bradley et al. (1975). Glycinin has a molecular mass of approximately 350, 300 Daltons. This fairly large molecule is assembled from several subunits as demonstrated by sodium dodecylsulfate-polyacrylamide gel electrophoresis (SDS-PAGE). There are six subunits, and each subunit contains two polypeptide chains, one acidic and one basic. Each subunit is synthesized as a single propeptide, which is then cleaved post-translationally into two distinct chains which are held together by a disulfide bridge. The six subunits are each very homologous, and can replace each other in the final assembled glycinin molecule. The actual composition of glycinin can, therefore, vary somewhat. The three-dimensional structure of glycinin has been investigated using both electron microscopy and X-ray diffraction. A model has been proposed consisting of a hollow cylinder made from two hexagons placed on top of each other. The distinct subunits then form the elements of the hexagons (Nielsen, 1984).

Experiments with Soy Proteins

We have tried to characterize the antigenic and allergenic properties of glycinin. The reaction of isolated subunit chains towards IgG and IgE have been investigated as a first attempt to identify the specific antigenic and allergenic epitopes on the molecule (Pederson and Djurtoft, 1989).

The sera used were from 10 atopic patients, 11-36 years of age, with a positive RAST value against soy. Some of the patients showed reactions against other legumes as well.

The acidic and basic chains from glycinin were purified using the procedure of Moreira et al. (1979). The acidic subunits, designated Ala, A2, A3, and A4 were fractionated, and one fraction containing all the basic subunits prepared.

Rabbit antisera, prepared by immunizing with purified glycinin, were used for a number of ELISA assays to evaluate the binding of individual subunit chains to the antibodies (Figure 1). All acidic chains showed approximately the same degree of reaction with rabbit anti-glycinin antiserum. The dilution curves are very similar with the exception of subunit Alb which reacted slightly more than the other subunits, but not as much as fractionated glycinin. In a capture ELISA (Figure 2), the reaction of A4 differed somewhat, indicating a different affinity between antibody and the peptide chain. Together the results indicate that none of the acidic subunits contains more dominating epitopes than the others. A4 may be slightly different from the other subunits concerning epitope structure.

Using a competitive ELISA with acidic subunits, it was confirmed that no acidic subunit dominates the immune response (Table 1). All acidic subunits were able to compete with each other for binding to the antiserum. Around 80% inhibition was observed. No single subunit could significantly displace the binding between native glycinin and the antiserum. This indicates that the isolated subunits do not contain the total antigenic repertoire of glycinin. Quite a large fraction must be considered as related to conformational epitopes, which are disrupted during separation of individual subunits.

The fraction containing the basic subunits did not show any antigenic reaction with the rabbit anti-glycinin antiserum, neither in the direct ELISA nor as competitor (data not shown). This confirms the hypothesis of Nielsen et al. (1989) that the basic subunits are located in the interior of the assembled molecule. For that reason, they will not be able to interact with the immune system of the animal.

Isolated acidic subunits and glycinin were coated on microtitre plates as well as preincubated with rabbit anti-glycinin antiserum. The complexes formed were transformed to the microtitre plated for final incubation. Detection was carried out with peroxidase-labeled pig anti-rabbit antiserum. Percentage inhibition is calculated by comparison with the inhibition by the buffer (Table 1).

The capability of the 10 patient sera to react with the individual subunits has been investigated. The level of IgG binding between the different sera varies very much (Figure 3). Six sera bound native glycinin preferably, and no sera seem to be directed specifically towards a single subunit. All subunits bound to the applied sera in very much the same way.

TABLE 1. Antigenic Homologies (% Inhibition) as Estimated from Competitive ELISA

Coating antigen Competing antigen	Glycinin	Alb	Ala	A2	A3	A4
Glycinin	60	62	67	71	67	76
Alb	0	85	78	83	80	65
Ala	0	83	77	82	72	53
A2	0	85	80	86	77	57
A3	6	76	74	78	81	81
A4	0	68	73	73	82	93

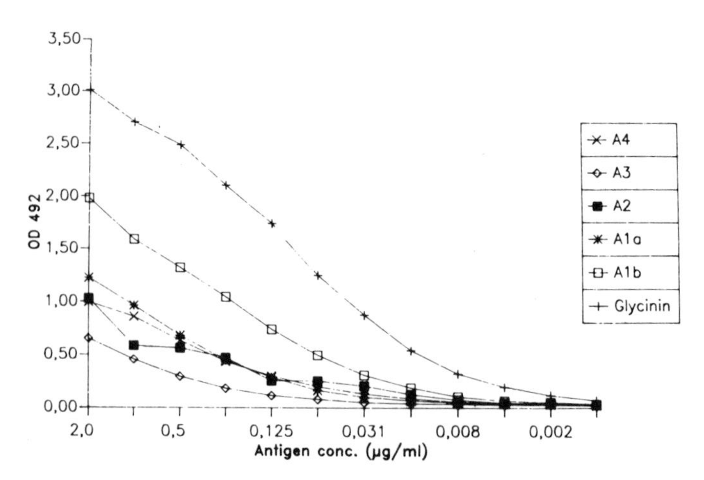

FIGURE 1. Direct ELISA with acidic subunits as coating antigen. The isolated acidic subunits and native glycinin were coated on the microtitre plates and detected with rabbit antiglycinin antiserum and peroxidase-labeled pig anti-rabbit antiserum.

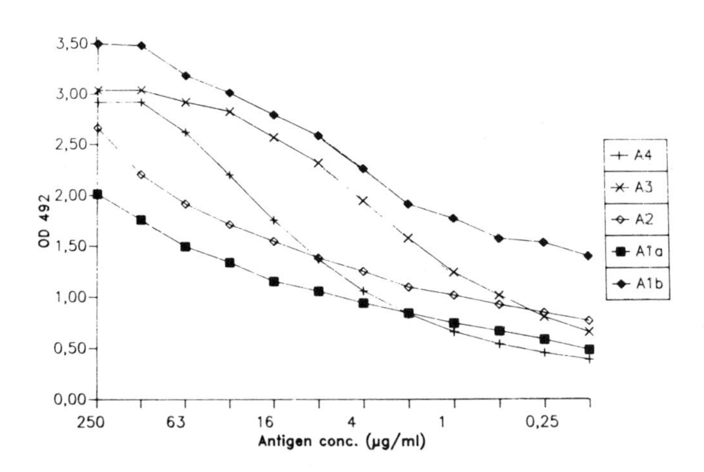

FIGURE 2. Capture ELISA for detection of biotin-labeled acidic subunits. Rabbit anti-glycinin antiserum, diluted 1: 2000 was coated on the microtitre plate and dilutions of the antigen, biotin-labeled acidic suunits were added. Bound antigen was detected with peroxidase-labeled avidin.

The results for the IgE binding are different (Figure 4). In four serum samples the binding to subunit A4 is the most pronounced and in only four serum samples the binding to native glycinin is more pronounced than the binding to the isolated subunits. One serum did not bind native glycinin at all, but still showed significant binding to all subunits. With the exception of subunit A4, all subunits reacted similarly .

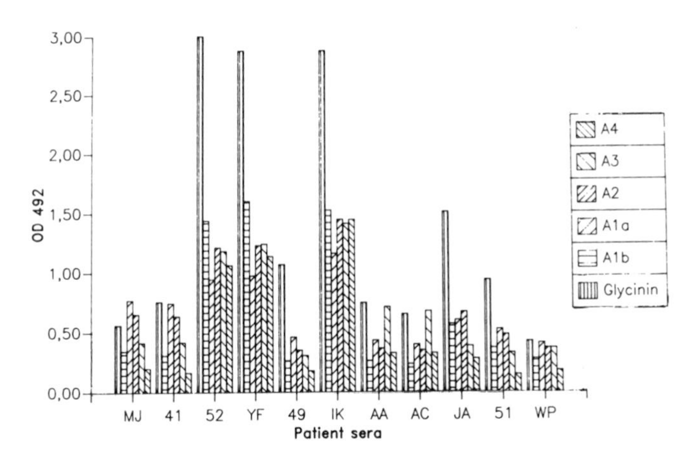

FIGURE 3. Direct ELISA for detection of patient IgG. The isolated acidic subunits and glycinin were coated on microtitre plates. Patient sera, diluted 1: 100 were added and incubated with all fractions. Detection was carried out using rabbit anti-human IgG antiserum and peroxidase-labeled pig anti-rabbit antiserum. Groups on the X-axis represent the individual patient sera, while the columns each represent the coated antigen.

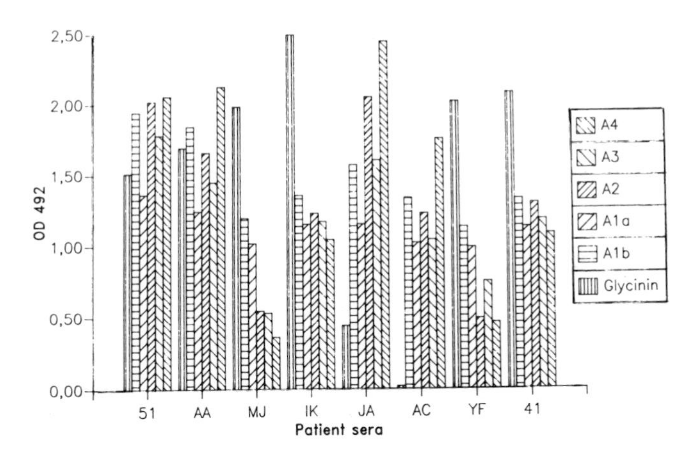

FIGURE 4. Direct ELISA detecting patient sera IgE. The isolated acidic subunits and glycinin were coated on microtitre plates. Patient sera, diluted 1: 100 were added and incubated with all fractions. Detection was carried out using rabbit anti-human IgE antiserum followed by biotin-labeled pig anti-rabbit antiserum and peroxidase-labeled avidin. Groups on the X-axis represent the individual patient sera, while the columns each represent the coated antigen.

EGG PROTEINS

Most of the work on the allergenicity of egg proteins has been performed on egg white (Langeland, 1983; Hoffmann, 1983; Anet et al., 1985). Albumen contains about 40 proteins, but most of them are minor components. IgE towards most of these proteins have been demonstrated (Langeland, 1982). We have concentrated on the four predominant proteins, making up 80% of the total protein content by weight: ovomucoid, ovotransferrin, ovalbumin, and lysozyme.

Ovomucoid (OM) accounts for 11% of the egg white protein. It consists of 186 amino acid residues in a single peptide chain including 18 cystines, all of which take part in disulfide bridges (Kato et al., 1987). Four or five asparagine linked carbohydrate chains make up about 20% of the total molecular mass of roughly 28,000 Daltons of the whole glycoprotein. The carbohydrate is the cause of extensive heterogeneity of OM preparations. OM is a serine protease inhibitor of the Kazal inhibitor family (Laskowski and Kato, 1980), and its function in egg white is supposed to be protease inhibition. OM is extremely stable towards degradation: not only will it inhibit an attack from serum proteases, and even very long incubation will result in only limited proteolysis (Kato et al., 1987). OM is very stable towards denaturation by heat as well as urea (Stevens and Feeney, 1963), most probably because the high content of disulfide bridges stabilizes the tertiary domains of OM (Laskowski, 1986).

Ovotransferrin (OT) accounts for 12% of the egg white protein. With a molecular weight of 78,000 Daltons, it consists of 683 amino acid residues in a single peptide chain (Williams et al., 1982) including 30 cysteines, all disulfide linked, and one asparagine linked carbohydrate chain (2% of the total mass). Microheterogeneity in the amino acid sequence as well as in the carbohydrate chain gives rise to heterogeneity of preparations of OT. It binds iron and other ions and is supposed thereby to act as an inhibitor of bacterial growth in the albumen. OT is homologous to other iron binding proteins, including serum transferrin. It consists of and N- and a C-terminal domain stabilized by disulfide bridges within the domains, but with no disulfide bridges between the domains. OT is more susceptible to denaturation by heat than other predominant egg white proteins (Burley and Vadhera, 1989).

Ovalbumin (OA) is the most abundant protein in hen's egg white, constituting 54% of the total. The molecular mass is 43,000 Daltons. It consists of 386 amino acid residues in a single peptide chain (Nisbet et al., 1981). It has one asparagine linked carbohydrate group representing 3% of the molecular mass. It contains six cysteines of which only two are connected in a disulfide bridge. The resulting high probability of disulfide rearrangement together with variations in post-translational modifications give rise to some heterogeneity of OA preparations. Heat denaturation of OA is accompanied by aggregation. Native ovalbumin is susceptible to cleavage by various enzymes. The function of OA is unknown, but it may be a protease inhibitor, as the amino acid sequence is homolgous to a group of serine protease inhibitors, the serpins.

Lysozyme (LY) constitutes only 3% of the content of proteins in egg white. It is a muramidase which can break up cell walls of gram-negative bacteria. The amino acid sequence of the 129 residues (Canfield, 1963) is homologous to x-lactalbumin from milk.

Experiments with Egg Proteins

Ovomucoid, ovotransferrin, and ovalbumin were the highest aavailable grade from Sigma, St. Louis, Missouri. OA was further purified by gel filtration on Superose (Pharmacia). OM and OT were purified by ion-exchange chromatography on Mono Q (Pharmacia) followed by gel filtration on Superose. Lysozyme (LY) was from Worthington.

For the selection of human sera containing IgE against the major egg white proteins, we have screened 32 sera from adults with a suspected allergy to egg: purified ovotransferrin, ovalbumin, ovomucoid, and lysozyme were applied as dots on a matrix of nitrocellulose. After incubation with the serum, bound IgE was detected with 125-I labeled rabbit antihuman IgE and subsequent chromatography.

In this group of sera we find IgE against all four egg white proteins: seven sera showed strong reaction with OT, three sera with OA, eleven sera with OM, and two with LY. Only two sera (No's 2 and 11) reacted to all four proteins. The individual reactions in this dot immunobinding assay are shown in Figure 5.

The reactivities of OM, OA, and OT towards human- and rabbit antibodies were measured by ELISA at three different stages of denaturation of the antigens/allergens:

1. Unmodified proteins (called native).
2. Cyanogen bromide cleaved proteins. The proteins were cleaved at methionine residues by cyanogen bromide, and the disulfide bridges were left intact.
3. Reduced and alkylated proteins. The disulfide bridges of the proteins were reduced and the liberated sulfhydryl groups alkylated with acetamide.

The success of cleavage and alkylation was controlled by SDS-PAGE and amino acid analysis. These modifications are expected to have very different effects on the three proteins, as the effect of cleavage of peptide bonds and the reduction of disulfide bridges depends very much on the arrangement of the disulfide bonds in individual proteins (Friedman, 1973).

Direct ELISA was performed as described by Pedersen and Djurtoft (1989). The substrate for peroxidase was 3,3',5,5'-tetramethylbenzidine, and the product formed was measured at 450 nm.

The reactivities of OM were tested towards the antisera against OM of four individual rabbits and towards IgE and IgG from six human sera with IgE binding to OM in the dot immunobinding assay. One human serum (No 12) which showed no IgE binding to OM, but did bind to OT, was included in the analyses as a control.

Figures 6 and 7 show that cyanogen bromide cleavage did not change the reaction of OM towards any of the antibodies in these analyses, whereas reduction and alkylation caused loss of activity in many cases. This is in accordance with a highly stabilizing effect of disulfides in OM, resulting in a complete change of structure upon reduction. Yet, the reactivity of reduced OM is retained towards IgE in two and IgG in three human sera. One of the rabbit sera also has substantial activity towards the denatured OM. It has been demonstrated that some of the denaturation resistant epitopes in OM contain carbohydrate (Gu et al., 1988; Matsuda et al., 1986).

The reactivities of OA were tested against three pools of rabbit antisera and against IgE and IgG of three human sera which showed IgE-binding to OA in the dot immunobinding assay. Two of the rabbit sera were raised against OA and one was raised against an extract of total egg.

The effect of modification of OA was different for reactions with IgG and IgE (Figures 8 and 9). As the structure of OA does not depend much on its only disulfide bridge, great loss in reactivity is not expected as a result of disulfide bond reduction. Indeed, the two rabbit antisera and the IgE of two humans were found to bind more effectively to the reduced OA than to the native form (assuming equal coating). Cyanogen bromide cleavage resulted in somewhat lower reactivities towards all antibodies in this trial.

The reactivities of OT were tested against IgG from two rabbits immunized with OT and IgE and IgG from three patients with IgE binding to OT in the dot assay (data not presented). Cyanogen bromide cleavage resulted in low, but still measurable reactivities towards all these antisera, whereas reduction caused more pronounced loss of reactivity. The reaction of IgE of one patient was reduced to below the detection limit.

DISCUSSION

The described experiments were designed to provide some fundamental information about different food proteins acting as antigens/allergens. The effect of different modifications on the ability to bind to antibodies was considered as a background for understanding antigenic/allergenic behaviour. In this way, the results could also be the basis for understanding and designing the effect of various production steps in the food industry.

A more profound understanding of food allergy requires a large range of investigations in several areas including clinical investigations, tests with laboratory animals, and experiments with T-cell activation. We find that the approaches presented here can be useful as part of these investigations on food allergy.

Our initial experiments show a good correlation between dot immunobinding assays and results from ELISA. We have also found a good correlation between ELISA with reduced alkylated proteins and immunoblotting after SDS-PAGE of reduced, alkylated fragments of the same proteins (data not presented). Furthermore, the results seem reliable in the sense that they are in agreement with the behaviour of the antigens/allergens expected from our physical and chemical knowledge of the proteins. From this limited data, it appears that there are very large differences even among sera with antibodies against the same protein. It may be possible to find a pool of rabbit antisera that mimic the sera of allergic people with regard to reaction towards processed foods. However, a careful selection is necessary.

In our investigations we have found that human serum IgE shows a stronger reaction towards the modified glycinin and ovalbumin than towards the native proteins. In contrast, human serum IgG reacted more strongly with the native proteins. A possible explanation is that production of IgE antibodies can be stimulated only when a certain, limited processing of the intact protein has taken place. If continued investigations confirm these results they deserve broader study for their significance.

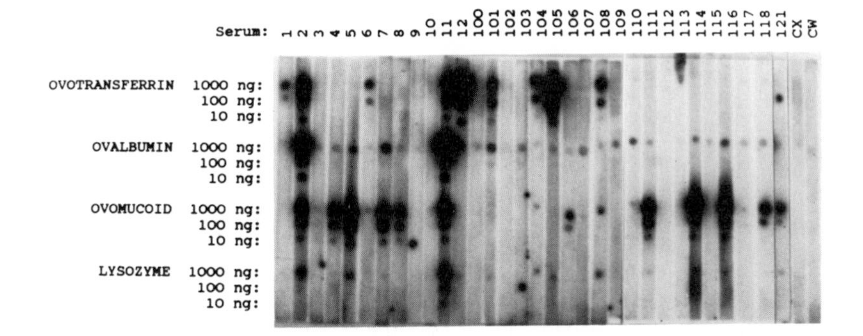

FIGURE 5. Detection of serum IgE against ovotransferrin (OT), ovalbumin (OA), ovomucoid (OM), and lysozyme (LY). Purified OT, OA, OM, and LY were applied at 3 concentrations as 1 ul dots on a matrix of nitrocellulose. After incubation with the serum, bound IgE was detected with 125-I labeled rabbit anti-human IgE (Phadebs RAST, Pharmacia Diagnostic AB, Sweden) and subsequent autoradiography. 1-121 referes to the sera tested, CX is a "non-egg allergic" control (644 IU IgE/ml serum) and CW is a pool of non-allergic sera.

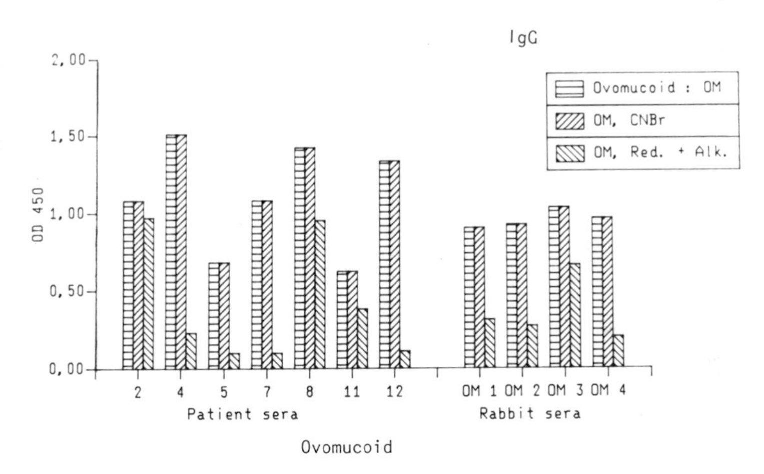

FIGURE 6. Direct ELISA for detection of serum IgG binding to ovomucoid, cyanogen bromide cleaved ovomucoid, and reduced-alkylated ovomucoid, coated on microtitre wells. The patient sera were diluted 1: 300 and IgG was detected by rabbit anti-human IgG antiserum followed by biotin-labeled pig anti-rabbit antiserum and peroxidase-labeled streptavidin. The analyzed rabbit antisera were diluted 1: 1000 and IgG was detected by peroxidase-labeled pig antiserum to rabbit immunoglobulins.

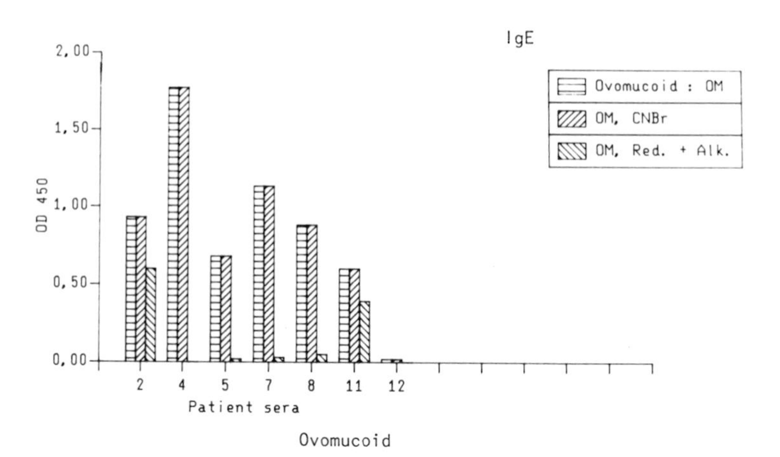

FIGURE 7. Direct ELISA for detection of serum IgE binding to ovomucoid, cyanogen bromide cleaved ovomucoids, and reduced-alkylated ovomucoid, coated on microtitre wells. The sera were diluted 1: 20 and binding of IgE was detected by rabbit anti-human IgE antiserum followed by biotin-labeled pig anti-rabbit antiserum and peroxidase-labeled streptavidin.

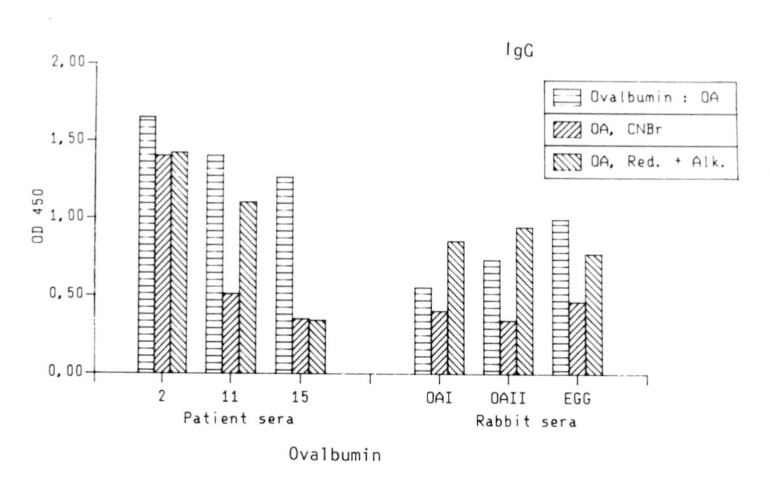

FIGURE 8. Direct ELISA for detection of serum IgG binding to ovalbumin, cyanogen bormide cleaved ovalbumin, and reduced-alkylated ovalbumin, coated on microtitre wells. The patient sera were diluted 1: 300 and IgG was detected by rabbit anti-human IgG antiserum followed by biotin-labeled pig anti-rabbit antiserum and peroxidase-labeled streptavidin. The analyzed rabbit antisera were diluted 1: 1000 and IgG was detected by peroxidase-labeled pig antiserum to rabbit immunoglobulins.

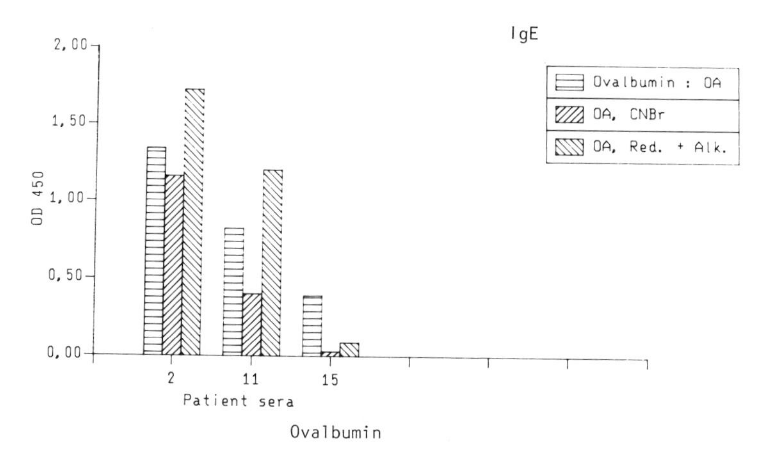

FIGURE 9. Direct ELISA for detection of IGE binding to ovalbumin, cyanogen bromide cleaved ovalbumin, and reduced-alkylated ovalbumin, coated on microtitre wells. The sera were diluted 1: 20 and binding of IgE was detected by rabbit anti-human IgE antiserum followed by biotin-labeled pig anti-rabbit antiserum and peroxidase-labeled streptavidin.

ACKNOWLEDGEMENTS

We are grateful to the Food Allergy Unit, Rigshospitalet, Copenhagen and Arhus Kommunehospital for making patient sera available to us, and to ALK, Horsholm, for rabbit antiserum towards total egg. Part of this work was supported by the Danish Council for Technical and Scientific Research.

REFERENCES

Anet, J., Back, J. F., Baker, R. S., Barnett, D., Burley, R. W., and Howden, M. E. H. (1985). Allergens in the white and yolk of hen's egg. Int. Archiv. Allergy Appl. Immunol. 77, 364-371.

Badley, R. A., Atkinson, D., Hauser, H., Oldani, D., Green, J. P., and Stubbs, J. M. (1975). The structure, physical, and chemical properties of soy bean protein glycinin. Biochim. Biophys. Acta, 412, 214-228.

Burks, A. W., Jr., Brooks, J. R., and Sampson, H. A. (1988).Allergenicity of major component proteins of soybean determined by ELISA and immunoblotting in children with atopic dermatitis and positive soy challenges. J. Allergy and Clin. Immunol. 81, 1135-1142.

Burley, R.W., and Vadehra, D. V. (1989). "The Avian Egg", John Wiley and Sons, New York.

Canfield, R. E. (1963). The amino acid sequence of egg white lysozyme. J. Biol. Chem. 238, 2698-2707.

Friedman, M. (1973). "The Chemistry and Biochemistry of the Sulfhydryl Group in Amino Acids, Peptides, and Proteins", Pergamon Press, Oxford, England.

Gu, J., Matsuda, T., and Nakamura, R. (1988). Isolation and properties of an immunoreactive glycopeptide connecting first and second domains of chicken ovomucoid. Agric. Biol. Chem. 52, 3029-3033.

Hoffmann, R. (1983). Immunochemical identification of the allergens in egg white. J. Allergy Clin. Immunol. 71, 481-486.

Husby, S. (1988). Dietary antigens: uptake and humoral immunity in man. Acta Pathol. Microbiol. Immunol. Scand. 96, Supplement 1.

Kato, I., Schrode, J., Kohr, W.J., and Laskowski, M., Jr. (1987). Chicken ovomucoid: determination of its amino acid sequence and of the trypsin reactive site, and preparation of all three of its domains. Biochemistry, 26, 193-201.

Langeland, T. (1982). A clinical and immunological study of allergy to hen's egg white. Allergy, 37, 323-333.

Langeland, T. (1983). A clinical and immunological study of allergy to hen's egg white. Allergy, 38, 493-500.

Laskowski, M., Jr. (1986). Protein inhibitors of serine proteases - mechanism and classification. In "Nutritional and Toxicological Signficance of Enzyme Inhibitors in Fodds", Friedman, M., Ed., Plenum Press, New York. pp. 1-17.

Laskowski, M., Jr., and Kato, I. (1980). Protein inhbitors of proteinasases. Annu. Rev. Biochem. 49, 593-626.

Matsuda, T., Nakashima, I., Nakamura, R., and Shimokata, K. (1986). Specificity to ovomucoid domains of human serum antibody from allergic patients: comparison with anti-ovomucoid antibody from laboratory animals. J. Biochem. 100, 985-988.

Moreira, M. A., Hermodson, M. A., Larkins, B. A., and Nielsen, N. C. (1979). Partial characterization of the acidic and basic polypeptides of glycinin. J. Biol. Chem. 254, 9921-9926.

Moroz, L. A., and Yang, W. H. (1980). Kunitz soybean trypsin inhibitor. A specific allergen in food anaphylaxis. New Engl. J. Med. 302, 1126-1128.

Nielsen, N. C. (1984). The chemistry of legume storage proteins. Phil. Trans. R. Soc. London, B 304, 287-296.

Nielsen, N. C., Dickinson, C. D., Cho, T. J., Thanh, V. H., Scallon, B.J., Sims, T. L., Fischer, R. L., and Goldberg, R. B. (1989). Characterization of the glycinin gnee family. J. Biol. Chem. submitted.

Nisbet, A. D., Saundry, R. H., Moir, A. J. G., Fothergill, L., and Fothergill, J. E. (1981). The complete amino acid sequence of hen ovalbumin. Eur. J. Biochem. 115, 335-345.

Pedersen, H. S. and Djurtoft, R. (1989). Antigenic and allergenic properties of acidic and basic peptide chains from glycinin. Food Agric. Immunol. 1, 101-109.

Reinhardt, D. and Schmidt, E. (eds.). (1988). "Food Allergy", Raven Press, New York.

Shibasaki, M., Suzuki, S., Tajima, S., Nemoto, H., and Kuruome, T. (1980). Allergenicity of major component proteins of soybean. Int. Archiv. Allergy Applied Immunol. 61, 441-448.

Stevens, F. C. and Feeney, R. E. (1963). Chemical modification of avian ovomucoids. Biochemistry, 2, 1346-1352.

Williams, J., Elleman, T. C., Kingston, B. I., Wilkins, A. G., and Kuhn, K. A. (1982). The primary structure of hen ovotransferrin. Eur. J. Biochem. 122, 297-303.

22

IDENTIFICATION OF SOY PROTEIN ALLERGENS IN PATIENTS WITH ATOPIC DERMATITIS AND POSITIVE SOY CHALLENGES; DETERMINATION OF CHANGE IN ALLERGENICITY AFTER HEATING OR ENZYME DIGESTION

A. Wesley Burks[1], Larry W. Williams[1], Ricki M. Helm[1], Wayne Thresher[2], James R. Brooks[3], Hugh A. Sampson[4]

[1]University of Arkansas for Medical Sciences and Arkansas Children's Hospital, Department of Pediatrics, 800 Marshall St., Little Rock, AR 72202

[2]Central Soya, Fort Wayne, Indiana
[3]Ross Laboratories, Columbus, Ohio
[4]John Hopkins University Medical School Baltimore, Maryland

INTRODUCTION

A common problem encountered by primary care physicians is the possible association between ingestion of a particular food and symptoms experienced by a child (Bock, 1980). The <u>Federal Register</u> in 1983 reported 15%, or approximately 34 million people, in the United States thought symptoms to be associated with ingestion of certain foods (USDA, 1983). A variety of adverse reactions have been reported to soy proteins (Cook, 1960; Duke, 1934; Sampson, 1983; Burks, 1988). Soy protein has been consistently one of the five major foods implicated in causing food hypersensitivity in over 90% of children with documented adverse food reactions (Sampson, 1988; Burks, 1988). However, there are very few studies on the allergenicity of the soybean protein fractions (Ratner, 1955; Shibasaki, 1980; Burks, 1988). Soy proteins are being used increasingly in a number of foods and food products (Anderson, 1984). Thus, the potential for possible sensitization is also increasing.

In some children with atopic dermatitis (AD), food hypersensitivity has been demonstrated to have a pathogenic role in the development of eczematous lesions (Sampson, 1984; 1985). Approximately 35% of children with AD undergoing double-blind placebo-controlled food challenges (DBPCFC) have food hypersensitivity reactions (Burks, 1988). Patients undergoing a DBPCFC were found to develop an eczematous, pruritic, morbiliform

---

Supported in part by grant R29 AI26629-02 from the National Institute of Health and a grant from the Arkansas Science and Technology Authority.

Nutritional and Toxicological Consequences of Food Processing
Edited by M. Friedman, Plenum Press, New York, 1991

rash within 2 hours of ingesting the offending food antigen. Soy, peanuts, milk, egg, wheat, and fish accounts for 90% of the foods involved in these reactions (Sampson, 1983; Burks, 1988).

An enzyme-linked immunosorbent assay (ELISA) was used to investigate allergen-specific IgE and IgG antibody to soy protein and its fractions in patients with AD and a positive DBPCFC to soy. Sera from these same individuals was used to examine the antigens in soy proteins by immunoblotting techniques. We then used pooled serum from similar patients to investigate the change in antibody response after heating or enzyme digestion of the crude soy protein and its fractions.

EXPERIMENTAL SECTION

Approval for this study was obtained from the Human Research Advisory Committee at the University of Arkansas for Medical Sciences and Johns Hopkins University Medical School. Eight patients (mean age 6.18 ± 5.12 years, range - 18 months to 16 years) with AD and allergic rhinitis and a positive immediate prick skin test to soy (wheal ≥ 3 mm) had a positive DBPCFC to soy. (Sampson, 1983) Control sera were obtained from normal children being followed at Arkansas Children's Hospital. These children were not atopic and reported previous use of soybean products without symptoms.

<u>Soybean proteins and fractions</u>

For the ELISA and immunoblotting investigations protein was fractionated into 7S, 11S and whey fractions in addition to a crude extract. The two major soy protein fractions, 7S and 11S were isolated and purified according to a modification of the procedure of Thanh and Shibasaki. (Thanh, 1976) Commercial defatted soy flakes were extracted with 20 parts (w/v) of 0.03M Tris-HCl buffer (pH 8.0) containing 0.001M dithioerythritol (DTE) at room temperature (20-22°C) for 1 hour. Following centrifugation at 35,000 x g for 20 minutes (20°C), the supernatant was recovered and adjusted to pH 6.4 with 1N HCl. A crude 11S protein precipitate was obtained by centrifuging this mixture at 4°C as above. Purified 11S protein was obtained by resolubilizing the precipitate in 50 parts (w/v) 0.035M phosphate buffer (pH 7.6) containing 0.4M NaCl, 0.001M DTE, and 0.05% sodium azide and passing through a concanavalin-A Sepharose 4B column to remove contaminating glycoproteins. To obtain the 7S protein fraction, commercial defatted soy flakes were extracted with 20 parts (w/v) of 0.03M Tris-HCl buffer (pH 6.2) containing 0.001M DTE at 9°C for 1 hour and centrifuged at 35,000 x g for 20 minutes (4°C). The supernatant was recovered, adjusted to pH 4.8 with 1N HCl and centrifuged as above. The precipitate, thus obtained, was dispersed in 10 parts Tris-HCl buffer, adjusted to pH 7.6 with 1N NaOH and stirred at 9°C until the precipitate dissolved. The solution was readjusted to pH 6.2 and centrifuged. Following the addition of 0.4M NaCl, the crude 7S supernatant was warmed to room temperature (20-22°C) and bound to a concanavalin-A Sepharose 4B column pre-equilibrated against 0.03M Tris-HCl buffer (pH 6.2) containing 0.4M NaCl and 0.001M DTE. After removing the unbound proteins, purified 7S protein was recovered by eluting the column with the same buffer containing 0.1M α-methyl-D-mannoside. The purified 7S solution was equilibrated against 0.035M phosphate buffer (pH 7.6) containing 0.4M NaCl, 0.001M DTE, 0.05% sodium azide by repeated (4 times) concentration in an Amicon stirred cell containing a YM-10 membrane followed by redilution (1:10 v/v).

To obtain the crude soy fraction commercial defatted flakes were extracted at room temperature for 45 minutes with 20 parts deionized water (w/v) adjusted to pH 8.0 with 1N NaOH. Following centrifugation as above the supernatant recovered is the crude soy preparation. The whey preparation is obtained by taking the crude supernatant and adjusting the pH to 4.0 with 1N HCl. The principle globulin proteins will precipitate when centrifuged as above. This supernatant is the whey fraction and is adjusted to pH 7.0 with dilute alkali.

ELISA for IgE and IgG

A biotin-avidin ELISA was developed to quantify IgE antisoy protein antibodies with modifications from an assay previously published. (Burks,1986) The upper two rows of a 96-well microtiter plate (Gibco, Santa Clara, CA) were coated with affinity purified goat anti-human IgE (Tago Diagnostics, Burlingame, CA). The remainder of the plate was coated with one of the soy products (crude, 7S, 11S, whey) at a concentration of 5 μg/ml in coating buffer (0.1 sodium carbonate-bicarbonate buffer, pH 9.5). The plate was incubated at 37°C for one hour and then washed five times with rinse buffer (phosphate-buffered saline, pH 7.4, containing 0.05% Tween 20; Sigma, St. Louis, MO) immediately and in between subsequent incubations. The upper two rows used a standard reference to generate a curve for IgE, ranging from 0.25 to 10 ng/ml.

The patient's serum samples were diluted (1:2, 1:10, 1:50, 1:100) and dispensed in duplicate in the lower portion of the plate. After incubation for one hour at 37°C and washing, biotinylated, affinity-purified goat anti-human IgE (Tago) was added to all wells. Plates were incubated again for one hour at 37°C and washed before horseradish peroxidase-avidin conjugate (Vector, Burlingame, CA) was added. After washing, the plates were developed by the addition of a buffer containing O-phenylenediamine (Sigma). The reaction was stopped by the addition of 2N-hydrochloric acid to each well, and absorbance was read at 492 nm (Titertek Multiscan; Flow Laboratories, McClean, VA). The standard curve was plotted on log-logit paper by means of simple linear regression, and values for each patient's sample were read from the curve as "nanogram-equivalent units" per milliliter (ng per milliliter).

IgG antisoy protein antibodies to the different fractions were determined by using a modification of a previously described enzyme-linked immunoassay. (Burks, 1986) The upper two rows of a 96-well microtiter plate (Gibco) were coated with a monoclonal antibody to human IgG (Miles Scientific; Naperville, IL). The remainder of the plate was coated with one of the soy products (crude, 7S, 11S, whey) at a standard concentration of 5 μg/ml in coating buffer. The plate was incubated at 37°C for one hour and then washed five times with rinse buffer immediately and in between subsequent incubations. The upper two rows used a standard reference to generate a curve for IgG, ranging from 5 to 250 μg/ml. The patient's serum samples were diluted (1:10, 1:50, 1:100, 1:500) and dispensed in duplicate in the lower portion of the plate. After incubation for one hour at 37°C and washing, alkaline phosphatase conjugated goat antihuman IgG (Tago) was added to all wells. After incubation for one hour at 37°C and washing, nitrophenyl phosphate in carbonate buffer was added and the plate was developed. Plates were then read at 405 nm (Titertek Multiscan). The standard curve was plotted on log-logit paper, and patient values were read from the curve in "nanogram equivalent units" per milliliter (ng per milliliter).

Immunoblotting: IgE and IgG

The electrophoresis procedure is a modification of Sutton et al. (Sutton, 1982) SDS-PAGE electrophoresis was carried-out with a 5-20% polyacrylamide separating gel and a stacking gel of 3%. The protein

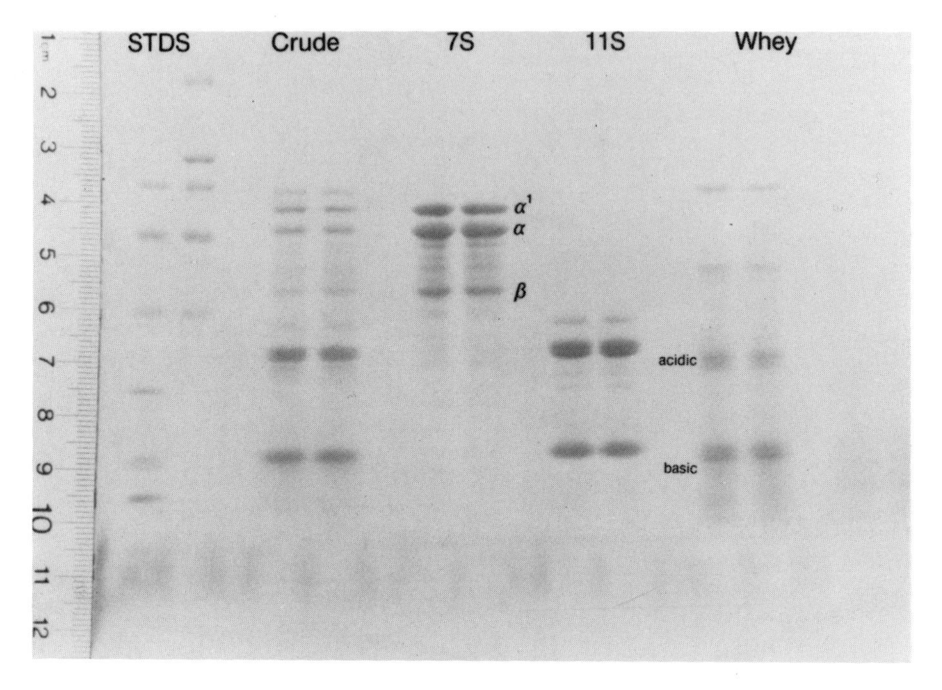

Fig. 1. Coomasie blue stain of 5% to 20% SDS-PAGE. Lanes are marked with low and high molecular weight standards, crude soy preparation, and 7S, 11S, and whey fractions of soy protein. The major subunits of 7S are α', α, and β. The 11S subunits are the acidic and basic fractions.

concentration of soy product added to each well in the gel was 1 mg/ml. Electrophoresis was performed overnight for 16 hours at 30 V (E-C Apparatus Corp., St. Petersbury, FL). To assure proper protein separation Coomassie Blue (Sigma) stains were done on preliminary gels.

Proteins were transferred from the separating gel to a nitrocellulose membrane in a transfer buffer with 10% SDS and methanol. The procedure was done in a transblot apparatus (Bio-Rad) for 5 hours (0.150 A). To assure transfer of the protein an Amido Black stain (Bio-Rad) was done on one portion of the nitrocellulose.

After removal from the transblot apparatus, the nitrocellulose was placed in blotto (25 gm non-fat dry milk, 0.5 gm sodium azide in 500 ml of phosphate buffered saline with EGTA added) and rocked for 1 hour. The nitrocellulose was then washed 3 times with TBS (2.42 gms Tris base, 29.42 gms NaCl in 1 liter of water and adjusted to pH 7.5). Next the nitrocellulose was incubated with the patient's serum

(1:10 dilution) overnight at 4°C with rocking. After washing again with TBS 3 times, goat anti-human IgE (1:100, Bio-Rad) was added and incubated at room temperature with rocking for 2 hours. Rabbit anti-goat alkaline phosphatase conjugated IgG (1:100, Bio-Rad) was then added after washing with TBS, at room temperature for 1.5 hours with rocking. After again washing with TBS 3 times, development was started by adding 250 μl of the NBT solution (30 mg nitro blue tetrazolium in 70% dimethylformamide; Bio-Rad) and 250 μl of the BCIP solution (15 mg 5-Bromo-4-Chloro-3-Indolyl phosphate in 70% dimethylformamide; Bio-Rad) into 25 ml of carbonate buffer at room temperature with rocking. The reaction is then stopped by decanting the NBT/BCIP solution and incubating the nitrocellulose for 10 minutes with distilled water. The paper is then air dried.

The procedure for specific IgG to the crude product is identical to IgE except in staining the first antibody added is goat anti-human IgG (1:500, Bio-Rad) and the second antibody is rabbit anti-goat alkaline phosphatase conjugated IgG (1:100, Bio-Rad).

HEATING AND ENZYME DIGESTION

The crude soy protein and the 7S, 11S and whey fractions were subjected to heat treatment. Each protein was subjected to either 80°C or 120°C for 60 minutes. Crude soy was subjected to enzyme digestion by the method of Thresher et al. (Thresher, 1989) The system is an immobilized digestive enzyme assay (IDEA). The assay consists of 2 bioreactors, one containing pepsin, and the other containing trypsin, chymotrypsin and intestinal mucosal peptidases.

ELISA INHIBITION

A competitive ELISA inhibition was done using the crude soy extract and the digested soy product (after passing over the IDEA system). One hundred μl of pooled serum (1:20) from the positive challenge patients was incubated with various concentrations of either the crude soy or digested soy (0.00005 to 50 ng/ml) for 18 hours. The inhibited pooled serum was then used in the ELISA described above. The percent inhibition was calculated by taking the food specific IgE value minus the incubated food-specific IgE value divided by the food-specific IgE value. This number is multiplied by 100 to get the percentage of inhibition.

DATA ANALYSIS

Statistical comparisons were done by the Mann-Whitney U test and the Wilcoxon signed-rank test.

RESULTS AND DISCUSSION

Preliminary gels were stained with Coomassie Blue to ensure adequate protein separation. One can see in Figure 1 the banding pattern for the crude, 7S, 11S and whey fractions. The major bands in the 7S fraction are the α', α and ß subunits. In the 11S fraction the acidic and basic subunits are also easily identified.

The ELISA results for specific IgE to crude soy, 7S, 11S and whey are shown in Table 1. Comparing the IgE in the atopic patients to the normal controls there was a statistically significant increase in the atopics to the crude extract ($p < 0.005$) and 7S ($p < 0.05$). The children with atopic dermatitis did not show a significant difference in the IgE response to whey ($p > 0.2$) or 11S ($p > 0.2$).

Table 2 gives the results of the IgG ELISA to crude soy, 7S, 11S and whey. The atopic patients did have a significant increase in IgG to the crude extract ($p < 0.02$) and 11S ($p < 0.005$). Their IgG specific for whey ($p > 0.2$) and 7S ($p > 0.05$) was not significantly different from the normal controls.

Comparisons were then made among the eight patients found to have soy protein hypersensitivity to determine if there was significantly more IgE specific for any of the three fractions. There was an increase in IgE specific for 7S versus whey ($p < 0.05$) but not when comparing 7S versus 11S ($p > 0.10$) or 11S versus whey ($p > 0.05$). A similar comparison was done for specific IgG. There was no significant difference in 7S versus 11S ($p > 0.10$), 7S versus whey ($p > 0.20$) and 11S versus whey ($p > 0.10$).

TABLE 1.

IgE ELISA (ng/ml) to crude soy protein, 7S, 11S, and whey fractions

Patients	Crude+	7S*	11S	Whey
1	30	70	25	30
2	10	15	3	0.25
3	60	220	120	40
4	2	4	4	2
5	60	70	25	60
6	15	40	40	15
7	3	0.72	3.1	7.5
8	3.3	2.5	1.3	0.6
Normals	7.69	11.42	9.87	22.83
(+/-S.D.)	10.62	19.45	9.94	19.75

+p<0.005
*p<0.05

Immunoblotting for specific IgE is shown in Figure 2. Numbers 5-8 are samples from the eight patients with positive DBPCFC's to soy (patients 1, 3, 5 and 6 from Table 1). These four patients had the highest specific IgE values and also had the strongest blotting pattern. All eight patients had immunoblots done. The intensity of their binding pattern correlated with their specific IgE values. Patient #5 had bands against the α', α and β subunits of 7S. Patient #6 had similar results against 7S but in addition had bands against the acidic and basic subunits of 11S. Patients #7 and #8 had bands representing 7S and 11S with the major band against the acidic subunit of 11S. Numbers 1-4 are representative samples from the sera of the control patients. Minimal bands can be seen in each strip with the strongest band in number 4 to the acidic subunit of 11S.

TABLE 2

IgG ELISA (ng/ml) to crude soy protein, 7S, 11S, and whey fractions

Patients	Crude+	7S	11S*	Whey
1	28,214	9,050	15,600	21,607
2	30,714	12,000	6,400	9,464
3	8,203	12,364	54,167	5,664
4	7,644	12,813	14,375	1,180
5	9,375	44,160	47,040	23,214
6	23,961	19,500	45,506	6,027
7	1,402	1,623	1,583	633
8	1,548	629	3,366	2,376
Normals (+/-S.D.)	2,766.11 2,948.76	4,024.44 4,948.76	2,016.44 1,874.86	4,150.67 4,281.81

+p<0.02
*p<0.005

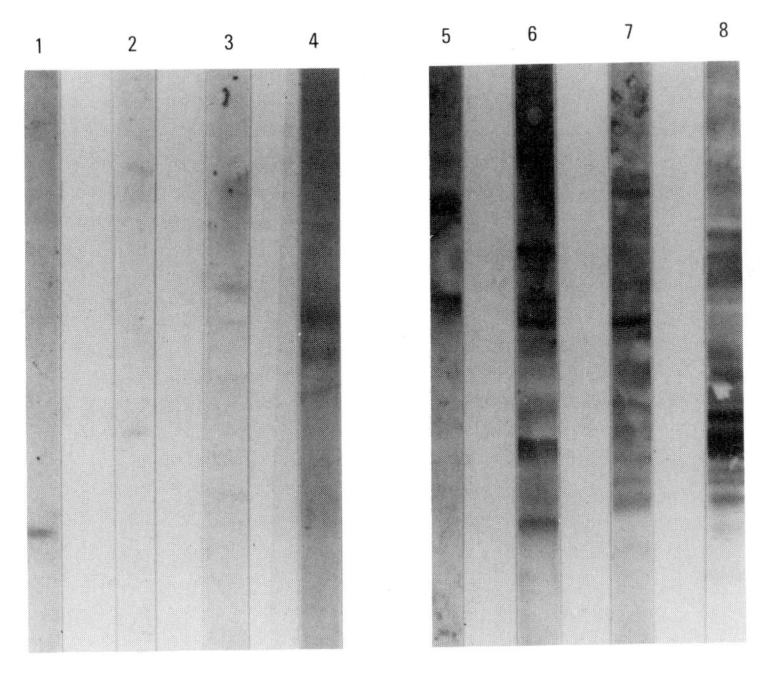

Fig. 2 IgE immunoblot of four normal control subjects (1 to 4) and four patients (5 to 8) with positive DBPCFC to soy.

The immunoblot for specific IgG to the crude product is shown in Figure 3. These are the same eight patients serum used in the IgE immunoblot. Numbers 1-4 are the control patients and numbers 5-8 are the patients with positive DBPCFC's to soy. Bands can be seen in each of the atopic's immunoblot to both the 7S and 11S fractions. As in the IgE immunoblots no one fraction is stained consistently darker.

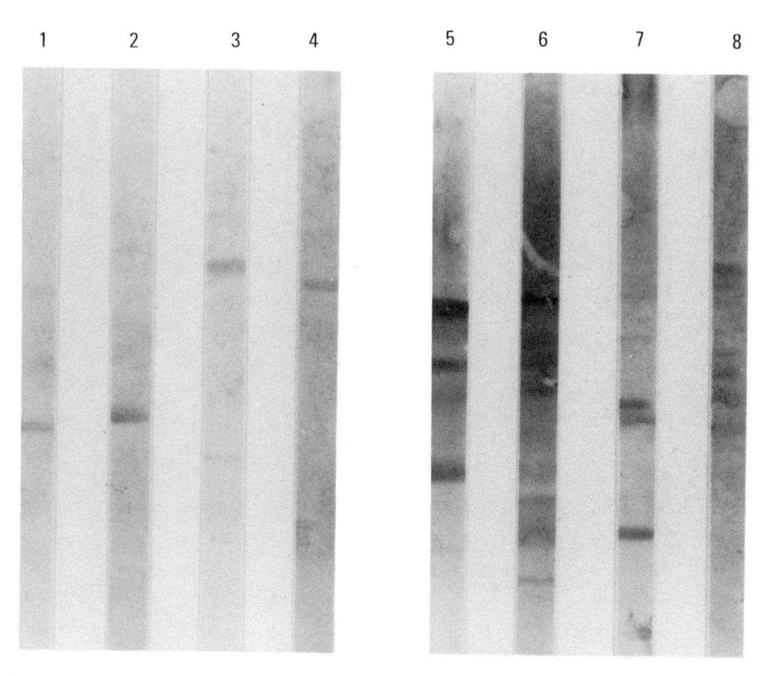

Fig. 3. IgG immunoblot of four normal control subjects (1 to 4) and four patients (5 to 8) with positive DBPCFC to soy.

To address the question of whether heating would change the allergenicity of the crude soy protein, and the 7S, 11S, and whey fractions. Heating of these proteins was done at 80°C and 120°C for 60 minutes. In the IgE specific ELISA there was a significant decrease in specific antibody for each protein except whey when heated to 80°C (Table 3) When the proteins were heated to 120°C there was a significant reduction in IgE specific for each one.

In the IgG specific ELISA heating to 80°C did not consistently decrease the immunoglobulin-specific binding as in the IgE ELISA. (Table 4) In fact, the IgG specific for the 11S component was higher when heated at 80°C. Heating to 120°C did significantly decrease the IgG binding to each protein fraction although not to the degree that IgE binding was affected.

TABLE 3

IgE (ng/ml) to heated soy proteins

Protein	Untreated	80°C	120°C
Crude Soy	8	0.4	0.005
Whey	4	4	0.2
7S	12	4	0.05
11S	24	10	0.05

TABLE 4

IgG (ng/ml) to heated soy proteins

Protein	Untreated	80°C	120°C
Crude soy	60	60	10
Whey	50	50	10
7S	200	100	10
11S	10	20	3

Crude soy was passed over the IDEA system to compare the specific IgE to the crude soy product versus the digested soy product. An ELISA inhibition assay was used for the comparison. We measured the IgE specific to crude soy with either crude soy or digested soy incubated overnight to inhibit IgE binding (data not shown). There was at least a 10,000 fold difference in the concentration required to inhibit the ELISA by 75%, suggesting a significant change in IgE binding of the digested soy product compared to the crude soy product.

Discussion

Allergic reactions to legumes including soybean and peanut are not uncommon. Soybean protein is used in many commercial foods. Approximately 90% of the seed proteins are salt-soluble globulins with the bulk of the remainder being water-soluble albumins. (Naismith, 1955; Wolf, 1972) They comprise a broad range of molecular sizes and display four major sedimentation components in the ultracentrifuge, designated 2, 7, 11 and 15S. (Wolf, 1956) Each fraction is a complex mixture of proteins. The 7S and 11S each make up approximately one-third of the extractable proteins followed by the 2S and 15S which comprise 20% and 10%, respectively. (Nielson, 1985)

Glycinin has a molecular weight of 320,000 to 360,000 and is made up of six subunits. Each subunit is comprised of one acidic and one basic polypeptide linked by a disulfide bond. Both the acidic and basic polypeptides are heterogeneous and possess molecular weights ranging from 34,000 to 45,000 and 19,000 to 22,000, respectively. (Brooks, 1985) One exception is an acidic polypeptide, designated A_5, which has a molecular weight of approximately 10,000. (Morreira, 1981) ß-conglycinin, a glycoprotein, has a molecular weight of approximately 180,000. (Koshiyama, 1976) It is a trimer composed of three major subunits, α', α and β, which have molecular weights of 76,000, 72,000 and 53,000, respectively. (Shattuck, 1985)

The allergenicity of the globulin fractions 2, 7 and 11S have previously been examined by RAST and RAST-inhibition experiments in patients with asthma and atopic dermatitis felt to be soybean-sensitive. (Shibasaki, 1980) There was considerable cross-reactivity among the 3 fractions with specific IgE antibody found against each one. But the 2S-globulin had the highest allergen potency to inhibit the RAST against 11S-, 7S- and 2S-globulin. Heat treatment of 80, 100 and 120°C for 30 minutes enhanced the RAST activity of the 2S-globulin (when heated at 80° for 30 minutes) while the activity of the others was reduced to 40 to 75% of the native globulin.

The results of the IgE and IgG ELISA's in these atopic patients with positive DBPCFC's to soy show a significant difference in both IgE and IgG specific for the crude protein extract compared to the normal controls. There was also an increase in IgE specific for 7S in the atopic population as demonstrated by IgE ELISA. The atopic group had more specific IgG for 11S, as shown by IgG ELISA.

The immunoblotting experiments reveal a great variation in the protein subunit polypeptides recognized by human antibodies. Variable patterns suggested that no one component of the soy fraction bands is more antigenic than others. There appeared to be IgE antibodies to both 7S and 11S fractions in varying amounts in most of the patients. The majority of protein bands identified in the IgG immunoblots also were detected in the IgE blots. It appears from this that most of the antigens are also allergens as well. The finding that IgE and IgG is present in varying amounts to the different soy fractions and that no one fraction is more antigenic than another is similar to findings in earlier studies with patients sensitive to wheat. (Theobald, 1986) These publications showed increased amounts of IgE in their population sensitive to these foods compared to their normal controls. They did not however, find any fraction of the food protein to be more consistently allergenic.

Very little is known about the alteration of the allergenicity of foods by heating. Simple heating of bovine gammaglobulin and alpha-lactoalbumin denatures these proteins while alpha-casein and beta-lactoglobulin are more heat stable (Anderson, 1984). From other studies with soy protein it is known that heating of the crude soy bean causes aggregation of the ß subunits of 7S with the basic subunits of 11S (Utsumi, 1984). Shibasaki et al (Shibasaki, 1980) examined heat treatment at 80°C, 100°C and 120°C of the crude soy fraction and its components and its effects on IgE binding. The IgE binding of the 2S fraction was enhanced with heating at 80°C, whereas the IgE binding to the other fractions were reduced 40% to 75% of the crude product.

Similarly, very few studies have determined the change in allergenicity of foods after enzyme digestion. In one study done to examine the difference in IgE activity to pepsin digests of β lactoglobulin, the patients had significantly lower levels of IgE antibodies to the digested proteins (Anderson, 1984)

CONCLUSION

This study demonstrates that children with AD and positive DBPCFCs to soy have significantly elevated levels of IgE and IgG antibodies to a crude soy protein extract. It also provided evidence of an IgE and IgG response to multiple fractions of the soy protein. Heating the crude soy and its fractions appears not to enhance IgE activity but significantly reduces its binding. The significant enzyme treatment provided by the IDEA system also appears to greatly reduce IgE binding.

Acknowledgements

We would like to thank Cathie Connaughton and Gael Cockrell for the laboratory assistance. We thank the C. V. Mosby Company for granting permission to include the 1988 article, Burks et al in the Journal of Allergy and Clinical Immunology.

REFERENCES

Anderson JA, Sogn DD eds: Adverse reactions to foods. Hyattsville, MD: NIH publication 1984 No. 84-2442.

Anet J, Bach JF, Baker RS, Barnett D, Burley RW, Howden MEH: Allergens in the white of yolk of hen's egg. Int Arch Allergy Appl Immunol 1985;77:364.

Bock, SA: Food sensitivity. Ann J Dis Child 1980;134:973.

Brooks JR, Morr CV: Current aspects of soy protein fractionization and nomenclature J Agri Chem Soc 1985;62:1347.

Burks AW, Sampson HA, Buckley RH: Anaphylactic reactions following gammaglobulin administration in patients with hypogammaglobulinemia: detection of IgE antibodies to IgA. N Engl J Med 1986;314:560.

Burks AW, Brooks JR, Sampson HA: Allergenicity of major component proteins of soybean determined by enzyme-linked immunosorbent assay (ELISA) and immunoblotting in children with atopic dermatitis and positive soy challenges. J Allergy Clin Immunol 1988;81:1135.

Burks AW, Butler HB, Brooks JR, Hardin J, Connaughton C: Identification and comparison of differences in antigens in two commercially available soybean protein isolates. J Food Sci 1988;53 (5):1456.

Burks AW, Mallory SB, Williams LW, Shirrell MA: Atopic dermatitis: Clinical relevance of food hypersensitivity reactions. J Pediatr 1988;113:447-451.

Conover WJ: Practical nonparametric statistics. New York: John Wiley 1971

Cook CD: Possible gastrointestinal reaction to soybean. N Engl J Med 1960;263:1076.

Duke WW: Soybean as a possible important source of allergy. J Allergy 1934:5:300.

Koshiyama I, Fukushima D: Identification of the 7S globulin with ß-conglycinin in soybean seeds. Phytochem 1976;15:157.

Morreira MA, Hermodson MA, Larkins BA, Neilsen NC: Comparison of the primary structure of the acidic polypeptides of glycinin. Arch Biochem Biophys 1981;210:633.

Naismith WEF: Ultracentrifuge studies of soybean protein. Biochem Biophys Acta 1955;16:203.

Nielson NC: Structure of soy proteins. In: Altschul AM, Wilcke HL eds., New protein foods vol 5. Orlanda, Academic Press 1985, pp 27-64.

Osterballe I, Ipsen H, Welke B, Lowenstein H: Specific IgE response toward allergenic molecules during perennial hyposensitization: a three-year prospective, double-blind study. J Allergy Clin Immunol 1983;71:40.

Ratner B: Allergenicity of modified and processed foodstuffs. V. Soybeans influence of heat on its allergenicity, use of soybeans preparations as milk substitute. J Allergy 1955;89:187.

Sampson HA: Role of immediate food hypersensitivity in the pathogenesis of atopic dermatitis. J Allergy Clin Immunol 1983;71:473.

Sampson HA, Jolie PL: Increased plasma histamine concentrations after food challenges in children with atopic dermatitis. N Engl J Med 1984;311:372.

Sampson HA, McCaskil CM: Food hypersensitivity and atopic dermatitis: evaluation of 113 patients. J Pediatr 1985;107:669.

Shattuck-Eidens DN, Beach RN: Degradation of ß-conglycinin in early stages of soybean embryogenesis. Plant Physiol 1985;78:895.

Shibasaki M, Suzuki S, Tajima S, Nemoto H, Kurome T: Allergenicity of major component proteins of soybeans. Int Arch Allergy Appl Immunol 1980;61:441.

Sutton R, Wrigley CW, Baldo BA: Detection of IgE and IgG binding proteins after electrophoretic transfer from polyacrylamide gels. J Immunol Methods 1982;52:183.

Thanh VH, Shibaski K: Major proteins of soybean seeds: A straight forward fractionization and their characterization. J Agri Food Chem 1976;24:1117.

Theobald K, Thiel H, Kallweit C, Ulmer W, Koniz W: Detection of proteins in wheat flour extracts that bind human IgG, IgE and mouse monoclonal antibodies. J Allergy Clin Immunol 1986;78:470.

Thresher WC, Swaisgood HE, Catignani GL: Digestibilition of the protein in various foods as determined in vitro by an immobilized digestive enzyme assay (IDEA). Plan Foods for Human Nutrition 1989;39:59.

USDA: Rules and Regulations. Federal Register 1983;48:32749.

Utsumi S, Damodaran S, Kinsella JE: Heat-induced interactions between soybean proteins: perferential association of 11S basic subunits and ß subunits of 7S. J Agric Food Chem 1984;32:1406.

Wolf WJ, Briggs DR: Ultracentrifugal investigation of the effect of neutral salts on the extraction of soybean proteins. Arch Biochem Biophys 1956;63:40.

Wolf WJ: Purification and properties of the proteins. In: Smith AK, Circle JJ eds. Soybeans: Chemistry and technology. Westport, CN: AVI, 1972, p 93.

23

REDUCTION OF WHEY PROTEIN ALLERGENICITY BY PROCESSING

R. Jost, J.C. Monti, and J.J. Pahud

Nestle Research Centre, Nestec Ltd.
Vers-ches-les-Blanc
CH-1000 Lausanne 26, SWITZERLAND

IMPORTANCE OF COW'S MILK ALLERGY IN INFANT FEEDING

Protein intolerance in the infant has many different clinical facets such as gastrointestinal disorders, respiratory troubles, and skin irritation such as eczema, urticaria, and others (Machtinger and Moss, 1986). Allergies may also cause disturbed sleep, as suggested by recent studies of Kahn et al. (1987).

The potential for cow's milk allergy is important in a normal population of newborns and very important in newborns with a family history in allergy. The percentage of all newborns showing allergic manifestations has been estimated by several authors. Careful European estimates are 2% or less (Jacobsson et al., 1979) whereas recent US estimates suggest a frequency of up to 5% in an unselected infant population (Bock, 1987).

Sensitization to cow's milk protein, frequently the first foreign protein a human newborn contacts in his diet, often occurs during the first weeks and months of life. Breast feeding undoubtedly minimizes the risk of such sensitization to bovine milk proteins, but sensitization in strictly breast-fed infants has been repeatedly reported and the transfer of dietary protein into human milk demonstrated (Jacobsson et al., 1985; Cant et al., 1985). Despite these (rare) cases, the early contact with an infant formula based on cow's milk is the most likely source of sensitization. For the infant formula producer, it is therefore imperative to ask whether there are means to reduce the allergenic potential of a starter formula.

Our paper considers two major technological alternatives which have been tested in the processing of milk with the aim of reducing or abolishing protein allergenicity.

REDUCTION OF ALLERGENICITY BY HEAT DENATURATION OF PROTEIN

Heat treatment of milk affects the *in vivo* allergenicity of bovine milk proteins, as shown by early studies of Ratner et al. (1958) and McLaughlan et al (1981). These studies showed a reduced sensitization capacity of more severely heated milks, such as canned sterilied evaporated milk or canned sterilized liquid infant formulae. Severe heating (120°C for 10-15 min) reduced or abolished β-lactoglobulin (BLG) or α-lactalbumin (ALA) specific allergic reactions but had little effect on the allergenicity of the caseins. The difference in heat sensitivity of casein and whey protein antigens was also shown in the guinea pig response to severely heated skimmed milk versus whey (Kilshaw et al, 1982) and led to the proposal of Heppel et al. (1984) to produce a whey-based hypoallergenic formula by adequate heat treatment of the whey.

It is well established by differential scanning calorimetry, that heat treatments above 70°C are required for efficient denaturation of most of the major whey proteins (de Wit, 1981). Exposing a whey concentrate to a heat treatment at a temperature of 90°C for 20 min resulted in complete denaturation of the protein according to DSC: while the initial protein had an enthalpy of denaturation of 9 ± 0.6 Joules/g, no enthalpy of denaturation could be measured following the heat treatment. Despite calorimetric information on the completeness of denaturation, immunochemical analysis showed that the heated protein reacted with specific antibodies raised against native BLG, ALA, and BSA (Figure 1). In an oral administration to specially bred guinea pigs (Pahud et al., 1985), such a heated whey sensitized 2 our of 5 test animals, while the untreated protein sensitized all of 5 test animals. We conclude that heat treatment considerably reduced oral sensitization capacity of the protein but did not completely abolish it.

Administered parenterally, by contrast, the heat denatured protein sensitized all animals yielding high serum titers, as revealed by passive cutaneous anaphylaxis (PCA) titrations (Table 1) Provocation with individual whey proteins showed that most of the reaginic antibodies were directed against determinants of BLG. The sera showed little or no reactivity with ALA or BSA. These results confirm the earlier findings of Ratner (1958), according to which BLG but not ALA retained sensitizing qualities in heated milk. The parenteral sensitization test showed that several proteins had preserved their allergenicity following the heat treatment.

Reduced antigenic response in an oral sensitization procedure might result from heat denaturation of protein facilitating its breakdown by proteolysis in the animal's intestine. The observed reduction in antigenicity would follow from this facilitated *in vivo* digestion.

The effect of heat denaturation on milk protein antigenicity was also tested by Baldo (1984), who measured binding of specific IgE from allergic infants to different milk protein preparations. Heating milk protein mixtures to 80-100°C markedly decreased IgE binding for BLG and BSA, but not for ALA nor for caseins. Heating at 100°C for 3 h still did not abolish IgE binding to the casein fraction. Also, the whey protein response of different sera was very variable. Baldo therefore concluded that significant reduction in milk protein allergenicity could not be achieved by mere heat treatment. This work measured the anaphylactic triggering capacity of the heated protein, not its capacity to induce sensitization. In order to evaluate this latter, we depend essentially on animal models.

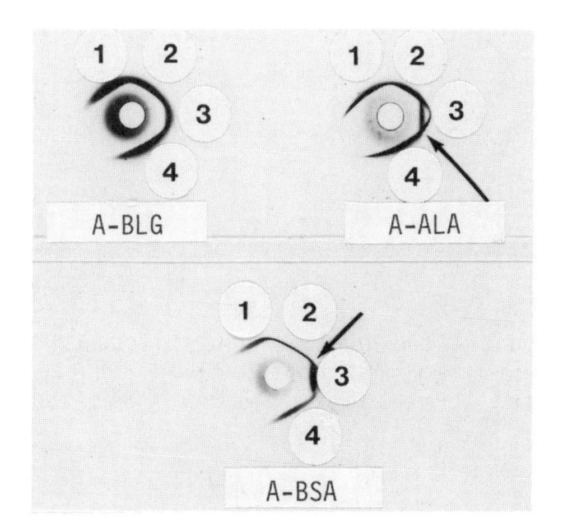

Figure 1. Reactivity of untreated and heat treated whey in the double immunodiffusion test. The test was performed on 3 rabbit antisera specific for β-lactoglobulin(A-BLG), α-lactalbumin (A-ALA and bovine serum albumin (A-BSA). In each test. Well 1 contained the pure protein antigen corresponding to the specificity of the antiserum. Wells 2 and 4 contained a dilution of untreated whey and Well 3 contained the heat treated (90°C/20 min) whey.

Table 1. Parenteral sensitization of guinea pigs with whey preparations

Sensitizing preparation	Sensitized animals/total	PCA Titrations Range titers	Mean (log_5) titers	Std. deviation
Untreated whey	5/5	1/625-1/15'625	4.06	0.33
Heat treated whey	5/5	1/625-1/5'625	4.78	0.39
Trypsin hydrolyzed whey	3/5	0-1/3'125	2.10	1.33
Tryptic hydrolysate after ultrafiltration	2/5	0-1/125	1.37	1.60

Sensitization protocol: 100 mg product with $Al(OH)_3$

day 1: intracutaneous injection

day 7: intraperitoneal injection

day 14: intraperitoneal injection

EFFECT OF CHEMICAL MODIFICATION AND CLEAVAGE ON ANTIGENICITY AND ALLERGENICITY OF SELECTED WHEY PROTEINS

Chemically modified BLG was tested in a quantitative immunoprecpitin reaction with antibody to native BLG. Modification of arginine side chains, trypotphan residues, or sulfhydyl groups had little effect on this antigenic reactivity. In contrast, modifications affecting protein net charge, such as succinylation of amino groups of lysine side chains led to a mrked decrease in antigenic reactivity (Otani et al., 1985).

Chemically denatured (reduced and S-carboxymethylated) BLG gave rise to a strong antibody response in rabbits. The resulting anti-SCM-BLG antibody weakly crossreacted with native BLG. This clearly shows that the denatured protein is antigenic on its own. Part of the antigenic determinants may be common to those displayed by the native protein. Fragmentation of S-CM-BLG with cyanogen bromide and with trypsin yielded a variety of fragments. Fragments corresponding to sequence segments 25-10, 108-145, and 146-162 showed strong antigenic reactivity as inhibitors of the immunoprecipitin reaction S-CM-BLG with homologeous antibody (Otani et al., 1987). The preptides mentioned must therefore participate in the global antigenic structure of S-CM-BLG. The represent sequential epitopes, but we do not know to what extent they have the capacity to induce an allergic response _in vivo_.

Tryptic cleavage of undenatured BLG resulted in the formation of at least 12 fragments (11 out of 18 possible T-sites were cleaved). Out of these fragments anti-BLG antibody was precipitated, only by the 2.7 KD peptide BLG-(61-69)-S-S(149-162), in which the C-terminal fragment 149-162 is linked vidisulfide bridge to the fragment 61-69 derived from the center part of the protein, was precipitated (Jost, 1988). Following repeated parenteral administration, the same peptide induced an allergic response to BLG in guinea pigs. None of the tryptic peptides triggered the PCA reaction in guinea pigs previously sensitized to BLG or whey protein. Enzymatic fragmentation of a relatively small and highly allergenic protein into a limited number of peptides with molecular weights ranging from 1-3 KD resulted in a complete loss of anaphylactic triggering capacity of the BLG molecule. The total suppression of sensitization capacity (on the parenteral route) appears to be a more difficult goal. It is conceivable that monovalent antigens can lead to sensitization. Their complete destruction may require a larger anumber of cleavages than those restricted to arginine and lysine residues.

The antigenic structure of bovine serum albumin (BSA) has been extensively studied (Peters, 1985). Cleavage of the protein with cyanogen bromide, pepsin, or trypsin, with maintenance of the disulfide loops, allowed the isolation of numerous fragments representing different disulfide loops of the molecule (Peters et al., 1977). Many of the fragments effectivey bound anti-BSA antibodies but only the largest ones gave immunoprecipitates, suggesting the presence of participation of more than a single determinant. Allergenic properties of such fragments were studied (Wahn et al., 1981) by sensitizing rabbits parenterally to BSA. Basophilic leucocytes were obtained from the blood of sensitized animals and assayed _in vitro_ for histamin release by BSA and BSA fragments. While the intact protein led to histamin discharge at 8×10^{-11} M concentrates, 10- to 100-fold higher concentrations of the active peptic fragments P-(1-306), P-(307-582), and T-(377-532) were required to trigger histamin release.

Peptic hydrolysis of BSA *in vitro* resulted in a rapid decrease in serologically defined antigenic determinants (Dosa et al., 1979). In addition to the drastically reduced immunogenicity observed with progressive stages of hydrolysis, components of the peptic digest were shown to be highly immunosuppressive. These peptides were characterized by a low density of antigenic determinants and were weakly immunogenic or not immunogenic at all. A strong suppression of a secondary IgE response to injected BSA was observed in mice primed with such peptides. The mechanism of this suppression is not established. It appears from this work, however, that peptides that maintain some antigenic reactivity may play a role in suppressing an allergenic response to protein and might play a role in the development of immunological tolerance to this protein.

SENSITIVITY OF DIFFERENT MILK PROTEINS TO IN VITRO PROTEOLYSIS

The frequency of cleavage sites for specific endopeptidases such as trypsin or chymotrypsin along the polypeptide chains of the major milk proteins is not very different from one protein to another. Thus, we find one average trypsin site per 7-10 residues within the primary structures of caseins and whey proteins (Jost, 1988). Naturally, the distribution of these sites along the polypeptide chain is not regular and there is a considerable variation in the size of the theoretical tryptic peptides. In practice we find striking differences in the susceptibility of individual milk proteins towards endopeptidases. While the breakdown of the caseins occurs quite smoothly, refractory behaviour of serum albumin and immunoglobulins is observed in the hydrolysis of skimmed milk or in whey. In concentrated, unheated whey, both serum albumin and IgG H- and L-chains persist after 4 h hydrolysis with pancreatic trypsin/chymotrypsin preparations (Jost et al., 1987). After heating at 90°C, both these proteins are rapidly hydrolyzed following the addition of fresh enzyme. Similarly, predenaturation by heat of the protein substrate leads to complete breakdown of the proteins in subsequent hydrolysis.

Studies with isolated whey proteins confirmed this protein-specific behaviour. Thus, the 18 KD BLG monomer was degraded to peptide fragments without prior heat denaturation. In contrast, BSA and IgG were not hydrolyzed to an important extent, unless denatured by heat (Figure 2). In the case of IgB, both H- and L-chains proved quite resistant to hydrolysis in the undenatured immunoglobulin. Protein fragmentation, which could be well followed by either HPLC exclusion chromatography or by SDS-PAGE, was clearly accompanied by a rapid decrease in antigenic reactivity. This is illustrated for BLG by means of drastically decreased antibody-binding capacity of the progressively hydrolyzed protein, as shown by solid phase radioimmunoassay (Pahud et al., 1985). Progressive hydrolysis of total whey protein is also characterized by the decreasing capacity of the hydrolysate to form specific immunoprecipitates. The final hydrolysate is unable to form precipitation arcs with specific rabbit sera against ALA, BLG, BSA, or IgG, indicating the destruction of polyvalent antigens (intact protein or large fragments thereof with more than one antigenic determinant).

ENZYMIC PROCESSING OF WHEY PROTEIN AND WHEY TO REDUCE ALLERGENICITY

While the low antigenicity of extensively hydrolyzed casein has been known for some time (Takase et al., 1979), whey protein hydrolysates were developed at a later stage. Ultrafiltrated whey protein was hydrolyzed with pancreatin to about 30% degree of hydrolysis (DH) defined as: (DH = % α-N x 100)/% N total. The resulting hydrolysate

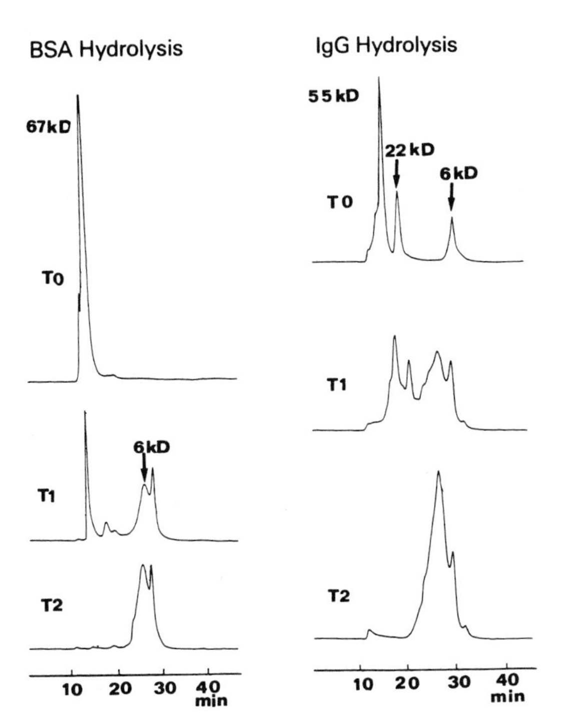

Figure 2. HPLC gell exclusion profiles of trypsin-hydrolysed bovine serum albumin (left side) and bovine immunoglobulin (right side), on TSK-2000 SW. Mobile phase was phosphate buffered (0.1 M., pH 6.8) guanidinium. HCl (3.5 M), 0.1% (w/v) of dithiothreitol. Detection was at 280 nm.
T0: immediately following start of hydrolysis with porcine trypsin
T1: after 3 h of hydrolysis at pH 7.5 and 50°C, after heat inactivation (10 min at 90°C)
T2: another 2 h of hydrolysis (pH 7.5, 50°C) following addition of trypsin (1% on total hydrolysate solids).

was then fractionated by an ultrafiltration process using an organic membrane with a nomimal cut-off near 10 KD (Oloffson et al., 1981). The purified hydrolysate, mainly composed of oligopeptides and free amino acids, had almost no parenteral sensitization capacity in guinea pigs, as demonstrated bu systemic provocation tests and PCA tests (Table 1). The major drawback of this kind of hydrolysate is the poor palatability and bad taste - a severe handicap. Although the bitterness is less drastic than in casein hydrolysates.

In addition, production costs for such oligopeptide products are very high. Debittering procedures are costly and result in losses of essential amino acids which have to be compensated by supplementation with synthetic ones. The hydrolysates produce a relatively high osmotic pressure and can cause osmotic discomfort or diarrhea. In order to avoid accentuation of this problem, lactose is usually not present in semielemental diets, the galenic form in which these hydrolystates are used. In long term nutrition of the young infant, lactose is essential for favouring a milk flora and calcium absoprtion in the intestine of the newborn. Consequently, we must attempt to achieve similar reduction in protein allergenicity by specific enzymatic hydrolysis, avoiding the formation of very small peptides and free amino acids and treat lactose rich substrates such as skimmed milk or partially demineralized whey.

An indication that limited enzymatic hydrolysis could substantially reduce allergenicity was found when we observed that tryptic hydrolysis of whey protein concentrates in vivo completely abolished the oral sensitization capacity of the protein in the guinea pig (Pahud et al., 1985). The tryptic hydrolysate also did not trigger anaphylaxis in guinea pigs previously sensitized by feeding them liquid skimmed milk or liquid whey. We then extended our studies to the hydrolysis of lactose-rich substrates such as demineralized whey (Jost et al., 1987). Protein hydrolysis proceeded equally well in such a substrate and the resulting hydrolysates were characterized by the same absence of oral sensitization capacity.

The complete absence of oral sensitization capacity is clearly what we are looking for in designing a hypoallergenic infant formula. Absence of oral sensitization capacity does not mean non-allergenic, however. If the product is, instead, administered by repeated intracutaneous and/or peritoneal injection, an allergic response can be induced. In a typical experiment with parenteral administration, 3 out of 5 guinea pigs became sensitized but the measured PCA titers were markedly lower than in the animals sensitized with untreated or heat treated whey (Table 1). Thus, the mean PCA titer was slightly higher than 1/25 in the hydrolysate animals, as compared with a mean PCA titer of 1/3125 in the protein group. In the hydrolysate animals, PCA could be elicited by whey protein as well as by purified BLG, but not with ALA nor with BSA. Preferential sensitization to BLG was observed in the guinea pigs.

The antigens in the hydrolysate responsible for the sensitization to BLG might have been either very small amounts of intact protein escaping hydrolysis, or fragments capable of inducing such a reaginic response. Following ultrafiltration of the hydrolysate on a 10 KD cut-off membrane, the remaining aggregates and largest polypeptides were removed but otherwise the peptide composition remained the same. The removal of a minor fraction of high molecular weight material was reflected in a strongly decreased in vivo allergenic potential: only 2 our of 5 animals were sensitized and the PCA titer measured by provocation with BLG or total whey protein was 1/125 naxunyn, Remarkably, provocation with the hydrolysate itself remained without effect, from which we deduced that if any sensitization to the peptides of the hydrolysate had occured, it remained undetected by a sensitive functional test such as PCA.

REACTIVITY OF ENZYME PROCESSED WHEY WITH IhE OF COW'S MILK ALLERGIC INFANTS

To test the sensitization potential of trypsin hydrolyzed whey in a formula in infants with a predisposition for allergy, long term clinical trials are required (Chandra et al, 1989). For process studies aimed at the production of hypoallergenic formula diets, we depend on a suitable animal model.

We can, however, test the reactivity of a formula with sera obtained from patients with a clinically established cow's milk allergy. This provides information about whether the formula contains antigenic proteins or peptides, recognizable by patient IgE. RAST discs can be impregnated with the hydrolysate and the binding of patient IgE to the disc measured via a (125-I)-labelled anti-IgE detecting antibody (Phadebas RAST kit of Pharmacia).

The capacity of the whey hydrolysate to bind cow's-milk-specific patient IgE was drastically reduced, although it did not reach the background level in most of the sera. Such a test could be used to characterize the extent of degradation of allergens in a hydrolysate based formula. High correlation of such RAST tests with skin-prick or oral provocation tests may be expected provided the intolerance is IgE mediated, but not if it is cell mediated hypersensitivity. Such a test is more likely to have predictive value in therapeutic use of protein hydrolysate formulae than in prophylactic use.

A more differentiated image of the remaining antigenic reactivity of hydrolysate formulae is obtained from an immunoblot of the hydrolysate drawn from polyacrylamide gels to nitrocellulose sheets, on which the antigen-antibody reaction with a patient serum is performed. Immunoblotting with a serum of a cow's-milk-allergic infant illustrates how, with a standard infant formula, IgE to practically all major milk proteins (whey proteins and caseins) are fixed in a high density (Figure 3). Compared with the standard formula, the hydrolysate-based formula was characterized by a very low density of IgE binding in one particular region (near 30 KD) of the blot. As the patient serum was shown to contain IgE against all types of milk protein antigens, the very low density of fixation of IgE with the hypoallergenic formula effectively documents its relatively low allergen content.

However, titers of cow's-milk-specific IgE in different patient sera may vary considerably. A formula would therefore have to be tested with a sufficient number of individual sera, in order to have a representative map of its residual antigens. Immunochemical data obtained from RAST tests, immunoblotting, and others have to be compared with skin-prick and oral-provocation tests to establish the biological significance of such immunochemical data.

It is unrealistic to believe that the low levels of antigenicity required for therapeutic formulas can be reached in an infant formula destinated for long-term feeding of the newborn. Extensive fractionation of milk-based hydrolysates involves the risk of losing nutritional factors. The total absence of larger peptides might be detrimental to normal secretory activity of intestinal enzymes. Finally, peptide antigens with a low density of antigenic determinants may play a role in the induction of immunological tolerance. What we attempt to achieve are hydrolysates which induce the lowest possible IgE response in predisposed individuals.

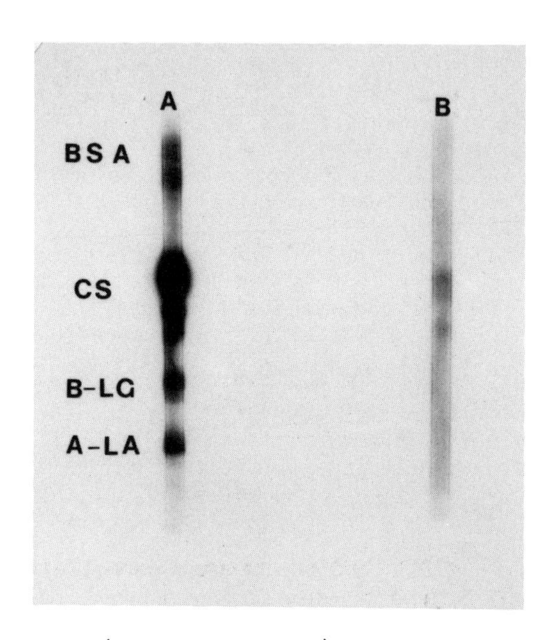

Figure 3. Immunoblot (autoradiographic) of a standard infant formula (A) and a hypoallergenic formula (B), the latter based on trypsin hydrolyzed whey. The test serum was from an infant with clinically established cow's milk allergy. Black zones represent patient IgE capable of recognizing antigenic polypeptides in the formula. Detecting antibody was 125-I labelled anti-human IgE.

A suitable IgE animal model, such as the rat, will prove a valuable tool in further research along these lines. Well controlled, long term clinical studies with hydrolysate formulae will soon become available to answer the question as to what extent hydrolysate formulae successfully contribute to prophylaxis of food allergy in infants.

Finally, in a relevant review Bounous and Kongshaven (1989) describe the influence of protein-type on the development of immunity and the immune-enhancing effect of whey protein concentrates.

ACKNOWLEDGEMENTS

We are indebted to Dr. A. Raemy for differential calorimetric analysis of our whey samples.

REFERENCES

Baldo, B.A. Milk allergies. Australian J. Dairy Technology, September 1984: 120-128.

Bock, S.A. Prospective appraisal of complaints of adverse reactions to foods in children during the first 3 years of life. Pediatrics 79, 683-688. 1987.

Bounous, G. and Kongshaven, P.A.L. Influence of protein type in nutritionally adequate diets on the development of immunity. In "Absorption and Utilization of Amino Acids", M. Friedman, Ed., CRC Press, Boca Raton, FL, Vol. 2, pp. 220-233. 1989.

Cant, A., Marsden, R.A. and Kilshaw, P.J. Egg and cow's milk hypersensitivity in exclusively breast fed infants with eczema and detection of egg protein in breast milk.

Chandra, R.K., Gurkirpal, S. and Bekal, S. Effect of feeding whey hydrolysate, soy and conventional cow's milk formulas on incidence. Annals of Allergy 63, 47-64. 1981.

De Wit, J.N. Structure and functional behaviour of whey proteins. Neth. Milk and Dairy J. 63, 47-64. 1981.

Dosa, s., Pesce, A.J., Ford, D.J., Muckerheide, A. and Michael, J.G. Immunological properties of peptic fragments of bovine serum albumin. Immunology 38, 509-517. 1979.

Heppel, L.M., Cant, A.J. and Kilshaw, P.J. Reduction in the antigenicity of whey proteins by heat treatment: a possible strategy for producing a hypoallergenic infant milk formula. Brit. J. Nutrition 51, 29-36. 1984.

Jacobsson, I., Lindberg, T., Benediktsson, B. and Hansson, B. Dietary bovine beta-Lactoglobulin is transferred to human milk. Acta Paediatr. Scand. 74, 342-345. 1985.

Jost, R. Physicochemical tratemtn of food allergens: application to cow's milk proteins. In "Food Allergy", D. Reinhardt and E. Schmidt, Eds., Raven Press, New York, Vol 17, pp. 187-197.

Jost, R., Monti, J.C. and Pahud, J.J. Whey protein allergenicity and its reduction by technological means. Food Technology 41, 118-121. 1987.

Kahn, A., Rebuffat, E., Blum, D., Casimir, G., Duchateau, J., Mozin, M.J. and Jost, R. Difficulty in initiating and maintaining sleep associated with cow's milk allergy in infants. Sleep 10, 116-121. 1987.

Kilshaw, P.J., Heppell, L.M. and Ford, J.E. Effects of heat treatment of cow's milk and whey on the nutritional quality and antigenic properties. Archives of Disease in Childhood 57, 842-847. 1982.

Machtinger, S. and Moss, K. Cow's milk allergy in breast fed infants. J. Allergy and Clinical Immunol. 77, 341-347. 1986.

McLaughlan, P., Anderson, K.J., Widdowson, E. and Coombs, R.A. Effect of heat on the anaphylactic sensitizing capacity of cow's milk, goat's milk, and various infant formulae fed to guinea pigs. Archives of Disease in Childhood 56, 165-171. 1981.

Oloffson, A., Buhler, M. and Wood, R. Process for the preparation of a purified protein hydrolysate. USA Patent 4, 293, 571. 1981.

Otani, H., Uchio, T. and Tokiat, F. Antigenic reactivities of chemically modified beta-lactoglobulins with antiserum to bovine beta-lactoglobulin. Agric. Biol. Chem. 49, 2531-2536. 1987.

Otani, H. and Hosono, A. Antigenic reactive regions of s-carboxy-methylated beta-lactoglobulin. Agric. Biol. Chem. 51, 531-536. 1987.

Pahud, J.J., Monti, J.C. and Jost, R. Allergenicity of whey protein: its modification by tryptic in vitro hydrolysis of the protein. J. Pediatric Gastroenterol. and Nutrition 4, 408-413. 1985.

Peters, T. Serumalbumin. Advances in Protein Chemistry 37, 161-245. 1985.

Peters, T., Feldhoff, R.C. and Reed, R.G. Immunochemical studies of fragments of bovine serum albumin. J. Biol. Chem. 252, 8464-8468. 1977.

Ratner, B., Dworetzky, M., Satako Oguri, B.A. and Aschheim, L. Studies on the allergenicity of cow's milk. Pediatrics 22, 648-657. 1958.

Takase, M., Fukuwatari, Y., Kawase, K., Kiyosawa, I., Ogasa, K., Suzuki, S. and Kuroume, T. Antigenicity of casein enzymatic hydrolysates. J. Dairy Sciences 62, 1570-1576. 1979.

Wahn, U., Peters, T. and Siraganian, R. Allergenic and antigenic properties of bovine serum albumin. Molecular Immunology 18, 19-28. 1981.

24

ELISA ANALYSIS OF SOYBEAN TRYPSIN INHIBITORS IN PROCESSED FOODS

David L. Brandon, Anne H. Bates,
and Mendel Friedman

Food Safety Research Unit, USDA Agricultural Research Service, Western Regional Research Center, 800 Buchanan Street, Albany, CA 94710

ABSTRACT

Soybean proteins are widely used in human foods in a variety of forms, including infant formulas, flour, protein concentrates, protein isolates, soy sauces, textured soy fibers, and tofu. The presence of inhibitors of digestive enzymes in soy proteins impairs the nutritional quality and possibly the safety of soybeans and other legumes. Processing, based on the use of heat or fractionation of protein isolates, does not completely inactivate or remove these inhibitors, so that residual amounts of inhibitors are consumed by animals and humans. New monoclonal antibody-based immunoassays can measure low levels of the soybean Kunitz trypsin inhibitor (KTI) and the Bowman-Birk trypsin and chymotrypsin inhibitor (BBI) in processed foods. The enzyme-linked immunosorbent assay (ELISA) was used to measure the inhibitor content of soy concentrates, isolates, and flours, both heated and unheated; a commercial soy infant formula; KTI and BBI with rearranged disulfide bonds; browning products derived from heat-treatment of KTI with glucose and starch; and KTI exposed to high pH. The results indicate that even low inhibitor isolates contain significant amounts of specific inhibitors. Thus, infants on soy formula consume about 10 mg of KTI plus BBI per day. The immunoassays complement the established enzymatic assays of trypsin and chymotrypsin inhibitors, and have advantages in (a) measuring low levels of inhibitors in processed foods; and (b) differentiating between the Kunitz and Bowman-Birk inhibitors. The significance of our findings for food safety are discussed.

Nutritional and Toxicological Consequences of Food Processing
Edited by M. Friedman, Plenum Press, New York, 1991

INTRODUCTION

Active protease inhibitors can limit the digestibility and bioavailability of soy protein. In addition, at least in the rat, consumption of dietary protease inhibitors can result in development of pancreatic adenomas (see reviews by Gallaher and Schneeman, 1986; Gumbmann et al., 1986; Grant, 1990). However, some inhibitors also have potentially beneficial effects. They have been shown to inhibit carcinogenesis in both _in vitro_ and _in vivo_ systems (Yavelow et al., 1983; Troll et al., 1986; St. Clair et al., 1990), and to stimulate human T cells (Richard et al., 1989).

In assessing the possible significance of protease inhibitors in human nutrition, Morgan et al. (1986) note that one group of humans, namely, infants fed soybean-containing formulas because of allergy to cow's milk, may be at risk. These authors also note that although it is possible that the period of continuous exposure to protease inhibitors is too short for any significant effect on the pancreas to occur, it would nevertheless be desirable to eliminate all trypsin inhibitor activity in these products.

In order to assess the impact of food processing on the nutritional quality and healthfulness of products containing soy protease inhibitors, it is necessary to measure these specific inhibitors accurately. In soy products, one complication is the occurrence of multiple soybean protease inhibitors, including the Kunitz trypsin inhibitor (KTI) and the double-headed Bowman-Birk trypsin and chymotrypsin inhibitor (BBI). Both KTI and BBI exist as several isoforms, which are derived from different genes or are produced by proteolysis; and other inhibitors are also present (Laskowski, 1986; Tan-Wilson and Wilson, 1987). It is therefore impossible to establish the exact protease inhibitor composition of a sample through enzymatic assay, especially in samples with low residual inhibitor activity, such as toasted soy flours or soy protein isolates. We have found that immunoassays using monoclonal antibodies offer the specificity and sensitivity necessary to analyze complex, processed food samples (Brandon et al., 1987b, 1988, 1989, 1990). These methods could be used to assess improved food processing strategies for optimizing the content of protease inhibitors in soy foods (Brandon et al., 1986a, 1987a; Friedman et al., 1989, 1990). In addition, we have found the immunoassays useful for screening soybean germplasm (Friedman et al., 1990). Finally, they can also be used, in model systems, to study the effects of processing conditions on food proteins.

METHODS AND MATERIALS

Antibodies and assays. Rabbit polyclonal antibodies and mouse monoclonal antibodies to KTI were described in detail previously (Brandon et al., 1986b, 1987b, 1988; Brandon and Bates, 1988). Antibodies to BBI were described by Brandon et al. (1989). The assays were described in detail in the above references, but will be summarized briefly here. In inhibition ELISA's, assays were conducted in plastic ELISA plates, coated with protease inhibitor. The analyte was mixed with purified, monoclonal, protease-specific immunoglobulin G at a concentration giving 50 to 90% of maximal antibody binding in the absence of analyte. After incubation and washing steps, the bound immunoglobulin was detected with horseradish peroxidase (HRP)-labelled rabbit anti-mouse IgG (Zymed Laboratories, South San Francisco, CA), using 2,2'-azinobis-3-ethylbenzthiazolinesulfonic acid (ABTS) plus H_2O_2 as substrate. In the competitive ELISA format, assay wells were conducted on monoclonal antibody-coated wells. The analyte was mixed with the HRP-conjugate of KTI (KTI-HRP) or BBI (BBI-HRP), before application to the assay wells. After incubation and washing, HRP conjugate bound to the assay wells was detected using ABTS plus H_2O_2. Standard curves were computed by fitting the data to a logistic model (Finney, 1978). The concentration of protease inhibitor analyte giving half-maximal response in the competitive or inhibition ELISA's is defined as the I_{50}. Inhibition of trypsin or chymotrypsin was determined by the methods of Hummel (1959), as described by Friedman et al., (1982).

Protease inhibitors and treatments with reagents. KTI and lima bean inhibitor (LBI) were obtained from Sigma Chemical Co. (St. Louis, MO). BBI (Birk, 1985) was provided by Prof. Y. Birk (Faculty of Agriculture, Hebrew University of Jerusalem, Rehovot, Israel). Treatment of inhibitors at elevated temperatures and with the disulfide-modifying reagents, sodium sulfite and N-acetylcysteine (NAC), was described and trypsin inhibitor units were defined previously (Friedman et al., 1982; Friedman and Gumbmann, 1986). Treatments of samples in a model system for non-enzymatic browning was conducted as described by Oste et al. (1990). Treatment of KTI with alkali was performed as described by Friedman and Liardon (1985).

RESULTS AND DISCUSSION

Antibodies and Immunoassays for the Kunitz Trypsin Inhibitor

The Kunitz trypsin inhibitor was characterized by Kunitz (1947) and the primary structure of isoform a (Ti^a) was reported by the Ikenaka group (Koide and Ikenaka, 1973a and b; Koide et al., 1973, Kim et al., 1985). Ti^a has 181 amino acids and two disulfide bonds. There are three closely related isoforms of KTI, encoded by codominant alleles in a multiple allelic system at one locus (Hymowitz and Hadley, 1972; Orf and Hymowitz, 1977). The $\underline{Ti}^c$ gene product, KTI isoform c (Ti^c), differs from the $\underline{Ti}^a$ product (Ti^a) in only one amino acid residue (Kim et al., 1985), a change from glycine to glutamic acid at residue 55. Ti^b retains glycine at position 55, but it differs at eight other positions from TI^a (Kim et al., 1985). KTI is also modified, apparently proteolytically, upon germination of the soybean (Orf et al., 1977).

We have previously described monoclonal antibodies that bind to KTI and have summarized earlier work using polyclonal antibodies (Brandon et al., 1986b, 1987b, 1988;

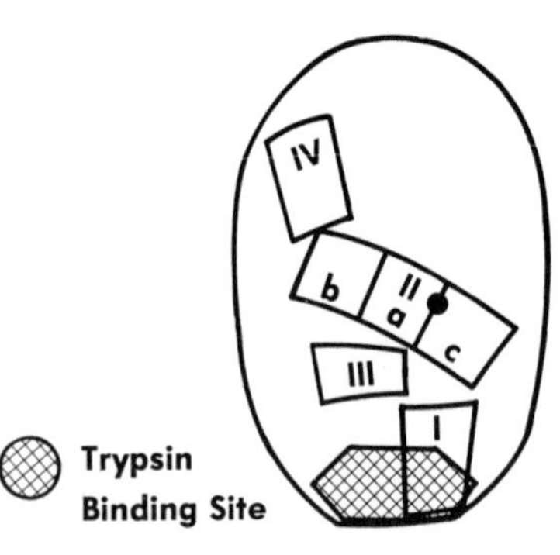

Figure 1. Schematic model of antibody-binding epitopes in relation to the trypsin-binding site (shaded area; surrounding residues 63 and 64). Roman numerals I-IV refer to epitopes, with epitope II further divided into sites a-c. The black dot represents the region of the molecular surface of KTI altered in isoform c due to the substitution of glutamic acid for glycine at residue 55.

Brandon and Bates, 1988). These studies led us to conceptualize the antigenic structure of KTI, as shown in Figure 1. The major antibody specificities are as follows:

Group 1. These antibodies bind poorly to the KTI-trypsin complex. Thus, epitope I is assumed to overlap the trypsin-binding site or be affected by allosteric changes which occur when KTI binds trypsin.

Group 2. These antibodies bind to several closely associated sites, denoted by the subdivision of epitope II into 3 sites. Sites IIa and IIc are altered when GLY-55 (isoform a) is replaced by GLU (as in isoform c). Site IIc is further distinguished from IIa and IIb by its sensitivity to heat and its proximity to epitope I.

Group 3. These antibodies bind to site III, which is distinct from site IIc, but close to sites IIa and IIb. Site III is moderately sensitive to heat and to substitution at residue 55, but not as sensitive as site IIc.

Group 4. These antibodies bind to a site which is highly conserved among the three isoforms of KTI. This site is unaffected by the binding of trypsin and is topographically close to site IIb.

ELISA provided the data which led to this conceptual model of KTI as an antigen. Figure 2 illustrates the competitive ELISA analysis of KTI isoforms, using 6 antibodies. Antibody 171 binds equally to the different isoforms, as illustrated by overlapping assay curves. Antibody 129 binds equally well to isoforms a and c, but not to isoform b (I_{50} = 78-fold greater than the I_{50} for isoform a). Antibody 180 binds better to isoform b than to isoform a. It does not bind to isoform c.

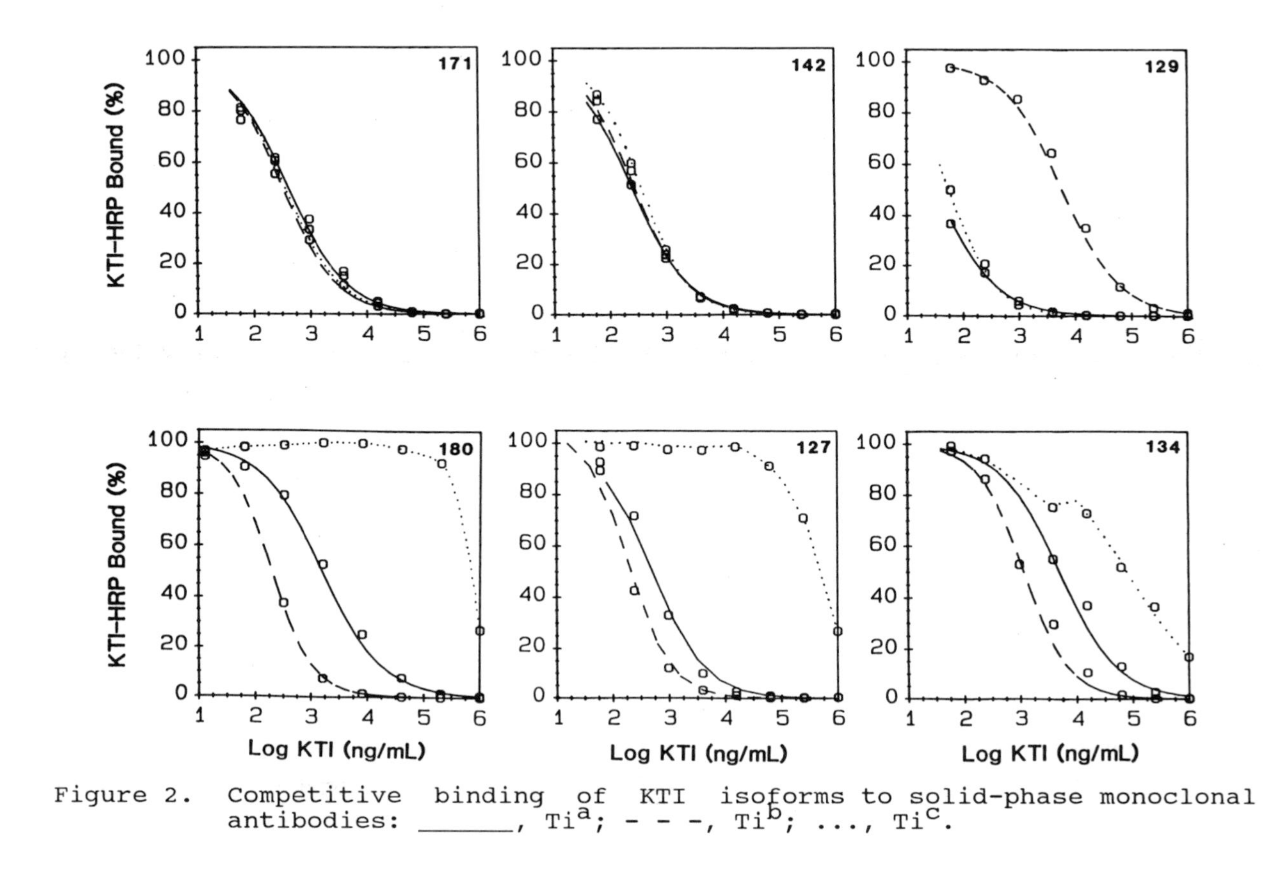

Figure 2. Competitive binding of KTI isoforms to solid-phase monoclonal antibodies: ______, Ti^a; - - -, Ti^b; ..., Ti^c.

Antibodies and Immunoassays for the Bowman-Birk Inhibitor

The second major protease inhibitor of soybeans is the low molecular weight Bowman-Birk inhibitor, discussed in detail by Birk (1985). This low molecular weight protein (M_r = 8000) was previously thought to be insufficiently antigenic to elicit usable antisera without crosslinking with glutaraldehyde, but we found that the native, uncrosslinked molecule is antigenic in mice. We prepared high affinity monoclonal antibodies (Brandon et al., 1989), and developed immunochemical methods for BBI which complement those for KTI and enable characterization of the major protease inhibitory activities in soybean cultivars and processed foods.

Antibody 238 has been used in most of our studies. The apparent high affinity of this monoclonal antibody resulted in an ELISA sensitivity 100-fold greater than could be obtained with polyclonal antibodies. The affinity was reduced by 97% for BBI denatured by treatment with sodium sulfite at 85 ^{o}C for 2 h. The antibody did not bind to the closely related BBI-type inhibitors from lima beans or chick peas. The antibody does not cross-react with KTI, though it will bind to some preparations of KTI, since they are contaminated with BBI.

A second BBI-specific monoclonal antibody, antibody 217, had some distinctly different characteristics from antibody 238. This antibody appears to bind to a very heat-labile epitope distinct from the binding site for antibody 238. This epitope is altered under relatively mild conditions which do not affect the trypsin- and chymotrypsin-reactive sites.

Antibodies Specific for Native Protease Inhibitors

The relationship between antigenic activity measured by competitive ELISA (Brandon et al., 1988) and trypsin inhibitory activity of KTI was studied, using KTI samples that had been treated for one hour at various temperatures. These assays were performed on plastic ELISA dishes coated with antibody 180. The concentration of KTI resulting in half-maximal binding of KTI-HRP conjugate was determined for each sample. These values (I_{50}'s) were then expressed as percentages relative to a control sample, as follows:

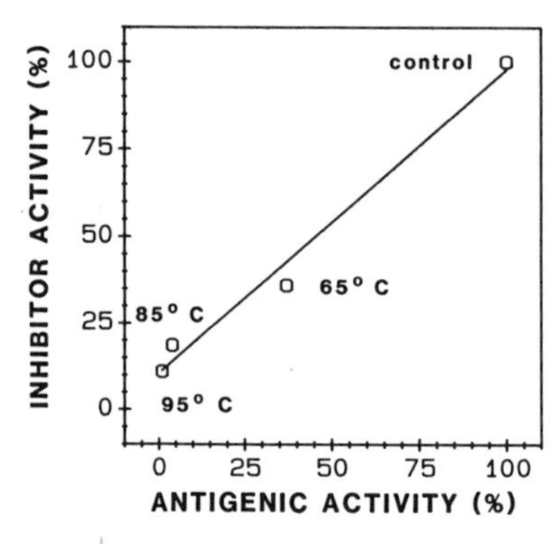

Figure 3. Correlation of antigenicity with the inhibitory activity of heat-treated KTI (r = 0.99).

Antigenic activity = [I_{50} (control) x 100]/I_{50} (sample). Thus, the sample treated at 65 ^{o}C produced 50% binding at 545 ng/ml, and was considered to have (201/545) X 100, or 37% antigenic activity. Figure 3 illustrates the strong correlation between inhibitory and antigenic activites.

The relationship between the enzyme inhibitory activities of BBI and ELISA was further studied using BBI treated at 85 ^{o}C in buffer only, or in the presence of NAC or sodium sulfite. The resulting samples were assayed for trypsin and chymotrypsin inhibition and for activity in the inhibition ELISA, using antibody 238. Relative ELISA activity was calculated from the I_{50} values as described above. Figure 4 shows that there is excellent agreement between the ELISA results and the enzymatic assays, especially in the area of low residual activity. Loss of activities was progressive over the 3-hour time course of the experiment. The NAC- and sodium sulfite-treated samples lost 85-90% of their inhibitory activities within 1 h. Samples treated at 85 ^{o}C without a disulfide-modifying reagent (not shown on graph) retained 69-87% of trypsin inhibitory activity, 60-75% of chymotrypsin inhibitory activity, and 100% of antigenic activity. It thus appears that the antibody recognizes the native structure of BBI, which is most affected by disruption of disulfides.

Inhibitors in Commercial Soy Concentrates and Infant Formulas

We have used the monoclonal antibody-based assays to analyze a variety of food products, employing antibody 142 to determine KTI and antibody 238 to determine BBI. The inhibitor content of commercial soy isolates, flours, and concentrates provided by the Archer Daniels Midland Company (Decatur, IL) were analyzed by ELISA and enzyme assays, and the data are summarized in Tables 1 and 2. The content of inhibitors varied markedly among various flours, isolates, and concentrates, depending on their prior processing. (Further information about these samples was not available from the company.) It is clear that the enzyme and ELISA

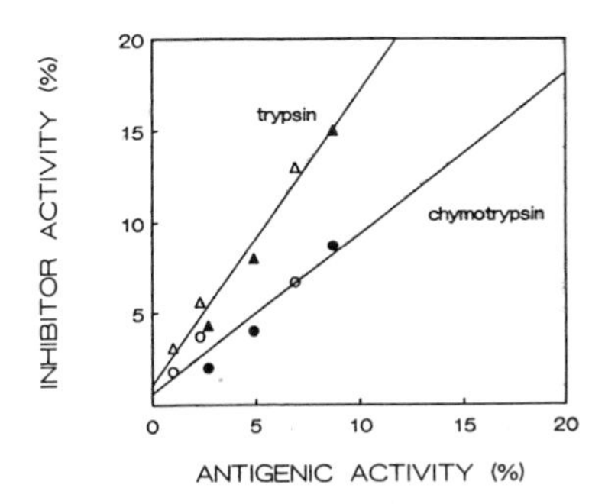

Figure 4. Correlation of BBI antigenic activity determined by ELISA and enzymatic activity determined by inhibition of chymotrypsin (r = 0.96) and trypsin (r = 0.98). Solid symbols depict N-acetylcysteine-treated (disulfide rearranged) samples, and open symbols depict samples treated sodium sulfite.

analyses correlate for most samples, with isolate A and concentrate F being notable exceptions which had much higher trypsin inhibitory activities than could be accounted for by their content of KTI measured by ELISA. ELISA analysis of prepared foods is presented in Table 3. Tofu, a precipitated fraction of a soy flour extract, and soy drinks containing protein isolates had 0.1 to 1% of the active KTI and BBI found in raw flours. Some products, such as that

Table 1. Enzyme-linked Immunosorbent Assay (ELISA) of Soy Isolates and Comparison with Enzymatic Assay[a]

Sample	Bowman-Birk Inhibitor (mg/g)		Kunitz Trypsin Inhibitor (mg/g)	
	ELISA	Enzymatic Assays[b]	ELISA	Enzymatic Assays[c]
Isolates				
A	2.1 ± 0.06	1.26 ± 0.04	0.30 ± 0.065	5.7 ± 0.49
B	0.05 ± 0.009	0	2.8 ± 0.62	1.3 ± 0.03
C	0.50 ± 0.07	0.32 ± 0.02	1.1 ± 0.09	1.5 ± 0.09
D	0.54 ± 0.14	0.19 ± 0.04	3.4 ± 0.20	1.2 ± 0.08
E	0.98 ± 0.11	0.15 ± 0.02	3.25 ± 0.58	3.8 ± 0.13

[a] Data are means ± standard deviation (n = 3).
[b] Calculated assuming BBI accounts for all of the chymotrypsin inhibition.
[c] "Enzymatic" determination of Kunitz trypsin inhibitor is the calculated amount of KTI which would account for all the trypsin inhibition not ascribable to the BBI (estimated from the chymotrypsin inhibition data).

Table 2. Enzyme-linked Immunosorbent Assay (ELISA) of Soy Concentrates and Flours and Comparison with Enzymatic Assay[a]

Sample	Bowman-Birk Inhibitor (mg/g)		Kunitz Trypsin Inhibitor (mg/g)	
	ELISA	Enzymatic Assays[b]	ELISA	Enzymatic Assays[c]
Concentrates				
F	0.64 ± 0.05	0.27 ± 0.05	0.48 ± 0.09	3.6 ± 0.15
G	0.19 ± 0.04	0.28 ± 0.06	2.8 ± 1.7	1.2 ± 0.10
H	0.02 ± 0.003	0.14 ± 0.03	0.25 ± 0.058	0.07 ± 0.04
Flours				
I	5.50 ± 0.51	3.85 ± 0.18	21.8 ± 7.4	22.8 ± 0.97
J	0.04 ± 0.002	0	1.6 ± 0.07	0.63 ± 0.01
K	0.04 ± 0.004	0	0.20 ± 0.061	0.29 ± 0.00

[a] Data are means ± standard deviation (n = 3).
[b] Calculated assuming BBI accounts for all of the chymotrypsin inhibition.
[c] "Enzymatic" determination of Kunitz trypsin inhibitor is the calculated amount of KTI which would account for all the trypsin inhibition not ascribable to the BBI (estimated from the chymotrypsin inhibition data).

Table 3. Enzyme-linked Immunosorbent Assay (ELISA) of Soy Products

Sample	BBI	KTI
Tofu	4 - 30	5 - 16
Soy protein powder	170	630
Soy drink 1	1.1	2.2
Soy drink 2	12	58
Soy drink 3	4.3	4.0
Soy sauce	n.d.	1.5

Data are typical values, expressed as µg/g for solids and µg/mL for liquids.

described in a local store as "soy protein powder" contained considerably higher levels. Soy protein isolates are used in many processed foods and can be produced by a variety of processes. As shown in Table 1, the BBI content of commercial isolates can vary from about 0.1 to 2 mg/g.

The impact of soy protease inhibitors, positive or negative, are likely to be most pronounced on infants receiving soy-based formula. The infant's lower gastric

Table 4. Analysis of Soy Infant Formula by Enzyme-linked Immunosorbent Assay (ELISA)

Inhibitor	µg/mL
BBI	6.9 ± 1.8 (n = 7)
KTI	6.8 ± 0.6 (n = 8)

Analyte was Isomil Concentrated Liquid (Ross Laboratories, Columbus, OH).

acidity and increased intestinal permeability could affect the fate of dietary protease inhibitors in the digestive tract. The ELISA analysis of one typical soy-based infant formula is summarized in Table 4. The results indicate that active KTI and BBI are lowered to about 0.1% of their activities compared to levels found in raw soy flour. Even this low level may be biologically significant, since an infant obtaining 100% nutriture from soy formula could consume about 10 mg of active KTI plus BBI per day.

Heat Inactivation of Inhibitors in Soy Meal

Soy flours are commercially "toasted," a steam-heating process which we model in the laboratory by use of an autoclave. In order to assess the protease inhibitor content of toasted flours, and to distinguish between KTI and other trypsin inhibitors, we studied soy meal prepared from two isolines of soybeans. The L81-4590 experimental

Table 5. Kunitz and Bowman-Birk Inhibitor Content of Unheated and Heated Soy Flours by ELISA

Soy Flour/ Heating Time	KTI (mg/g)	KTI % of control	BBI (mg/g)	BBI % of control
Williams 82				
0 min	7.6 ± 0.7 (3)[a]	100	3.3 ± 0.3 (3)	100
10 min	4.6 ± 0.6 (2)	59	0.7 ± 0.2 (3)	22
20 min	1.5 ± 0.2 (3)	19	0.04 ± 0.03 (3)	1.2
30 min	1.9 ± 0.1 (2)	25	0.04 ± 0.03 (3)	1.2
L81-4590				
0 min	0.008 ± 0.003 (3)		3.3 ± 0.2 (3)	100
10 min	n.d.[b]		1.2 ± 0.3 (3)	35
20 min	<0.005 (3)		0.02 ± 0.01 (3)	0.6
30 min	n.d.[b]		0.02 ± 0.01 (3)	0.6

[a] Values in parentheses are number of separate determinations.
[b] Not determined.

Table 6. Trypsin and Chymotrypsin Inhibitor Content of Unheated and Heated Soy Flours[a,b]

Soy Flour/ Heating time	TIU/g	% of control	CIU/g	% of control
Williams 82				
0 min	7140 ± 100	100	144 ± 6.4	100
10 min	3930 ± 50	55	28 ± 6.0	19
20 min	1060 ± 40	15	0	0
30 min	1030 ± 20	14	0	0
L81-4590				
0 min	3860 ± 50	100	113 ± 3	100
10 min	1840 ± 60	60	24 ± 0	21
20 min	110 ± 10	4	0	0
30 min	100 ± 0	3	0	0

[a] Samples were heated in an autoclave for the indicated time at 121°C.
[b] Values are averages from two separate determinations ± average deviation from the mean.

line (Hymowitz, 1986) lacks KTI, while the Williams 82 cultivar contains the Ti^a isoform. The effect of the autoclave treatment on inhibitor content is shown in Tables 5 and 6. BBI appears to have been rapidly inactivated, as indicated by the elimination of chymotrypsin inhibitory activity and BBI antigenicity in the samples heated for 20 min. About 14% of the trypsin inhibitory activity remained in the Williams sample, even after 30 min of autoclaving. The ELISA analysis indicates that 25% of KTI activity was still present after this treatment. The relative stability of KTI under these processing conditions was surprising, and was not apparent from consideration of the enzymatic data alone. These results confirmed and extended the findings by Liener and Tomlinson (1981) and DiPietro and Liener (1989), which suggested that matrix effects may make BBI in a flour more heat sensitive than purified inhibitor. We postulate that much of the residual chymotrypsin inhibitory activity sometimes found in soy protein-containing products is due to nonspecific inhibitors such as phytate and fat. Some of the residual TI activity is probably due to other minor protease inhibitors. Other studies in our laboratory suggest that the rate of heating may influence the stability of BBI.

The Ti-null isoline could be used to prepare flours with near-zero TI levels in toasted samples, with less heating than than commercially used varieties. Most strikingly, despite the greater heat stability of purified BBI compared to KTI (e.g., Obara and Watanabe, 1971) it is KTI that is responsible for the heat-stable trypsin-inhibitory activity of toasted soy products. Thus, the matrix in which protease inhibitors are found appears to influence their stability (Friedman et al., 1991).

Chemical Changes During Food Processing

As part of a study of the effects of food processing conditions on the antigenicity of food proteins, we investigated the reactions of alkali and carbohydrates with KTI. Treatment of plant proteins such as corn or soy storage proteins or animal protein such as casein with alkali brings about desirable changes in flavor, texture, and solubility and is used by the food industry to prepare protein isolates. Chemical changes which may accompany such treatment include crosslinking, degradation, browning reactions, and racemization. While these treatments may inactivate undesirable protein activities (such as lectins), they can also cause nutritional damage to the protein and may induce the formation of toxic components such as lysinoalanine. The biological effects of alkali-treated protein were discussed by Gould and MacGregor (1977).

Effects of Alkali on KTI

Alkali, especially at elevated temperatures, can inactivate KTI, as determined by enzymatic assay and by ELISA (Brandon et al., 1988). Most likely, several mechanisms are involved in this inactivation. We considered whether the antigenic changes could be related to lysinoalanine (LAL) formation (one of the several crosslinks which are induced by alkali). Figure 5 shows the antigenicity of KTI treated with alkali at 75°C (relative to the control of pH 7) plotted vs. the predicted number of LAL crosslinks (based on the study of soy protein by Friedman et al., 1984). Over 90% of the antigenic change occurs under conditions expected to induce crosslinking of essentially all KTI molecules (0.5 crosslink per molecule). LAL or other crosslinks and the concurrent racemization of L-amino acids could contribute significantly to the altered antigenicity of KTI or other proteins. This possibility remains to be tested directly.

Effects of Carbohydrates on KTI

Carbohydrates react with proteins to form non-enzymatically browned products, especially under conditions of high temperature and low moisture content that occur during toasting of soy flour. The reactions include the Maillard reaction of reducing carbohydrates as well as reactions of non-reducing carbohydrates such as sucrose (Smith and Friedman, 1984). Oste et al. (1990) studied the effects of carbohydrate-protein reactions on the antigenicity of KTI, using monoclonal antibodies. Table 7 shows the results of heating KTI as a dry powder in the presence of various carbohydrates for 50 min at 120°C. When KTI was heated in the presence of reducing carbohydrates, the antigenicity decreased up to 90% compared to control samples lacking carbohydrate. Non-reducing carbohydrate had a lesser effect. Analysis of the protein by gel electrophoresis showed that the KTI, in the presence of reducing carbohydrate, underwent a small mobility change within 10 min of heating.

The time course of the glucose-induced antigenic changes indicated that there is an initial rapid reaction of carbohydrate with protein (probably lysine residues) which results in antigenic changes. Further rearrangement of protein-bound carbohydrate appears to increase the browning of the protein (change in absorbance at 420 nm) without significant further change in antigenicity. These experiments suggest that conditions might be developed to exploit the beneficial effects of non-enzymatic browning. For example, it may be possible to inactivate undesirable components, such as KTI, selectively by food processing strategies. Further work is needed to evaluate the positive as well as negative consequences of such an approach.

Table 7. Relative Antigenicity of Native Kunitz Trypsin Inhibitor (KTI) Heated with Carbohydrate.

Sample	Relative Antigenicity %
KTI, unheated	100
KTI, heated	100
KTI + glucose, heated	10
KTI + maltose, heated	34
KTI + lactose, heated	27
KTI + starch, heated	55

Samples were heated for 50 min at 121°C.

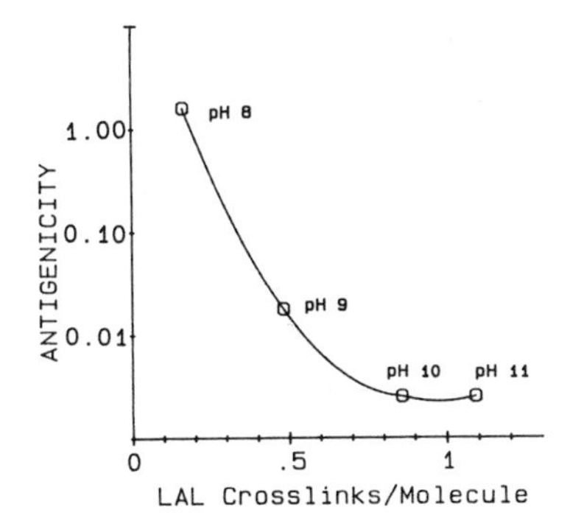

Figure 5. Effects of alkali on KTI at 75°C. Antigenicity determined by ELISA with antibody 180 is plotted against the expected degree of LAL crosslinking.

CONCLUSIONS

Monoclonal antibodies which recognized selected epitopes were used to develop ELISA's which can analyze native inhibitors in the presence of heat-treated and other denatured forms, both in pure samples and in extracts of soybean cultivars and processed foods. The analysis of processed soy products and the model studies conducted in our laboratory indicate that food processing conditions could be designed to optimize soy products to minimize protease inhibitor content. These strategies should take into account the nutritional, toxicological, and potentially beneficial pharmacological effects of these naturally occurring compounds. Together with genetic approaches for modifying the protease inhibitor content of soybeans (Hymowitz, 1986), such food processing research could help provide more nutritious, healthful soy products for human and animal consumption.

REFERENCES

Birk, Y. The Bowman-Birk inhibitor, Int. J. Peptide Protein Res. 25, 113-131. 1985.

Brandon, D. L., and Bates, A. H. Definition of functional and antibody-binding sites on Kunitz soybean trypsin inhibitor isoforms using monoclonal antibodies, J. Agric. Food Chem. 36, 1336-1341. 1988.

Brandon, D. L., Bates, A. H., Friedman, M., and Corse, J. W. Monitoring nutritional and toxicological changes in processed foods using monoclonal antibodies. In: Food Processing, Online International, New York, pp. 27-37. 1986a.

Brandon, D. L., Haque, S., and Friedman, M. Antigenicity of native and modified Kunitz soybean trypsin inhibitors. In: Nutritional and Toxicological Significance of Enzyme

Inhibitors in Foods, (Friedman, M., Ed.), Plenum Press, New York, pp. 449-467. 1986b.

Brandon, D. L., Bates, A. H., and Friedman, M. Immunoassays for measuring beneficial and adverse changes in food proteins. In: Biotech USA 1987, Online International, London, pp. 308-317. 1987a.

Brandon, D. L., Haque, S., and Friedman, M. Interaction of monoclonal antibodies with soybean trypsin inhibitors. J. Agric. Food Chem. 35, 195-200. 1987b.

Brandon, D. L., Bates, A. H., and Friedman, M. Enzyme-linked immunoassay of soybean Kunitz trypsin inhibitor using monoclonal antibodies. J. Food Sci. 53, 97-101. 1988.

Brandon, D. L., Bates, A. H., and Friedman, M. Monoclonal antibody-based enzyme immunoassay of the Bowman-Birk protease inhibitor of soybeans. J. Agric. Food Chem. 37, 1192-1196. 1989.

Brandon, D. L., Bates, A. H., and Friedman, M. Monoclonal antibodies to soybean Kunitz trypsin inhibitor and immunoassay method. U. S. Patent No. 4,959,310. 1990.

DiPietro, C. M., and Liener, I. E. Heat inactivation of the Kunitz and Bowman-Birk soybean protease inhibitors. J. Agric. Food Chem. 37, 39-44. 1989.

Finney, D. J. Statistical Method in Biological Assay, 3rd ed., MacMillan Publishing, New York. 1978.

Friedman, M., and Gumbmann, M. R. Nutritional improvement of soy flour through inactivation of trypsin inhibitors by sodium sulfite. J. Food Sci. 51, 1239-1241. 1986.

Friedman, M., and Liardon, R. Racemization kinetics of amino acid residues in alkali-treated soybean proteins. J. Agric. Food Chem. 33, 666-672. 1985.

Friedman, M., Brandon, D. L., Bates, A. H., and Hymowitz, T. Comparison of a commercial soybean cultivar and an isoline lacking the Kunitz trypsin inhibitor: composition, nutritional value, and effects of heating. J. Agric. Food Chem. 39, 327-335. 1991.

Friedman, M., Grosjean, O.-K., and Zahnley, J. C. Inactivation of soya bean trypsin inhibitor by thiols. J. Sci. Food Agric. 33, 165-172. 1982.

Friedman, M., Levin, C. E., and Noma, A. T. Factors governing lysinoalanine formation in soy proteins. J. Food Sci. 49, 1282-1288. 1984.

Friedman, M., Gumbmann, M. R., and Brandon, D. L. Nutritional, toxicological, and immunological consequences of food processing, Front. Gastrointest. Res. 14, 79-90. 1988.

Friedman, M., Gumbmann, M. R., Brandon, D. L., and Bates, A. H. Inactivation and analysis of soybean inhibitors of digestive enzymes. In: Food Proteins (Kinsella, J. E., and Soucie, W. G., Eds.), American Oil Chemists' Society, Champaign, IL, pp. 296-328. 1989.

Gallaher, D., and Schneeman, B. O. Nutritional and metabolic response to plant inhibitors of digestive enzymes. In: Nutritional and Toxicological Significance of Enzyme Inhibitors in Foods, (Friedman, M., Ed.) Plenum Press, New York, pp. 167-184. 1986.

Gould, D. H., and MacGregor, J. T. Biological effects of alkali-treated protein and lysinoalanine: an overview. In: Protein Crosslinking: Nutritional and Medical Consequences (M. Friedman, Ed.), Plenum Press, New York, pp. 29-48. 1977.

Grant, G. Antinutritional effects of soybean: a review. Progr. Food Nutr. Sci. 13, 317-348. 1990.

Gumbmann, M. R., Spangler, W. L., Dugan, G. M., Rackis, J. Safety of trypsin inhibitors in the diet: effects on the rat pancreas of long-term feeding of soy flour and soy protein isolate. In: Nutritional and Toxicological Significance of Enzyme Inhibitors in Foods, (Friedman, M., Ed.) Plenum Press, New York, pp. 33-79. 1986.

Hummel, B. C. A modified spectrophotometric determination of chymotrypsin, trypsin, and thrombin. Can J. Biochem. Physiol. 37, 1393-1399. 1959

Hymowitz, T. Genetics and breeding of soybeans lacking the Kunitz trypsin inhibitor. In: Nutritional and Toxicological Significance of Enzyme Inhibitors in Foods (M. Friedman, Ed.), Plenum Press, New York, pp. 291-298. 1986.

Hymowitz, T., and Hadley, H. H. Inheritance of a trypsin inhibitor variant in seed protein of soybeans. Crop Sci. 12, 197-198. 1972.

Kim, S., Hara, S., Hase, S., Ikenaka, T., Toda, H., Kitamura, K., and Kaizuma, N. Comparative study on amino acid sequences of Kunitz-type soybean trypsin inhibitors, Ti^a, Ti^b, and Ti^c. J. Biochem. 98, 435-448. 1985.

Koide, T., and Ikenaka, T. Studies on soybean trypsin inhibitors. 1. Fragmentation of soybean trypsin inhibitor (Kunitz) by limited proteolysis and by chemical cleavage. Eur. J. Biochem. 32, 401-407. 1973a.

Koide, T., and Ikenaka, T. Studies on soybean trypsin inhibitors. 3. Amino-acid sequence of the carboxyl-terminal region and the complete amino-acid sequence of soybean trypsin inhibitor (Kunitz). Eur. J. Biochem. 32, 417-431. 1973b.

Koide, T., Tsunasawa, S., and Ikenaka, T. Studies on soybean trypsin inhibitors. 2. Amino-acid sequence around the reactive site of soybean trypsin inhibitor (Kunitz). Eur. J. Biochem. 32, 408-416. 1973.

Kunitz, M. Crystalline soybean trypsin inhibitor. II. General properties, J. Exp. Med. 30, 291-310. 1947.

Laskowski, L., Jr. Protein inhibitors of serine proteases - mechanism and classification. In: Nutritional and Toxicological Significance of Enzyme Inhibitors in Foods, (Friedman, Ed.), Plenum Press, New York, pp. 1-17. 1986.

Liener, E. I., and Tomlinson, S. Heat inactivation of protease inhibitors in a soybean line lacking Kunitz trypsin inhibitor, J. Food Sci. 46, 1354-1356. 1981.

Morgan, R.G.H., Crass, R.A., and Oates, P. S. Dose effects of raw soyabean flour on pancreatic growth. In: Nutritional and Toxicological Significance of Enzyme Inhibitors in Foods, (Friedman, M., Ed.), Plenum Press, New York, pp. 81-89. 1986.

Obara, T., and Watanabe, Y. Heterogeneity of soybean trypsin inhibitors, II. Heat inactivation. Cereal Chem. 48, 523-527. 1971.

Orf, J. H., and Hymowitz, T. Inheritance of a second trypsin inhibitor variant in seed protein of soybeans, Crop Sci. 17, 811-813. 1977.

Orf, J. H., Mies, D. W., and Hymowitz, T. Qualitative changes of the Kunitz trypsin inhibitor in soybean seeds during germination as detected by electrophoresis, Bot. Gaz. 138, 255-260. 1977.

Oste, R. E., Brandon, D. L., Bates, A. H., and Friedman, M. Effects of nonenzymatic browning reactions of the Kunitz soybean trypsin inhibitor on its interaction with monoclonal antibodies, J. Agric. Food Chem. 38, 258-261. 1990.

Richard, K. A., Speciale, S. C., Staite, N. D., Berger, A. E., Daibel, M. R., Finzel, B. C. Soybean trypsin inhibitor. An IL-1-like protein. Agents and Actions 27, 265-267. 1989.

Smith, G. A., and Friedman, M. Effect of carbohydrates and heat on the amino acid composition and chemically available lysine content of casein. J. Food Sci. 49, 817-820, 843. 1984.

St. Clair, W. H., Billings, P. C., Carew, J. A., Keller-McGandy, C., Newberne, P., and Kennedy, A.R. Suppression of dimethylhydrazine-induced carcinogenesis in mice by dietary addition of the Bowman-Birk protease inhibitor. Cancer Res. 50, 580-586. 1990.

Tan-Wilson, A. L., Cosgriff, S. E., Duggan, M. C., Obach, R. S., and Wilson, K. Bowman-Birk proteinase isoinhibitor complements of soybean strains, J. Agric. Food Chem. 33, 389-393. 1985.

Tan-Wilson, A. L., Chen, J. C., Duggan, M. C., Chapman, C., Obach, R. S., and Wilson, K. A. Soybean Bowman-Birk trypsin isoinhibitors: Classification and report of a glycine-rich trypsin inhibitor class. J. Agric. Food Chem. 35, 974-980. 1987.

Troll, W., Frankel, K., and Wiesner, R. Protease inhibitors: their role as modifiers of carcinogenic processes. In: Nutritional and Toxicological Significance of Enzyme Inhibitors in Foods, (Friedman, M., Ed.) Plenum Press, New York, pp. 153-165. 1986.

Yavelow, J., Finley, T.H., Kennedy, A.R., and Troll, W., 1983, Bowman-Birk soybean protease inhibitor as an anticarcinogen. Cancer Res. 43, 2454s-2459s. 1983

25

EFFECT OF HEAT ON THE NUTRITIONAL QUALITY AND SAFETY OF SOYBEAN CULTIVARS

Mendel Friedman, David L. Brandon, Anne H. Bates, and Theodore Hymowitz[1]

Western Regional Research Center, U.S. Department of Agriculture, Agricultural Research Service, 800 Buchanan Street, Albany, CA 94710. [1]Department of Agronomy University of Illinois, Urbana, IL 61801

ABSTRACT

To evaluate whether soybean strains with reduced levels of trypsin inhibitors have enhanced nutritional and safety characteristics, we measured protease inhibitor content of a standard cultivar (Williams 82) and an isoline (L81-4590) lacking the Kunitz trypsin inhibitor, using enzyme inhibition assays and enzyme-linked immunosorbent assays (ELISA). Less heat was needed to inactivate the remaining trypsin inhibitory activity of the isoline than that of the standard soybean cultivar. In fact, autoclaving (steam heating at 121°C) of the isoline for 20 min resulted in a near zero level of trypsin inhibitor activity, while 20% remained in the Williams 82 sample. Feeding studies with rats showed that the raw soy flour prepared from the isoline was nutritionally superior to the raw flour prepared from the standard variety, as measured by PER and pancreatic weights. Since the content of amino and fatty acids of the flours from both strains was identical and the hemagglutinating activities were within a factor of 2, the increased PER was likely due to the lower level of trypsin inhibitory activity in the isoline. Steam heating the flours for up to 30 min at 121°C progressively increased the PER for both strains. Preliminary screening of several accessions from the USDA Soybean Germplasm Collection showed considerable variation in the content of trypsin inhibitors, sulfur amino acids, and lectins. The BBI content of these cultivars, determined by chymotrypsin inhibition assays, was identical to that found by ELISA. The results indicate that further screening studies could lead to the discovery of soybeans which yield flour that is safe and nutritious, with minimal need for heating.

INTRODUCTION

Adverse effects following short- and long-term ingestion of raw soybean meal by mammals and birds have been attributed to the presence of soybean protease inhibitors and lectins (Gumbmann et al., 1986; Liener, 1989). These inhibitors include the Kunitz trypsin inhibitor (KTI) (Rackis et al., 1986), the double-headed Bowman-Birk inhibitor (BBI) of trypsin and chymotrypsin (Birk, 1985), a glycine-rich trypsin inhibitor structurally unrelated to either KTI or BBI, and other minor inhibitors and proteolytically modified forms (Tan-Wilson et al., 1987). To minimize possible human health hazards and to improve the nutritional quality of soy foods, inhibitors are generally inactivated by heat treatment during food processing or removed by fractionation (Wolf and Cowan, 1975). Most commercially heated soy flours retain 5 to 20% of the original trypsin and chymotrypsin inhibitory activity (Rackis et al., 1986). The more protracted heating required to destroy all inhibitor activity would damage the nutritive value of soy proteins.

Another approach to producing low trypsin- and chymotrypsin-inhibitor foods is to develop soybean cultivars lacking major inhibitors. Orf et al. (1977) used polyacrylamide gel electrophoresis under nondenaturing conditions to determine the phenotype of soybean seeds with respect to KTI. Hymowitz (1986) discusses the genetics of KTI expression in soybeans and the development of KTI-null isolines on three genetic backgrounds. These isolines lack a functioning Ti allele, and are denoted ti/ti.

Earlier studies indicated that the KTI-lacking soybeans had lower trypsin inhibitor levels (Hymowitz, 1986) and supported the growth of pigs better than isolines expressing KTI (Cook et al., 1988). In this study, we wished to further analyze the protease inhibitor and lectin content of normal and KTI-lacking isolines and to study the nutritional properties of soy flour derived from these strains. Since soybeans contain multiple protease inhibitors and because measurements of enzyme inhibition can be inaccurate at low levels, we used ELISA's specific for KTI and BBI (Brandon et al., 1987, 1988, 1989, 1990; Oste et al., 1990) to characterize the inhibitor content in raw and heated flours. For comparison we also evaluated the sulfur amino acid, inhibitor, and lectin content of soybean flours derived from seeds in the USDA Soybean Germplasm Collection.

MATERIALS AND METHODS

Materials. Trypsin, chymotrypsin, N-α-tosyl-arginine methyl ester (TAME), N-benzoyl-L-tyrosine ethyl ester (BTEE), other reagents, and soybean Kunitz trypsin inhibitor were obtained from Sigma Chemical Co., St. Louis, MO. The Bowman-Birk soybean inhibitor was kindly provided by Prof. Y. Birk (Faculty of Agriculture, Hebrew University of Jerusalem, Rehovot, Israel).

Commercial soybeans (Williams 82) and an isoline with highly reduced Kunitz trypsin inhibitor activity (L81-4590) were grown by the Illinois Foundation Seeds, Inc., Champaign, IL, under the sponsorship of the Department of Agronomy, University of Illinois. The experimental line was grown in the field or isolated in a greenhouse to minimize cross-pollination.

Sample Preparation. Soybean seeds were ground in a Udy mill and sieved through a No. 60 mesh sieve. The meal was then defatted by extraction with ether in a Soxhlet apparatus for 16 h (AOAC, 1980) and air-dried. Samples (600 g) in porcelain dishes covered with aluminum foil were heated at 121°C in an autoclave for 10, 20, or 30 min.

Soybean samples from the USDA Soybean Germplasm Collection were selected for analysis of trypsin and chymotrypsin inhibitor, lectin, and amino acid content. Samples were milled but not defatted. Soybean meals were extracted by stirring the meal for 1 h in 0.5 M Tris-HCl buffer, pH 8.5 (500 mg meal/15 mL).

Trypsin Assays. The following conditions were used: buffer, 46 mM TRIS-HCl containing 11.5 mM $CaCl_2$ pH 8.1; substrate, 0.03 mM TAME (37.9 mg/10 mL H_2O); enzyme, 1 mg/mL, 1 mM HCl. The enzyme solution was diluted to 10-20 mg/mL. In the absence of inhibitor, 2.6 mL buffer and 0.3 mL TAME were added to a 3 mL cuvette followed by 0.1 mL diluted enzyme solution. The absorption was then recorded at 247 nm (A_{247}) for 3 min on a Perkin-Elmer Lambda 6 spectrophotometer. The increase in absorbance was then determined from the initial linear portion of the curve. In the presence of inhibitors, 2.6 mL buffer, 0.1 mL enzyme solution, and 20 µL of inhibitor solution (prepared to give 50% inhibition) were pre-incubated for 6 min. The reaction was started by adding 0.3 mL TAME and recording at A_{247} for 3 min. Values were based on sample dilutions yielding 40 to 60% inhibition (Worthington, 1982):

$$\text{Units/mg} = (\Delta A_{247}/\text{min} \times 1000 \times 3)/540 \times \text{mg enzyme used.}$$

Chymotrypsin Assays. The following conditions were used: buffer, 80 mM Tris-HCl containing 100 mM $CaCl_2$, pH 7.8; substrate, 1.07 mM BTEE (8.4 mg/25 mL 50% methanol); enzyme, 1 mg/mL, 1 mM HCl. The enzyme solution was diluted to a concentration of 10-20 µg/mL. In the absence of inhibitor, 1.5 mL buffer, 1.4 mL BTEE, and 0.1 mL enzyme solution were added and the increase in absorbance at 256 nm (A_{256}) was recorded for 3 min. The ΔA_{256}/min was then calculated from the initial linear portion of the curve. In the presence of inhibitor, 2.6 mL buffer, 0.1 mL enzyme solution, and 20 µL inhibitor were incubated for 6 min before adding 1.4 mL BTEE and recording as above. Values were based on sample dilutions yielding 40 to 60% inhibition (Worthington, 1982):

$$\text{Units/mg} = (\Delta A_{256}/\text{min} \times 1000 \times 3)/964 \times \text{mg enzyme used.}$$

Buffer plus substrate served as controls for all measurements (Friedman and Gumbmann, 1986; Friedman et al., 1984).

A trypsin unit (TU) is defined as the amount of trypsin that catalyzes the hydrolysis of 1 µmol of substrate/min. A trypsin inhibitor unit (TIU) is the reduction in activity of trypsin by 1 TU.

One chymotrypsin unit (CU) is defined as the amount of chymotrypsin that catalyzes the hydrolysis of 1 µmol of substrate/min. A chymotrypsin inhibitor unit (CIU) is the reduction in activity of chymotrypsin by 1 CU.

Pure KTI or BBI was used as a standard with each assay of the soy flours. The calculated values, which are averages from 2 to 3 separate determinations, are based on the individual control values.

ELISA Methods. Inhibition ELISA for KTI was performed using monoclonal antibodies as described by Brandon et al. (1987, 1988), or by competition ELISA with a KTI conjugate of horseradish peroxidase (HRP) as described by Brandon et al. (1988). Antibodies used (129, 142, 171, and 180) bind to four distinct epitopes of KTI (Brandon and Bates, 1988). ELISA for BBI was performed using monoclonal antibody 238 and BBI-conjugated HRP as described previously (Brandon et al., 1989).

Concentrations were determined by reference to a standard curve or by computation using samples spiked with an internal standard as follows:

$$\frac{\text{KTI (sample)} + \text{KTI (spike)}}{\text{KTI (sample)}} = \text{Antilog}\,[I_{50}(\text{sample}) - I_{50}\,(\text{sample + spike})]$$

where I_{50} is the concentration of the unknown sample yielding 50% inhibition of maximal binding of labeled ligand, KTI (sample) is the concentration of KTI in the sample, and KTI (spike) is the KTI added as an internal standard. The relative concentration of two samples was computed as follows:

$$\frac{\text{KTI (sample 2)}}{\text{KTI (sample 1)}} = \text{Antilog } [I_{50}(\text{sample 1}) - I_{50}(\text{sample 2})]$$

Hemagglutination Assay. The sample (soy flour, 150 mg) was mixed with 1.5 mL of phosphate buffered saline, pH 7.2. Lectin was extracted by stirring for 1 h at room temperature. After extraction, the resulting slurry was immediately chilled and centrifuged at 9000 g for 5 min in a Beckman Microfuge (Beckman Instruments, Palo Alto, CA). When necessary, the extracts were diluted with isotonic phosphate buffer (0.05 M NaH_2PO_4 and 0.15 M NaCl, pH 7.2) before plating so that incipient activity would fall midrange in the plated series.

Fifty µL aliquots of glutaraldehyde-stabilized human group A red blood cells diluted with a buffer to 3.3% hematocrit were added to equal volumes of serially diluted extracts and a buffer blank. Agglutination was observed visually after 1 h (Liener, 1974, 1989; Wallace and Friedman, 1985).

Activity is calulated as the reciprocal of the minimum amount of soy flour required to cause agglutination of blood cells under these test conditions. This value is derived from the minimum experimental value (µg/50 mL) which produces hemagglutination. The results of four separate assays conducted on each sample were averaged.

Amino Acid Composition. Three analyses with flour containing about 5 mg of protein (N X 6.25) were used to establish the amino acid composition of the soybean protein (Friedman et al., 1979): (a) standard hydrolysis with 6 HCl for 24 h in evacuated sealed tubes; (b) hydrolysis with 6 N HCl after performic acid oxidation to measure half-cystine and methionine content as cysteic acid and methionine sulfone, respectively; (c) basic hydrolysis by barium hydroxide to measure tryptophan content (Friedman and Cuq, 1988). The reproducibility of these analyses is estimated to be $\pm$ 3%, based on past experience (Friedman et al., 1979).

Fatty Acid Composition. Fatty acids were methylated using boron trifluoride-methanol, and subsequently analyzed by GLC analysis utilizing a Varian 5050A Gas Chromatograph modified for capillary injection using the following conditions: A fused silica capillary column from J&W Scientific Inc., Rancho Cordova, CA; liquid phase, DBWAX + 0.25 micron thickness, 30 m X 0.245 mm; initial temperature 150°, final temperature 220°, program 2°/min for 20 min, total time 55 min, injector temperature 220°, FID 300°, helium carrier gas flow as 30 cm/s, split ratio 25:1. The estimated reproducibility of these analyses is $\pm$ 10% (Crawford, 1989).

Animal Feeding Studies. Protein efficiency ratios (PER) were determined by the official method using weanling, male Sprague-Dawley derived rats from Simonsen Laboratories, Inc., Gilroy, California. Each assay ran 28 days (AOAC, 1980). Food consumption and PER (weight gain/ protein intake) were evaluated. Pancreata were excised at the end of the feeding period and weighed (wet weight).

RESULTS

Amino Acid Composition. Table 1 compares the amino acid composition of a standard variety of defatted soy flour and of the isoline lacking KTI. This analysis is compared to the values of the scoring pattern of the essential amino acids for an ideal protein, as defined by the Food and Agricultural Organization of the United Nations (FAO, 1973). Neither strain meets the provisional requirements for the sulfur amino acids (Cys + Met). Table 2 compares the amino acid composition of the unheated and heated flours. There was no significant difference between unheated samples (g/16 g N) as evidenced by the standard normal test. Linear regression showed no significant linear change in the amino acid composition between unheated and heated samples (Madansky, 1959).

ELISA Analysis of L81-4590. ELISA was performed in both the inhibition and competition formats as described by Brandon et al. (1988). Initial results using the inhibition format indicated that isoline L81-4590 expressed KTI at less than 2% of the Williams 82 isoline levels. To ascertain the nature of this low level of expression, additional samples were analyzed. These samples were grown on two experimental plots (Swine Farm and South Farm) or in a greenhouse, and were analyzed using antibody 180, with specificity for native KTI. The results, shown in Figure 1, analyzed as described in Methods and Materials, indicated that the Swine Farm samples contained 1.8% of control levels of KTI and the South Farm sample, 0.97% compared to Williams 82 isoline. Assay of the greenhouse-grown sample at very high concentration produced 25% inhibition of antibody binding. Although the KTI level is too low to assess accurately by ELISA using this antibody, extrapolation of the steepest section of the assay curve leads to an upper estimate of the KTI content as about 3 µg/g, or less than 0.1% of the level in Williams 82 cultivar expressing KTI. Direct reading from the assay curve (82% inhibition at a log concentration value of 3.9) indicates a shift of 4.0 log units from the control (Williams 82) sample, equivalent to a 10,000-fold reduction of KTI concentration.

Antibody 180 binds to only one epitope of KTI, and is highly specific for the native molecule. Therefore, to examine whether the L81-4590 experimental line expressed a truncated form of the KTI molecule with altered electrophoretic mobility, extracts of soybeans were examined for the presence of each of the major epitopes of KTI (Brandon and Bates, 1988) using competition ELISA and three other antibodies. The results of these assays agreed qualitatively with the low level (< 5 µg/g) detected using antibody 180: antibody 129, 11 µg/g; antibody 142, 5 µg/g; antibody 171, <10 µg/g. Thus, the amount of protein possessing the structural or functional properties of KTI is less than 0.2% of the level found in Williams 82.

Analysis of Protease Inhibitors in Raw and Heated Flours. Table 3 and Figure 2 presents the protease inhibitor content derived from enzymatic measurements. This flour had 54% of the trypsin inhibitory activity and 79% of the chymotrypsin inhibitory activity of flour from Williams 82. Less heat was needed to inactivate trypsin inhibitory activity in the L81-4590 experimental line than in the standard variety. In contrast, the chymotrypsin inhibitory activity of both strains was equally susceptible to inactivation by heat, reaching zero after 20 min. Table 4 shows the ELISA analysis of the two major inhibitors, KTI and BBI. KTI was present at a very low, but measurable level in L81-4590. In Williams 82, the KTI content decreased during the first 20 min of heating to about 20% of the initial value. No further reduction was observed at 30 min. BBI levels were identical in the two strains. The time course for heat inactivation of BBI was similar for both isolines.

Table 1. Comparison of Amino Acid Composition of Flour from a Standard Variety of Soy (Williams 82) and of an New Experimental Strain lacking the Kunitz Trypsin Inhibitor (L81-4590)

AMINO ACID	Williams 82		KTI-free		FAO
	g/100 g	g/16 g N	g/100 g	g/16 g N	g/16 g N
Asp	5.19	11.16	5.67	11.42	
Thr	1.86	3.99	1.99	4.01	4.0
Ser	2.45	5.28	2.70	5.45	
Glu	8.77	18.87	9.72	19.58	
Pro	2.61	5.61	2.80	5.64	
Gly	1.93	4.15	2.09	4.22	
Ala	1.94	4.17	2.12	4.27	
Cys	0.65	1.39	0.68	1.37	
Met	0.65	1.39	0.66	1.32	3.5
Val	2.02	4.35	2.15	4.34	5.0
Ileu	2.05	4.42	2.13	4.30	4.0
Leu	3.65	7.84	3.91	7.87	7.0
Tyr	1.75	3.76	1.91	3.86	6.0
Phe	2.19	4.70	2.38	4.80	
His	1.18	2.53	1.31	2.63	
Lys	2.88	6.20	3.15	6.35	5.5
Arg	3.33	7.16	3.72	7.49	
Trp	0.33	0.71	0.38	0.76	
Total	45.42	97.68	49.48	99.70	

Table 2. Effect of Heating in an Autoclave at 121°C on the Amino Acid Composition of a Standard Soybean Variety (Williams 82) and of an Isoline Lacking KTI (L81-4590). (in g/16g N).

AMINO ACID	Williams 82 Heated for (in min):				KTI-free heated for (in min):			
	0	10	20	30	0	10	20	30
Asp	11.16	11.79	11.19	11.48	11.42	11.15	11.24	11.52
Thr	3.99	4.26	4.01	4.10	4.01	3.96	3.97	4.01
Ser	5.28	5.60	5.27	5.38	5.45	5.34	5.38	5.40
Glu	18.87	20.13	18.87	19.39	19.58	19.54	19.35	19.69
Pro	5.61	4.81	5.18	5.98	5.64	5.36	5.65	5.87
Gly	4.15	4.40	4.18	4.30	4.22	4.18	4.17	4.19
Ala	4.17	4.42	4.20	4.32	4.27	4.24	4.21	4.27
Cys	1.39	1.42	1.32	1.39	1.37	1.36	1.30	1.41
Val	4.34	4.63	4.36	4.50	4.34	4.39	4.28	4.42
Met	1.39	1.40	1.34	1.43	1.32	1.33	1.32	1.36
Ileu	4.42	4.55	4.21	4.31	4.30	4.24	4.16	4.28
Leu	7.84	8.22	7.69	7.89	7.87	7.83	7.76	7.91
Tyr	3.76	4.16	3.74	3.91	3.86	3.85	3.87	3.93
Phe	4.70	5.05	4.75	4.92	4.80	4.80	4.81	4.87
His	2.53	2.68	2.54	2.62	2.63	2.57	2.55	2.62
Lys	6.20	6.67	6.19	6.37	6.35	6.27	6.04	6.26
Arg	7.16	7.67	7.11	7.43	7.49	7.43	7.26	7.53
Trp	0.71	1.06	0.69	0.42	0.76	0.87	0.98	0.858
Total	97.68	102.92	96.83	100.14	99.70	98.84	98.34	100.39

Table 3. Trypsin and Chymotrypsin Inhibitor Content Determined by Enzyme Assays of Unheated and Heated Soy Flours[a,b]

Sample	Trypsin inhibitor			Chymotrypsin inhibitor		
		Trypsin inhibited			Chymotrypsin inhibited	
	TIU/g	mg/g	% remaining	CIU/g	mg/g	% remaining
Williams 82 flour						
Unheated	7136 ± 96	36.0 ± 0.5	100	144 ± 6.4	4.2 ± 0.2	100
Heated-10 min	3933 ± 52	25.0 ± 0.3	69	28 ± 6.0	0.8 ± 0.2	19
Heated-20 min	1058 ± 39	7.0 ± 0.2	19	0	0	0
Heated-30 min	1030 ± 16	6.0 ± 0.0	17	0	0	0
L81-4590 flour						
Unheated	3858 ± 53	20.0 ± 0.4	100	113.0 ± 2.8	3.3 ± 0.1	100
Heated-10 min	1838 ± 59	12.0 ± 0.4	60	24.0 ± 0	0.7 ± 0	21
Heated-20 min	111 ± 13	0.8 ± 0.1	4	0	0	0
Heated-30 min	100 ± 0	0.6 ± 0	3	0	0	0

[a] Samples were heated in an autoclave for the indicated time periods at 121°C.
[b] Listed numbers are averages from two separate determinations ± standard deviations.

Table 4. Kunitz (KTI) and Bowman-Birk (BBI) Content of Unheated and Heated Soy Flours by ELISA

Soy Flour Soy Flour	KTI (mg/g)	% of Control	BBI (mg/g)	% of Control
Williams-82				
Unheated	7.6 ± 0.65 (3)[a]	100	3.31 ± 0.34 (3)	100
Heated-10 min	4.6 ± 0.55 (2)	59	0.725 ± 0.19 (3)	22
Heated-20 min	1.5 ± 0.24 (3)	19	0.04 ± 0.025 (3)	1.2
Heated-30 min	1.9 ± 0.065 (2)	25	0.04 ± 0.03 (3)	1.2
KTI-free (L81-4590)				
Unheated	0.008 ± 0.0025 (3)		3.31 ± 0.185 (3)	100
Heated-10 min	n.d.[b]		1.165 ± 0.30 (3)	35
Heated-20 min	<0.005 (3)		0.02 ± 0.005 (3)	0.6
Heated-30 min	n.d.[b]		0.02 ± 0.005 (3)	0.6

[a] Values in parentheses are number of separate determinations.

[b] Not done.

Hemagglutinating Activity. Table 5 shows that the hemagglutinating activities of the raw flours derived from Williams 82 and L81-4590 were within a factor of 2. Since a visual endpoint was used to assess the agglutination in a two-fold dilution series, further analysis is needed to assess whether this difference is significant (Lotan and Sharon, 1977; Nachbar and Oppenheim, 1980; Pusztai et al., 1981). Autoclave treatment resulted in rapid and parallel decreases in activities in both varieties.

Fatty Acid Content. Gas chromatograhy showed identical distribution of fatty acid isomers in the two soybean strains (Table 6).

Nutritional Quality and Effects on the Pancreas. Table 7 shows the results of feeding studies in rats of the raw and heated soy flours. Raw Williams 82 flour had a negative PER (-0.14), while the isoline had a PER of 0.46. Total food consumption, body weight gain, and PER increased with heating time. By 20 min, the isoline had a significantly higher PER than Williams 82.

The improvement in nutritional quality following heat treatment was accompanied by a decrease in pancreatic weights. However, the KTI-free soy flour required only 10 min of heating for the pancreatic hypertrophy to be normalized, compared to 20 min for the Williams 82 flour.

Screening for Cultivars with Modified Protease Inhibitor Content. Since the KTI-free soybean cultivars still contain inhibitors of the Bowman-Birk type, it would be useful to develop soybean cultivars which lack both KTI and BBI. We carried out a feasibility study by screening 13 of the approximately 14,000 soybean cultivars from the USDA Soybean Germplasm Collection for BBI content by enzymatic methods. All of the samples inhibited both trypsin and chymotrypsin (Table 8). The values ranged from 4.7 to 13.2 mg of chymotrypsin and 37.2 to 61.4 mg of trypsin inhibited per gram, about a two-fold range in inhibitory activity. Table 8 and Figure 3 also show that the BBI content of nine of these samples calculated from the chymotrypsin inhibition assay agrees with that obtained by ELISA. Error-in-variables regression assuming an error variance ratio of 1 was used to compare the two assays (Madansky, 1959). The calculated regression equation is: BBI (enzyme assay) = -0.44 + 1.11 (BBI ELISA assay), with a 95% confidence interval of upper and lower limits on the slope of: (0.87, 1.42). The fact that the confidence interval includes the value of 1 implies that the same variable is being measured by both assays.

These results show that other varieties contain more or less KTI or BBI than Williams 82. These samples were also evaluated for their content of sulfur-containing amino acids and lectins (Table 9). The half-cystine content ranged from 0.60 to 0.90 g/100 g and the methionine content from 0.51 to 0.73 g/100 g. There was not a strong correlation between the sulfur-containing amino acid levels and the content of BBI. The hemagglutinating activity varied over a six-fold range.

DISCUSSION

Amino Acid Analysis. The fact that the two strains have identical amino acid compositions is surprising, in view of the fact that they differ genetically at the locus of a major protein. At this time we do not know whether the amino acids otherwise used in the synthesis of KTI are free or whether they are incorporated into other proteins. However, the result suggests that the Ti-null isoline maintains normal amino acid distribution despite reduced expression of a significant structural gene.

Table 5. Effect of Heat on Lectin Activity of a Standard Soybean Variety (Williams 82) and a KTI-free Cultivar (L81-4590)

Heating Time (min)	Activity[a]	
	Williams 82	L81-4590
0	31.5	65
10	8.0	8.0
20	3.2	6.5
30	0.85	1.65
60	0.16	0.32

[a] Activity is defined as the reciprocal of the minimum concentration of sample (expresed as mg/mL) which causes agglutination of human red blood cells.

Table 6. Fatty Acid Composition (g/100 g oil) of a Standard Soybean Variety (Williams 82) and an Isoline Lacking the Kunitz Trypsin Inhibitor (L81-4590)

Fatty Acid	Williams 82	L81-4590
C16:0	11.1	11.2
C18:0	4.2	4.0
C18:1ω9	20.9	19.9
C18:1ω7	1.3	1.3
C18:2ω6	55.2	56.2
C18:3ω3	7.3	7.4

Table 7. Food Consumption, Body Weight Gain, Protein Efficiency Ratio, and Pancreas Weights of Rats Fed Unheated and Heated Soy Flours[a,b]

Diet	Total food consumption, g	Total body weight gain, g	PER	Pancreas weights: Absolute (g)	Pancreas weights: Relative to body weight
Williams 82 flour					
Unheated	156.8 ± 9.6[a]	-1.8 ± 1.7[f]	-0.14[f]	0.421[f]	0.806[a]
Heated-10 min	282.4 ± 13.1[c]	40.2 ± 3.2[c]	1.42[d]	0.537[cde]	0.572[c]
Heated-20 min	386.4 ± 10.5[a]	82.4 ± 4.2[b]	2.13[b]	0.581[bcd]	0.427[de]
Heated-30 min	395.8 ± 12.4[a]	88.0 ± 3.4[b]	2.22[b]	0.635[b]	0.446[de]
L81-4590 flour					
Unheated	189.0 ± 14.9[d]	9. 2 ± 2.6[e]	0.46[e]	0.458[ef]	0.721[b]
Heated-10 min	325.8 ± 6.4[b]	40. 6 ± 5.0[c]	1.63[c]	0.525[de]	0.503[d]
Heated-20 min	390.0 ± 11.2[a]	87. 4 ± 1.0[b]	2.25[b]	0.609[bcd]	0.430[e]
Heated-30 min	398.0 ± 14.0[a]	90. 8 ± 3.4[b]	2.28[b]	0.624[bc]	0.431[e]
Casein control	405.0 ± 3.3[a]	132. 6 ± 1.0[a]	3.27[a]	0.870[a]	0.466[de]

[a] Defatted soy flours were placed in autoclave and heated at 121°C for the indicated time periods.

[b] Duncan's Multiple Range Test (Duncan, 1955). Means without the same superscript letter for food consumption, weight gain, PER and pancreatic weight are significantly different, $P < 0.05$; 6 rats per group fed 28 days. PER (protein efficiency ratio) is defined as grams body weight gain per gram soybean protein ingested.

Table 8. Trypsin and Chymotrypsin Inhibition by Cultivars of Soybeans Evaluated as Milled Flours (not defatted)

Soybean variety	Inhibitor Activity (mg/g)	
	Trypsin	Chymotrypsin
COB 1983 MISS	37.2 ± 2.2[a]	4.87 ± 0.2
HOL 1983 MISS	63.8 ± 1.1	13.0 ± 0.2
JAC 1983 MISS	47.1 ± 0.7	5.71 ± 0.1
ROA 1983 MISS	44.0 ± 2.8	4.71 ± 0.2
MM610 SF83	58.8 ± 1.0	13.2 ± 0.2
T908 SF83	61.4 ± 0.9	10.3 ± 0.2
BSA1004 SF84	56.2 ± 0.9	8.3 ± 0.2
FUJ822 SF84	44.9 ± 1.6	4.8 ± 0.1
ML1014 SF84	48.6 ± 0.7	6.8 ± 0.01
MLB1015 SF84	58.1 ± 1.0	8.77 ± 0
CAY408 SF85	53.4 ± 0.9	9.03 ± 0.4
POL814 SF85	44.0 ± 2.8	7.02 ± 0
BSC803 SF86	56.2 ± 0.9	11.2 ± 0.2
Raw soy flour[b] (control)	49.0 ± 0.6	7.22 ± 0.2

[a] Number of separate determinations = 2.
[b] Prepared from soybeans purchased in a local store.

Table 9. Lectin Activity and Sulfur Amino Acid Content of Different Soybean Cultivars

Soybean Cultivar	Hemagglutinating Activity[b]	Half-cystine		Methionine	
		g/100 g	g/16g N	g/100g	g/16g N
Williams 82	36.0 ± 15 (3)[b]	0.65	1.39	0.65	1.39
L81-4590	78.1 ± 13 (4)	0.68	1.37	0.66	1.32
COB 1983 MISS	64.9 ± 7.6 (2)	0.60	1.62	0.52	1.41
HOL 1983 MISS	13.7 ± 2.8 (2)	0.72	1.72	0.60	1.44
JAC 1983 MISS	82.0 ± 16.1 (2)	0.75	1.58	0.63	1.32
ROA 1983 MISS	64.1 ± 30 (2)	0.60	1.58	0.51	1.35
MM 610 SF83	25.0 ± 0 (2)	0.63	1.70	0.52	1.41
T 908 SF83	27.5 ± 5.6 (2)	0.80	1.67	0.70	1.46
BSA 1004 SF84	54.9 ± 11 (2)	0.90	1.89	0.75	1.59
FUJ 822 SF84	27.3 ± 5.6 (2)	0.57	1.55	0.54	1.45
ML 1014 SF84	24.0 ± 0 (2)	0.75	1.49	0.65	1.28
MLB 1015 SF84	13.7 ± 2.8 (2)	0.86	1.86	0.75	1.62
CAY 408 SF85	38.5 ± 10.9 (2)	0.76	1.54	0.67	1.36
POL 814 SF85	48.0 ± 0 (2)	0.76	1.47	0.66	1.28
BSC 803 SF86	24.0 ± 0 (2)	0.80	1.71	0.73	1.56

[a] Activity is defined as the reciprocal of the minimum concentration of sample (expresed as mg/mL) which causes agglutination of human red blood cells.
[b] Values in parenthesis are number of separate determinations.

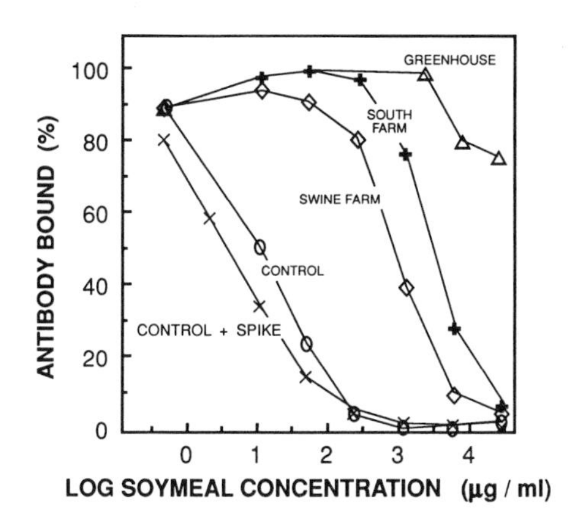

Figure 1. Analysis of KTI in soybean meal by inhibition ELISA using antibody 180. Williams 82 (control, with or without addition of an authentic KTI spike of 10 mg/g) was analyzed, along with L81-4590 grown on two different experimental plots or isolated in a greenhouse. The standard curve is given in Brandon et al. (1988).

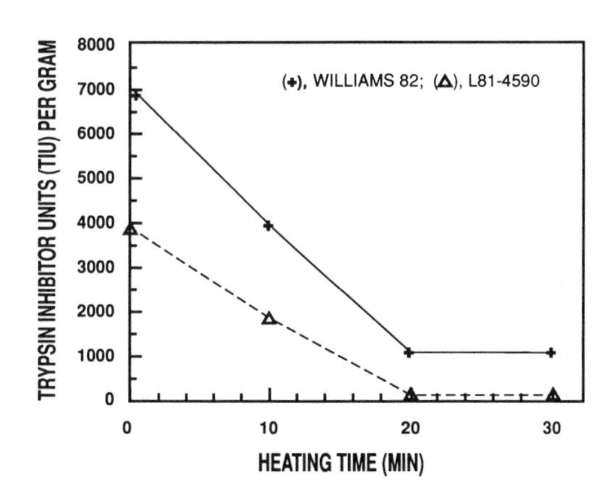

Figure 2. Effect of autoclaving time on trypsin inhibitory activity of soybean meals from Williams 82 and L81-4590 determined by enzyme assay.

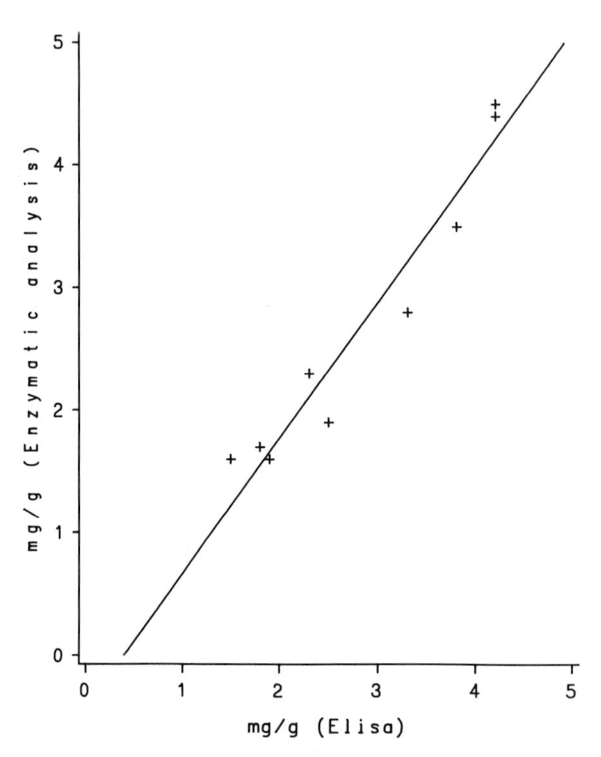

Figure 3. Linear relationship between BBI content of soybean accessions of the USDA Soybean Germplasm Collection, determined by enzyme assays and ELISA.

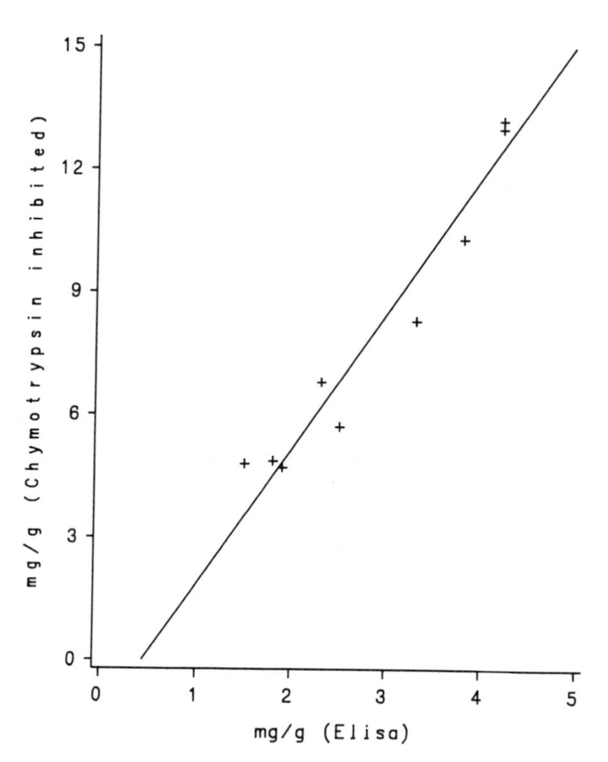

Figure 4. Linear relationship between extent of chymotrypsin inhibition of soybean accessions of the USDA Soybean Germplasm Collection determined by enzyme assays and BBI content determined by ELISA.

Protease Inhibitors. In agreement with electrophoretic, enzymatic, and genetic data (Orf and Hymowitz, 1979; Hymowitz, 1986; Jofuku et al., 1989), ELISA shows that isoline L81-4590 produces a low level of KTI. Jofuku et al (1989) show that the reduction in the KTI is a consequence of three mutations resulting in a frameshift in the ti allele. This frameshift causes lower mRNA levels and the premature termination of the mRNA translation, producing low levels of a truncated protein. We found a residual activity of less than 0.2% of the wild type which may reflect the low level of expression of truncated KTI or the product of a related KTI gene. Another inference from the ELISA studies is that KTI can increase when field-grown samples are cross-pollinated by KTI-expressing cultivars. This may explain the significantly higher KTI content of the samples from Swine Farm and South Farm. If the KTI-null isolines are to be used commercially, the ELISA method could be an important quality control test.

L81-4590 still contains 54% of the TI activity of Williams 82. Based on our experimentally determined values for BBI (81 CIU and 545 TIU/mg), the results in Tables III and IV indicate that 49% of the TI activity of Williams 82 can be ascribed to non-KTI inhibitors, including BBI and other specific and nonspecific inhibitors of trypsin. The residual 54% TI activity of L81-4590 is thus in the range expected for a KTI-lacking isoline. It is, therefore, unlikely that truncated KTI molecules or genetically related minor forms contribute significantly to the total TI activity.

In agreement with our previous report (Brandon et al., 1989), the BBI content of L81-4590 is identical to that of Williams 82, as determined by ELISA. The results permit us to estimate BBI from the chymotrypsin inhibition data as well, based on the assumption that all of the chymotrypsin inhibitory activity is due to BBI. Using the empirically determined relationship that 1 mg of chymotrypsin is inhibited by 0.34 mg BBI, we calculate the BBI content of the flours as 1.4 mg/g for Williams 82 and 1.1 mg/g for L81-4590. Thus, the ELISA gave a higher estimate of BBI than the enzymatic assay of soybean meal for these varieties. In contrast, the analysis of non-defatted flours from nine other varieties indicated close agreement between enzymatic and ELISA estimates of BBI. These results suggest that some of the BBI in defatted soybean meal is not fully active as a chymotrypsin inhibitor, perhaps due to processing-induced changes in BBI structure or its interaction with other soybean constituents. Since antibody 238 binds to BBI outside the region of the chymotrypsin-reactive site (Brandon et al., 1990), it would still detect BBI which had formed a complex with and was inhibited by another component of the soybean meal.

Effects of Heating. Heat treatments for up to 30 min at 121°C, though sufficient to inactivate trypsin and chymotrypsin inhibitors, did not appear to affect the amino acid composition of the soy flours. Such heat treatments often damage lysine, arginine, and the sulfur amino acids (Smith and Friedman, 1984). However, when measured as g/16 g N, the content of these amino acids was unchanged under the experimental conditions of this study.

The ELISA data indicate that KTI is responsible for the heat-stable trypsin inhibitory activity for commercial soy flour, as illustrated in Figure 2, which shows that the trypsin inhibitory activity of L81-4590 can be 100% inactivated, without the residual activity found in the Williams 82 sample. This could result from the stability of KTI itself, or from an effect of KTI on other inhibitors. However, it appears that the BBI content of both varieties is similar, with equal susceptiblity to inactivation under the moist heating conditions of the study.

Previous studies have indicated the sensitivity of KTI and BBI to inactivation by heat and chemical modification (Friedman et al., 1980, 1982, 1984 and 1989; Sanderson et al., 1982; Brandon et al., 1988). In contrast, BBI was unaltered antigenically by a one-hour treatment at 95°, and also retained most of its enzyme inhibitory activities (Brandon et al., 1989). These data indicate that KTI in solution is more susceptible than BBI to inactivation by heat. However, pure KTI in the dry state appears highly stable to heat inactivation (Oste et al., 1990).

The microenvironment in the soy flour appears to catalyze heat inactivation of BBI to a greater extent than KTI. Other studies have also found that the matrix has a strong influence on the susceptibilities of inhibitors to heat inactivation (Begbie and Pusztai, 1989; DiPietro and Liener, 1989 a, b; Hancock et al., 1989; Liener and Tomlinson, 1981; Paik, 1988; Peters and Czukor, 1989; Tanahashi et al., 1988). A possible explanation of these results is that sulfhydryl groups found in the proteins of soy flour participate in sulfhydryl-disulfide interchange with the disulfide-rich BBI (Friedman et al., 1984). The disruption of disulfide bonding facilitates further denaturation of the BBI. The influence of disulfide-disrupting agents on KTI and BBI has been described previously (Friedman and Gumbmann, 1986; Friedman et al., 1989; Brandon et al., 1988 and 1989).

In addition to moisture and the presence of agents which can induce changes in disulfide bonds, interaction with other constituents such as carbohydrates appears to contribute to the denaturation of inhibitors (Oste et al., 1990). It seems likely that the thermal gradients to which the inhibitors are exposed could also influence heat-induced denaturation.

Pancreatic Effects. According to the biofeedback hypothesis (Gumbmann et al., 1986; Liener et al., 1988), complexation between proteolytic enzymes and enzyme inhibitors such as the Kunitz and Bowman-Birk types in the intestinal tract creates a deficiency of proteolytic enzymes. This deficiency triggers an endocrine sensing mechanism involving cholecystokinin and gastrin. The endocrine system then induces increased protein synthesis in the pancreas, which can result in pancreatic enlargment (hypertrophy) followed by hyperplasia and development of adenoma. The results of the present study are in agreement with this hypothesis. The two soy flours, whether raw or heated, did not differ from each other in amino acid composition and their lectin contents were within a factor or two. Therefore, the differences in both the PER values and pancreatic weights are probably due to differences in inhibitor content.

The nutritional value of the raw KTI-free soy flour, although greater than that of the standard variety, is still too low to sustain normal growth and development. While both samples needed to be heated to achieve this objective, a shorter heating time sufficed for the KTI-free flour. A flour with a trypsin inhibitor content of about 1000 TIU/g supported growth well, without pancreatic weight gain. Our results suggest that this level of trypsin inhibitor could be achieved by a 15 min heat treatment of the KTI-free flour, compared to 20 min for Williams 82 (Figure 2). Table VII shows that a flour with 1800 TIU/g did not produce pancreatic hypertrophy. Figure 2 shows that this level of trypsin inhibitor activity could be achieved by an estimated heating time of 18 min for Williams 82, compared to only 10 min for the KTI-free sample. The precise conditions to achieve these parameters, with minimal use of heat, could readily be determined using the ELISA for BBI. An important difference between flours from the two cultivars is that the KTI-free flour could be processed to achieve near zero levels of TI after 20 min, well below the level achievable with standard varieties (Figure 2).

Accessions of the USDA Soybean Germplasm Collection

The preliminary screening of 13 soybean cultivars indicated that there is considerable variation in the content of protease inhibitors and sulfur amino acids. We found a linear correlation between the BBI content determined by the enzyme and ELISA assays (Figure 3). The pancreatic response to protease inhibitors is altered by the quality and quantity of the protein, as well as by the sulfur amino acid content of the diet (Gumbmann and Friedman, 1987). Therefore, the amino acid composition of soybean cultivars should be taken into account when they are being evaluated for commercial use. There was no apparent correlation between sulfur amino acid content and the chymotrypsin inhibitor activity, presumably because the inhibitors contribute only about 15-20% to the total sulfur amino acid content. Therefore, our preliminary data suggest that it may be possible to discover, through screening of the Soybean Germplasm Collection, varieties low in KTI and BBI but with a sulfur amino acid content which meets the nutritional requirements for an ideal protein (FAO, 1973; Gumbmann and Friedman, 1987).

OUTLOOK

While this and other studies (Cook et al., 1988) demonstrate that low-KTI soybeans may offer nutritional advantages, it would be desirable to develop varieties with low levels of BBI as well. The genetic diversity of soybeans with respect to the major seed lectin has been studied (Stahlhut et al., 1981), and double nulls for the lectin and KTI have been produced (Prischmann and Hymowitz, 1988). If the low-BBI characteristic can be bred into a lectin-KTI double null, soybean meal from these varieties could possibly be used as animal feed with no processing. In addition, flours from such varieties might be processed into human foods more readily than current commercial flours.

The potential beneficial effects of protease inhibitors should also be taken into account in establishing criteria for improved varieties. There is substantial evidence that wound-induced inhibitors of digestive enzymes may play a role in protecting potato and tomato plants against insects and other phytopathogens (Brown et al., 1986; Johnson et al., 1990). The protection presumably arises directly from inactivation of insect digestive enzymes or indirectly by the over-stimulation of trypsin secretion, resulting in depletion of sulfur amino acids (Broadway and Duffus, 1986). It has been suggested that such inhibitors, in combination with other toxicants, could form a battery of defenses in the *Leguminosae* (Janzen et al., 1986). Evidence presented thus far suggests that KTI is not agronomically important in soybeans, since TI-nulls appear equivalent to their parental strains (Hymowitz, 1986). It is therefore important to determine whether the elimination of additional inhibitors from soybeans by classical or molecular genetic methods such as anti-sense RNA methodology (Delauney et al., 1988) would have adverse agronomic consequences.

Soy protease inhibitors have been shown to inhibit carcinogenesis, in both *in vitro* and *in vivo* systems (Yavelow et al., 1983; Troll et al., 1986; St. Clair et al., 1990) and to stimulate human T cells (Richard et al., 1989). Therefore, more data are also needed on the health effects of protease inhibitors to guide further development of soybean varieties with enhanced health-promoting qualities.

An alternative approach to optimizing the content of protease inhibitors in soybean protein is to use food processing to achieve the desired mix. This study and others have noted that temperature, moisture, and interaction with other nutrients can influence the inactivation of protease inhibitors. A comparison of the present results to those from a previous study (Brandon et al., 1989) suggests that sample size and defatting of the flour may influence the relative susceptibilities of KTI and BBI to heat inactivation. Food processing could be combined with genetic approaches. For example, inhibitors modified by site-directed mutagenesis to produce heat-labile molecules could protect the plants, but be eliminated from foods and feeds by mild processing. High resolution techniques, such as the monoclonal antibody-based ELISA's for measuring KTI and BBI, could help optimize production of the most nutritious and healthful soy protein product.

REFERENCES

AOAC Official Methods of Analysis: 12th ed., Association of Official Analytical Chemists, Washington, D.C. p. 857. 1980.

Begbie, R. and Pusztai, A. The Resistance to Proteolytic Breakdown of Some Plant (Seed) Proteins and Their Effects on Nutrient Utilization and Gut Metabolism. In "Absorption and Utilization of Amino Acids",, M. Friedman (Ed.). CRC Press, Boca Raton, Florida. pp. 265-295, Volume 3. 1989.

Birk, Y. The Bowman-Birk Inhibitor. Int. J. Peptide Protein Res. 25, 113-131. 1985

Brandon, D.L. and Bates, A.H. Definition of Functional and Antibody-binding Sites of Kunitz Soybean Trypsin Inhibitor Isoforms Using Monoclonal Antibodies. J. Agric. Food Chem. 36, 1336-1341. 1988.

Brandon, D.L., Haque, S. and Friedman, M. Interaction of Monoclonal Antibodies with Soybean Trypsin Inhibitors. J. Agric. Food Chem. 34, 195-200. 1987.

Brandon, D.L., Bates, A.H. and Friedman, M. Enzyme-Linked Immunoassay of Soybean Kunitz Trypsin Inhibitor Using Monoclonal Antibodies. J. Food Sci. 53, 97-101. 1988.

Brandon, D.L., Bates, A.H. and Friedman, M. Monoclonal Antibody-Based Enzyme Immunoassay of Bowman-Birk Protease Inhibitor of Soybeans. J. Agric. Food Chem. 37, 1192-1196. 1989.

Brandon, D.L., Bates, A.H. and Friedman, M. Antigenicity of Soybean Protease Inhibitors. In "Protease Inhibitors as Potential Cancer Chemopreventive Agents", W. Troll and R.A. Kennedy (Eds.). Plenum, New York. 1990.

Broadway, R.M. and Duffey, S.F. Plant Proteinase Inhibitors Mechanism of Action and Effect on the Growth and Digestive Physiology of Larval Heliothis zea and Spodoptera exigua. J. Insect Physiol. 32, 827-834. 1986.

Brown, W.F., Graham, J.S., Lee, J.S. and Ryan, C.A. Regulation of Proteinase Inhibitor Genes in Food Plants. In "Nutritional and Toxicological Significance of Enzyme Inhibitors in Foods", M. Friedman (Ed.). Plenum, New York. pp. 281-190. 1986.

Cook, D.A., Jensen, A.H., Fraley, J.R. and Hymowitz, T. Utilization by Growing and Finishing Pigs of Raw Soybeans of Low Kunitz Trypsin Inhibitor Content. J. Anim. Sci. 66, 1686-1691. 1988.

Crawford, L. The Effects of Oxidation and Partial Hydrogenation of Dietary Soy Oil on Rat Liver Microsomal Fatty Acid Distribution and the Resulting Influence on the Mixed Function Oxidase System. Nutr. Res. 9, 173-181. 1989.

Delauney, A.J., Tabaeizadeh, Z. and Verman, D.P.S. A Stable Bifunctional Antisense Transcript Inhibiting Gene Expression in Transgenic Plants. Proc. Natl. Acad. Sci. USA 85, 4300-4304. 1988

DiPietro, C.M. and Liener, I.E. Soybean Protease Inhibitors in Foods. J. Food Sci. 54, 606-609. 1989a.

DiPietro, C.M. and Liener, I.E. Heat Inactivation of the Kunitz and Bowman-Birk Soybean Protease Inhibitors. J. Agric. Food Chem. 37, 39-44. 1989b.

Duncan, D.B. Multiple Range and Multiple F Tests. Biometrics 1-42. 1955.

FAO, Energy and Protein Requirements. FAO Nutritional Meetings Report Series No. 52, Food and Agricultural Organization of the United Nations, Rome. 1973.

Friedman, M. and Cuq, J.L. Chemistry, Analysis, Nutritional Value, and Toxicology of Tryptophan in Food. A Review. J. Agric. Food Chem. 36, 1079-1093. 1988.

Friedman, M. and Gumbmann, M.R. Nutritional Improvement of Soy Flour through Inactivation of Trypsin Inhibitors by Sodium Sulfite. J. Food Sci. 51, 1239-1241. 1986a.

Friedman, M. and Gumbmann, M.R. Nutritional Improvement of Legume Proteins Through Disulfide Interchange. Adv. Exp. Med. Biol. 199, 357-390. 1986b.

Friedman, M., Gumbmann, M.R., Brandon, D.L. and Bates, A.H. Inactivation and Analysis of Soybean Inhibitors of Digestive Enzymes. In "Food Proteins", J.E. Kinsella, and W.G. Soucie (Eds.). The American Oil Chemists Society, Champaign, IL. pp. 296-328. 1989.

Friedman, M., Gumbmann, M.R. and Grosjean, O.K. Nutritional Improvement of Soy Flour. J. Nutr. 114, 2241-2246. 1984.

Friedman, M., Grosjean, O.K. and Zahnley, J.C. Inactivation of Soya Bean Trypsin Inhibitors by Thiols. J. Sci. Food Agric. 33, 165-172. 1982a.

Friedman, M., Grosjean, O.K. and Zahnley, J.C. Cooperative Effects of Heat and Thiols in Inactivating Trypsin Inhibitors from Legumes in Solution and in the Solid State. Nutr. Repts. International 25, 743-751. 1982b.

Friedman, M. Grosjean, O.K. and Zahnley, J.C. Effect of Disulfide Bond Modification on the Structure and Activities of Enzyme Inhibitors. In "Mechanisms of Food Protein Deterioration", J.P. Cherry (Ed.). ACS Symposium Series, Washington, D.C. pp. 359-407. 1982c.

Friedman, M., Zahnley, J.C. and Wagner, J.R. Estimation of the Disulfide Content of Trypsin Inhibitors as S-β-(2-pyridylethyl)-L-cysteine. Analytical Biochem. 106, 27-34. 1980.

Friedman, M., Noma, A.T. and Wagner, J.R. Ion-exchange Chromatography of Sulfur Amino Acids on a Single-Column Amino Acid Analyzer. Analytical Biochem. 98, 293-305. 1979.

Gumbmann, M.R. and Friedman, M. Effect of Sulfur Amino Acid Supplementation of Raw Soy Flour on the Growth and Pancreatic Weights of Rats. J. Nutr. 17, 1018-1023. 1987.

Gumbmann, M.R., Spangler, W.L., Dugan, G.M. and Rackis, J.J. Safety of Trypsin Inhibitors in the Diet: Effects on the Rat Pancreas of Long-term Feeding of Soy Flour and Soy Protein Isolate. In "Nutritional and Toxicological Significance of Enzyme Inhibitors in Foods," M. Friedman (Ed.). Plenum Press, New York. pp. 33-79. 1986.

Hancock, J.D., Lewis, A.J., and Peo, E.R., Jr. Effects of Ethanol Extraction on the Utilization of Soybean Protein by Growing Pigs. Nutr. Rep. Int. 39, 813-821. 1989.

Hymowitz, T. Genetics and Breeding of Soybeans Lacking the Kunitz Trypsin Inhibitor. In "Nutritional and Toxicological Significance of Enzyme Inhibitors in Foods," M. Friedman (Ed.). Plenum Press, New York. pp. 291-298. 1986.

Janzen, D.H., Ryan, C.A., Liener, I.E. and Pearce, G. Potentially Defensive Proteins in Mature Seeds of 59 Species of Tropical Leguminosae. J. Chem. Ecol. 12, 1469-1480. 1986.

Jofuku, K.D., Schipper, R.D. and Goldberg, R.B. A Frameshift Mutation Prevents Kunitz Trypsin Inhibitor Messenger RNA Accumulation in Soybean Embryos. Plant Cell 1, 427-436. 1989.

Johnson, R., Lee, J.S. and Ryan, C.A. Regulation of Expression of Wound-Inducible Tomato Inhibitor I Gene in Transgenic Nightshade Plants. Plant Mol. Biol. 14, 349-356. 1990.

Liener, I.E., Goodale, R.L., Deshmukh, A., Satterberg, T.L., DiPietro, C.M., Bankey, P.E. and Borner, J.W. Effect of Bowman-Birk Trypsin Inhibition from Soybeans on the Secretory Activity of the Human Pancreas. Gastroenterology 94, 419-427. 1988.

Liener, I.E. The Nutritional Significance of Lectins. In "Food Proteins", J.E. Kinsella, and W.G. Soucie (Eds.) The American Oil Chemists Society, Champaign, IL. pp. 239-353. 1989.

Liener, I.E. Phytohemagglutinins: Their Nutritional Significance. J. Agric. Food Chem. 22, 17-22. 1974.

Liener, I.E. and Tomlinson, S. Heat Inactivation of Protease Inhibitors in Soybean Line Lacking the Kunitz Trypsin Inhibitor. J. Food Sci. 46, 1354-1356. 1981.

Lotan, R. and Sharon, N. Modification of the Biological Properties of Plant Lectins by Chemical Crosslinking. In "Protein Crosslinking: Biochemical and Molecular Aspects", M. Friedman (Ed.). Plenum Press, New York. pp. 149-168. 1977.

Mandansky, A. The Fitting of Straight Lines when both Variables are Subject to Error. 1959. J. Am. Statist. Assn. 54, 173-206. 1959.

Nachbar, M.S. and Oppenheim, J.D. Lectins in the United States Diet: A Survey of Lectins in Commonly consumed Foods and A Review of the Literature. Amer. J. Clin. Nutr. 33, 2338-2345. 1980.

Orf, J.H. and Hymowitz, T. Inheritance of the Absence of the Kunitz Trypsin Inhibitor in Seed Protein of Soybeans. Crop Sci. 19, 107-109. 1979.

Orf, J.H., Mies, D.W. and Hymowitz, T. Qualitative Changes of the Kunitz Trypsin Inhibitor in Soybean Seeds During Germination as Detected by Electrophoresis. Bot. Gaz. 138, 255-260. 1977.

Oste, R.E., Brandon, D.L., Bates, A.H. and Friedman, M. Effect of Maillard Browning Reactions of the Kunitz Soybean Trypsin Inhibitor on its Interaction with Monoclonal Antibodies. J. Agric. Food Chem. 38, 258-261. 1990.

Paik, I.K. Review of the Antinutritional Factors and Heat Treatment Effects of Soybean Products. Korean J. Anim. Nutr. Feed 12, 284-291. 1988.

Peters, J. and Czukor, B. Effect of Extrusion Cooking on Trypsin-Inhibitor Activity. Nahrung 33, 275-281. 1989.

Pusztai, A., Clarke, E.M.W., Grant, G. and King, T.P. The Toxicity of Phaseolus vulgaris Lectins. Nitrogen Balance and Immunochemical Studies. J. Sci. Food Agric. 32, 1037-2046. 1981.

Prischmann, J.A. and Hymowitz, T. Inheritance of Double Nulls for Protein Components of Soybean seed. Crop Sci. 28, 1010-1012. 1988.

Rackis, J.J., Wolf, W.J. and Baker, E.C. Protease Inhibitors in Plant Foods: Content and Inactivation. In "Nutritional and Toxicological Significance of Enzyme Inhibitors in Foods", M. Friedman (Ed.). Plenum Press, New York. pp. 299-347. 1986.

Richard, K.A., Speciale, S.C., Staite, N.D., Berger, A.E., Daibel, M.R. and Finzel, B.C. Soybean trypsin inhibitor. An IL-1-like protein. Agents and Actions 27, 265-267. 1989.

St. Clair, W.H., Billings, P.C., Carew, J.A., Keller-McGandy, C., Newberne, P. and Kennedy, A.R. Suppression of Dimethylhydrazine-induced Carcinogenesis in Mice by Dietary Addition of the Bowman-Birk Protease Inhibitor. Cancer Res. 50, 580-586. 1990.

Sanderson, J.E., Freed, R.C. and Ryan, D.S. Thermal Denaturation of Genetic Variants of the Kunitz Soybean Trypsin Inhibitor. Biochim. Biophys. Acta 701, 237-241. 1982.

Smith, G.A. and Friedman, M. Effects of Carbohydrates and Heat on Amino Acid Composition and Chemically Available Lysine Content of Casein. J. Food Sci. 49, 817-820. 1984.

Stahlhut, R.W. and Hymowitz, T. Screening the USA Department of Agriculture Glycine Soja Collection for Presence or Absence of a Seed Lectin. Crop Sci. 21, 110-112. 1981.

Tan-Wilson, A.L., Chen, J.C., Duggan, M.C., Chapman, C., Obach, R.S. and Wilson, K.A. Soybean Bowman-Birk Trypsin Isoinhibitors: Classification and Report of Glycine-rich Trypsin Inhibitor Class. J. Agric. Food Chem. 35, 974-981. 1987.

Tanahashi, K., Takano, K., Matsumoto, S., Kamoi, I. and Obara, T. Effects of Soybean Protein on Thermal Stability of Soybean Trypsin Inhibitor. Nippon Shokuhin Kogyo Gakkaishi 35, 534-540. 1988.

Troll, W., Frankel, K. and Wiesner, R. Protease Inhibitors: Their Role as Modifiers of Carcinogenic Processes. In "Nutritional and Toxicological Significance of Enzyme Inhibitors in Foods", M. Friedman (Ed.). Plenum, New York. pp 153-165. 1986.

Wallace, J.M. and Friedman, M. Inactivation of Hemagglutinins in Lima Bean (Phaseolus lunatus) Flour by N acetyl-L-cysteine, pH, and Heat. Nutr. Rep. Int. 32, 743-748. 1985.

Wolf, W. J. and Cowan, J. C. Soybeans as a Food Source. CRC Press, Boca Raton, Florida. 1975.

Worthington Biochemical Products. Worthington Diagnostic System. Freehold, New Jersey. 1982.

Yavelow, J., Finley, T. H., Kennedy, A. R. and Troll, W. Bowman-Birk Soybean Protease Inhibitor as an Anticarcinogen. Cancer Res. 43,2454-2459. 1983.

Zahnley, J. C. and Friedman, M. Absorption and Fluorescence Spectra of S-quinolylethylated Kunitz Soybean Trypsin Inhibitor. Journal Prot. Chem. 1, 225-240.

Additional References

Brandon, D. L., Bates, A. H. and Friedman, M. The Impact of Food Processing on Soybean Trypsin Inhibitors. 1991. This Volume.

Belitz H. D. and Weder, J. K. P. Protein Inhibitors of Hydrolases in Plant Foodstuffs. Food Reviews International 6 (2), 151-211. 1990.

Friedman, M., Brandon, D. L., Bates, A. H. and Hymowitz, T. Comparison of a Soybean Cultivar and an Isoline Lacking the Kunitz Trypsin Inhibitor: Composition, Nutritional Value, and Effects of Heating. J. Agric. Food Chem. 39, 327-335. 1991.

Grant, G. Anti-Nutritional Effects of Soyabean: A Review.Progress in Food and Nutritiona Science 13, 317-348. 1989.

26

UTILIZATION OF EARLY MAILLARD REACTION PRODUCTS BY HUMANS

Helmut F. Erbersdobler, Michael Lohmann and Karin Buhl

Institute for Human Nutrition and Food Science, Christian-Albrechts- University, 2300 Kiel 1, Fed. Rep. of Germany

ABSTRACT

The paper reports results from balance trials with a total of 42 volunteers testing glycated casein samples containing the Maillard-(Amadori-) product fructoselysine (= FL, analysed as furosine). On a almost FL free diet only traces of FL were excreted. If test meals with 0.8-5.0 g FL were given, only 2.0- 1.2 % were found in the urine. In the feces of 3 persons eating 0.96 g FL in a single meal 2.6-5.6 % were excreted. It is concluded that digestion is the main limiting factor for the uptake of FL. Possibly also the transit time of the ingesta through the gastro intestinal tract is important. Since there is no indication from animal studies for a utilization of FL it can be assumed that the microorganism in the hind gut decompose the main part of the not recovered more than 90 % of FL. Obviously the bacterial flora is more active and able to attack such components as assumed until now.

INTRODUCTION

Man is eating a lot of processed food and the main changes during processing are due to the Maillard reaction. The metabolic transit of Maillard products in humans is very uncertain. The metabolism of the Amadori products with lysine, however, has been tested in laboratory animals by several authors (e.g. Finot 1973, Finot 1983, Erbersdobler et al. 1981). This applies in particular to ε-fructoselysine, the Amadori product formed by the reaction of the ε-amino group of protein bound lysine with glucose in the initial stage of Maillard condensation, which is found in considerable amounts in many heated foods (e.g. Erbersdobler 1989). Figure 1 shows the initial stage of the Maillard reaction with the formation of fructoselysine and its indicator furosine.

The present study deals with the influence of different doses of protein bound FL on the urinary excretion in humans and includes a preliminary balance study with the collection of urine and feces. The results are discussed in connection with earlier results obtained in balance experiments with rats and in vitro studies with several rat tissues by using radioactive labeled material.

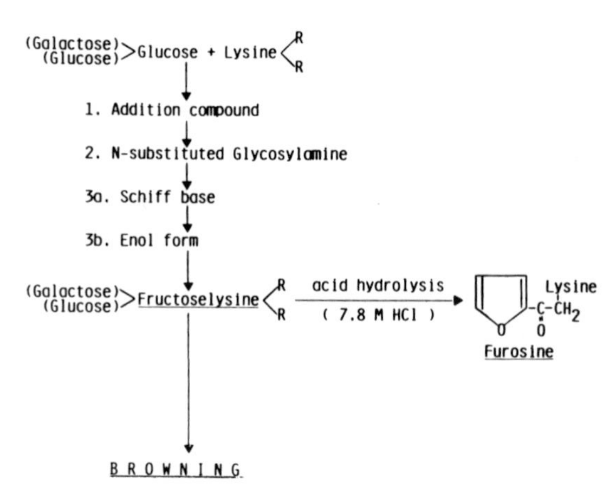

Figure 1. Initial steps of the Maillard reaction with the formation of furosine

MATERIALS AND METHODS

Fructoselysine (FL), which itself is very unstable during acid hydrolysis was estimated by analysing furosine on an amino acid analyser after hydrolysing 500 mg of the foods, 1 g of the feces or 100 ml of the urines with 7.8 M HCL for 20 h under reflux as described elsewhere (Erbersdobler et al. 1987). For the calculation of FL from furosine the factor of 2.5 was used. The test materials consisted in casein samples (acid precipitated), which were heated together with glucose and tap water in order to get quite high values of FL.

For the studies on the urinary excretion of FL in human volunteers a total of 40 persons were fed a defined diet extremely low in (FL) two days prior to the test meal and during the time of urine collection. At the second day a pooled blank sample of urine was collected 24 hours for each person. After that, the persons consumed the FL-rich test proteins in form of 25 to 56 g of heated casein samples containing 0.8; 1.6; 3.0; and 5.0 g FL. After the test meal the volunteers collected their urines for 24 hours in single portions. After one experiment 20 volunteers, who had been just on a normal diet for a while, were asked again to collect their pooled urines for 24 h. In a preliminary balance study two male persons collected urine and feces after a test meal with 0.96 g of FL up to 48 hours. The test was repeated with one person after a week. The experimental details are described elsewhere (Erbersdobler et al. 1989, 1990).

The animal experiments were described earlier (Erbersdobler et al. 1981, 1989). In contrast to the studies with humans the animal experiments were conducted by using ^{14}C-labeled free and protein bound FL.

RESULTS AND DISCUSSION

Compared with the intake the amount of FL excreted was very low as shown in figure 2. Only 1.2-2.5 % were found in the urine which is considerably lower than found in earlier results on rats. In that experiments proteins containing ^{14}C-labeled FL were fed and about 30% of the ingested FL were excreted (Erbersdobler, 1977, Erbersdobler et al. 1981, 1989, Mori and Nakatsujii, 1977). The results of Niederwieser et al. (1975) demonstrated a 16 % absorption of FL in infants given normal or moderately heat damaged formula diets based on a lactose-deprived glucose containing milk.

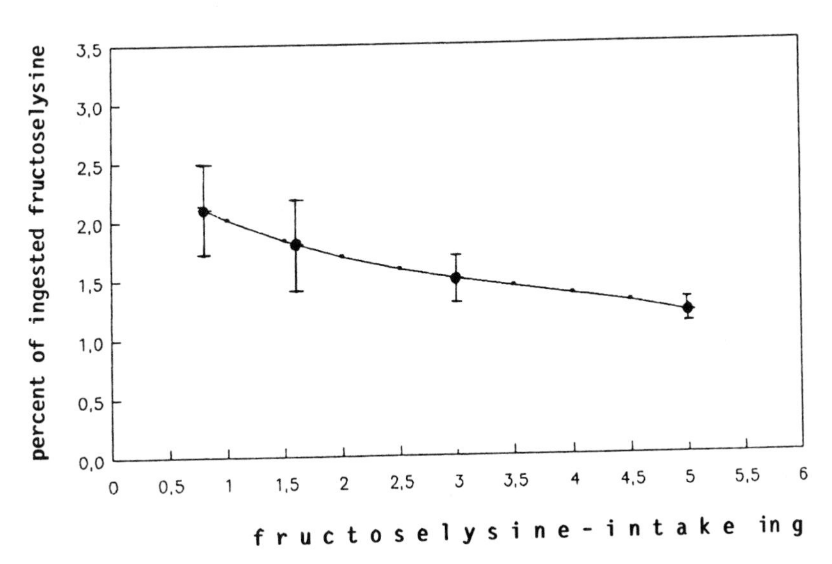

Figure 2. Percentage of ingested fructoselysine (FL) excreted in the urine as influenced by the amount of FL intake.

This relative high absorption rate can be explained by the higher permeability of the young childś intestinal tract. The discrepancy to the rat values remains unexplained.

Figure 3 shows two examples of the course of FL excretion after the ingestion of 800 and 1600 mg of FL respectively. Depending on the time of urination different profiles for the individual volunteers can be observed. In all persons, however, the peak of excretion was reached in the time between 4 and 8 hours after the meal, which is rapid considering the transit time of the ingesta through the gastro intestinal tract. In rats a similar excretion profile was observed as is described elsewhere (Erbersdobler et al. 1981). It appears from the early decrease of the FL levels after 8-10 hours that only in the upper part of the digestive tract an absorption of FL is possible and that FL is not absorbed to a significant degree if there would be a release by microbial digestion in the hind gut as was proposed by Finot from experiments with rats (1973).

As was demonstrated by figure 2, the dose response of FL excretion is not linear increasing with decreasing amount of ingested FL. This indicates a saturation phenomenon either in the liberation by digestion, in the intestinal absorption or in the renal excretion. The rapid increase of the FL concentrations after the test meal and the early and relatively sharp decline (figure 3) suggest that not the renal excretion is responsible for that phenomenon. It was shown in earlier experiments with rats that ^{14}C-FL was rapidly excreted after intravenous injection (Erbersdobler et al. 1981). A slow absorption of FL only by diffusion and its accumulation at the intestinal wall was demonstrated in rats with the everted sac technique and with ligated jejunal loops (Erbersdobler et al. 1981). On the other hand Finot (1973) demonstrated that 60 -70 % of free FL were absorbed and excreted in the urine after oral administration. For this reason it becomes clear that particularly the enzymic release of FL and possibly also absorption in connection with the passage of the ingesta through the gastro intestinal tract are limiting for the uptake.

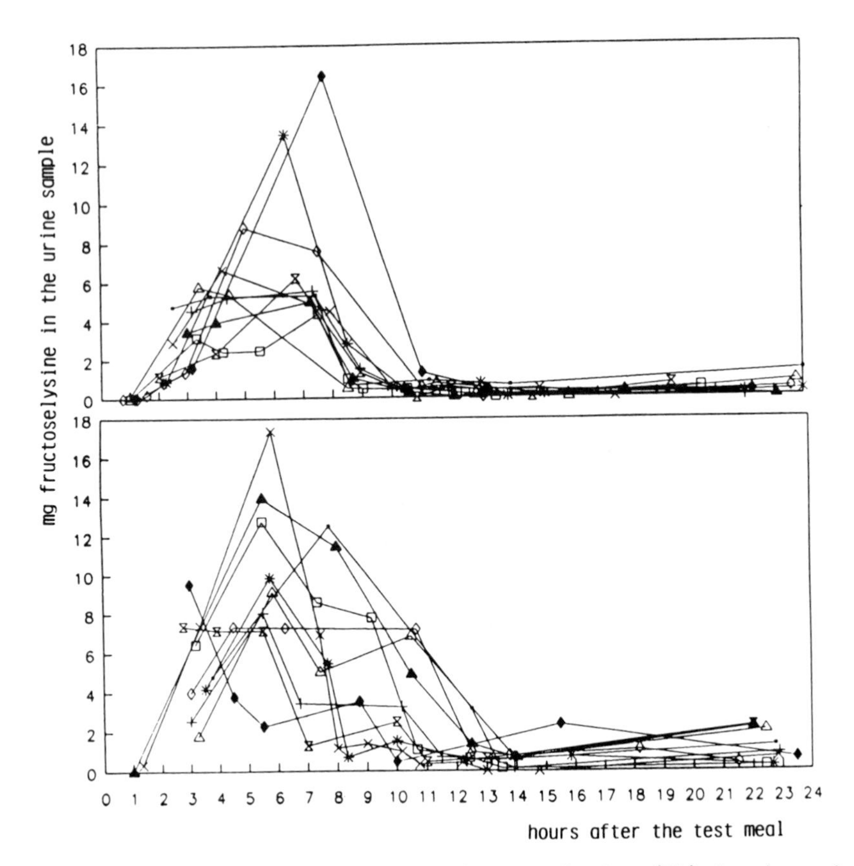

Figure 3. Profiles of the excretion of fructoselysine (FL) in the urine of 20 volunteers after test meals of 0.8 g or 1.6g of FL respectively.

In one trial with 20 volunteers the average value of FL excretion under normal conditions of eating was 3.3 ± 1.4 mg FL per day. By assuming an excretion rate of 2.5 % for relatively low amounts of intake this represents a daily uptake of about 130 ± 60 mg FL. This corresponds to about 60 mg of unavailable lysine or 1-2 % of the daily intake of lysine in adults. One person, however, excreted 6.9 mg, which was explained by the consumption of high amount of a breakfast cereal, rich in FL. In this way the method may be useful to expose the intake of high amounts of heat damaged foods eg. by unusual eating habits.

After the ingestion of the "FL free" diet only traces of FL were measured in the urine, which may result from residues of FL in the diet or from the degradation of endogenously glycosylated proteins. In earlier studies (Erbersdobler et al. 1986) somewhat higher "zero levels" in non diabetic persons were obtained. This could be attributed to the fact, that the specificity of the furosine determination was meanwhile improved by now separating off a small unknown peak. In contrast to that Reindl (1983) found much higher values in normal persons and in persons after a starvation of 12 hours. It appears that his HPLC method for furosine determination elaborated for the use in purified blood proteins was not specific enough for the application on urine samples. Most recently Knecht et al. (1990) found similar low values for furosine (and in that way in FL), but they could easily quantify her data by using the more sensitive gaschromatography/ mass spectroscopy.

Table 1 shows the results of preliminary experiments measuring also the FL excretion in the feces. In the feces only 3-6 % of the ingested FL were recovered too, which is also low compared to preliminary results with rats (Lee and Erbersdobler, unpublished).

Table 1. Excretion of fructoselysine (FL) in feces and urine within 48 hours in 3 volunteers. All values are given in % of the intake of FL (0.96 g in a single meal)

FL in the feces (48h)		FL in the urine (24h)	
in mg	in %	in mg	in %
25-54	2.6-5.6	14-55	1.5-5.7

CONCLUSIONS

In summarizing the results obtained hitherto the unsatisfactory phenomenon that more than 90 % of the ingested fructoselysine were not recovered has to be recognized. The question arises whether the other 90 % are metabolized or destroyed e.g. by the intestinal flora.

There is no evidence fromout all data in the literature that fructoselysine would be available as source of lysine in any way. Results with ^{14}C-labeled FL on rats have shown no indication for a metabolism or utilisation (Erbersdobler 1977, Erbersdobler et al. 1981, Finot 1973, 1983). On the other hand there are some reports for a limited utilization of related compounds like lysinoalanine, which shows an availability of 35 % in growing chicken (Robbins et al., 1980) or glutamyllysine, which is available even by rats (Hurrell and Carpenter, 1977).

In overheated dried skim milks, however, the lactuloselysine moiety was not or at least very poorly available as source of lysine. This is demonstrated by experiments with growing rats as is shown in figure 4. Rats receiving a diet with a severely heat damaged dried skim milk containing only 2.2 % available lysine in the protein (instead of about 8.5) did not grow sufficiently. A supplementation of lysine in the form of lysine-HCl restored the protein quality almost to the level of the control sample of good quality (Erbersdobler, 1986). Similar results were obtained in experiments with heat damaged caseins and also by performing balance studies and tests measuring the concentrations of lysine in the portal plasma of rats after single meals of the heat damaged proteins.

The up to now first and only human study indicating a growth retardation with browned milk proteins was reported in nutrition Reviews, (1978). In this study school children in Surinam (6-12 years old) showed a poor growth rate compared to controls if they received a lactose hydrolyzed dried milk. It was suggested that this milk product experienced an increased browning due to the more aggressive monosaccharides and the poor storage conditions (the experiment lasted one year). The difference could not be confirmed in animal studies the chemically determined (FDNB) content in available lysine, however, was lower in the hydrolyzed milk.

Anyway, it appears from the results that the main part of the protein-bound FL was decomposed in the gastro intestinal tract. Namely the microorganisms in the hind gut are obviously active enough to attack and

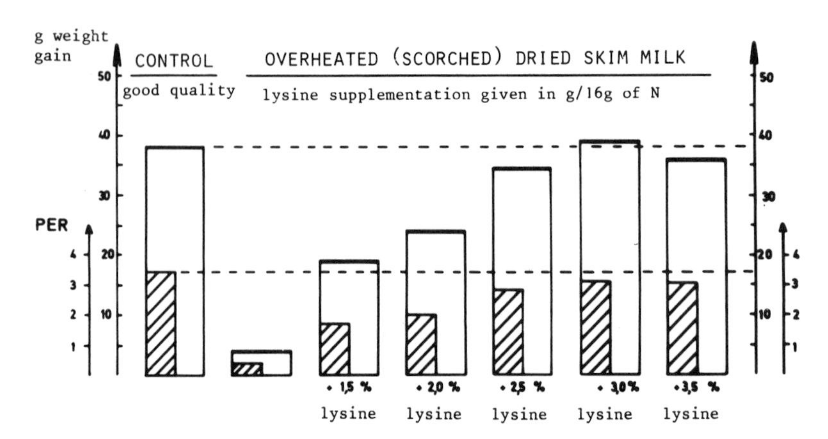

Figure 4. Effects of lysine supplementation of a severely heat damaged dried skim milk on weight gains and protein efficiency ratio (PER=g weight/g of protein intake) of rats. (From Erbersdobler, 1986, see ref. Nr. 2).

even utilize this type of unavailable lysine. Figure 5 shows results of earlier in vitro experiments in which the contents of the hind gut of rats were incubated together with a FL rich casein (Erbersdobler et al. 1970). Similar studies were performed later on with hind gut contents of pigs (Erbersdobler, 1986). The results showed that the intestinal flora was able to decompose FL and that at the same time ammonia was produced.

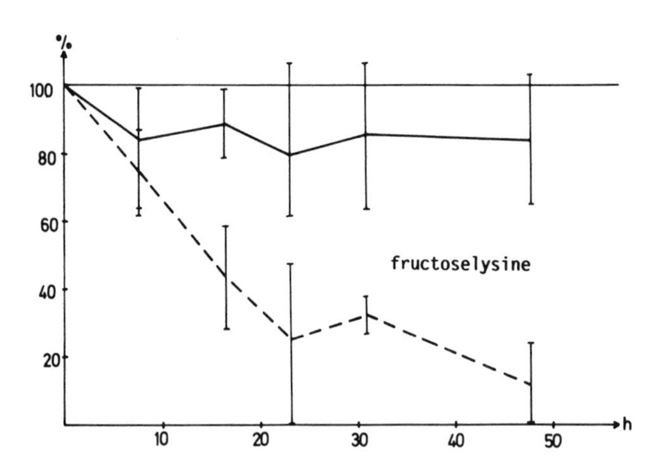

Figure 5. Microbial degradation of fructoselysine during the in vitro incubation of a heat damaged protein together with the contents of the hind gut of rats.
- - - = Incubation without additive; ——— = addition of $HgCl_2$.

REFERENCES

Erbersdobler, H.F. (1977). The biological significance of carbohydrate-lysine crosslinking during heat treatment of food proteins. In: "Protein Crosslinking. Nutritional and Medical Consequences", Adv. Exp. Med. Biol., 86b, M. Friedman, ed., Plenum Press, New York, pp. 367-387.

Erbersdobler, H.F. (1986). Loss of nutritive value on drying. In: "Concentration and Drying of Foods", D. MacCarthy ed., Elsevier Applied Science Publishers, London and New York, pp. 69-87.

Erbersdobler, H.F. (1989). Protein reactions during food processing and storage - Their relevance to human nutrition. In: "Nutrition Impact of Food Processing", Bibliotheca Nutritio et Dieta, J.C. Somogyi and H.R. Müller eds., S. Karger, Basel, pp. 140-155.

Erbersdobler, H.F., Brandt, A., Scharrer, E. and v. Wangenheim, B. (1981). Transport and metabolism studies with fructose amino acids. Prog. Fd. Nutr. Sci. 5, 257-263.

Erbersdobler, H.F., Buhl, K. and Klusmann, U. (1990). Balance studies with glycosylated proteins on human volunteers. In: "The Maillard Reaction in Food Processing, Human Nutrition and Physiology", P.A. Finot, H.U. Aeschbacher, R.F. Hurrell and R. Liardon, eds., Birkhäuser, Basel, Boston and Berlin, pp. 273-278.

Erbersdobler, H.F., Dehn, B., Nangpal, A. and Reuter, H. (1987). Determination of furosine in heated milk as a measure of heat intensity during processing. J. Dairy Research, 54, 147-151.

Erbersdobler, H.F., Groß, A., Klusmann, U. and Schlecht, K. (1989). Absorption and metabolism of heated protein carbohydrate mixtures in humans. In: "Absorption and Utilization of Amino Acids", Vol. III., M. Friedman, ed., CRC Press, INC, Boca Raton, pp. 91-102.

Erbersdobler, H.F., Gunsser, I., and Weber, G. (1970). Abbau von Fruktoselysin durch die Darmflora. Zentralbl. Veterinaermed., Reihe A., 17, 573.

Erbersdobler, H.F., Purwing, U., Bossen, M. and Trautwein, E. (1986). Urinary excretion of fructoselysine in human volunteers and diabetic patients. In: "Amino-Carbonyl Reactions in Food and Biological Systems", M. Fujimaki, M. Namiki, and H. Kato, eds., Elsevier, Amsterdam, Oxford, New York, Tokyo and Kodansha, LTD, Tokyo, pp. 503-508.

Finot, P.A. (1973). Non-enzymatic browning. In: "Proteins in Human Nutrition". J.W.G. Porter, and B.A. Rolls, eds., Academic Press, London and New York, pp. 501-514.

Finot, P.A. (1983). Chemical modifications of the milk proteins during processing and storage. Nutritional, metabolic and physiological consequences. Kieler Milchwirtschaftliche Forschungsberichte 35, 357-369.

Hurrell, R.F. and Carpenter, K.J. (1977). Nutritional significance of cross-link formation during food processing. In: "Protein Crosslinking. Nutritional and Medical Consequences", Adv. Exp. Med. Biol., 86b, M. Friedman, ed., Plenum Press, New York, pp. 225-238.

Knecht, K.J., Dunn, J.A., McFarland, K.F., Thorpe, S.R. and Baynes, J.W. (1990). Oxidative degradation of glycated proteins: Effect of diabetes and aging on carboxymethyllysine levels in human urine. Diabetes, in press.

Mori, N. and Nakatsujii, H. (1977). Utilization in rats of ^{14}C-L-lysine labeled casein browned by amino carbonyl reaction. Agric. biol. Chem. 14, 345-350.

Niederwieser, A., Giliberti, P. and Matasovic, A. (1975). N-ε-desoxyfructosyl-lysine in urine after ingestion of a lactose free glucose containing milk formula. Pediatric Res. 9, 867.

Nutrition Reviews, (1978). Nutritional significance of lactose intolerance. 36, 133-134.

Reindl (1983). Freies und proteingebundenes Fruktoselysin im Blutplasma und Urin und seine Bedeutung zur Stoffwechselüberwachung bei Diabetikern. Med. Diss. München.

Robbins, K.R., Baker, D.H. and Finley, J.W. (1980). Studies on the utilization of lysinoalanine and lanthionine. J. Nutrition 110, 907-915.

27

DIGESTIBILITY OF PROCESSED FOOD PROTEIN

Rickard E. Öste

Department of applied nutrition and food chemistry
Chemical Center
University of Lund, S-221 00 Lund, Sweden

ABSTRACT

An overview is given on the effects of food processing on the protein digestibility. Beneficial effects of food processing are primarily observed in a range of plant foods containing toxic substances and/or anti-nutrients (legumes, cereals, some seed food). Digestibilities improve by heating, soaking, germination and fermentation. These processing steps reduce the amount of active enzyme inhibitors through extraction, inactivation by heat or microorganisms, or by compositional modification through germination. Reduced protein digestibility is primarily associated with excessive heat, exemplified by the comparatively low digestibility of ready-to-eat breakfast cereals. Experiments with model systems indicate that some caution should be observed with the use of alkalis in food processing, and with products prone to the Maillard reaction.

INTRODUCTION

Industrial and domestic processing of food products includes energy and mass transfer operations such as blanching, boiling, sterilizing, microwave heating, drying and deep-freezing. The objectives are to increase product palatability and/or shelf life. By necessity, these treatments affect the chemical and physical nature of the foodstuff. From the nutritional point of view, the treatment may be beneficial or detrimental. When the digestibility of food proteins is examined, both gains and losses are frequently observed. The subject was reviewed by Hurrel and Finot (1985). The present paper gives an overview of published work on the effect of food processing on protein digestibility, with the emphasis on contributions from 1984 and onwards.

Guidelines for the correction of protein intake by quality were given by the joint UNU/FAO/WHO expert consultation in 1985. The quality aspects of food proteins include the amino acid composition and the true digestibility. To calculate effective protein intake of children up to the age of 12, both protein digestibility and amino acid composition have to be considered. For adolescents and adults, only digestibility is of relevance, since their need for essential amino acids is considered to be met by most food protein.

Nutritional and Toxicological Consequences of Food Processing
Edited by M. Friedman, Plenum Press, New York, 1991

Further, the expert consultation suggested that the overall digestibility of a mixed diet should be the weighted average in digestibility of the different proteins in the diet. This means that the low digestibility of a protein may not be compensated by protein supplementation or complementation to the diet, as is the case when the low protein quality is due to an imbalanced amino acid composition (Bressani, 1977). Accordingly, losses in protein digestibility upon food processing should be of more concern to the food technologist than, for example, the reduced availability of a single limiting amino acid, since the latter effect may be efficiently eliminated by deliberate formulation or through the contributions from other dietary components.

Data on the digestibility of protein in man have been reviewed by Hopkins (1981) and by the 1985 UNU/FAO/WHO expert consultation. Typical values of digestibilities of mixed diets of various ethnicity are given in table I. They range from 78 to 96, which is roughly the span of digestibilities obtained with individual protein sources. The differences are in general attributed to the following circumstances: intrinsic differences in the nature of food protein, including cellular organization (Bradbury et al., 1984) and protein structure, and modifying dietary factors, in particular dietary fibre, tannins and other polyphenols, fytate and specific enzyme inhibitors. Obviously, food processing may alter all these factors, and the processing may be beneficial for digestion or it may hinder it.

Table I. TRUE DIGESTIBILITY BY ADULTS OF PROTEIN IN MIXED DIETS

American mixed diet	96
Chinese mixed diet	96
Filipino mixed diet	88
Indian rice + beans diet	78
Brazilian mixed diet	78

From FAO/WHO/UNU (1985).

Digestibility in this context is that of crude protein, which means the digestibility of nitrogen. It has been argued that this may not necessarily be the digestibility of the individual amino acids. Significant differences between crude protein digestibility and individual amino acid digestibility have been reported for some plant foods and for such foods this fact may have to be taken into account (Sarwar, 1987). However, few data exist on the digestibility of individual amino acids from different foods, and even less data are available which reflect the effect of food processing in this context. Therefore, the present overview will essentially be based on reports on digestibility measured by nitrogen balance techniques, or by enzymatic in vitro methods.

For obvious practical reasons, enzymatic methods have been used in most studies on protein digestibility and food processing. In some cases rat nitrogen balance techniques were used. However, since we are interested in the protein digestibility of man the question arises: how well do these methods reflect human protein digestibility?. Correlations between published data from animal assays and apparent digestibilities for human subjects were found to be generally poor, ranging from about 0.1 to 0.6 (Ritchey and Taper, 1981). These were data collected from the literature, which means that at least part of the variation comes from the fact that non-identical samples were compared. A few studies have been made in which the digestibility of the same proteins have been compared between humans, rats and in vitro enzyme assays (Bodwell et al., 1980, Rich et al., 1980). The results gave correlations from 0.40 to 0.98 between true digestibility of rats and calculated values of true digestibility of adult humans (Bodwell et al., 1980), and correlations between 0.96 and 1.0 for apparent digestibilities (Rich, 1980). Enzyme assays and results from humans correlate less well, but the results improve when data are grouped with respect to protein origin (Satterlee et al., 1981). This is also true for the rat assays (Bowell et al., 1980; Rich, 1980). Other studies show that enzymatic methods correlate fairly well with rat assays, but that individual values may deviate considerably (Kan and Shipe, 1984; El Malik, 1986; Mitchell and Grundel, 1986; Thresher et al., 1989; Chang et al., 1990).

Although not the objective of most human nitrogen balance studies, the effect of food processing on digestibility may be illustrated by grouping the data (from different sources) given by Hopkins (1981), as shown in table II. Two important observations may be made. The heating of cereals, in the production of ready-to-eat breakfast cereals, is concomitant with a considerable reduction in protein digestibility. Secondly, processing of soybean increases its digestibility. These effects have often been observed with rats and are found in enzyme experiments as will be seen below. We may thus feel confident that studies on the effect of food processing on a least these types of substrates bear relevance to human nutrition, and that the methods used are valid.

Table II. TRUE DIGESTIBILITY BY ADULTS OF PROTEIN IN SOME COMMON FOOD. EFFECT OF PROCESSING

	Digestibility	
	Mean	Range
Corn, whole	87	84-92
Corn, ready-to-eat cereal	70	62-78
Rice, polished	89	82-91
Rice, ready-to-eat cereal	75	63-85
Oatmeal	86	76-92
Oats, ready-to-meat cereal	72	63-89
Soy flour	86	75-92
Soy protein, isolate	95	93-97
Soy protein, isolate, spun	104	101-107

Ranges of values from different reports. From Hopkins,D.T.(1981).

Studies on methodology indicate that both rat assays and enyme assays may be suitable for the evaluation of the effect of food processing on protein digestibility, when the same types of protein are examined before and after some form of treatment (see also Hung et al., 1984). The methods are useful to identify suitable processing conditions, but they clearly do not allow the prediction of the magnitude of the change in digestibiliy that may be expected in humans.

With the above mentioned limitations in mind, published studies on the effect of food processing on protein nutrition were grouped into those experiments showing an increase in digestibility and those showing a decrease in digestibility. This simplified basis for the evaluation of published results was supported by the fact that a wide range of different enzyme methods has been used in these studies, which made comparisons difficult. Nevertheless, the ambition was to specify characteristics of the two groups.

EFFECTS OF FOOD PROCESSING ON PROTEIN DIGESTIBILITY

Enhanced protein digestibility

Table III shows a number of different foods that have been shown to benefit from processing, when assayed for protein digestibility. They include various legumes (African yam beans, fafa beans, moth beans, mucuna beans, mung beans, soybeans), cereals (wheat, sorghum) and seeds (buckwheat seed, fenugreek seed, sunflower seed) and a few food item of animal origin (cheese, pork-loin). Notably, enzymes have been used in most of the studies referred to in table III. The effects of food processing that have been observed in these studies must thus have been directly related to the action of the enzymes in the assay.

A number of possible mechanisms for these enzyme interaction exists. The processing conditions may alter the food structure on a cellular level, enabling the enzyme to gain physical access to the protein. That such an effect may be of importance has been shown by Bradbury et al. (1984). They found that the in vitro protein digestibility of the aleurone layer and grain coat was about 25 % in raw rice, but increased to 65% after cooking. With the help of electron micrographs they concluded that the increase in digestibility was due to the disruption of the cellulosic cell walls at 100 °C (Bradbury et al., 1984). Further, Iskander et al. (1987) showed that the in vitro digestibility of milled wheat varied with particle size. Obviously, particle size and cell structure should be considered in relation to processing and digestibility.

Protein structure is susceptible to food processing. It is generally agreed that some native legume proteins resist proteolysis but are degraded after heating (Liener and Thompson, 1980; Nielsen et al.,1988). The major storage protein phaseolin contributes significantly to this effect, and the conformational changes in the protein molecule induced by heating are directly related to its digestibility (Deshpande and Srinivasan-Damodaran, 1989). Indirect observations also support the relationship between protein structure and digestibility. Strains of maize that differ in amino acid composition also differ in in vitro protein digestibility (Gupta et al., 1987). Germinated fafa beans show increased in vitro protein digestibility and altered functional properties of the protein (Rahma et al., 1987). Fermentation of sorghum and green gram increases protein solubility and protein digestibility in vitro (Chavan et al., 1988). Fermentation of lentils was found to change protein composition and digestibility (Ragaee et al., 1986). There is thus considerable evidence illustrating the importance of plant protein structure in relation to digestibility, and protein structure is a factor that is affected by food processing.

However, the beneficial effect of processing on plant food is more often attributed to the reduction of anti-nutrients and toxicants. Some of these compounds are believed to directly or indirectly affect the protein digestion. In principle, they may interfere with the enzyme activity through complex formation with either the enzyme or the substrate. Compounds that have been ascribed to such effects include specific protease inhibitors, polyphenols (including tannins and saponins), hemagglutinins (lectins, Thompson et al., 1986), phytates (phytic acids and salts) and dietary fiber (Moron et al., 1989). The subject has recently been extensively covered (Friedman, ed., 1986, see Friedman and Gumbman, 1986).

Table III. BENEFICIAL EFFECTS OF FOOD PROCESSING ON PROTEIN DIGESTIBILITY

Food	Process	Digestion assay	References
African yam beans	soaking + cooking/autoclaving	enzyme	Abbey & Berezi, 1988
Buckwheat seed	germination	enzyme	Ikeda, 1984
Buffalo meat /soybean mix	frying	enzyme	Ibrahim et al., 1985
Cheese, processed	emulsifying + heating	enzyme	Ahn and Lee, 1985
Cowpea	boiling	enzyme	Lamena et al., 1987
Cowpea	storage, 20 °C, 7 mo	enzyme	Akinyele and Onigbinde, 1988
Fafa beans	soaking + germination	enzyme	Rahma et al., 1987
Fenugreek seeds	germination	enzyme	Abd El Aal and Rahma, 1986
Fish(dried)/ wheat flour mix	extrusion cooking	enzyme	Battachanya et al., 1988
Guar meal	acidification + roasting	enzyme rat	Kochar, 1985
Maize	cooking	enzyme	Hamaker et al., 1986
Millet flour italian	acid treatment	enzyme	Monteiro et al., 1988
Moth bean	soaking, sprouting, cooking, preassure cooking	enzyme	Kokhar and Chauhan, 1986
Mucuna bean seeds	cooking	enzyme	Ravindran and Ravindran, 1988
Mung beans	boiling	enzyme	Barroga et al., 1985
Mung beans	soaking + cooking	enzyme	Kataria et al., 1989
Pearl millet (Rabach)	fermentation	enzyme	Neerja-Dhankher and Chauhan, 1987
Pork-loin cured	storage, -3°C, 0-12 week	enzyme	Szmanko et al., 1988

(continued)

Table III, (Continued)

Food	Process	Digestion assay	References
Sorghum	steeping + germination	enzyme	Bhise et al., 1988
Sorghum	steaming; fermentation	enzyme	Moharram, 1987
Sorghum	extrusion cooking	enzyme human	Fapojuwo,et al., 1987 MacLean et al., 1983
Sorghum/ green gram	fermentation	enzyme	Chavan et al., 1988
Soybean	germination, autoclaving	rat	Morou-Jimenez, 1985
Soybean (miso)	cooking	enzyme	Nikkuni et al., 1988
Sunflower seed	extraction + protein precipitation	enzyme	Brueckner et al., 1986
Sunflower seeds	roasting 120 °C, 5 min	enzyme	Madhusudhan et al., 1986
Wheat	extrusion cooking	enzyme	Dorokohvich et al., 1987
Wheat, white flour	extrusion cooking, drum-drying	rat	Håkansson et al., 1987
Wheat, whole grain	steam flaking, autoclaving popping	rat	ibid.

The beneficial effect of food processing on digestibility will depend on the relative importance of the different potential enzyme inhibitors in the specific plant food. While most protease inhibitors, lectins and resistant protein structures may be modified (with regard to molecular structure) by relatively mild processing conditions and with the aid of mild agents, including disulfides (Friedman and Gumbman, 1986), it is less likely that polyphenols and phytates are reduced by heat alone. This may however, not be obvious in a rat feeding experiments with processed plant foods containing toxicants, since the elimination of the toxic effect may mask the increase in digestibility. If we are primarily concerned with the enzymatic process of protein digestion and factors that affect this process, enzyme experiments in vitro provide useful data. Enzyme studies by Ikeda et al. (1986) identified the protease inhibitors of buckwheat to be more influential on protein digestibility than tannins, phytates and dietary fibers which were present. Accordingly, in buckwheat, and most likely also in soybeans and dry beans *(Phaseolus vulgaris* L.), such as Great Northern beans (Eicher and Satterlee, 1988), processing which alters the protein structure is beneficial.

On the other hand, the in vitro digestibility of cowpeas was found to be enhanced by soaking, and the effect was assumed to be the result of tannin extraction (Laurena et al., 1986, 1987). Mung beans, which are low in protease inhibitors and lectins, contain tannins that are reduced by simple cooking in water, with a concomitant increase in enzyme digestibility (Baroga et al., 1985). Further, it has been reported that tannins present in black beans (*Phaseolus vulgaris*, L.) affect the in vitro digestibility and *Tetrahymena*-PER (Aw and Swanson, 1985). It has been argued that tannins may not be present in sufficient amounts in plant food to exert an effect in humans (Oh and Hoff, 1986). This assumption is not supported by two studies performed on rats: reduction of polyphenols by acidification did in fact increase protein digestibility in sorghum (Bach-Knudsen et al., 1988), and the addition of the bean broth containing polyphenols to a diet containing the cooked beans did reduce the protein digestibility (Braham and Bressani, 1985).

Phytic acid interacts with positively charged ions and functional groups, and may bind strongly to proteins (Cheryan, 1980). It has been speculated that phytates reduce protein digestibility, and model studies show an effect on pepsin activity (Knuckles et al., 1985; Inagawa et al., 1987). Other studies contradict this. Reddy et al. (1988) found no effect from phytates in the digestibility of proteins of Great Northern bean. Serraino et al. (1985) found no effect from phytic acid in rapseed flour on digestibility in vitro with a pepsin-pancreatin digestion, but observed a reduced rate of release of many essential amino acids. When the same rapseed flours were fed to rats, however, no effect on protein utilization was observed (Thompson and Serraino, 1986). The present evidence indicates that phytates do not contribute significantly to the low digestion that may be observed with plant foods.

Table IV. FOOD PROCESSING FOLLOWED BY REDUCED PROTEIN DIGESTIBILITY

Food	Process	Digestion assay	References
Black beans	storage, 30-40 °C	enzyme	Sievwright and Shipe, 1986
Breakfast cereals	extrusion puffing; flaking + toasting	enzyme	McAuley et al., 1987
Cereals	roasting 220 -280 °C, 2-2.5 min	rat	Nupur-Chopra and Hira, 1986
Chickpea	frying	enzyme	El Faki et al., 1984
Duck	Frozen storage	enzyme	Panov, 1986
Hake meat	heating, >100 °C	enzyme	Seidler, 1987
Parma ham	curing	enzyme	Chizzolini et al., 1985
Peanut protein	radiation	enzyme	Mostafa, 1987
Sorghum	cooking	enzyme	Hamaker et al., 1986
Sorghum (Njali)	cooking/acidification	rat	Bach-Knudsen et al., 1988
Winged bean	oven heating	rat	Kadam, 1987

Reduced protein digestibility

Table IV show a summary of some recent studies that show reduced protein digestibility after food processing. The studies by McAuley et al. (1987) and Nupur-Chopra and Hira (1986) confirm that the processing of ready-to-eat cereals includes heat treatment that is strong enough to decrease digestibility. Other plant foods also suffer from excessive heat or radiation (El Faki, 1984; Mostafa, 1987). It is now well known that strong heat induces crosslinks in protein, either through the formation of isopeptide bonds, or the formation of disulfide bonds, thereby reducing the access of proteolytic enzymes to the protein core (recent review: Erbersdobler, 1989, see also Kella et al., 1986).

Sorghum supplies a large portion of the protein for many people living in semi-arid tropical regions. It has been known for some time that normally cooked sorghum is poorly digested by children, in contrast to other heated grains such as wheat, maize and rice (MacLean, 1981). It is believed that the cooking process itself is responsible for this effect (Bookwalter et al., 1987). The reason may partly be attributed to a rearrangement of certain heat sensitive protein fractions, with the formation of disulfide bonds and reduced digestibility as the result (Hamaker et al., 1984, 1986; Hamaker, 1987). Sorghum contains tannins, and studies in vitro and on rats suggest that they do affect its digestibility (Moharran, 1987; Bach-Knudsen et al., 1988).

Food processing may be used to alleviate the problem of sorghum. Extrusion cooking was found to be an excellent way to significantly increase sorghum protein digestibility in preschool children (MacLean et al., 1983). It is noteworthy that in vitro experiments on extruded sorghum indicate that detoxification of tannins is not the effective mechanism (Fapojuwo et al., 1987). Evidently, the true nature of the sorghum digestibility remains obscure. Other beneficial processing forms for sorghum include steeping followed by germination (Bhise et al., 1988), or steaming and fermentation (Moharram, 1987).

Model studies

A number of model experiments have shown other possible effects of food processing. These are experiments that do not necessarily imitate conditions to be found in real food processing, but they may give the food technologist guidelines when designing new processes. Alkalis are used as a processing aid in the pre-treatment of plant food or in protein isolation (e.g. Brueckner et al., 1986; Yu et al., 1987, Abbey and Berezi, 1988). It is believed that alkalis may affect protein digestibility in at least two different manners: through the formation of racemic amino acids, and through the formation of intra- or intermolecular crosslinks (reviews: Friedman and Gumbman, 1984; Man and Bada, 1987). The effect might be directly related to impaired enzymic hydroysis of alkali-treated proteins. D-amino acids present in the peptide chain obstruct protein digestion in vitro, as does the formation of lysino-alanine (Abe et al., 1984; Savoie, 1984; Friedman, 1985). Some recent experimental evidence supports the notion that severe alkali treatment affects digestibility both in vitro and in rats (Chung et al., 1986; Chang et al., 1990).

The seeds of the Cruciferae family (e.g. rapseed) contain many kinds of glucoseinolates, that may be degraded to alkyl isothiocyanates by enzymes (Tookey et al., 1980). This is of some concern, since isothiocyanates may form adducts with compounds having nucleophilic functional groups. Kawakishi and Namiki (1982) found that allyl isothiocyanate reacts with cystine and possibly other sites on the protein molecule under mild conditions.

Digestibilities with trypsin and chymotrypsin were markedly reduced (Kawakishi and Kaneko, 1987). The reaction proceeds under relatively mild conditions (40 °C up to about 100 hours), and may pose a problem in the utilization of Brasscia seed meal.

Another possible reason for a reduced digestibility in heated samples containing carbohydrates is the formation of enzyme inhibitors in the Maillard reaction (Öste and Sjödin, 1984). Such compounds might inhibit the action of, in particular, aminopeptidase N, which is a microvilli enzyme that may be essential in the final protein hydrolysis in the small intestine (Öste et al., 1986). Some aromatic, heterocyclic Maillard reaction compounds were found to be inhibitors (Öste et al., 1987). One of them, 2-formyl-5(hydroxymethyl)pyrrole-1-norleucine, was recently identified in some heated foods in concentrations ranging from 1 - 200 ppm (Chiang, 1988). In the study by Öste et al. (1987), a concentration of 3,000 ppm of this compound in the diet was found to exert an effect on the protein utilization of rats. Chiang (1988) performed the analysis on the free compound, which must have been formed from free lysine. In food, protein-bound amino groups, in particular the epsilon-amino group of lysine, should be the dominant amino reactant in the Maillard reaction. It is reasonable to assume that additional 2-formyl-5(hydroxymethyl)-pyrrole-1-norleucine is formed with protein-bound lysine, and that the total content in heated foods may be considerably higher than those reported by Chiang (1988). Considering that more than one inhibitor may be formed (Öste et al., 1987), the combined effect of such compounds may be sufficient to affect the protein digestion of severely heated foods.

Protein crosslinks are formed by some intermediate Maillard reaction products as the dicarbonyl 3-deoxyglucosone (Kato et al., 1989). Such crosslinks reduce in vitro digestibility (Kato et al., 1986) and may contribute to the impaired digestion of strongly heated food. A naturally occurring dicarbonyl, gossypol (a dialdehyde), is present in cottonseed flours and may, upon heating, crosslink protein and reduce the digestibility in rats (Anderson et al., 1984). Another natural bifunctional compound is chlorogenic acid, which may be found in crude leaf protein concentrates. Barbeau and Kinsella (1985) have demonstrated that if chlorogenic acid is covalently bound to proteins, the in vitro digestibility is decreased.

Chemical modification of proteins may be used to improve the functional properties, such as emulsifying and foaming capacities, and thereby extend the use of some food protein and protein sources (Kinsella, 1982). Further, the nutritional quality may be improved through fortification by covalently attaching limiting amino acids (Noack and Hajos, 1984; Tokioka et al., 1985; Gärtner and Puigserver, 1986; Iwamin and Yasumoto, 1986; Bercovici et al., 1989).

Although these techniques have not gained practical applications to any degree so far, they may be an attractive choice for the food technologist in the future. These forms of treatment induce changes not only on the tertiary and secondary structure of the protein, but also on the primary structure, that is, they affect the covalent bonds of the amino acids of the peptide chain. Obviously, there is a chance that this might affect the digestibility of the protein, and the necessity to check digestibility seems important.

The addition of methionine into the peptide chain of proteins has been performed with enzymes, employing the plastein reaction (Noack and Hajos, 1984; Tokioka et al., 1985). Not surprisingly, the reversed reaction, hydrolysis by enzymes, was easily achieved, indicating good nutritional value of the substrate. Another technique that may be used to modify protein structure is the transglutaminase reaction. The carboxamide group of protein-bound glutamine is used by the enzyme to form isopeptide bonds with added amino acids. Iwamin and Yasumoto (1986) used this method to enrich wheat gliadin with lysine. Rat feeding experiments with the modified gliadin showed no adverse effects on digestibility. Gaertner and Puigserver (1986) used N-carboxy-methionine anhydride to covalently attach methionine to casein. This did not decrease the enzymatic hydrolysis of the casein moiety, and the attached polymethionine chain was readily hydrolyzed, although the release of methionine was inversely correlated to chain length. The same method was used to form a branched chain polypetide, multi-oligo(L-methionyl)poly-L-lysine (Bercovici et al., 1989.). This purely synthetic substrate was completely bioavailable, as confirmed by rat feeding experiments. These results show promising new approaches for the nutritional enrichment of food of lower quality. By using polymerized amino acids instead of the same amount of monomers, the heat stability of the enriched food should increase, since the amount of reactive amino groups is considerably reduced.

Techniques that are employed with the objective to improve functional properties also affect basic protein structure, and the digestibilities of modfied proteins should be checked. Seidler at al. (1985) found that the spinning and additional crosslinking with dialdehyde starch did not reduce the digestibility of casein-*Vicia faba* protein solutions by rats. Also, phosphorylation of soy protein with $POCl_3$ (Hirotsuka et al., 1984) and the succinylation of field bean protein (Proll and Schwenko, 1983) can be done conveniantly and without a reduced nutritional value. Obviously, chemical protein modification, which will manipulate basic structures in the protein, may be performed without adverse effects on digestion. This should, of course, always be experimentally confirmed.

CONCLUSIONS

Most data on the effect of food processing on protein digestion arise from in vitro studies using enzymes. The validity of such assays, when it comes to tha prediction of human protein digestion, may be questioned. However, when the data are examined together, and occasionally are supported by rat assays and even nitrogen balance assay on humans, some general principles may be perceived.

Processing techniques that have proven to be beneficial for the diges tion of plant food containing antinutrients include one or more of the following steps:

- a direct extraction of the antinutrient, when necessary with the addition of salt and proper adjustment of pH.
- elimination of the antinutrient by induced, altered plant metabolism i.e. germination.
- elimination of the antinutrient through consumption by added or naturally present microorganisms, i.e. fermentation.
- inactivation of the antinutrient through structural modification by heat or other means.

In addition, the processing may facilitate protein digestion by:

- modifying cellular and/or protein structure through heat
- modifying cellular and/or protein composition through germination or fermentation.

Comparatively few studies show a reduced protein digestibility after pro cessing. When it happens, evidence indicates that altered protein structure is the main reason. This may be due to:

- strong heating which induces crosslinks. This may be enhanced by the presence of carbohydrates.
- mild heating of certain sensitive proteins (e.g. kafirins in sorghum)
- storage which alters protein structure in an unfavorable way through metabolism, or trough the formation of disulfide crosslinks

Finally, model studies indicate that the food technologist also should be aware of the detrimental effects of:

- conditions favoring the Maillard reaction. Protein crosslinking may occur before browning is plainly visible.
- severe or prolonged alkaline conditions.
- some chemically reactive, naturally occurring plant constituents.

REFERENCES

Abbey, B. W., and Berezi, P. E. (1988). Influence of processing on the digestibility of the African yam bean (Sphenostylis stenocarpa) flour. Nutr. Repts. Int., 37 (4), 819-827.

Abd-El-Al, M. H., and Rahma, E. H. (1986). Changes in gross chemical composition with emphasis on lipid and protein fractions during germination of fenugreek seeds. Food Chem., 22 (3), 193-207.

Abe, K., Homma, S., Fujimaki, M., and Arai, S. (1984). Administration of lysinoalanine-containing proteins to rats and the characterization of their small intestinal contents. Agric. Biol. Chem., 48 (3), 573-578.

Ahn, H. I., and Lee, B. O. (1985). Studies on the fabrication of processed cheese made from milk powder. I. Melting characteristics. (In: "Proceedings of the 3:rd AAAP Animal Science Congress", Vol 2, (see FSTA (1987) 19 3S66).), 1146-1149.

Akinyele, I. O., and Onigbinde, A. O. (1988). Stability of protein digestibility and composition of cowpea (Vigna unguiculata L. Walp.) during sealed storage at different temperatures. Int. J. Food Sci. Tech., 23 (3), 293-296.

Andersson, P. A., Sneeed, S. M., Skurray, G. R. and Carpenter, K. J. (1984). Measurement of Lysine Damage in Porteins Heated with Gossypol. J. Agric. Food Chem., 32, 1048-1053.

Aw, T. L., and Swanson, B. G. (1985). Influence of tannin on Phaseolus vulgaris protein digestibility and quality. J. Food Sci., 50 (1), 67-71.

Bach-Knudsen, K. E. B., Munck, L., and Eggum, B. O. (1988). Effect of cooking, pH and polyphenol level on carbohydrate composition and nutritional quality of a sorgum (*Sorghum bicolor*, (L.) Moench) food, ugali. Br. J. Nutr., 59(1), 31-47.

Barbeau, W. E., and Kinsella, J. E. (1985). Effect of free and bound chlorogenic acid on the in vitro digestibility of ribulose bisphosphate carboxylase from spinach. J. Food Sci., 50 (4), 1083-1087.

Baroga, C. F., Laurens, A. C., and Mendoza, E. M. T. (1985). Effect of condensed tannins on the in vitro protein digestibility of mung bean (*Vigna radiata* (L.) Wilczek). J. Agric. Food Chem., 33 (6), 1157-1159.

Bercovici, D., Gaertner, H. F., and Puigserver, A. J. (1989). Poly-L-lysine and Multioligo(L-methionyl)poly-L-lysine as Nutritional Sources of Essential Amino Acids. J. Agric. Food Chem., 37, 873-877.

Bhattacharya, S., Das, H., and Bose, A. N. (1988). Effect of extrusion process variables on in-vitro protein digestibility of fish-cheat flour blends. Food Chem., 28 (3), 225-231.

Bhise, V. J., Chavan, J. K., and Kadam, S. S. (1988). Effects of malting on proximate composition and in vitro protein and starch digestibilites on grain sorghum. J. Food Sci. Tech. India, 25 (6), 327-329.

Bodwell, C.E., Satterlee, L.D. and Hackler, L.R. (1980). Protein digestibility of the same protein preparations by human and rat assays and by in vitro enzymic digestion methods. Am. J. Clin. Nutr., 33(3), 677-686.

Bookwalter, G. N., Kirleis, A. W., and Mertz, E. T. (1987). In vitro digestibility of protein in milled sorhum and other processed cereals with and without soy-fortification. J. Food Sci., 52 (6), 1577-1579.

Bradbury, H. H., Collins, J. G., and Pyliotis, N. A. (1984). Digestibility of proteins of the histological components of cooked and raw rice. Br. J. Nutr., 52 (3), 507-513.

Braham, J. E., and Bressani, R. (1985). Effect of bean broth on the nutritive value and digestibility of beans. J. Sci. Food Agric., 36 (10), 1028-1034.

Bressani, R. (1977). Protein Supplementation and Complementation. In: "Evaluations of proteins for humans", C. E. Bodwell, ed., Avi, Westport, Connecticut, pp. 204-232.

Brueckner, J., Proll, J., Noack, J., Mieth, G., and Schadereit, R. (1986). (Studies on the quality of protein-enriched products from sunflower seed. I. Influence of different production processes on the nutritive value of meals and protein isolates.) Untersuchung zur Qualitaet proteinangereicherter Produkte aus Sonnenblumensamen. I. Einfluss unterschiedlicher Gewinnungsverfahren auf den nutritiven Wert von Mehlen und Proteinisolaten. Nahrung, 30 (7), 693-699.

Chang, H. I., Catignani, G. L. and Swaisgood, H. E. (1990). Protein Digestibility of Alkali- and Fructose-Treated Protein by Rat True Digestibility Assay and by the Immobilized Digestive Enzyme Assay System. J. Agric. Food Chem., 38 (4), 1016-1018.

Chiang, G. H. (1988). High-Performance Liquid Chromatographic Determination of epsilon-Pyrrole-lysine in Processed Food. J. Agric. Food Chem., 36, 506-509.

Chung, S. Y., Swaisgood, H. E., and Catignani, G. L. (1986). Effects of alkali treatment and heat treatment in the presence of fructose on digestibility of food proteins as determined by an immobilized digestive enzyme assay (IDEA). J. Agric. Food Chem., 34 (3), 579-584.

Chavan, U. D., Chavan, J. K., and Kadam, S. S. (1988). Effect of fermentation on soluble proteins and in vitro protein digestibility of sorghum, green gram and sorghum-green gram blends. J. Food Sci., 53 (5), 1574-1575.

Cheryan, M. (1980). Phytic acid interactions in food systems. CRC Crit. Rev. Food Sci. Nutr., 13, 297-335.

Chizzolini, R., Dazzi, G., Parolari, G., and Bellatti, M. (1985). (Physical and chemical changes in proteins during maturation of Parma ham. III Nutritional aspects.). Viandes Produits Carnes, 6 (6), 226-228.

Deshpande, S. S., Srinivasan-Damodaran. (1989). Structure-digestibility relationship of legume 7S proteins. J. Food Sci., 54 (1), 108-113.

Dorokhovich, A. N., Ostrik, A. S., and Tsirik, N. A. (1987). (Extruded groats - the new non-traditional raw material in a confectionery production.). Pishchevaya Promychlennost', 2, 34-35.

Eicher, N. J., and Satterlee, L D. (1988). Nutritional quality of Great Northern bean proteins processed at varying pH. J. Food Sci., 53 (4), 1139-1143.

El Faki, H. A., Venkataraman, L. V., and Desikachar, H. S. R. (1984). Effect of processing on the in vitro digestibility of proteins and carbohydrates in some Indian legumes. Qualitas Plantarum Plant Foods Human Nutr., 34 (2), 127-133.

Elmalik, M. M. (1986). Studies on the nutritional quality of sorghum grain. Dissertation Abstr. Int. B, 47 (1), 16

Erbersdobler, H. F. (1989). Protein Reactions During Food Processing and Storage - Their Relevance to Human Nutrition. In: "Nutritional Impact of Food Processing", J. C. Somogyi, H. R. Muller, eds., Karger, Basel, pp. 140-155.

Fapojuwo, O. O., Maga, J. A., and Jansen, G. R. (1987). Effect of extrusion cooking on in vitro protein digestibility of sorghum. J. Food Sci., 52 (1), 218-219.

Friedman, M., Grosjean, O.-K., and Zahnley, J.C. (1985). Carboxypeptidase Inhibition by Alkali-Treated Food Proteins. J. Agric. Food Chem., 33, 208-213.

Friedman, M., and Gumbman, M.R. (1986). Nutritional improvement of legume protein through disulfide interchange. In: "Nutritional and toxicological significance of enzyme inhibitors in foods", M. Friedman, ed., Plenum, New York, pp.357-387.

Friedman, M., Gumbman, M.R. and Masters, P. M. (1984). Protein-alkali reactions: chemistry, toxicology, and nutritional consequences. In: "Nutritional and toxicological aspects of food safety", M. Friedman, ed., Plenum, New York, pp.367-412.

Gaertner, H. F., and Puigserver, A. J. (1986). Hydrolysis of poly-L-methionyl proteins by some enzymes of the digestive tract. J. Agric. Food Chem., 34 (2). 291-297.

Gupta, H. O., Singh, J., Ram, P. C., and Singh, R. P. (1987). Chemical and biochemical investigations on normal, chalky and modified opaque-2 strains of maize. J. Food Sci. Tech. India, 24 (4), 184-186.

Håkansson, B., Jägerstad, M., Öste, R., Åkesson, B., and Jonsson, L. (1987). The Effects of Various Thermal Processes on Protein Quality, Vitamins and Selenium Content in Whole -grain Wheat and White Flour. J. Cereal Sci. 6, 269-282.

Hamaker, B. R. (1987). Effect of heat and reducing agents on in vitro digestibility of sorghum proteins. Dissertation Abstr. Int. B, 48 (1), 13.

Hamaker, B. R., Kirleis, A. W., Mertz, E. T., and Axtell, J. D. (1986). Effect of cooking on the protein profiles and in vitro digestibility of sorghum and maize. J. Agric. Food Chem., 34 (4), 647-649.

Hamaker, B. R., Mertz, E. T., and Axtell, J. D. (1984). Landry-Moureaux protein profile of sorghum and its pepsinindigestible residue. Federation Proc., 43 (3), 466.

Hirotsuka, M., Taniguchi, H., Narita, H., and Kito, M. (1984). Functionality and digestibility of a highly phosphorylated soybean protein. Agric. Biol. Chem., 48 (1), 93-100.

Hopkins, D. T. (1981). Effects of variation in protein digestibility. In: "Protein Quality in Humans:Assessment and in vitro estimation", C. E. Bodwell, J. S. Adkins, and D. T. Hopkins, eds., Avi, Westport, Connecticut, pp. 169-190.

Hung, N. D., Cseke, E., Vas, M., and Szabolcsi, G. (1984). Processed protein foods characterized by in vitro digestion rates. J. Food Sci., 49 (6), 1543-1546.

Hurrel, R. F.; Finot, P.A. (1985). Effects of food processing om protein digestibility and amino acid availability. In: "Digestibility and amino acid availability in cereals and oilseeds", J. W. Finley , and D. T. Hopkins, eds., Am. Ass. Cereal Chem., St. Paul, pp. 233-246.

Ibrahim, A. A., Ibrahim, M. A., El-Hashimy, F. S. A., and Morsi, M. K. S. (1985). Protein quality of soybean-meat blends. Egyptian J Food Sci., 13 (2), 193-199.

Ikeda, K., Arioka, K., Fujii, S., Kusano, T., and Oku, M. (1984). Effect on buckwheat protein quality of seed germination and changes in trypsin inhibitor content. Cereal Chem., 61 (3), 236-238.

Ikeda, K., Oku, M., Kusano, T., and Yasumoto, K. (1986). Inhibitory potency of plant antinutrients towards the in vitro digestibility oc buckwheat protein. J. Food Sci., 51 (6), 1527-1530.

Inagawa, J., Kiyosawa, I., Nagasawa, T. (1987). (Effects of phytic acid on the digestion of casein and soybean protein with trypsin, pancreatin and pepsin.). J. Jap. Soc. Nutr. Food Sci., 40(5), 367-373.

Iskander, F. Y., Morad, M. M., Klein, D. E., and Bauer, T. L. (1987). Determination of protein and 11 elements in six milling fractions of two wheat varieties. Cereal Chem., 64 (4), 285-287.

Iwami, K., and Yasumoto, K. (1986). Amine-binding capacities of food proteins in transglutaminase reaction and digestibility of wheat gliadin with epsilon-attached lysine. J. Sci. Food Agric., 37 (5), 495-503.

Kadam, S. S., Smithard, R. R., Eyre, M. D., and Armstrong, D. G. (1987). Effects of heat treatments of antinutritional factors and quality of proteins in winged bean. J. Sci. Food Agric., 39 (3), 267-275.

Kan, T. M., and Shipe, W. F. (1984). Enzyme diafiltration technique for in vitro determination of protein digestibility and availability of amino acids of legumes. J. Food Sci., 49 (3), 794-798.

Katarina, A., Chauhan, B. M., and Darshan-Punia. (1989). Antinutrients and protein digestibility (in vitro) of mungbean as affected by domestic processing and cooking. Food Chem., 32 (1), 9-17.

Kato, H., Chuyen, N. van, Utsunomiya, N., and Okitani, A. (1986). Changes of amino acids composition and relative digestibility of lysozyme in the reaction with alpha-dicarbonyl compounds in aqueous system. J. Nutr. Sci. Vitaminology, 32 (1), 55-65.

Kato, H., Hayase, F., Dong Bum Shin, Onimoni, M., and Shigeaki, S. (1989). 3-dexyglucosone, an intermediate product of the Maillard reaction. In: "The Maillard Reaction in Aging, Diabetes, and Nutrition", J. W. Baynes and V. M. Monnier, eds., Alan R. Liss, New York, pp. 69-84.

Kawakishi, S., and Kaneko, T. (1987). Interaction of proteins with allyl isothiocyanate. J. Agric. Food Chem., 35 (1), 85-88.

Kawakishi, S., and Namiki, M. (1982). Oxidative cleavage of the disulfide bond of cysteine by allyl isothiocyanate. J. Agric. Food Chem. 30, 618-620.

Kella, N. K. D., Barbeau, W. E., and Kinsella, J. E. (1986). Effect of oxidative sulfitolysis of disulfide bonds of glycinin on solubility, surface hydrophobicity, and in vitro digestibility. J. Agric. Food Chem., 34 (2), 251-256.

Khokhar, S., and Chauhan, B. M. (1986). Effect of domestic processing and cooking on in vitro protein digestibility of moth bean. J. Food Sci., 51 (4), 1083-1084.

Kinsella, J. E. (1982). Relationship between structure and functional properties of food proteins. In: "Food Proteins", P. F. Fox, J. J. Condon, eds., Applied Sciences, London, pp. 51-103.

Knuckles, B. E., Kuzmicky, D. D., and Betschart, A. A. (1985). Effect of phytate and partially hydrolyzed phytate on in vitro protein digestibility. J. Food Sci., 50 (4), 1080-1082.

Kochar, G. K. (1985). Trypsin-inbihibion and protein digestibility of raw and detoxified guar meal (Cyamopsis tetragonoloba). Indian J. Animal Res., 19 (1), 45-47.

Laurena, A. C., Garcia, V. V., and Mendoza, E, M. T. (1986). Effects of soaking in aqueous acidic and alkali solutions on removal of polyphenols and in-vitro digestibility of cowpea. Qualitas Plantarum Plant Foods Human Nutr., 36 (2), 107-118.

Laurena, A. C., Garcia, V. V., and Mendoza, E. M. T. (1987). Effects of heat on the removal of polyphenols and in vitro protein digestibility of cowpea. Qualitas Plantarum Plant Foods Human Nutr., 37 (2), 183-192.

Liener, I. E., Thompson, R. M. (1980). *In vitro* and *in vivo* studies on the digestibility of the mayor storage protein of the navy bean (*Phaseolus vulgaris*). Qual. Plant. Plant Foods Hum. Nutr. 30, 13-25.

MacLean, W.C., Lopez de Romana, G., Gastanaduy, A., and Graham, G.C. (1983). The Effect of Decortication and Extrusion on the Digestibility of Sorghum by Preschool Children. J. Nutr., 113, 2071-2077.

MacLean, W.C., Lopez de Romana, G., Placko, R.P., and Graham, G.C. (1981). Protein Quality and Digestibility of Sorghum in Preschool Children: Balance Studies and Plasma Free Amino Acids. J. Nutr., 111, 1928-1936.

Madhusudhan, K. T., Shamanthaka-Sastry, M. C., and Srinivas, H. (1986). Effect of roasting on the physico-chemical properties of sunflower proteins. Lebensmittel Wissenschaft Tech., 19 (4), 292-296.

Man, E. H., and Bada, J.L. D-Amino Acids. In: "Annual Review of Nutrition", R. E. Olson, E. Beutler, and H. P. Broquist, eds., Annual Reviews, Palo Alto, pp. 209-225.

McAuley, J. A., Kunkel, M. E., and Acton, J. C. (1987). Relationships of available lysine to lignin, color and protein digestibility of selected wheat-based breakfast cereals. J. Food Sci., 52 (6), 1580-1582.

Mitchell, G.V., and Grundel, E. (1986). Nutritional value of proteins powdered infant formula: In vitro and in vivo methods. J. Agric. Food Chem., 34, 650-653.

Moharram, Y. G. (1987). Sorghum: properties and tannin deactivation treatments. Alexandria J. Agric. Res., 32 (1), 215-226.

Monteiro, P. V., Hara-Gopal, D., Virupaksha, T. K., and Geeta-Ramachandra. (1988). Chemical composition and in vitro protein digestibility of Italian millet (Setaria italica). Food Chem., 29 (1), 19-26.

Moron, D., Melito, C., and Tovar, J. (1989). Effect of indigestible residue from foodstuffs on trypsin and pancreatic alfa-amylase activity in vitro. J. Agric. Food Chem., 47(2), 171-179.

Moron Jimenez, M. J., Elias, L. G., Bressani, R., Navarrete, D. A., Gomez-Brenes, R., and Molina, M. R. (1985). (Biochemical and nutritional studies on germinated soybeans.). Archivos Latinoamericanos Nutr., 35 (3), 480-490.

Mostafa, M. M. (1987). Nutritional aspects of thermal and irradiation processing of peanut kernels and their oil. Food Chem., 26 (1), 31-35.

Neerja-Dhankher, and Chauhan, B. M. (1987). Effect of temperature and period of fermentation on protein and starch digestibility (in vitro) of Rabadi - a pearl millet fermented food. J. Food Sci., 52 (2), 489-490.

Nielsen, S. S., Deshpande, S. S., Hermodson, M. A., and Scott, M. P. (1988). Comparative digestibility of legume storage proteins. J. Agri. Food Chem., 36 (5), 896-902.

Nikkuni, S., Okada, N., and Itoh, H. (1988). Effect of soybean cooking temperature on the texture and protein digestibility of miso. J. Food Sci., 53 (2), 445-449.

Noack, J., and Hajos, G. (1984). The enzymic in vitro digestibility of mehionine-enriched plastein. Acta Alimentaria, 13 (3), 205-213.

Nupur-Chopra, and Hira, C. K. (1986). Effect of roasting on protein quality of cereals. J. Food Sci. Tech. India, 23 (4), 233-235.

Öste, R. E., Dahlqvist, A., Sjöström, H., Noren, O., and Miller, R. (1986). Effect of Maillard reaction products on protein digestio. In vitro studies. J. Agric. Food Chem., 34, 355-358.

Öste, R.E., Miller, R., Sjöström, H., and Noren, O. (1987). Effect of Maillard Reaction Products on Protein Digestion. Studies on Pure Compounds. J. Agric. Food Chem., 35, 938-942.

Öste, R.E., Sjödin, P. (1984). Effect of Maillard Reaction Products on Protein Digestion. In vivo studies on rats. J. Nutr., 114, 2228-2234.

Oh, H. I., and Hoff, J. E. (1986). Effect of condensed grape tannins on the in vitro activity of digestive proteases and activation of their zymogens. J. Food Sci., 51 (3), 577-580.

Panov, V. P. (1986). The influence of storage time of frozen duck meat on its biological value. Proc. European Meeting Meat Res. Workers, 32, Vol. III, 8:21, 72-74.

Proll, J., and Schwenke, K. D. (1983). Chemical modification of proteins. IX. Nutritive evaluation of succinylated faba bean proteins (Vicia faba L.). Nahrung, 27 (3), K9-K11.

Ragaee, S. M. El-Banna, A. A., Damir, A. A., Mesallam, A. S., and Mohamed, M. S. (1986). Effect of natural lactic acid fermentation on amino acids content and in-vitro digestibility of lentils. Alexandria Sci. Exch., 7 (2), 217-224.

Rahma, E. H., El-Bedawey, A. A., El-Adawy, T. A., and Goma, M. A. (1987). Changes in chemical and antinutritional factors and functional properties of faba beans during germination. Lebensmittel Wissenshaft Tech., 20 (6), 271-276.

Ravindran, G., and Ravindran, V. (1988).Nutritional and Anti-nutritional Characteristics of Mucuna (*Mucuna utilis*) Bean Seeds. J. Sci. Food Agri., 46 (1), 71-79.

Reddy, N. R., Sathe, S.K., and Pierson, M. D. (1988). Removal of Phytate from Great Northern Beans (*Phaesolus vulgaris* L.). J. Food Sci. 53(1), 107-110.

Rich, N., Satterlee, L.D., and Smith, J. L. (1980). A comparison of *in vivo* apparent protein digestibility in man or rat to *in vitro* protein digestibility as determined using human and rat pancreatins and commercially available proteins. Nutr. Rep. Int., 21, 285-300.

Richey, S. J., and Taper, L. J. (1981). Estimating protein digestibility for humans from rat assays. In: "Protein Quality in Humans:Assessment and in vitro estimation", C. E. Bodwell, J. S. Adkins, and D. T. Hopkins, eds., Avi, Westport, Connecticut, pp. 306-315.

Sarwar. G. (1987). Digestibility of protein and bioavailability of amino acids in foods. Effects on protein quality assessment. World Review Nutr. Dietetics, 54, 26-70.

Satterlee, L. D., Kendrick, J. G., Jewell, D. K., and Brown, W. D. (1981). Estimating apparent protein digestibility from in vitro assays. In: "Protein Quality in Humans:Assessment and in vitro estimation", C. E. Bodwell, J. S. Adkins, and D. T. Hopkins, eds., Avi, Westport, Connecticut, pp. 316-339.

Savage, G. P. (1989). Influence of tannin-binding substances on the quality of yellow and brown sorghum. Nutr. Repts. Int., 39 (2), 359-366.

Savoie, L. (1984). Effect of protein treatment on the enymatic hydrolysis of lysinoalanine and other amino acids. In: "Nutritional and toxicological aspects of food safety", M. Friedman, ed., Plenum, New York, pp.413-422.

Seidler, T. (1987). Effects of additives and thermal treatment on the content of nitrogen compounds and the nutritive value of hake meat. Nahrung, 31 (10), 959-970

Seidler, W., Bergner, H., Simon, O., Schmandke. H. (1985). Effect of Spinning and Additional Cross-Linking with Dialdehyde Starch or Aluminum Ions on the Digestibility and Quality of Casein-*Vicia faba* Protein Isolate Mixtures. Ann. Nutr. Metab. 29, 184-188.

Serraino, M. R., Thompson, L. U., Savoie, L., and Parent, G. (1985). Effect of phytic acid on the in vitro rate of digestibility of rapeseed protein and amino acids. J. Food Sci., 50 (6), 1689-1692.

Sievwright, C. A., and Shipe, W. F. (1986). Effect on storage conditions and chemical tratments on firmness, in vitro protein digestibility, condensed tannins, phytic acid and divalent cations of cooked black beans (Phaseolus vulgaris). J. Food Sci., 51 (4), 982-987.

Szmanko, T., Duda, Z., Kajdan, L., and Kubis, B. (1988). Storage of selected sort of processed meat product at cryoscopic temperature - an attempt at energy conservation. Changes in proteins, amino acids balance and in vitro digestibility of cured smoked raw pork-loin. Acta Alim. Pol., 14 (2), 145-146.

Thompson, L. U., and Serraino, M. R. (1986). Effect of phytic acid reduction on rapeseed protein digestibility and amino acid absorption. J. Agric. Food Chem., 34 (3), 468-469.

Thompson, L. U., Tenebaum, A. V., and Hui, H. (1986). Effect of lectins and the mixing of proteins on rate of protein digestibility. J. Food Sci., 51 (1), 150-162.

Tookey, H. L., VanEtten, C. H., and Daxenbichler, M. E. (1980). Glucosinolates In: "Toxic constituents of plant foodstuffs",I. E. Liener, ed., Academic, New York, pp. 103-142.

Tresher, W.C.; Swaisgood, H. E. and Catignani, G. L. (1989). Digestibility of protein and amino acids in various foods as determined by an immobilized digestive enzyme assay (IDEA). Plant Foods Hum. Nutr., 39, 59-62.

Tokioka, J., Takahashi, M., and Kametaka, M. (1985). (In vitro digestibility of suplsup5N-methionine invorporated into soy protein), J. Japanese Soc. Nutr. Food Sci., 38 (6), 435-445.

WHO, (1985). FAO/WHO/UNU Joint Expert Consultation: Energy and protein requirements. WHO techn. rep. 724, Geeneva.

Yu, R. S. T., Kyle, W. S. A., Hung, T. V., and Zeckler, R. (1987). Characterisation of aqueous extracts of seed proteins of Lupinus albus and Lupinus angustifolius. J. Sci. Food Agric., 41 (3), 205-218.

28

AMINO ACID RATINGS OF DIFFERENT FORMS OF INFANT FORMULAS BASED ON VARYING DEGREES OF PROCESSING

G. Sarwar

Bureau of Nutritional Sciences, Food Directorate
Health Protection Branch, National Health and Welfare
Tunney's Pasture, Ottawa, Ontario, Canada K1A 0L2

ABSTRACT

Amino acid profiles, protein digestibility and/or amino acid bioavailability for the various forms (powder, liquid concentrate, ready-to-use, etc.) of infant formulas (involving varying degrees of heat processing during preparation) have been determined. Amino acid scores (based on the single most limiting amino acid) were calculated by comparing the essential amino acid data with that of human milk. Amino acid scores were multiplied by total protein (g/100 kcal) to obtain amino acid ratings, which take into account both quality and quantity of protein. Amino acid scores for milk- and soy-based formulas ranged from 49 to 90 and 59 to 81%, respectively, due to deficiencies in methionine plus cystine and/or tryptophan. The deficiency in the limiting amino acids was more marked in liquid concentrate than powder prepared by the same manufacturer. Because of significantly higher total protein contents (g/100 kcal) of soy- (2.65-3.68) and milk-based (2.20-2.95) formulas compared to human milk (1.5), the relative amino acid ratings (human milk= 100) of all formulas except two milk-based liquid concentrates and one ready-to-feed (with values of 77-87%) were > 100%. When corrected for protein digestibility, the relative amino acid ratings for all four liquid concentrates were < 100%. Lower levels of digestible protein and bioavailable amino acids in liquid concentrate compared with powder (prepared by the same manufacturer) would suggest that inferior protein quality of liquid concentrates may be due to more severe heat treatment involved in their preparation.

INTRODUCTION

When foods containing high levels of lysine-rich proteins, lactose (and other reducing sugars) and unsaturated fats (such as milk-based infant formulas) are subjected to moderate heat treatment (such as drying of milk), some reduction in amino acid availability and protein digestibility may occur due to the formation of Maillard compounds, oxidized forms of sulfur amino acids and cross-linked peptide chains such as lysinoalanine (Cheftel, 1979; Finot, 1983; Hurrell and Carpenter, 1981; Pompei et al., 1987; Sarwar et al., 1988).

The loss in protein quality of different forms of milk-based infant formulas would depend on the extent and severity of heat treatment involved in their preparation, and on the conditions of storage. (Pompei et al., 1987; Sarwar et al., 1988). The preparation of liquid concentrates requires more heat treatment than that of powders (Packard, 1982). Moreover, the liquid concentrates are formulated to be concentrated sources of protein, lactose and fat containing unsaturated fatty acids. The thermal and concentration effects may result in more protein damage by Maillard and oxidation reactions in liquid concentrates than in powders of the same manufacturer. Information on the comparative nutritional value of the different forms of milk-based infant formulas is limited.

Nutritional and Toxicological Consequences of Food Processing
Edited by M. Friedman, Plenum Press, New York, 1991

Pompei et al. (1987) published an excellent paper comparing protein nutritional value of various forms of milk-based formulas sold in several Western European countries. In their study, the formation of lysinoalanine was found to be one of the most sensitive predictor of protein damage in infant formulas caused by heat processing. They found that powder forms contained only very small amounts of lysinoalanine, while liquid (ready-to-feed) forms of milk-based formulas contained up to ten times more lysinoalanine (more than 1 g/kg protein). This suggested more severe heat processing in the preparation of liquid forms compared to powders. Amino acid bioavailability and protein quality of the various forms of infant formulas have been studied at the Health Protection Branch Laboratories (Sarwar et al., 1988; Sarwar et al., 1989a; Sarwar et al. 1989b). Values for true digestibility of protein and essential amino acids in liquid concentrates were up to 13% lower than those in powders (Sarwar et al., 1989a). Similarly, protein quality (as predicted by rat growth methods) of liquid concentrates was up to 25% lower than that of powders.

The protein efficiency ratio (PER) method is still widely used for evaluating protein quality of foods, including infant formulas, but it is the poorest of current animal tests in meeting criteria of a valid routine test. A detailed examination of existing animal assays and more promising amino acid scoring methods was carried out by an Ad Hoc Working Group on Protein Quality Measurement for the Codex Committee on Vegetable proteins during the last five years (Codex Alimentarius Commission, 1989). Factors such as inadequacies of PER and other animal assays, progress made in standardising methods for amino acid analysis and protein digestibility, information on digestibility of protein and bioavailability of amino acids in a variety of foods, and reliability of human amino acid requirements and scoring patterns were evaluated.

Based on this examination and review, amino acid score, corrected for true digestibility of protein, was recommended to be the most suitable routine test for predicting protein quality of foods for humans. Amino acid scores adjusted for true digestibility of protein (as determined by rat balance or fecal method) were termed "protein digestibility-corrected amino acid scores". More recently, a Joint FAO/WHO Expert Consultation on Protein Quality Evaluation agreed that the protein digestibility-corrected amino acid score is the most suitable approach for routine evaluation of protein quality for humans, and recommended the adoption of this method as an official method at the international level (FAO/WHO, 1990). The Consultation also recognized that the amino acid scoring pattern proposed by FAO/WHO/UNU (1985) for children of preschool age is at present the most suitable pattern for calculating scores for all age groups except infants. It was further recommended that the amino acid composition of human milk should form the basis of the scoring pattern to evaluate protein quality in foods for infants under 1 year of age. The protein digestibility-corrected amino acid score method has been described in detail by Sarwar and McDonough (1990).

Amino acid scores of infant formulas, using the amino acid composition of human milk as the reference, have been calculated (Sarwar et al., 1989b). Amino acid scores for the milk- and soy-based formulas ranged form 59 to 90 and form 59 to 81%, respectively, due to deficiencies in sulfur amino acids and/or tryptophan. In practice, the amino acid deficiencies of the infant formulas are, however, compensated for by the higher level of protein in infant formulas compared to human milk, resulting in no evidence of amino acid deficiencies in clinical studies. Amino acid adequacy of infant formulas (a sole source of nutrition) should, therefore, be assessed by a method that takes into account both quality and quantity of protein. One such method, termed "amino acid rating: amino acid score X total protein, g/100 kcal" has been developed (Sarwar et al., 1989b). Amino acid ratings were also corrected for true digestibility of protein, as determined by the rat balance method. This review examines available information on protein and amino acid composition, and on protein and amino acid digestibility of various forms (e.g. powders, liquid concentrates, ready-to-use, etc.) of infant formulas, and discusses their comparative protein nutritional value in terms of amino acid scores and amino acid ratings.

PROTEIN CONTENTS OF INFANT FORMULAS

Total protein contents (N X 6.25) of milk- and soy-based infant formulas sold in Canada are shown in Tables 1 and 2, respectively. All milk-based formulas except the low-birth-weight formulas contained about 2.2 g protein/100 kcal (Table 1). The low-birth-weight formulas contained 2.75-2.95 g protein/100 kcal. A wide variation existed in protein contents of soy-based formulas (Table 2). The formulas of manufacturer 1 contained lower protein (2.65-2.66 g/100 kcal) than formulas of other manufacturers (3.00-3.68 g/100 kcal, Table 2). All formulas, especially the soy-based formulas and the low-birth-weight milk-based formulas, contained higher protein than mature human milk (1.5 g protein/100 kcal) (USDA, 1976). The differences between true protein content of infant formulas and human milk will even be larger when consideration is given to the higher level of nonprotein nitrogen in human milk (about 25% of total nitrogen) (Raiha, 1985) compared to that in infant formulas (about 8-13% of total nitrogen) (Sarwar et al., 1989a).

Table 1. Protein contents (g/100 kcal) and protein sources used in the preparation of milk-based infant formulas sold in Canada[a]

Formula	Protein source	Protein content
Manufacturer 1		
Powder	nonfat milk	2.20
Powder-whey	whey protein concentrate, caseinate	2.22
Liquid concentrate	nonfat milk	2.22
Liquid concentrate-whey	nonfat milk, demineralized whey solids	2.20
Low-birth-weight	nonfat milk, demineralized whey solids	2.72
Manufacturer 2		
Powder	skim milk powder, whey powder (demineralized)	2.28
Liquid concentrate	whey powder (demineralized), skim milk powder	2.24
Low-birth-weight	whey powder (demineralized), skim milk powder	2.95
Manufacturer 3		
Powder	skim milk, reduced minerals whey	2.21
Liquid concentrate	skim milk, reduced minerals whey	2.22
Low-birth-weight	skim milk, reduced minerals whey	2.95
Manufacturer 4		
Liquid concentrate	skim milk, skim milk powder	2.28

[a] Data of Sarwar et al. (1989a).

Total protein contents of milk-based formulas sold in some Western European countries (Pompei et al. 1987) have been reported to vary from 2.24 to 2.83 g/100 kcal (Pompei, C. and Rossi, M., personal communication).

The question of upper limit of the safe range of protein intake, with particular reference to the protein content of infant formulas used for feeding healthy term infants, was discussed by Young and Pelletier (1989). Based on evaluation of growth and blood biochemical data, an arbitrary upper limit of about 3.5 g protein/100 kcal was suggested. To address concerns for renal solute load, the need for a further desirable refinement in the proposed safe upper limit for protein content of infant formulas was also identified (Young and Pelletier, 1989).

An evaluation of present infant formulas revealed that all milk-based infant formulas contained considerably lower protein (Table 1) than the proposed upper limit of 3.5 g/100 kcal. Similarly, all soy-based formulas except one contained less protein than the proposed upper limit (Table 2). A further reduction in the protein content of milk-based formulas to simulate human milk more closely has been suggested (Jarvenpaa et al., 1982; Raiha, 1985). This would make the quality of milk proteins in infant formulas more critical; suggesting that any effort to further reduce protein content of formulas should be accompanied by a critical examination of their amino acid adequacy.

Table 2. Protein contents (g/100 kcal) and protein sources used in the preparation of soy-based infant formulas sold in Canada[a]

Formula	Protein source	Protein content
Manufacturer 1		
Powder	soy protein isolate, L-Met, Tau	2.66
Liquid concentrate	soy protein isolate, L-Met, Tau	2.65
Manufacturer 2		
Powder	soy protein isolate, L-Met, Tau	3.00
Ready-to-use	soy protein isolate, L-Met, Tau	3.68
Manufacturer 3		
Powder	soy protein isolate, L-Met	3.19
Manufacturer 4		
Powder	aqueous extract from soybeans, L-Met	3.14
Liquid concentrate	soy protein isolate, L-Met	3.01

[a] Data of Sarwar et al. (1989a).

AMINO ACID COMPOSITION OF INFANT FORMULAS

Essential amino acid profiles of milk-based infant formulas sold in Canada and Western Europe are shown in Tables 3 and 4, respectively. The amino acid data for human milk, as specified by the FAO/WHO/UNU (1985) report is included in Tables 3 and 4 for comparison purposes.

The levels of essential amino acids in the milk-based formulas sold in Canada were affected by protein source (s) containing different proportions of casein and whey proteins, and by varying degrees of heat processing involved in their preparation (Table 1 and 3). The whey-predominant formulas (based on combination of whey products and skim or nonfat milk) contained high levels of threonine while the casein-predominant formulas (based on nonfat or skim milk and skim milk powder) contained high levels of phenylalanine plus tyrosine (Tables 1 and 3). Liquid concentrates contained lower levels of lysine and sulfur amino acids than other forms of formulas prepared by the same manufacturer (Table 3). Compared to human milk, all milk-based infant formulas sold in Canada, especially liquid concentrates, were lower in sulfur amino acids (Table 3). Most of these formulas also contained lower levels of tryptophan than human milk (Table 3).

In general, the essential amino acid profiles of the milk-based formulas sold in Western Europe (Table 4) were similar to those sold in Canada (Table 3). The liquid (ready-to-feed) forms contained lower levels of tryptophan and several other essential amino acids compared to powder forms prepared by the same manufacturer (Table 4). All formulas contained lower levels of tryptophan and sulfur amino acids than human milk (Table 4).

Essential amino acid profiles of the soy-based infant formulas sold in Canada are shown in Table 5. Despite being supplemented with L-methionine (Table 2), the sulfur amino acid content of the formulas was still lower than that of human milk (Table 5). The soy-based formulas were also marginally lower in threonine, tryptophan, lysine, leucine and valine when compared to human milk (Table 5).

DIGESTIBILITY OF PROTEIN AND BIOAVAILABILITY OF AMINO ACIDS IN INFANT FORMULAS

Data on true digestibility of protein and selected essential amino acids, which are likely to be deficient, in powder and liquid concentrate forms of milk-based infant formulas sold in Canada are shown in Table 6. The digestibility values were determined by the rat balance method. In the case of manufacturer 1, 2 or 3, the values for true digestibility of protein, lysine, methionine, cystine, and threonine in liquid concentrate were 5-13 percentage units lower than the values for protein and

Table 3. Amino acid profiles (g/100 g protein) of milk-based formula sold in Canada[a]

Formula	His	Ile	Leu	Lys	Met + Cys	Phe + Tyr	Thr	Trp	Val
Manufacturer 1									
Powder	2.58	4.92	9.52	7.90	3.35	10.13	4.45	1.23	6.56
Powder-whey	2.30	6.41	11.10	8.11	3.78	8.57	6.57	1.66	6.37
Liquid concentrate	2.82	5.09	9.76	6.01	2.48	9.76	4.29	1.38	6.62
Liquid concentrate-whey	2.44	5.77	10.28	6.61	3.11	8.46	5.84	1.59	6.05
Low-birth-weight	2.16	5.79	10.53	8.42	3.68	7.91	6.27	1.63	6.24
Manufacturer 2									
Powder	2.15	5.44	10.66	8.20	3.82	8.31	5.61	1.44	6.26
Liquid concentrate	2.00	5.61	10.24	7.16	3.16	8.29	5.35	1.48	6.21
Low-birth-weight	2.28	5.52	10.39	7.75	3.52	7.94	5.89	1.57	5.89
Manufacturer 3									
Powder	2.30	5.46	9.94	8.04	3.23	8.34	5.52	1.44	6.17
Liquid concentrate	2.28	5.31	9.89	7.31	2.99	8.46	5.57	1.37	6.02
Low-birth-weight	2.32	6.11	11.37	7.71	3.22	9.92	6.04	1.72	6.14
Manufacturer 4									
Liquid concentrate	2.60	5.14	9.77	6.71	2.41	10.32	4.35	1.22	6.39
Human milk[b]	2.60	4.60	9.30	6.60	4.20	7.20	4.30	1.70	5.50

[a] Data of Sarwar et al. (1989a)

[b] Data for human milk were abstracted from the FAO/WHO/UNU 1985 report.

these amino acids in powder(s) (Table 6). Although the powder form of manufacturer 4 was not available for analysis at the time of the study, the true digestibility values of protein and most amino acids for the liquid concentrate of manufacturer 4 were similar to the lower values for liquid concentrates prepared by the other three manufacturers (Table 6).

Lysine availability values in the various forms of the milk-based infant formulas sold in Canada were also determined by a rat growth response method using regression analysis (weight gain vs. lysine consumed). A wheat gluten basal diet adequate in all nutrients for rat growth except lysine (0.26%) was supplemented with graded levels of crystalline lysine (0.08- 0.64%) or infant formulas providing 0.16% supplemental lysine. Simple linear regression for weight gain (1 wk) on total lysine consumed (food intake X dietary lysine consumed) was calculated for the standard diets (Figure 1). The expected weight gains for the test diets were calculated by substituting the total lysine consumption for the test diets in the regression equation. The lysine bioavailability was then determined by the following equation:

Lysine bioavailability = (actual weight gain/expected weight gain) X 100.

Although limited to the bioavailability determination of a single amino acid at a time, the bioavailability results of the properly processed protein sources obtained by the animal growth method (not affected by the modification in the large intestine) are considered (in theory) more accurate than those obtained by the balance method. Lysine in condensed skim milk (used as the main source of protein in several milk-based infant formulas) was 95% bioavailable (Table 7). Values for bioavailability of lysine in milk-based formulas were up to 18 percentage units lower than those in condensed skim milk. In the case of manufacturer 1 and 2, liquid concentrates had the lowest lysine bioavailability values (78-81%). In the case of manufacturer 4, one ready-to-use product was lower in lysine bioavailability (81%) than the other two products (89%) (Table 7).

Table 4. Essential amino acid profiles (g/100 g protein) of some milk-based infant formulas sold in Western Europe[a]

Formula	His	Ile	Leu	Lys	Met + Cys	Phe + Tyr	Thr	Trp	Val
Manufacturer 1									
Powder	4.00	5.72	11.14	9.06	3.58	8.19	6.21	1.36	5.79
Liquid	2.20	5.46	10.65	8.59	3.42	7.03	5.69	1.41	5.59
Manufacturer 2									
Powder	2.32	5.45	10.47	8.82	3.49	7.39	5.83	1.24	5.30
Liquid	2.60	5.06	9.67	7.69	3.63	6.82	5.10	0.84	5.17
Manufacturer 3									
Powder	2.45	5.61	10.64	8.58	3.23	7.61	5.73	1.42	4.46
Liquid	2.24	4.83	9.49	7.99	3.33	7.46	4.67	1.22	5.09
Manufacturer 4									
Powder	2.82	5.25	10.64	8.24	2.97	9.67	4.56	1.21	5.96
Liquid	2.54	4.82	9.93	7.32	3.09	8.39	4.10	0.99	5.22
Manufacturer 5									
Powder	2.42	5.74	11.22	9.13	3.71	7.96	5.91	1.29	5.57
Manufacturer 6									
Powder	2.42	5.17	10.15	8.08	3.01	8.34	5.16	1.27	5.42
Manufacturer 7									
Liquid	2.25	4.40	8.87	7.40	3.19	6.99	3.85	1.02	4.85
Manufacturer 8									
Liquid	2.65	5.69	11.06	9.38	3.20	8.09	5.57	1.11	5.63
Manufacturer 9									
Liquid	2.02	5.22	9.73	7.79	3.30	6.73	5.37	1.16	4.98
Human milk[b]	2.60	4.60	9.30	6.60	4.20	7.20	4.30	1.70	5.50

[a] Data were abstracted from Pompei et al. (1987).
[b] Data for human milk were abstracted from the FAO/WHO/UNU 1985 report.

Table 5. Amino acid profiles (g/100 g protein) of soy-based formula sold in Canada[a]

Formula	His	Ile	Leu	Lys	Met + Cys	Phe + Tyr	Thr	Trp	Val
Manufacturer 1									
Powder	2.75	4.77	8.06	5.56	2.96	9.16	3.68	1.08	4.83
Liquid concentrate	2.72	4.66	7.89	5.28	2.98	8.73	3.70	1.32	4.78
Manufacturer 2									
Powder	2.32	4.62	7.87	5.73	2.97	8.65	3.61	1.22	4.69
Ready-to-use	2.19	4.80	8.12	5.75	2.82	9.20	3.80	1.16	5.00
Manufacturer 3									
Powder	2.55	4.69	8.13	5.60	3.33	9.02	3.76	1.00	4.73
Manufacturer 4									
Powder	2.39	4.49	7.61	5.32	3.48	8.91	3.89	1.48	4.68
Liquid concentrate	2.45	4.56	7.74	5.70	3.22	8.51	3.67	1.20	4.85
Human milk[b]	2.60	4.60	9.30	6.60	4.20	7.20	4.30	1.70	5.50

[a] Data of Sarwar et al. (1989a).
[b] Data for human milk were abstracted from the FAO/WHO/UNU 1985 report.

Table 6. True digestibility of (%) crude protein and selected amino acids in milk-based infant formulas sold in Canada[a]

Formula	Protein	Lys	Met	Cys	Thr	Trp
Casein + methionine (control)	98[a]	98[a]	100[a]	97[ab]	97[a]	100[a]
Manufacturer 1						
Powder	94[c]	96[b]	93[c]	92[c]	93[bc]	95[bc]
Powder-whey	95[bc]	95[b]	98[b]	98[a]	92[c]	96[b]
Liquid concentrate	88[de]	87[e]	86[e]	85[e]	90[d]	92[d]
Manufacturer 2						
Powder	93[c]	93[c]	98[b]	96[b]	90[d]	96[b]
Liquid concentrate	88[e]	85[f]	89[d]	89[d]	87[e]	96[b]
Manufacturer 3						
Powder	97[ab]	98[a]	99[ab]	96[b]	94[b]	96[b]
Liquid concentrate	90[a]	85[ef]	92[c]	91[c]	87[e]	94[c]
Manufacturer 4						
Liquid concentrate	90[d]	89[d]	89[d]	89[d]	93[b]	92[d]

[a] Data of Sarwar et al. (1989b). Means in each column without a common superscript differ significantly ($P<0.05$).

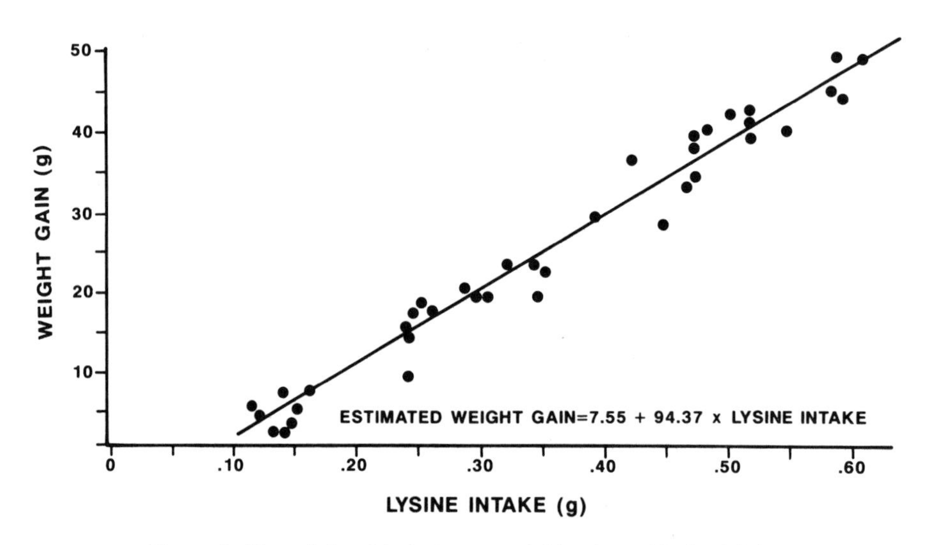

Figure 1. The relationship between weight gain and lysine intake.

Table 7. Data on lysine availability (%) and contents of bioavailable lysine in milk-based infant formulas sold in Canada[a]

Product	Lysine bioavailability[b]	Lysine availability chemical	Bioavailable lysine[c]
Condensed skim milk	95 ± 2	97	7.75
Formulas of manufacturer 1			
Powder	87 ± 1	83	7.10
Liquid concentrate	78 ± 2	75	5.55
Ready-to-use	83 ± 2	87	6.46
Low-birth-weight	82 ± 1	83	6.40
Formulas of manufacturer 2			
Powder	87 ± 3	79	6.38
Liquid concentrate	81 ± 2	82	5.47
Ready-to-use	85 ± 2	85	6.46
Low-birth-weight	83 ± 2	92	6.99
Formulas of manufacturer 3			
Powder	83 ± 2	76	6.80
Liquid concentrate	88 ± 3	81	6.11
Ready-to-use	89 ± 2	82	6.55
Low-birth-weight	86 ± 2	86	6.44
Formulas of manufacturer 4			
Ready-to-use	89 ± 2	91	6.29
Ready-to-use	81 ± 2	83	5.61
Ready-to-use	89 ± 3	94	6.26

[a] Data were abstracted from Sarwar et al. (1988).
[b] Mean ± SEM.
[c] g/100 g protein, total lysine x lysine bioavailability.

The values for lysine availability estimated chemically by using the sodium, borohydride method (Couch and Thomas, 1976) were included for comparison with the bioavailability data in Table 7. In general, the chemical values for lysine availability were comparable to the bioavailability values; the differences between the values obtained by the two methods were less than 10% (Table 7).

The data for total lysine were corrected for lysine bioavailability to obtain estimates of bioavailable lysine in various forms of milk-based infant formulas (Table 7). Liquid concentrates contained lower levels of bioavailable lysine than present in powders and/or other forms of formulas prepared by the same manufacturer. The ready-to-use formulas of manufacturer 4 were lower in bioavailable lysine (5.61-6.29 g/100 g protein) than similar forms of other manufacturers (6.44-6.55 g/100 g protein) (Table 7).

PROTEIN QUALITY OF INFANT FORMULAS AS PREDICTED BY RAT GROWTH METHODS

The relative PER (RPER) and relative net protein ratio (RNPR) values for some milk-based formulas sold in Canada are shown in Table 8. The RPER and RNPR values were obtained by using a modified method (2-wk PER or NPR including the use of 8% dietary protein and methionine-supplemented casein control). There were significant differences between various forms (powder and liquid concentrate) of the same manufacturer, and among formulas prepared by different manufacturers. In the case of manufacturer 1, 2 or 3, the RPER and/or RNPR values for the liquid concentrate were significantly lower than those for powders (Table 8). Among liquid concentrates, the products of manufacturer 1 and 4 had the lowest RPER (64-65%) and RNPR (78%) values. Among powders, the product of manufacturer 2 had higher RPER and RNPR values than the products of manufacturer 1 and 3. In the case of manufacturer 1, the powder (based on whey protein concentrate + caseinate) was superior to that based on nonfat milk in RPER or RNPR value (Table 8).

The lower RPER and RNPR values of liquid concentrates compared with powders of the same manufacturer could be explained by the lower levels of digestible (bioavailable) lysine and methionine plus cystine in liquid concentrates (Table 8). Similarly, the differences between the two powders of manufacturer 1, and among formulas of different manufacturers could be related to the contents of bioavailable methionine plus cystine and/or lysine.

The protein quality results obtained with rats (Table 8) should only be considered of predicatory nature, and should be confirmed with infants because of differences in protein and amino acid requirements of rats and infants. Moreover, the use of RPER data reported in Table 8 would not be appropriate in evaluating regulatory compliance of infant formulas because the official method (4-wk PER, 10% dietary protein and unsupplemented casein control) was not used.

AMINO ACID SCORES AND AMINO ACID RATINGS OF INFANT FORMULAS

Amino acid scores and amino acid ratings of milk-based formulas sold in Canada are shown in Table 9. Amino acid contents of formulas were used in calculating amino acid ratios (mg of an essential amino acid in 1.0 g of test protein/mg of the same amino acid in 1.0 g of reference protein X 100) for 9 essential amino acids. The suggested FAO/WHO/UNU (1985) requirement pattern for infants (based on amino acid composition of human milk) was used as the reference protein. The lowest amino acid ratio (%) was termed amino acid score. Amino acid ratings were calculated in the following manner (Sarwar et al., 1989a):

Amino acid rating (AAR)
= amino acid score X total protein (g/100 kcal)

Relative amino acid rating (RAAR, %)
= (AAR of test formula)/(AAR of human milk) X 100,
AAR of human milk was considered to be 1.50 (i.e. amino acid score of 100% X 1.5 g protein/100 kcal).

True protein digestibility (TPD)-corrected AAR
= TPD X AAR of test formula

TPD-corrected RAAR (%)
= (TPD-corrected AAR of test formula)/(AAR of human milk) X 100,

TPD of human milk was considered to be 100%.

In the case of manufacturer 1, liquid concentrate and liquid concentrate-whey had considerably lower amino acid scores than powder and powder-whey, respectively; suggesting higher protein damage in the preparation of liquid concentrates compared to powders (Table 9). Similarly, liquid concentrate had lower amino acid score than powder prepared by manufacturer 2 or 3. The liquid concentrate of manufacturer 4 also had a low amino acid score of 57%. All formulas, except the powder of manufacturer 1, were first limiting in methionine plus cystine. The powder of manufacturer 1 was first limiting in tryptophan.

The higher protein content of the low-birth-weight formulas (Table 1) was mainly responsible for their higher amino acid ratings compared to other forms of the same manufacturer (Table 9). All formulas except the liquid concentrates of manufacturer 1 and 4 had amino acid ratings of more than 1.5 (1.58-2.39) or above 100% of that of human milk (relative amino acid ratings of 105-159%) (Table 9). The liquid concentrates of manufacturer 1 and 4 had an amino acid rating of about 1.3 or relative amino acid rating of 87%. The correction for true protein digestibility reduced the relative amino acid ratings of liquid concentrates of manufacturer 1 and 4 by about 10%, and also lowered the relative amino acid ratings of the other two liquid concentrates (manufacturer 2 and 3) to 95-97% (Table 9).

Table 8. Relative protein efficiency ratio (RPER) and relative net protein ratio (RNPR) values (casein + methionine = 100) and contents of digestible amino acids in milk-based formulas sold in Canada[a]

Formula	RPER[b]	RNPR[b]	Digestible Lys[c]	Digestible Met + Cys[c]
Manufacturer 1				
Powder	76 ± 3[d]	88 ± 2[c]	7.6	3.1
Powder-whey	90 ± 3[b]	95 ± 2[b]	7.7	3.7
Liquid concentrate	65 ± 2[e]	78 ± 1[d]	5.2	2.1
Manufacturer 2				
Powder	98 ± 2[a]	100 ± 1[a]	7.6	3.7
Liquid concentrate	85 ± 2[bc]	94 ± 2[b]	6.1	2.8
Manufacturer 3				
Powder	88 ± 1[bc]	94 ± 1[b]	7.9	3.1
Liquid concentrate	83 ± 2[cd]	86 ± 2[c]	6.3	2.7
Manufacturer 4				
Liquid concentrate	64 ± 2[e]	78 ± 1[d]	6.0	2.1

[a] Data of Sarwar et al. (1989b).
[b] Mean ± SEM (n=8). Means in each column without a common superscript differ significantly ($P<0.05$).
[c] Digestible amino acid (g/100 g protein) = total amino acid x true amino acid digestibility.

The amino acid data of Pompei et al. (1987) (Table 4) were used in calculating amino acid scores for the milk-based formulas sold in Western Europe. The amino acid scores for formulas of nine manufacturers varied from 49 to 80% (Table 10). In the case of manufacturer 2, 3 or 4, the liquid form had a lower amino acid score than the powder form. Most infant formulas were first limiting in tryptophan while some were co-limiting in methionine plus cystine and tryptophan (Table 10). In most cases, the liquid forms contained higher levels of lysinoalanine than the powder forms; suggesting higher protein damage in the case of liquid formulas (Table 10). All formulas except the liquid forms of manufacturer 2, 4 and 7 had amino acid ratings of more than 1.5 (1.57-1.99) or above 100% of that of human milk (relative amino acid ratings of 105-133%) (Table 10). The liquid form of manufacturer 2 had a low relative amino acid rating, 77%, while the values for the liquid forms of manufacturer 4 and 7 were nearly 100% (97-99%).

Amino acid scores and amino acid ratings of soy-based formulas sold in Canada are shown in Table 11. The powder of manufacturer 3 had the lowest amino acid score, 59%, with tryptophan being the first limiting amino acid. Amino acid scores for other soy-based formulas varied from 63 to 81% due to deficiencies in methionine plus cystine and/or tryptophan (Table 11). Because of substantially higher protein in soy-based formulas compared to human milk, amino acid ratings for all of these products were> 100% (Table 11). Even after adjustment for protein digestibility, amino acid ratings for all soy-based formulas were > 100% (104-159%).

The protein and amino acid digestibility, protein quality (amino acid score, RPER, RNPR), and amino acid adequacy (amino acid rating) data for the milk-based formulas suggested that liquid concentrates may be inferior to powders prepared by the same manufacturer. This convincing evidence, about the inferior protein quality of liquid concentrates compared to powders, supports the need to investigate the effects of processing used in the preparation of various forms of milk-based formulas on their protein digestibility and quality in infants.

Table 9. Amino acid score (AAS), true protein digestibility (TPD), amino acid rating (AAR), and relative AAR (RAAR, human milk = 100) values for milk-based infant formulas sold in Canada[a]

Formula	AAS, %	AAR[b]	RAAR, %	TPD, %	TPD-corrected RAAR, %
Manufacturer 1					
Powder	72	1.58	105	94	99
Powder-whey	90	2.00	133	95	127
Liquid concentrate	59	1.31	87	88	77
Liquid concentrate-whey	74	1.63	109	90	98
Low-birth-weight	88	2.39	159	94	149
Manufacturer 2					
Powder	85	1.94	129	93	120
Liquid concentrate	75	1.68	112	87	97
Low-birth-weight	84	2.48	165	93	154
Manufacturer 3					
Powder	77	1.70	113	97	110
Liquid concentrate	71	1.58	105	90	95
Low-birth-weight	77	2.27	151	96	145
Manufacturer 4					
Liquid concentrate	57	1.30	87	90	78

[a] Data of Sarwar et al. (1989a).
[b] Amino acid rating of human milk was considered to be 1.50 (i.e., amino acid score of 100% x 1.5 g protein/100 kcal).

Table 10. Lysinoalanine (LAL) contents, amino acid scores (AAS), amino acid rating (AAR), and relative AAR (RAAR, human milk=100) values of some milk-based infant formulas sold in Western Europe

Formula	LAL (mg/kg protein)[a]	AAS (%)[b]	AAR[c]	RAAR(%)
Manufacturer 1				
Powder	n.d.	80	1.79	119
Liquid	445	81	1.99	133
Manufacturer 2				
Powder	215	73	1.71	114
Liquid	556	49	1.16	77
Manufacturer 3				
Powder	107	77	1.71	114
Liquid	343	72	1.61	107
Manufacturer 4				
Powder	n.d.	71	1.57	105
Liquid	1032	58	1.45	97
Manufacturer 5				
Powder	115	76	1.95	130
Manufacturer 6				
Powder	n.d.	72	1.88	125
Manufacturer 7				
Liquid	379	60	1.48	99
Manufacturer 8				
Liquid	190	65	1.84	123
Manufacturer 9				
Liquid	660	68	1.65	110

[a] Data of Pompei et al. (1987), n.d., not detected.
[b] Calculated from the amino acid data of Pompei et al. (1987).
[c] Calculated by using the protein data of Pompei, C. and Rossi, M. (personal communication).

Table 11. Amino acid score (AAS), true protein digestibility (TPD), amino acid rating (AAR), and relative AAR (RAAR, human milk = 100) values for soy-based infant formulas sold in Canada[a]

Formula	ASS, %	AAR[b]	RAAR, %	TPD, %	TPD-corrected RAAR, %
Manufacturer 1					
Powder	63	1.67	111	94	104
Liquid concentrate	71	1.88	125	92	115
Manufacturer 2					
Powder	71	2.13	142	95	135
Ready-to-use	67	2.46	164	93	152
Manufacturer 3					
Powder	59	1.88	125	95	119
Manufacturer 4					
Powder	81	2.54	169	94	159
Liquid concentrate	70	2.10	140	92	129

[a] Data of Sarwar et al. (1989a).
[b] Amino acid rating of human milk was considered to be 1.50 (i.e., amino acid score of 100% x 1.5 g protein/100 kcal).

ACKNOWLEDGEMENTS

The author acknowledges the assistance of R.W. Peace and H.G. Botting in the preparation of this manuscript.

REFERENCES

Cheftel, J.C. (1979). Proteins and amino acids. In: "Nutritional and safety aspects of food processing", S.R. Tannenbaum, ed., Marcel Decker Inc., New York, p. 153-215.

Codex Alimentarius Commission (1989). Document Alinorm 89/30, working group's report of the fifth session of CCVP on protein quality measurement, Food and Agriculture Organization, Rome, Italy, p. 2-3.

Couch, J.R. and Thomas, M.C. (1976). A comparison of chemical methods for the determination of available lysine in various proteins. J. Agr. Food Chem. 24, 943-946.

FAO/WHO (1990). Report of the joint expert consultation on protein quality evaluation, Food and Agriculture Organization, Rome, Italy (In press).

FAO/WHO/UNU Joint Expert Consultation (1985). Energy and protein requirements, World Health Organization Techn. Rep. Ser. No. 724, geneva, Switzerland.

Finot, P.A. (1983). Chemical modification of the milk proteins during processing and storage. Nutritional, metabolic and physiological consequences. Kieler Milchwirtschaftliche Forschungsberitchte 35, 357-369.

Hurrell, R.F. and Carpenter, K.J. (1981). The estimation of available lysine in foodstuffs after Maillard reactions. Prog. Food Nutr. Sci. 5, 159-176.

Järvenpää, A-L, Räihä, N.C.R., Rassin, D.K. and Guall, G.E. (1982). Milk protein quantity and quality in the term infant 1. Metabolic responses and effects on growth. Pediatrics 70, 214-220.

Packard, V.S. (1982). In: "Human milk and infant formulas", Academic Press, New York, p. 157-162.

Pompei, C., Rossi, M. and Mare, F. (1987). Protein quality in commercial milk-based infant formulas. J. Food Quality, 10, 375-391.

Räihä, N.C.R. (1985). Nutritional proteins in milk and the protein requirement of normal infants. Pediatrics 75(suppl), 136-141.

Sarwar, G., Botting, H.G. and Peace, R.W. (1989a). Amino acid rating method for evaluating protein adequacy of infant formulas. J. Assoc. Off. Anal. Chem. 71, 622-626.

Sarwar, G., and McDonough, F.E. (1990). Evaluation of protein digestibility-corrected amino acid score method for assessing protein quality of foods. J. Assoc. Off. Anal. Chem. 73, 347-356.

Sarwar, G., Peace, R.W. and Botting, H.G. (1988). Bioavailability of lysine in milk-based infant formulas as determined by rat growth response method. Nutr. Res. 8, 47-55.

Sarwar, G., Peace, R.W. and Botting, H.G. (1989b). Differences in protein digestibility and quality of liquid concentrate and powder forms of milk-based infant formulas fed to rats. Am. J. Clin. Nutr. 49, 806-813.

USDA (1976). Composition of foods: diary and egg products, raw-processed-prepared. United States Department of Agriculture Handbook No. 8-1 (item no. 01-041), U.S. Government Printing Office, Washington, D.C.

Young, V.R. and Pelletier, V.A. (1989). Adaptation to high protein intakes, with particular reference to formula feeding and the healthy, term infant. J. Nutr. 119, 1799-1809.

29

NUTRITIONAL VALUE OF PROCESSED RAPESEED MEAL

In-Kee Paik

Department of Animal Science, Chung-Ang University

Ansung-Kun, Kyonggi-Do, South Korea

ABSTRACT

Supplementation of iodine at the level of 3.5 ppm reduced weight gain of the rats fed rapeseed oil meal(ROM) diets. Treatment of ROM with ammonia at the level of 2 or 4% tended to increase metabolizable energy value and availability of dry matter, crude protein and crude ash of ROM in the chicken. Potential goitrin level of ROM was reduced by ammoniation at 6% level. On the other hand, level of potential isothiocyanates increased by ammoniation.

Treatment of ROM with ammonia at the level of 3% and above reduced weight gain of the chickens fed treated ROMs. Weight of thyroid glands of the birds increased as the level of ammoniation of ROM increased. Supplementation of Avoparcin to the diets containing ROM improved weight gain and dressing percentage of the broiler chickens.

INTRODUCTION

Rapeseed oil meal(ROM) is fairy good source of protein and its amino acid composition is similar to other protein supplements such as soybean meal(SBM). However, some difficulties have been experienced when ROM is used in the diets of monogastric animals. One of the main problems encountered has been related to the presence of glucosinolates in the meal. Glucosinolates are hydrolyzed by enzyme thioglucoside gluco-hydrolase(myrosinase) to yield various hydrolysis products. Aglucone products formed by hydrolysis of glucosinolates are more or less anti-thyroid substances(VanEtten, 1969). Enzymatic hydrolysis of progoitrin (2-hydroxy-3-butenyl glucosinolate) yields 2-hydroxy-3-butenyl isothio-cyanate which is unstable and cyclizes to 5-vinyl-2-thiooxazolidone. This cyclic compound is also called goitrin due to its strong goitrogenic effect(Kjaer, 1960). Other isothiocyanates may exert a goitrogenic effect by formation of thiourea derivatives(Greer, 1950, 1962) or by formation of thiocyanate ion(VanEtten, 1969).

Supplementation of iodinated casein to the diet containing ROM reduced weight of thyroid glands of the chickens(Klain et al. 1956) and pigs(Nordfeldt et al. 1959) but did not improve growth rate. Supplementation of iodinated casein(50-200 mg/kg) or calcium iodate (0.14-0.28 ppm iodine) to the pig diets containing canola meal did not produced significant responses in gain and feed efficiency(Bell et al. 1980)

Nutritional and Toxicological Consequences of Food Processing
Edited by M. Friedman, Plenum Press, New York, 1991

Blake(1983) produced ammoniated canola meal by treating 5%(w/w) anhydrous ammonia at 105℃. Ammonia treatment significantly increased nitrogen content by 8.3% and reduced glucosinolates content approximately 40%. When the ammonia treated canola meal was included at level of 20% in the chicken diet, availability of lysine, body weight, feed conversion and thyroid size were not significantly affected(Goh et al. 1983). In a rat feeding trial, however, plasma lysine concentrations were lower in rats fed the ammoniated meals indicating decreased absorption of lysine(Keith and Bell, 1984). Anmmoniation of mustard meal reduced glucosinolates by over 80% and reduced lysine by 20% as well but increased crude protein from 44.6 to 51.1%, dry basis(Bell et al. 1984).

Jensen and Thomsen(1980) reported that Avoparcin supplementation to the diet containing ROM improved performance of the broiler chickens. Growth promoting effect of Avoparcin increased when the level of protein and lysine in the diet was lowered(Kirchgessner and Roth, 1981).

The objectives of this study were (1) to assess the influences of feeding diets containing ROMs hydrolyzed with mustard seed powder, ammoniated or supplemented with iodine on the performance of the rats(Experiment 1); (2) to study chemical characteristics of ROM and effects of ammoniation on the metabolizable energy value and levels of hydrolysis products of glucosinolates in ROM(Experiment 2); (3) to study effects of feeding diets containing ammoniated ROMs on the performance of broiler chickens(Experiment 3); (4) to examine the effects of supplementation of Avoparcin in the broiler diet with or without ROM(Experiment 4).

MATRIAL AND METHODS

Experiment 1. Thirty six weanling male Wistar rats were divided into 6 groups and housed randomly in the individual cages. Each group was placed on one of the following six experimental diets ; Basal+Korea ROM(ROM), Basal+Canadian ROM(Canola), Basal+Korean ROM hydrolyzed with mustard seed powder(Hydrolyzed ROM), Basal+ammoniated Korean ROM(Ammoniated ROM), Basal+Korean ROM+iodine(Iodized ROM), and Basal+Korean ROM hydrolyzed with mustard seed powder+iodine(Iodized, hydrolyzed ROM).

Basal diet was formulated with corn, soybean meal, fish meal, animal fat, D,L-methionine, dicalcium phosphate, salt, vitamin and mineral premix. Rapeseed meal was included at the level of 20% replacing the basal diet in each treatment.

For the hydrolysis of ROM, mustard seed powder was added to ROM at the level of 5% and then thoroughly mixed along with distilled water of twice the weight of ROM. The mixture was incubated for 24 hrs at 37℃ and then dried. Ammoniation of ROM was conducted by spraying diluted aqueous ammonia(12.5% NH_3 concentration) at the level of 10% of ROM treated, followed by incubation at 37℃ for one week. Iodized diets were supplemented with KI at the level of 3.5 ppm iodine.

Experiment 2. General compositions of ROM were analyzed by A.O.A.C.(1984) methods. Analysis of ammino acids and minerals were conducted with automatic amino acid analyzer and atomic absorption spectrometer respectively. Energy utilization of ROM was determined by apparent metabolizable energy(AME) method(Hill and Anderson, 1958) and true metabolizable energy(TME) method(Sibbald, 1986). Twelve week old cockerels of White Leghorn breed were employed for the determination. Ammoniation of ROM was conducted at 0, 2, 4 and 6% ammonia level using the procedure described in Experiment 1. Nutrients availability of treated ROMs were determined by using the procedure employed in TME determination. Hydrolysis products of glucosinolates in ROM were determined by the method of Appelqvist & Joseffson(1967) using UV-spectrophotometer.

Experiment 3. Two broiler feeding trials were conducted for 6 weeks each. In Experiment 3-A, 160 hatched male broiler chickens were fed 4 experimental diets; 0, 1.5, 3 and 4.5% ammoniated ROM diet, in which ammoniated ROMs were included at level of 15%.

In Experiment 3-B, 24 chickens were fed 3 experimental diets; 0, 3 and 6% ammoniated ROM diet, in which ammoniated ROMs were used at the level of 20%. Nutrients availabilities of the experimental diets were determined by a metabolic trial at 7 weeks of age of the birds.

Experiment 4. Two hundred fifty hatched male broiler chickens were divided into 25 groups of 10 birds each and fed in raised floor batteries. Five groups were placed on one of the following five experimental diets; Soybean meal(SBM) diet, SBM diet+Avoparcin 10 ppm, ROM diet, ROM diet+Avoparcin 10 ppm, and ROM diet+Avoparcin 15 ppm.

Diets were formulated using computer to have isocalorie and isonitrogen, and ROM diets contained Indian ROM at the level of 8%. Dressing percentage of the birds was measured after removing head, feet, feathers and intestinal organs.

For the statistical analysis of the results, analysis of variance and Duncan's Multiple Range Test(Steel and Torrie, 1980) were used to test for significance between treatment means.

RESULTS AND DISCUSSION

Experiment 1. Weight gain, feed intake and feed efficiency of the rats are shown in Table 1. Weight gains of the rats fed Canola, Ammoniated ROM, and ROM were significantly(P<0.01) greater than that of Iodized, hydrolyzed ROM. Weight gain of the rats fed Hydrolyzed ROM and Iodized ROM were not significantly different from others. Feed intake of the rats fed Ammoniated ROM, Canola, and ROM were significantly(P<0.01) greater than that of Iodized, hydrolyzed ROM. Those of Hydrolyzed ROM and Iodized ROM were not significantly different from others. The differences of feed efficiency among treatments were not statistically significant.

Table 1. Weight gain, feed intake, and feed efficiency of the rats fed for 6 weeks(Exp. 1).

Treatments	Weight gain, g	Feed intake, g	Feed efficiency
ROM	218.6[a]	662.7[a]	3.04
Canola	225.7[a]	663.3[a]	2.95
Hydrolyzed ROM	206.4[ab]	615.6[ab]	2.98
Ammoniated ROM	219.5[a]	683.8[a]	3.12
Iodized ROM	194.9[ab]	610.8[ab]	3.14
Iodized, hydrolyzed ROM	173.7[b]	542.1[b]	3.13
S.E.M.[1]	9.8	27.2	0.07

S.E.M.[1]: Standard error of means, [a,b]: Values with different superscript in the same column are significantly different at P<0.01.

Weight of thyroid glands, liver and kidneys per 100 g body wt of rats are shown in Table 2. The weight of thyroid glands of the rats fed Hydrolyzed ROM and Iodized, hydrolyzed ROM were significantly(P<0.01) heavier than those of rats fed ROM, Canola, Ammoniated ROM and Iodized ROM. The weight of liver and kidneys were not significantly different among treatments.

Table 2. Weight of thyroid glands, liver and kidneys per 100 g of body weight(mean±S.D.) (Exp. 2)

Treatment	Thyroid glands, mg	Liver, g	Kidneys, g
ROM	6.61±2.05[a]	4.96±0.42	0.958±0.047
Canola	4.89±0.74[a]	4.52±0.58	0.869±0.028
Hydrolyzed ROM	10.89±1.91[b]	4.90±0.52	0.849±0.076
Ammoniated ROM	5.22±1.71[a]	4.47±0.21	0.894±0.069
Iodized ROM	5.07±1.39[a]	4.25±0.21	0.876±0.082
Iodized, hydrolyzed ROM	11.57±3.51[b]	4.73±0.54	0.868±0.067

a-c: Values with different superscript in the same column are significantly different at P<0.01.

Results of the experiment indicate that treatments of ROM influence weight gain and feed intake without significant influence on feed efficiency. Canadian ROM(Canola) showed best weight gain and feed efficiency probably due to low glucosinolates content and properly controlled processing conditions. Mustard seed powder has been known to be a rich source of myrosinase. Thus, hydrolysis of ROM with mustard seed powder might have produced goitrin and isothiocyanates from glucosinolates of ROM. Enlarged thyroid glands of the rats fed hydrolyzed ROMs are the evidence of goitrin production(Paik et al. 1980). Hydrolysis of ROM tended to reduce weight gain and feed intake probably due to the production of goitrin and isothiocyanates. Nordfeldt(1959) and Ochetim(1980) indicated that iodine supplementation alleviates goiter. Supplementation of iodine to ROM and Hydrolyzed ROM diets did not reduce the size of thyroid glands but reduced weight gain. Present results indicate that excessive supplementation of iodine to ROM diets may be detrimental. Ammoniation of ROM at the present level(1.25% in NH_3 basis) did not significantly influence performance or weight of organs of the rats.

Experiment 2. General composition, amino acids and mineral contents of ROM are shown in Table 3. The chemical compositions of Korean ROM were similar to those of Canadian ROM. Metabolizable energy contents of ROMs are shown in Table 4.

Average metabolizable energy content of Korean ROM was 1,628 kcal in AME, 1,552 kcal in AMEn, 2,414 kcal in TME and 2,350 kcal in TMEn per kg of dry matter(DM). Average metabolizable energy content of Canadian ROM was 1,933 kcal in AME, 1,846 kcal in AMEn, 2,955 kcal in TME and 2,833 kcal TMEn per kg DM.

The potential goitrin levels of ROMs are shown in Table 5. The level of goitrin was highest in Chilean ROM being 6.40 mg and lowest in Indian ROM being 1.05 mg. The level of isothiocyanate was highest in local mustard seed powder being 4.46 mg and lowest (not detectable) in Canadian ROM(mash). Table 5 also shows that goitrin level of Korean ROM(4.37 mg) was not affected by ammoniation at the level of 2 or 4% but reduced to 2.52 mg at 6% ammoniation. The level of isothiocyanate in ROM tended to increase by ammoniation.

TME values of ammoniated Korean ROM are shown in Table 6. Ammoniation of ROM increased TME showing the highest value at 4% ammoniation. Nutrients availability of ammoniated ROMs, which were determined with samples obtained from TME bioassay procedure, are shown in Table 7. Availability of crude protein of ROMs ammoniated at 2 or 4% was significantly(P<0.05) higher than that of untreated ROM. Availability of dry matter and crude ash tended to be higher in ROMs ammoniated at 2 or 4% than in ROM ammoniated at 6% or untreated ROM.

Metabolizable energy value of ROM varies widely depending on the source of reports. Reported ME values of ROMs were 1,670 kcal/kg DM(Sibbald and Slinger, 1963), 1,230, 1,313 and 1,782 kcal/kg DM(Lodhi

Table 3. Chamical composition of rapeseed oil meal(Exp. 2)

Composition	Korean ROM	Canadian ROM
General composition, %		
Moisture	13.09	11.70
Crude protein	34.20	33.59
Crude fat	2.09	3.22
Crude fiber	10.03	6.90
Crude ash	7.91	6.90
N.F.E.	32.68	37.66
Amino acids, %		
Lysine	1.85	1.64
Histidine	0.92	0.92
Arginine	1.44	1.67
Aspartic acid	2.32	2.02
Threonine	1.43	1.45
Serine	1.50	1.32
Glutamic acid	5.99	6.04
Proline	2.80	1.73
Glycine	1.59	1.44
Alanine	1.49	1.36
Cystine	0.23	0.63
Valine	1.48	1.66
Methionine	0.42	0.62
Isoleucine	1.13	1.29
Leucine	2.23	2.12
Tyrosine	0.47	0.77
Phenylalanine	1.32	1.10
Minerals, %		
Ca	0.60	0.73
P	0.94	1.06
K	3.00	3.25
Na	0.30	0.27
Mg	0.47	0.46
Fe	0.035	0.016
Mn	0.008	0.004
Zn	0.007	0.008
Cu	0.001	0.001

Table 4. Metabolizable energy content of rapeseed oil meals determined with cockerels(kcal/kg, mean±S.D.) (Exp. 2)

Items	Korean ROM	Canadian ROM
AME, DM	1,877±332	1,933±335
AME, 90% DM	1,689±299	1,740±335
AMEn, DM	1,818±333	1,846±342
AMEn, 90% DM	1,400±327	1,661±342
TME, DM	2,414±195	2,955±52
TME, 90% DM	2,173±195	2,660±52
TMEn, DM	2,350±200	2,833±65
TMEn, 90% DM	2,115±200	2,595±65

Table 5. Content of hydrolysis products of glucosinolates in rapeseed oil meal, mg/g (Exp. 2)

ROMs	Goitrin	Isothiocyanate
Korean ROM	4.37	0.46
Korean ROM (2% ammoniation)	4.47	0.78
Korean ROM (4% ammoniation)	4.35	0.72
Korean ROM (6% ammoniation)	2.52	0.84
Canadian ROM (mash)	3.12	ND
Canadian ROM (pellet)	1.48	0.99
Indian ROM	1.05	1.51
Mustard seed powder	1.71	4.46
Chinese ROM	2.41	0.48
Chilean ROM	6.40	0.86

Table 6. True metabolizable energy content of ammoniated Korean ROMs determined with cockerels(kcal/kg, mean±S.D.) (Exp. 2)

TME	ROM	2% Ammoniated	4% Ammoniated	6% Ammoniated
TME, DM	2,599±191	2,771±320	2,824±222	2,647±106
TME, 90% DM	2,303±191	2,494±320	2,542±222	2,382±106
TMEn, DM	2,475±180	2,514±309	2,566±102	2,482±125
TMEn, 90% DM	2,228±180	2,263±309	2,309±102	2,234±125

Table 7. Nutrients availability[1] of ammoniated Korean rapeseed oil meals, % (Exp. 2)

Treatments	DM	Crude protein	Crude fat	Crude fiber	Crude ash	N.F.E.
0% ammoniated ROM	40.1	21.7[b]	63.7	28.5	18.0	99.3
2% ammoniated ROM	44.9	50.4[a]	55.1	33.6	29.7	89.0
4% ammoniated ROM	48.4	48.4[a]	69.1	29.0	30.8	98.1
6% ammoniated ROM	41.7	29.6[ab]	63.6	28.5	21.5	100.0
S.E.M.[2]	3.44	6.81	5.66	1.68	4.03	4.51

[1].Determined with samples obtained from TME assay procedure using cockerels, [2].S.E.M. Standard error of means, [a,b]: Values with different superscripts are significantly different at P<0.05

et al. 1969), and 2,295 kcal/kg ADM(Sell, 1966). AMEn values of ROM were 1,530 kcal/kg DM(Sibbald and Slinger, 1963), 2,120 kcal/kg ADM(Sell, 1966), 1,682 kcal/kg ADM(Lee et al. 1973) and 1,843 kcal/kg ADM(Paik et al. 1975). TME and TMEn values of ROM were reported to be 2,270-3,060 kcal and 2,055-2,533 kcal/kg DM, respectively(Sibbald, 1986).

Except TMEn value of Canadian ROM, other metabolizable energy values of ROM determined in the present experiment are with in the range of the published data. Increase of TME values of ammoniated ROMs, especially 2 and 4% ammoniated, may be due to the improved availability of dry matter and crude protein.

Giotrin level of ROM was reduced only at 6% ammoniation. Glucosinolates content of Canadian ROM could be reduced by the treatment with heat(105℃), steaming and 5% anhydrous ammonia(Blake, 1983). Increase of isothiocyanate level in the ammoniated ROM seems to be related to the formation of thiourea derivatives as a result of ammoniation. Isothiocyanates react with ammonia to form thiourea derivatives which is measured by UV-spectrophotometer in the quantitative analysis of isothiocyanates.

The results of Experiment 2 indicates that 2 or 4% ammination is proper to improve ME value and nutrients availability of ROM although 6% ammoniation may be required to reduce potential goitrin level of high glucosinolate ROMs.

Experiment 3. Weight gain, feed intake and feed efficiency of the broiler chickens of Experiment 3-A are shown in Table 8. Performance of the birds fed for 6 weeks were not statistically different among treatments. Numerically, however, weight gain and feed defficiency of the birds showed quardratic response to the level of ammoniation. Birds fed 1.5% ammoniated ROM diet performed better than other treatments while those fed 4.5% ammoniated ROM diet perfomed worst. Weekly weight gain data showed that birds fed 1.5% ammoniated ROM diet weighed less than those fed 0% ammoniated (untreated) ROM diet up to 4 weeks of age but weighed more after 5 weeks of age.

Nutrients availability of the experimental diets(Exp. 3-A) are shown in Table 9. Availability of crude protein, crude fat and crude fiber tended to be higher in 1.5% ammoniated ROM diet but availability of NFE was higher in 4.5% ammoniated ROM diet than other treatments.

Table 8. Weight gain, feed intake, feed efficiency of broiler chickens fed for 6 wks(Exp. 3-A)

Treatments	Weight gain, g	Feed intake, g	Feed efficiency
0% ammoniated ROM diet	1,735.0	3,394.9	1.96
1.5% ammoniated ROM diet	1,742.3	3,344.3	1.92
3% ammoniated ROM diet	1,676.0	3,312.4	1.98
4.5% ammoniated ROM diet	1,618.5	3,224.9	1.99
S.E.M.[1]	35.49	82.58	0.02

[1] Standard error of means

Table 9. Nutrients availability of the experimental diets, %(Exp. 3-A)

Treatments	Dry matter	Crude protein	Crude fat	Crude* fiber	N.F.E.**
0% ammoniated ROM diet	71.11	51.98	87.39	6.97[ab]	85.15[ab]
1.5% ammoniated ROM diet	71.61	54.10	89.29	15.50[a]	85.36[ab]
3% ammoniated ROM diet	69.46	50.06	87.10	5.78[b]	83.88[b]
4.5% ammoniated ROM diet	71.22	49.26	88.78	6.99[ab]	86.38[a]
S.E.M.[1]	0.87	2.31	1.71	2.63	0.32

[1] Standard error of means, [a,b]: Values with different superscript in the same column are significantly different at P<0.01(**) or P<0.05(*)

Weight gain, feed intake and feed efficiency of broiler chickens of Experiment 3-B are shown in Table 10. Birds fed 3 or 6% ammoniated ROM diet gained significantly(P<0.01) less weight than those fed 0% ammoniated(untreated) ROM diet. Birds fed 6% ammoniated ROM diet consumed significantly(P<0.05) less feed than those fed untreated ROM diet but there was no significant difference in feed efficiency.

Table 10. Weight gain, feed intake, feed efficiency of broiler chickens fed for 6 wks(Exp. 3-B)

Treatments	Weight gain,** g	Feed intake,* g	Feed efficiency
0% ammoniated ROM diet	1,768.0[a]	3,350.0[a]	1.90
3% ammoniated ROM diet	1,631.2[b]	3,180.0	1.95
6% ammoniated ROM diet	1,613.7[b]	3,107.5[b]	1.93
S.E.M.[1]	11.03	55.06	0.029

[1] Standard error of means, [a,b]: Values with different superscript in the same column are significantly different at P<0.01(**) or P<0.05(*)

Table 11 shows contents of potential hydrolysis products of glucosinolates in ammoniated ROMs which were used in Experiment 3. Goitrin level of untreated ROMs were rather low(2.34 mg in Exp. 3-A and 2.45 mg in Exp. 3-B) and ammoniation did not lower them to any considerable extent. On the other hand, level of isothiocyanates of ROM increased as the level of ammoniation increased.

Table 11. Contents of hydrolysis products of glucosinolates in ammoniated ROMs(mean±S.D.) (Exp. 3-A and B)

Treatments	Goitrin, mg/g		Isothiocyanate, mg/g	
	Exp. 3-A	Exp. 3-B	Exp. 3-A	Exp. 3-B
0% ammoniated ROM	2.34±0.28	2.45±0.33	0.91±0.03	1.18±0.13
1.5% ammoniated ROM	2.32±0.17	-	1.16±0.15	-
3% ammoniated ROM	2.51±0.18	2.35±0.27	1.21±0.04	1.58±0.16
4.5% ammoniated ROM	2.27±0.11	-	1.39±0.28	-
6% ammoniated ROM	-	2.26±0.20	-	1.56±0.07

Weight of thyroid glands of the birds fed experimental diets(Exp. 3-A and B) are shown in Table 12. Weight of thyroid glands in mg per 100 g body weight increased as the level of ammoniation of ROM in the diet increased.

Table 12. Weight of thyroid glands(Exp. 1 and 2)

Treatments	Weight, mg/100g body wt.	
	Exp. 3-A	Exp. 3-B
0% ammoniated ROM diet	6.75[b]	6.91
1.5% ammoniated ROM diet	6.80[b]	-
3% ammoniated ROM diet	9.13[ab]	7.73
4.5% ammoniated ROM diet	10.95[a]	-
6% ammoniated ROM diet	-	8.14
S.E.M.[1]	0.358	1.08

[1] Standard error of means, [a,b]: Values with different superscript in the same column are significantly different at $P<0.01$

Results of Experiment 3 shows that treatment of ROM with 3% and above level of ammonia lowers feeding value of ROM although 2 or 4% ammoniation increased TME value and crude protein availability, and 6% ammoniation decreased potential goitrin level of ROM in Experiment 2. Keith and Bell(1984) and Bell et al.(1984) reported that ammonia treatment reduces availability of lysine in the meal.

It is assumed that treatments of ROM with 3% and above level of ammonia reduced the lysine availability excessively. The fact that birds fed 1.5% ammoniated ROM diet weighed less up to 4 weeks of age but weighed more after 5 weeks of age than those fed untreated ROM diet indicates that lysine availability of ROM was marginally reduced by 1.5% ammoniation. It is well known that lysine requirement of the broiler chikens decreases sharply after 4 weeks of age. The ROM used in Experiment 2 and 3 came from same lot but Experiment 3 was conducted 2 years after Experiment 2. Stortage of ROM under room temperature for such a long period might have influenced chemical nature of ROM. As a probable consequence, the ROM used in Experiment 3 was lower in potential goitrin but higher in isothiocyanates content than the ROM used in Experiment 2. As was in Experiment 2, ammoniation increased the level of isothiocyanates in ROM.

In earlier discussion, it was mentioned that isothiocyanates combine with ammonia to form thiourea derivatives which have a goitrogenic effect(Greer, 1950, 1962). Increase in weight of thyroid glands might be the result of formation of thiourea derivatives in the ammoniated ROM. Goh et al.(1983) showed slightly enlarged thyroid glands when the birds had been fed ammonia treated canola meal. Factors reponsible for the hypertrophy of thyroid glands may also have contributed to the poor performance of the birds fed diets containing ROM treated with high level of ammonia.

Experiment 4. Table 13 shows the experimental diets which were least-cost formulated to compare SBM diet and ROM diet under practical condition. The Indian ROM used in this Experiment contained 0.83 mg of goitrin and 1.51 mg of isothiocyanates per g meal.

Results of the feeding trial of the broiler chickens are shown in Table 14. Supplemention of Avoparcin in the ROM diets significantly ($P<0.05$) improved weight gain at 3 weeks of age. At 4 weeks of age, however, significant differences among treatments disappeared but dose related tendency still existed both in SBM and ROM diets. Gain of the birds fed the ROM diets supplemented with Avoparcin were not significantly different from that of the birds fed SBM diet with or without Avoparcin supplementation in all ages. Feed efficiency of ROM diet was significantly($P<0.01$) poorer than other treatments.

Table 13. Formula and chemical composition of the experimental diets

Ingredients, %	SBM diet	ROM diet
Corn, yellow	56.078	57.083
Soybean meal	36.453	24.359
Rapeseed meal	-	8.000
Fish meal	0.726	4.551
Animal fat	3.500	3.500
Limestone	0.545	0.397
Calcium phosphate(18%)	1.776	1.299
Salt	0.288	0.219
Methionine(45%)	0.434	0.393
Broiler premix[1]	0.200	0.200
Total	100.000	100.000
Chemical composition, %		
ME, kcal/kg	3000.00	3000.00
Crude protein	21.50	21.50
Ca	0.95	0.95
P	0.47	0.48
Lysine	1.21	1.17
Methionine+Cystine	0.87	0.87

[1] Premix provides following amounts of micronutrients per kg of diet; Vit.A 10000IU, Vit.D_3 2000IU, Vit.E 8IU, Vit.K 1mg, Vit.B_1 0.5mg, Vit.B_2 5mg, Vit.B_6 0.5mg, Vit.B_{12} 0.01mg, Niacin 25mg, Ca·phantothenate 10mg, Folic acid 0.5mg, Choline 300mg, Ethoxyquin 1mg, I 0.6mg, Zn 50mg, Mn 55mg, Fe 40mg, Cu 4mg, Co 0.3mg

Table 14. Cumulative weight gain, feed intake, feed efficiency of broiler chickens fed for 4 wks

Items	Weeks	T_1	T_2	T_3	T_4	T_5	S.E.M.[1]
Weight gain, g	0	41.8	41.9	41.9	41.8	41.8	0.07
	0-1	109.8	102.3	101.7	106.6	110.9	3.98
	0-2	328.4	331.3	311.8	333.2	337.8	7.45
	0-3*	683.2[ab]	698.6[b]	657.8[a]	696.6[b]	702.3[b]	9.78
	0-4	1114.2	1125.9	1101.2	1127.1	1138.3	10.59
Feed intake, g	0-1	127.4	116.1	127.3	130.6	128.8	4.60
	0-2	430.5	421.1	437.6	444.5	446.7	10.78
	0-3	957.6	951.1	975.5	978.7	978.2	15.33
	0-4	1675.8	1671.3	1719.3	1703.9	1696.8	22.61
Feed efficiency	0-1**	1.16[a]	1.13[a]	1.25[b]	1.22[b]	1.16[a]	0.02
	0-2**	1.31[ab]	1.27[a]	1.40[c]	1.33[b]	1.32[b]	0.02
	0-3*	1.40[b]	1.36[a]	1.48[c]	1.41[b]	1.39[ab]	0.01
	0-4**	1.50[a]	1.48[a]	1.56[b]	1.51[a]	1.49[a]	0.01

T_1: Soybean meal diet, T_2: T_1+Avoparcin 10ppm, T_3: Rapeseed meal diet, T_4: T_3+Avoparcin 10ppm, T_5: T_3+Avoparcin 15ppm

[1] Standard error of means, [a-c]: Values with different superscript in the same row are significantly different at $P<0.01$(**), $P<0.05$(*)

Mortality, dressing percentage and weight of thyroid glands are shown in Table 15. Mortality was high in birds fed ROM diet(T_3) and low in those fed SBM diet but they were not significantly different. Dressing percentage of the birds fed ROM diet(T_3) was lower than those fed other diets. Weight of thyroid glands of the birds were not significantly different among treatments.

Table 15. Mortality, dressing percentage and weight of thyroid glands

Item	T_1	T_2	T_3	T_4	T_5	S.E.M.[1]
Mortality[2], %	0.00	8.00	12.00	2.00	10.00	4.05
Dressing %[3]	69.02	68.56	66.95	69.89	69.20	-
Thyroid glands (mg/100g body wt)	6.57	8.43	7.73	8.99	7.50	1.04

T_1: Soybean meal diet, T_2: T_1+Avoparcin 10ppm, T_3: Rapeseed meal diet, T_4: T_3+Avoparcin 10ppm, T_5: T_3+Avoparcin 15ppm

[1] Standard error of means, [2] Natural death and culled from 0 to 4 wks, [3] Mean of 15 birds in each treatments

Improvement of weight gain and feed efficiency by Avoparcin supplementation to ROM diets agree with the report of Jensen and Thomsen(1980). Extent of improvement of feed efficiency also agrees with the report of Spoerl and Kirchgessner(1978), in which supplementation of Avoparcin at the level of 7.5-15.0 ppm improved feed efficiency of broiler diet by 3.6%.

It is interesting to note that poor dressing percentage of ROM diet was improved by supplementation of Avoparcin. Since the Indian ROM was low in potential goitrin content(0.83 mg/g) and isothiocyanates are known be less goitrogenic compared to goitrin, 8% inclusion of the Indian ROM in the broiler diet might not have influenced weight of thyroid glands. Therefore, improvement of performance of the birds in Avoparcin supplemented diets should be explained by general growth promoting effects of antibiotics.

However, greater response of supplementation in ROM diets than in SBM diet may warrant further studies.

CONCLUSIONS

Supplementation of iodine to the diets containing high level of ROM did not improve performance of the rats. Treatment of ROM with ammonia at 2 or 4% increased metabolizable energy content and improved availability of dry matter, crude protein and crude ash. Potential goitrin level of ROM did not decrease at 2 or 4% ammoniation but considerably decreased at 6% ammoniation. On the other hand, potential isothiocyanates level of ROM increased by ammoniation.

Although the ammoniation of ROM showed beneficial effects in some of the parameters studied, ammoniation at the level of 3% and above decreased weight gain of the broiler chickens probably due to reduced lysine availability of ROM. Moreover, weight of thyroid glands increased as the level of ammoniation increased. It may not be advisable to ammoniated ROM at the level of 3% and above when the treated ROM should be used in the monogastric animal diets which is marginally deficient in lysine content.

Improvement of growth rate and dressing percentage of broilers were observed when the ROM diets had been supplemented with Avoparcin. It is noteworthy that greater response from Avoparcin supplementation was shown with ROM diet than with SBM diet.

REFERENCES

A.O.A.C. 1984. Official Methods of the Association of Official Analytical Chemist(14th ed.). A.O.A.C., Washington, D.C.

Appelqvist, L. Å. and E. Joseffson. 1967. Method for Quantitative Determination of Isothiocyanates and Oxazolidinethiones in Digests of Seed Meals of Rape and Turnip Rape. J. Sci. Food Agric. 18:516-519.

Bell, J. M., L. W. McCuaig and A. Shires. 1980. Effect of supplementary iodine, iodinated casein, lysine and methionine on the nutritive value of Tower canola meal for swine. 6th Progress Report. Research on Canola Seed, Oil, Meal and Meal Fractions. Canola Council of Canada. pp.182-185.

Bell, J. M., M. O. Keith, J. A. Blake, and D. I. McGregor. 1984. Nutritional evaluation of ammoniated Mustard Meal for use in swine feed. Can. J. Anim. Sci. 64:1023-1033

Blake, J. A. 1983. The production of ammoniated canola meal. 7th Progress Report. Research on Canola Seed, Meal and Meal Fractions. Canola Council of Canada. pp. 123-127

Goh, Y. K., A. Shires, A. R. Robblee, and D. R. Clandinin. 1983. The effect of ammoniation on the nutritive value of canola meal for chicks. 7th Progress Report. Research on Canola Seed, Meal and Meal Fractions. Canola Council of Canada. pp. 128-132.

Greer, M. A. 1950. Nutrition and goiter. Physiol. Rev. 30:513-518

Greer, M. A. 1962. II. Thyroid Hormones. The natural occurrence of goitrogenic agents. Recent Progr. Hormone Res. 18:187-219

Hill, F. W. and D. L. Anderson. 1958. Comparison of metabolizable energy and productive energy determinations with growing chicks. J. Nutr. 64:587.

Jensen, J. F. and M. G. Thomsen. 1980. Rapsskrå og avoparcin i slagtekyllingefoder. Statens Husdyrbrugsforsøg. Meddelese.

Keith, M. O. and J. M. Bell. 1984. Effects of ammoniation of canola(low glucosinolate raprseed) meal on its nutritional value for the rat. Can. J. Anim. 64:997-1004.

Kirchgessner, M. and F. X. Roth. 1981. Nutritional-Physiological prerequisites for supplementing feed with growth promotants. Proc. of the International Conference on Feed Additives, Hungarian Society of Agr. Sci. Vol. 1:11-20.

Kjaer, A. 1960. Naturally derived isothiocyanates(mustard oil) and their parent glucosides. Fortschr. Chem. Org. Naturstoffe. 18:122-176.

Klain, G. J., D. C. Hill, H. D. Branion, and J. A. Gray. 1956. The value of rapeseed oil meal and sunflower seed oil meal in chick starter rations. Poult. Sci. 35:1315.

Lee, N. H., C. S. Kim, and J. L. Yuk. 1973. Studies on the metabolizable energy of some oil seed meals for broiler chickens. Korean J. Anim. Sci. 15(1):29-44.

Lodhi, G. H., R. Renner, and D. R. Clandinin. 1969. Studies on the metabolizable energy of rapeseed meal for growing chicks and laying hens. Poult. Sci. 49:289-294.

Nordfeldt, S., N. Gellerstedt, and S. Falkmer. 1959. Studies on rapeseed meal and its goitrogenic effect on pigs. Microbiological. Scandinavica. 35:217-231.

Ochetim, S., J. A. Bell, C. E. Doige, and C. G. Youngs. 1980. Can. J. Anim. Sci. 60:407-421

Paik, I. K., I. K. Han, and C. S. Kim. 1975. Studies on the nutritive values of locally produced oil meals. II. Comparative studies on the feeding values of various oil meals in broiler chickens. Korean J. Anim. Sci. 17(4):348-358.

Paik, I. K., A. R. Robblee, and D. R. Clandinin. 1980. The effect of sodium thiosulfate and hydroxocobalamin on rats fed nitrile-rich or goitrin-rich rapeseed meals. Can. J. Anim. Sci. 60:1003-1013.

Sell, J. L. 1966. Metabolizable energy for rapeseed meal for the laying hen. Poult. Sci. 45:854-856.

Sibbald, I. R. and S. J. Slinger. 1963. Factors affecting the metabolizable energy content of poultry feeds. 12. Protein Quality. Poult. Sci. 42:707-710.

Sibbald, I. R. 1986. The TME system of feed evaluation. Animal Research Centre, Ottawa, Canada. pp. 46-47.

Spoerl, V. R. and M. Kirchgessner. 1978. Arch. Geflugelk. 42:52-55.

Steel, R. G. D. and J. H. Torrie. 1980. Principles and Procedures of Statistics, 2nd ed. McGraw-Hill Book Co. Inc., New York, N. Y.

VanEtten, C. H. 1969. Goitrogens. Toxic Constituents of Plant Foodstuffs. Food Sci. and Technology. A Series of Monographs. Academic Press. 103-142.

30

IMPROVEMENT IN THE NUTRITIONAL QUALITY OF BREAD

Mendel Friedman[1] and Paul-Andre Finot[2]

[1]Western Regional Research Center, U.S. Department of Agriculture, Agricultural Research Service, 800 Buchanan Street, Albany, CA 94710

[2]Nestlé Research Centre, Nestec Ltd., Vers-chez-les-Blanc 1000 Lausanne, 26 Switzerland

ABSTRACT

To assess whether the dipeptide N-ε-(γ-L-glutamyl)-L-lysine (glutamyl-lysine) can serve as a nutritional source of lysine, we compared the growth of mice fed (a) an amino acid diet in which lysine was replaced by six dietary levels of glutamyl-lysine; (b) wheat gluten diets fortified with lysine; (c) a wheat bread-based diet (10% protein) supplemented before feeding with lysine or glutamyl-lysine (0, 0.75, 1.50, 2.25, and 3% lysine HCl-equivalent in the final diet), not co-baked and (d) bread diets co-baked with these levels of lysine or glutamyl-lysine. With the amino acid diet, the relative growth response to glutamyl-lysine was about half that of lysine. The effect of added lysine on the nutritional improvement of wheat gluten depended on both lysine and gluten concentrations in the diet. With 10 and 15% gluten, 0.37% lysine HCl produced a marked increase in weight gain. Further increase in lysine HCl to 0.75% proved deterimental to weight gain. Lysine HCl addition improved growth at 20 and 25% gluten in the diet and did not prove detrimental at 0.75%. For whole bread, glutamyl-lysine served nearly as well as lysine to improve weight gain. The nutritive value of bread crust fortified or not was markedly less than that of crumb or whole bread. Other data showed that lysine or glutamyl-lysine at the highest level of fortification, 0.3%, improved the protein quality (PER) of crumb over that of either crust or whole bread, indicating a possible greater availability of the second-limiting amino acid, threonine, in crumb. These data and additional metabolic studies with U-^{14}-C glutamyl-lysine suggest that glutamyl-lysine, co-baked or not, is digested in the kidneys and utilized *in vivo* as a source of lysine; it and related peptides merit further study as a sources of lysine in low-lysine foods.

INTRODUCTION

Wheat gluten, the major protein in many baking formulations, is considered a poor quality protein, primarily because it has insufficient amounts of two essential amino acids: lysine, the first limiting amino acid, and threonine, the second-limiting one (Jansen,

Nutritional and Toxicological Consequences of Food Processing
Edited by M. Friedman, Plenum Press, New York, 1991

1981). To compensate for the poor quality of most cereal proteins such as gluten, the minimum recommended daily allowance (RDA) for these proteins has been set at 65 g compared to 45 g for good-quality proteins such as casein (Hegsted, 1977; Sarwar et al., 1983, 1984, 1985).

As noted by Ziderman and Friedman (1985), during baking, the mixture of protein, carbohydrate, and water plus additives in dough is exposed to two distinct transformations. Desiccation of the surface on exposure to temperatures reaching 215°C produces the crust. The crust encloses part of the dough in steam phase at approximately 100°C, resulting in the formation of the crumb.

Because lysine's ε-amino group interacts with food constituents to make it nutritionally less available (Bjorck et al., 1983; Finley and Friedman, 1973; Friedman, 1977; Friedman, 1982; Geervani and Devi, 1986; Gumbmann et al., 1983; Tsen and Reddy, 1977; Tsen et al., 1977; Sherr et al., 1989), the baking process further reduces the dietary availability and utilization of lysine, especially in the crust, which makes up about 40% of the bread by weight. Many such interactions have been described including (a) reaction of the amino group with carbonyl groups of sugars and fatty acids to form Maillard browning products; (b) formation of cross-linked amino acids such as lanthionine, lysinoalanine, and glutamyl-lysine; (c) interaction with tannins and quinones; and (d) steric blocking of the action of digestive enzymes by newly introduced crosslinks, as well as native ones such as disulfide bonds (Friedman, 1977, 1982; Otterburn, 1989; Otterburn et al., 1977). Because these reactions of lysine with other dietary components may lead to protein damage and to the formation of physiologically active compounds, an important objective of food science and nutrition is to overcome these effects.

In principle, it is possible to enhance the nutritional quality of bread by either (a) creating new wheat varieties through plant genetic manipulations with improved protein quality characteristics (Johnson and Mattern, 1978); (b) by covalent attachment of lysine residues to food proteins (Chan et al., 1979; Ikura et al., 1985; Iwami and Yasumoto, 1986); and (c) by amino acid fortification (Betschart, 1978). A major problem encountered when fortifying with free lysine is that the added amino acid can itself participate in browning and other side reactions. Because ε-acyl-lysine derivatives are less susceptible to Maillard reactions than free lysine (Finot et al., 1978), the main objective of this study was to compare the effectivness of lysine and glutamyl-lysine to serve as a nutritional source of lysine for mice fed bread crust, crumb, and whole bread co-baked with these amino acids. Since the ε-NH_2 group of N-ε-(γ-L-glutamyl)-L-lysine (glutamyl-lysine) is blocked in the form of an isopeptide bond with the γ-COOH group of glutamic acid, as illustrated, expectations were that it should also undergo less damage than lysine during baking.

$$\overset{\varepsilon}{NH_2}CH_2CH_2CH_2CH_2CH(NH_2)COOH \quad HOOCCH(NH_2)CH_2\overset{\gamma}{CH_2}\text{-}CO\text{-}NH\text{-}\overset{\varepsilon}{CH_2}CH_2CH_2CH_2CH(NH_2)COOH$$

<u>L-lysine</u> <u>N-ε-(γ-L-glutamyl)-L-lysine</u>

MATERIALS

Unbleached, unbrominated, malted, and enriched white wheat flour (Mellow Judith) was obtained from Con Agra Inc., Oakland, CA. Fresh yeast was obtained every two weeks as a gift from Red Star Yeast,

Oakland, CA. L-lysine HCl and commercial wheat gluten were obtained from U.S. Biochemical Corp., Cleveland, Ohio, and (U-^{14}C)-L-lysine came from Amersham, England. The other lysine compounds were synthesized as previously described (Finot et al., 1978). The fermentation chamber and the roller sheeting/loaf shaping machine were made by National MFG Co., Lincoln, NE. The humidity in the chamber was monitored with a wet and dry bulb hygrometer made by Premium Instruments, Chicago, IL. The gas oven was an "EZ" by E.J. Chubbuck, Co., Inc., Oakland, CA. Oven racks rotated during baking to maintain uniformity. The dough kneader was a Hobart MFG Co. model #C100 (Troy, Ohio). The pans were EKCO Baker's Secret, size 14.6 X 7.6 X 5.4 cm, with a non-stick coating, obtained in a local market.

Lysine derivatives, N-ε-acetyl-L-lysine, N-ε-formyl-L-lysine, N-ε-benzylidene-L-lysine were synthesized as previously described (Finot et al., 1978). N-ε-(γ-L-glutamyl)-L-lysine used for the feeding tests was synthesized according to the patent mentioned in the same paper. The synthesis of N-α, N-ε-diformyl-L-lysine was adapted from that of N-ε-formyl-L-lysine using an excess of acetic anhydride. Radioactive lysine derivatives were obtained using uniformly labelled (U)-^{14}C-L-lysine. The specific activities were 0.38, 0.38, 0.38, and 0.89 mCi/mmole for ^{14}C-L-lysine, N-α-formyl-^{14}C-L-lysine, N-ε-formyl-^{14}C-L-lysine, and N-ε-(γ-L-glutamyl)-L-lysine respectively.

The purity of the lysine compounds listed in the tables was confirmed by previously described analytical procedures (Finot et al., 1978). In addition, ion-exchange chromatography of N-ε-(γ-L-glutamyl)-L-lysine on an amino acid analyzer (Friedman et al., 1979) produced a single peak eluting in the same position as methionine. The same peak resulted from chromatography of a commercial sample of the glutamyl-lysine obtained from Sigma (St. Louis, MO).

METHODS

<u>Baking Experiments</u>. The recipe for one loaf of bread consisted of: 183.1 gm flour; 106.8 g water; 3.5 g NaCl; 6.1 g yeast; and lysine or glutamyl-lysine as applicable. The level of lysine fortification was as follows: L-lysine-HCl was added to equal 0, 0.75, 1.50, 2.25, and 3.0% of the protein in the flour. Glutamyl-lysine fortification levels were the molar equivalent of the lysine-HCl series. The water and flour were brought to 30°C and maintained at this temperature in an incubator overnight. The salt, yeast, and lysine additives were solubilized in the water and then added to the flour. The mixture was kneaded on a Hobart at speed #1 for 30 sec, then at speed #2 for a total of 8 min. The dough was divided into workable balls, approximately 3 loaves per ball. The balls were placed in large stainless steel bowls, smooth side up, and placed in a fermentation chamber (37°C, 90% humidity) for 45 min. The dough was then turned out on a board and divided into loaves. The loaves were made into oval balls and de-gassed by running through the roller sheeting machine. Each sheet was rolled into a loaf, using pressure to exclude air bubbles. The loaves were placed in the pans, smooth side up and returned to the fermentation chamber (37°C, 90% humidity) for 45 min. They were baked for 35 min at 215°C, removed from the oven and the pans, and left to cool on open racks. Using an electric knife, 2/3 of the loaves were trimmed of crust. Crumb is the soft part of the bread produced during removal of the crust using the electric knife. Crust is the brown portion which contains a small

fraction of the crumb that sticks to it during cutting. Crumb, crust, and whole bread were lyophilized and ground in a Wiley mill with a 1 mm screen. The resulting flours were analyzed for nitrogen on an Erba model #1400 automatic nitrogen analyzer.

Bioassays. Biological utilization of several lysine derivatives as nutritional sources of lysine was tested by a 14-day growth assay in mice (Swiss Webster strain, Simonsen Laboratories, Inc., Gilroy, CA) by using an amino acid diet (Table 1) in which lysine was omitted, to be partly or fully replaced by the derivative to be tested as described for each experiment (Friedman and Gumbmann, 1981; 1988). Mice were housed singly or two per cage. The cages were polycarbonate with stainless steel wire tops and pine shavings for litter. Feed and water were provided *ad libitum*. The temperature of the animal room was 22 ± 1°C and humidity was maintained at 50 ± 10%. The light cycle was 0600-1800 h light and 1800-0600 h dark, as regulated by an automatic timer. Animals were assigned so that all treatment groups had nearly the same initial body weight.

The effect of lysine fortification on the weight gain and protein efficiency ratio (PER), defined as weight gain/protein intake, of wheat gluten was examined by feeding mice 10, 15, 20, and 25% wheat gluten in the diet (Table 2) fortified with 0, 0.37, and 0.75% lysine HCl for 21 days (Friedman et al., 1987).

Bread (whole, crust, or crumb), with and without lysine or glutamyl-lysine fortification, was fed to weanling mice for 14 days as the sole source of protein in the diet (Table 2). All diets contained 10% protein (N X 6.25) plus various levels of lysine HCl or glutamyl-lysine either co-baked or not co-baked (supplemented at time of feeding).

Cholesterol Assays. At the end of the 14-day feeding study, serum samples obtained from the brachial artery by axillary space incision were analyzed for total cholesterol content by a modification of the Lieberman-Burchard reaction on a Technicon AutoAnalyzer II.

Metabolic Studies. These were carried out as described previously (Finot et al., 1978). The metabolic transit of the radioactive lysine derivatives was studied in rats (80-100 g) kept for 24 hours in a metabolic cage designed to collect $^{14}CO_2$, urine, and feces, seprarately. The animals received by stomach tube after one night fasting, 5 µCi of each lysine derivative solubilized in 1 mL water. The radioactivity excreted during the 24 hour experiment was expressed in % of the ingested dose.

Statistics. The relative potencies of the lysine derivatives as a nutritional source of lysine were calculated as the slope ratios or horizontal distances between the growth curves (weight gain after 14 days per unit of dietary concentration) for which linearity was approximated. Slope ratio and parallel line analyses of growth data were performed as described by Finney (1978) with SAS using the General Linear Model (GLM) procedure (SAS, 1987). Confidence intervals of potencies relative to lysine were estimated using Fieller's theorem (Zerbe, 1978). Individual body weights for the generally six mice per diet group were used in the bioassay analyses. All comparisons are based on lysine equivalents.

For Table 3, data for relative dietary levels of amino acid diets greater than 50% was deleted for the L-lysine diet (1.35% lysine HCl)

Table 1. Composition of Amino Acid Basal Diet

Ingredient	%
L-Ala	0.35
L-Arg.HCl	1.35
L-Asn	0.60
L-Asp	0.35
L-Glu	3.50
Gly	2.33
L-His.HCl	0.41
L-Ile	0.82
L-Leu	1.11
L-Lys-HCl	1.35
L-Met	1.17
L-Phe	1.51
L-Pro	0.35
L-Ser	0.35
L-Trp	0.174
L-Thr	0.82
L-Val	0.82
Alphacel[a]	3.00
Corn Oil	8.00
Cornstarch	20.00
Dextrose	38.33
Salts USP XIV[b]	5.00
Sodium acetate	1.31
Water (added)	5.00
Complete vitamin mixture[b]	2.00
Total	100.00

[a]Nutritional Biochemical Corp., Cleveland, OH.

[b]Friedman and Gumbmann, 1988.

Table 2. Composition of Protein Basal Diets

Ingredient	Amount
	%
Protein	Variable, at the expense of cornstarch
Alphacel	3.00
Corn oil	8.00
Cornstarch	(21.7), variable depending on protein level
Dextrose	43.3
AIN Mineral Mixture[a]	5.00
Water (added)	5.00
Complete vitamin mixture[a]	2.00
Total	100.00

[a] Friedman and Gumbmann, 1988.

since it exceeded the linear portion of the response curve. The slope ratio analysis requires two assumptions: linearity and common intercepts. The test for curvature is non-significant, P = 0.22, but the test for departure from common intercepts is significant, P = 0.002. Despite this violation of assumption, the slope ratio estimates still provide useful approximations of potencies in this case since the reduction in R^2 (coefficient of determination) is only from 0.937 to 0.926 from forcing the common intercept. (In looking at the graph, the potencies will be slightly under-estimated.) The most useful information in the output are the slopes and standard errors, and the potency ratios and their 95% confidence intervals.

The data in Tables 5, 6, 7, and 9 were subjected to standard parallel line analyses. In each case, the tests for lack of parallelism are non-significant (P < 0.10). For this type of assay, the potencies represent the distances between lines in the abscissa direction (concentration or % gluten for Table 9).

RESULTS AND DISCUSSION

Utilization of Lysine Derivatives in Mice. To assess the potential value of several lysine derivatives as nutritional sources of lysine, growth response to N-ε-acetyl-, N-ε-benzylidene, N-α-, N-ε-diformyl-, and glutamyl-lysine was determined in mice. Dose-response data were obtained as permitted by the supply of test material on hand. Table 3 shows that benzylidene-lysine produced the greatest weight gain after lysine, followed by glutamyl-lysine. Acetyl- and formyl-lysines were equivalent at a relative concentration of about 30 to 50%, depending on the method used to calculate potency; however, multi-level feeding of formyl-lysine was not done. Diformyl-lysine barely supported growth.

The data are best visualized when plotted as shown in Figure 1. The dose-response curves show that relative responses varied with the dietary concentrations. It is also possible to rank the amino acid derivatives according to a relative growth response. This was defined as follows: percent relative weight gain equals the net growth response between 0 and 50% relative concentration of amino acid in the diet (Figure 1) divided by the growth response of L-lysine times 100. The calculated values in Table 4 show a range of values from 16.8 for diformyl-lysine to 85.4% for benzylidene-lysine, approximately of the same order as observed with rats (Finot et al., 1978). Note that glutamyl-lysine, the main object of our study, had a lysine-equivalent value of 51.1%. Thus, any assessment of the relative damaging effects of baking on lysine or glutamyl-lysine should take into account the relative potencies of these two compounds in the amino acid diets. Note also that the slope ratio statistical analysis gave only a 45% lysine-equivalent for the lysine part in glutamyl-lysine. The corresponding value for N-ε-acetyl-lysine was 30%.

A surprising result is that the lysine-equivalent value of glutamyl-lysine is about 45-50% when evaluated as part of an all-amino-acid diet (Table 4) but nearly 100% when the dipeptide is added to diets containing wheat gluten (Tables 5-8). A possible explanation is that provision of maintenance lysine levels by gluten in the bread-based diets tends to reduce immediate need for lysine derived from added glutamyl-lysine. Since glutamyl-lysine has to reach the rat kidneys before it can be hydrolyzed to lysine, its utilization is delayed about 2 hr compared to free or protein- bound lysine There is an apparent difference in the bioavailability of lysine from the two different diets. A more likely explanation is that, because glutamyl-lysine is transported across the intestinal tract by passive

Table 3. Utilization of Lysine Derivatives Compared to that of L-Lysine in Mice in an Amino Acid Diet[a]

		Body Weight Gain (g)[b]				
Relative dietary level(%) L-lysine	L-Lysine	N-ε-(γ L-glutamyl) L-lysine	N-ε-acetyl-	N-ε-Formyl-	N-ε-benzylidene-L-lysine	N-α,N-ε-Diformyl-L-lysine
300	11.7					
200	13.0					
100	13.8	10.5	6.3			
75	13.5	7.8				
50	12.2	5.5	4.0	4.0	10.2	0.8
25	5.8		3.2	1.5		
12.5	1.8	0.7				
6.25	0.3					
0	-1.5					
Slope ratio to L-lysine		0.45	0.30			
95% C. I.[c]		(0.42, 0.50)	(0.25, 0.34)			

[a] Lysine derivatives were fed on an equal molar basis to L-Lysine.

[b] Six mice per group; initial body weight, 10.5 g. Weight gain at 14 days.

[c] C.I. = confidence interval; values in parentheses show lower and upper limits.

Table 4. Relative Growth Response to L-Lysine Derivatives in Mice[a]

Amino Acid	Relative weight gain[b]	Slope ratio analysis[c]
	%	%
L-lysine . HCL	100.0	
N-ε-benzylidine-L-lysine	85.4	
N-ε-(γ-L-glutamyl)-L-lysine	51.1	45
N-ε-acetyl-L-lysine	40.0	30
N-ε-formyl-L-lysine	40.0	
N-α, N-ε-diformyl-L-lysine	16.8	

[a]Comparisons made at 50% relative concentration in the diet, which is equal molar to 0.68% L-lysine HCl.

[b]Percent relative weight gain equals the net growth response between 0 and 50% relative concentration of amino acid in the diet divided by that of L-lysine HCl times 100.

[c]Percent weight gain equals slope ratio to L-lysine (Table III) times 100.

Table 5. Weight Gain and PER in Mice Fed Whole Bread with L-Lysine or Glutamyl-lysine Supplementation in the Diet (not co-baked)

Added amino acid concentration[a] %	Weight gain[b]		PER[c]	
	Lysine	Glutamyl-lysine	Lysine	Glutamyl-lysine
0	5.0	5.0	1.0	1.0
0.075	7.8	6.2	1.4	1.0
0.150	8.8	6.7	1.6	1.4
0.225	9.7	9.3	1.8	1.8
0.300	11.0	9.5	1.9	1.8
Standard error			± 0.7	± 0.3
Potency difference[d]		0.069		
95% C.I.		(0.026, 0.114)		

[a]L-lysine HCl concentration in the diet as shown; glutamyl-lysine added to achieve molar equivalent levels (factor = 275.3/182.6 = 1.51 x lysine concentration).

[b]Mean weight gain at 14 days, 6 mice per group.

[c]PER based on 2 mice per cage, N = 3.

[d]Potency difference = added % glutamyl-lysine concentration required to achieve equivalent weight gain as with L-lysine.

Table 6. Weight Gain and PER in Mice Fed Whole Bread, Bread Crumb, and Bread Crust Supplemented with Lysine Before and After Baking

Lysine HCl concentration	Weight gain (g)[a]			PER[a]		
%	Whole bread	Bread crumb	Bread crust	Whole bread	Bread crumb	Bread crust
			Co-baked			
0	4.0	4.5	2.7	0.9	0.9	0.5
0.075	6.5	8.8	2.8	1.3	1.7	0.6
0.150	7.7	9.8	4.0	1.4	2.1	0.9
0.225	9.3	9.7	5.8	1.8	1.9	1.1
0.300	8.8	10.2	6.5	1.6	1.9	1.3
Standard error		± 0.6			+ 0.1	
Potency difference[b]	- 0.085	0.185				
95% C.I.		(- 0.148, - 0.029)	(0.124, 0.263)			
			Not co-baked			
0	5.0	6.3	2.2	1.0	1.2	0.6
0.075	7.8	8.8	2.8	1.4	1.9	0.7
0.150	8.8	9.5	6.0	1.6	1.9	1.5
0.225	9.7	9.8	6.7	1.8	2.2	1.4
0.300	11.0	9.8	7.7	1.9	3.3	1.7
Standard error		± 0.7			+ 0.3	
Potency difference		- 0.055	0.207			
95% C.I.		(- 0.135, 0.016)	(0.128, 0.318)			

[a]Mean weight gain at 14 days, 6 mice per group. Casein control = 12.0 g.

[b]Potency difference X 100 = % additional lysine HCl needed to achieve weight gain as with whole bread.

Table 7. Weight Gain and PER in Mice Fed Whole Bread, Bread Crust, and Bread Crumb Co-baked with Glutamyl-Lysine[a]

Glutamyl-lysine concentration (lysine equivalent) %	Whole bread	Bread crumb	Bread crust
		Weight gain	
0	5.2	7.5	3.0
0.075	6.5	9.0	3.5
0.150	7.3	9.8	3.7
0.225	8.5	9.7	3.7
0.300	9.8	9.5	5.8
Standard error		$\pm$ 0.5	
Potency difference[b]		- 0.201	0.376
95% C.I.		(- 0.325, - 0,116)	(0.265, 0.560)
		PER	
0	0.92	1.2	0.6
0.075	1.3	1.6	0.6
0.150	1.2	1.6	0.7
0.225	1.3	1.6	0.6
0.300	1.5	1.7	1.0
Standard error		$\pm$ 0.1	

[a]Mean PER at 14 days, 2 mice per cage, N = 3. Mean weight gain at 14 days, 6 mice per group.

[b]Potency difference X 100 = % additional glutamyl-lysine needed to achieve weight gain as with whole bread.

Table 8. Percent of radioactivity Excreted in CO_2 Urine and Feces 24 h after Intragastric Ingestion of U-^{14}C-lysine Derivatives (rat studies)

	Nε-L-formyl-Lysine (n = 2)	L-lysine (n = 4)	Nα-formyl-L-lysine (n = 2)	N-ε-(α-L-glutamyl)-(L-lysine (n = 2)
CO_2	18.5 $\pm$ 0.9	17.3 $\pm$ 2.1	12.7 $\pm$ 3.0	28.0 $\pm$ 4.0
Urine	5.5 $\pm$ 0.6	25.1 $\pm$ 2.2	73.6 $\pm$ 3.6	2.75 $\pm$ 0.5
Feces	7.6 $\pm$ 1.1	3.2 $\pm$ 0.6	5.6 $\pm$ 0.8	35.7 $\pm$ 10.0

transport which is disturbed by the high osomolarity of the free amino acids in the amino acid diets, its absorption is reduced compared to glutamyl-lysine-fortified protein diets (Reichl, 1989; Scharrer, 1989).

Metabolic Studies. The metabolic fates of the lysine derivatives are relevant to their possible nutritional value and safety. Preliminary indications (Finot et al., 1978; Hurrell and Carpenter, 1978) suggest that the derivatives go through different metabolic pathways than does lysine. The following are some additional findings in rats. L-lysine, N-α-formyl-L-lysine, N-ε-formyl-L-lysine, glutamyl-lysine, all labelled with ^{14}C-lysine, were fed to rats by stomach tube. The resulting $^{14}CO_2$, urine, and feces were collected for 24 hr in metabolic cages. The N-α-formyl-lysine, which is not utilized as a source of lysine, was mainly excreted in the urine (Table 8), showing that this molecule was not deformylated by the kidneys. The N-ε-formyl-lysine, which is 50% utilized as a source of lysine, was also mainly excreted in the urine. Evidently, this molecule cannot be completely deformylated by the kidney. In contrast, glutamyl-lysine, which is a good source of lysine, was not excreted in the urine. The low level of digestibility as measured by the high level of radioactivity (35.7%) in the feces is probably due to route of administration through tubing inserted into an empty stomach. When given in a diet for several days in a bioavailability trial, this molecule was more efficiently absorbed by the gut than when it was given by stomach tubing in fasted animals.

The incorporation of the labelled compounds into intestnal mucosa, liver, and kidney tissues 24 h after ingestion was found to be the same for both lysine and glutamyl-lysine. In contrast, when the two compounds were administered intravenously, the $^{14}CO_2$ peak in the expired air appeared approximately between 165 and 195 min after administration of glutamyl-lysine, and only after about 30 min following injections of lysine. These and additional studies with tissues homogenates suggest that glutamyl-lysine has to reach the kidneys before it is hydrolyzed to lysine.

The $^{14}CO_2$ comes from the catabolism of lysine in the gut and from its catabolism in the body. The N-α-formyl-^{14}C-lysine was found to be excreted in the urine as N-α-formyl-lysine only. This molecule was not transformed in the body.

N-ε-formyl-^{14}C-lysine was detected in traces in the urine. The main radioactive product found in the urine had an acidic behaviour on ion-exchange chromatography and on paper electrophoresis. On acid hydrolysis, this mole- cule regenerated lysine. The main urinary catabolite is probably N-ε-formyl-N-α-acetyl-lysine, formed by the kidney to detoxify N-ε-formyl-L-lysine by acylation of the α-amino group.

These metabolic studies show that the bioavailability of the acyl derivatives of lysine depends on the absorption rates (passive diffusion mechanism) by the intestine and by the capacity of the kidneys to liberate lysine. The kidneys have the ability to liberate ε- but not α-amino groups from acylated lysine derivatives.

L-Lysine Fortification of Wheat Gluten. In addition to the already-mentioned studies on the utilization of lysine derivatives as part of an all-amino-acid diet, we studied the availability of lysine when added to wheat gluten (Table 9; Figures 2-3) in order to design quantitative studies of the effect of fortifying bread with lysine (Table 3).

As expected, weight gain and PER were improved by the addition of lysine. The effect of any given level of added lysine however, depended on the level of gluten in the diet. With 10% and 15% gluten, 0.37% lysine produced a marked increase in weight gain and improvement in protein quality as measured by PER (Figures 2 and 3). Further increase in lysine to 0.75% proved to be somewhat detrimental. Gluten at 20% and 25% in the diet produced better growth than gluten at 10% or 15% without lysine supplementation. Lysine addition improved growth at the higher gluten levels, and surprisingly did not prove detrimental at 0.75%. At lower gluten levels, the ratio of added lysine to total protein at 0.75% lysine apparently unbalanced the essential amino acids causing one other than lysine to become nutritionally-limiting, and thereby reducing protein nutritional quality and growth.

The absolute PER of gluten was markedly superior at 10% gluten plus 0.37% lysine than at higher levels of gluten (Figure 3). Except for unsupplemented gluten, increasing the amount of gluten in the diet decreased PER; i.e. less growth occurred per unit of protein eaten. This is a well known characteristic of protein nutrition (Owens and Pettigrew, 1989): after sufficient protein is supplied to meet basic demands, increased amounts of protein allow the animal to use it for purposes other than growth, thus decreasing the efficiency of utilization for growth (PER).

Another interesting observation is that plots of body weight gain versus percent of gluten in the diet supplemented with 0, 0.37 or 0.75% lysine HCl produced two parallel lines (Figure 3). These plots show (a) a linear increase in weight gain with gluten levels in the diet ranging from 10 to 25%; (b) a dramatic enhancement in weight gain with the gluten diets containing added 0.37% lysine HCl; and (c) weight gains with the 0.37% lysine HCl diets were essentially the same as with the 0.75% lysine diets (Table 9).

These phenomena merit further analysis, first in terms of the amino acid composition of gluten and the nutritional requirements of mice. For gluten, lysine is the first-limiting amino acid (N ≈ 1.4 g/16 g N) and threonine the second (N~2.5 g/16 g N). For mice, the minimum lysine and threonine requirements are about 0.4% each in the diet (Bell and John, 1981). If we calculate the levels of lysine and threonine in the diets, it can be shown that lysine is limiting in only a few diets (10% gluten diets, for example) and that threonine is limiting only in diets containing about 10 to 15% gluten, as illustrated in Table 10. These considerations suggest that no improvement in growth would be expected for diets fortified with 0.37% lysine HCl and 0.75% lysine HCl because lysine is not limiting.

Second, amino acids are used both metabolically as building blocks for protein biosynthesis and catabolically as energy sources. Catabolism for most amino acids proceeds through transamination pathways; the exceptions are lysine and threonine (Lougnon and Kiener, 1989; Khan-Siddiqui, 1989; Milner, 1989). These two amino acids (which are nutritionally limiting in wheat gluten) are catabolized by non-amino transferase specific enzymes: threonine dehydratase acts on threonine and lysine ketoglutarate reductase on lysine (Hegsted, 1977).

Table 9. Effect of Unsupplemented and Lysine-supplemented Gluten on Weight Gain (g) in Mice after 21 Days

Gluten, %	Lysine, % 0	0.37	0.75
10	3.5	11.3	10.5
15	6.2	15.5	15.2
20	9.7	17.5	18.7
25	12.8	19.0	19.3
Potency difference from 0 lysine		13.5	13.7
95% C.I.		(10.6, 17.3)	(19.7, 17.4)

Chu and Hegsted (1976) showed that the concentrations of these enzymes in the liver of rats are subject to adaptive responses that control the utilization of these two amino acids. Although both enzymes are induced by feeding diets high in protein, rats differ in the mechanism of the adaptive response to high-protein diets and to diets whose threonine or lysine content is less than needed for growth. Thus, reductase falls to very low levels in the liver of rats fed wheat

Table 10. Calculated Nutritionally Limiting Amino Acids as a Function of Gluten and Lysine Concentration

	No Lysine added		0.37% Lysine HCl added		0.75% Lysine HCl added	
	Lysine %	Threonine $	Lysine %	Threonine %	Lysine %	Threonine %
10% Gluten	0.14[a]	0.25[b]	0.44	0.25[b]	0.74	0.25[b]
15% Gluten	0.21[a]	0.37[b]	0.51	0.37[b]	0.81	0.37[b]
20% Gluten	0.28[a]	0.50	0.58	0.50	0.88	0.50
25% Gluten	0.35[a]	0.62	0.65	0.62	0.95	0.62

[a] Limiting in lysine.

[b] Limiting in threonine.

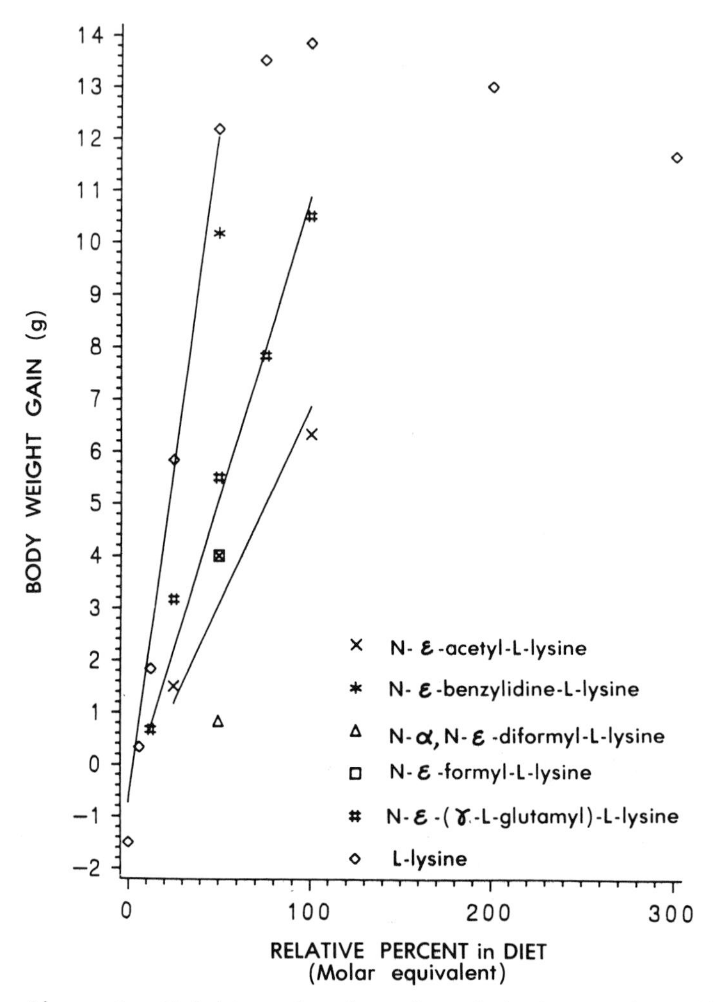

Figure 1. Weight gain in mice fed increasing dietary levels of lysine and lysine derivatives as part of an amino acid diet

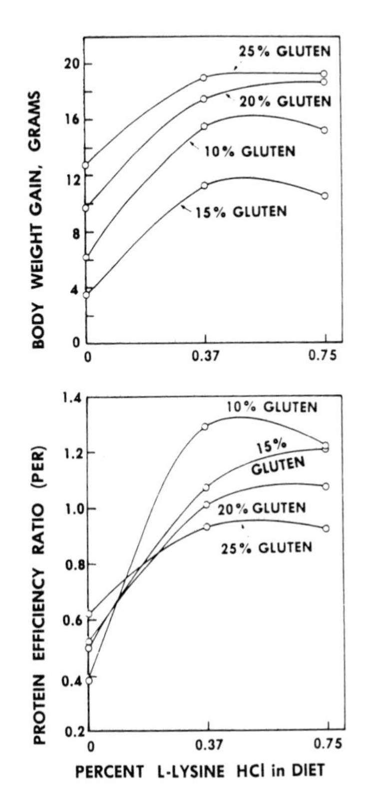

Figure 2. Effect of gluten and lysine in diets on body weight gain and PER in mice after 21 days.

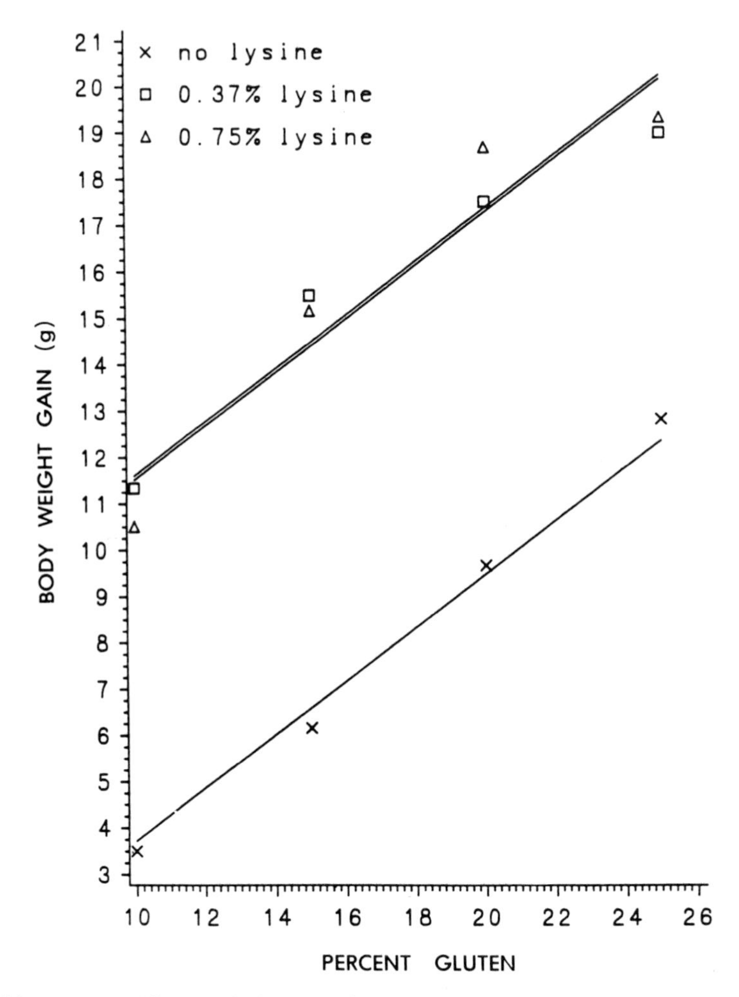

Figure 3. Plots of body weight gain in mice after 21 days as a percent of gluten supplemented with 0, 0.37 or 0.75% of lysine HCl.

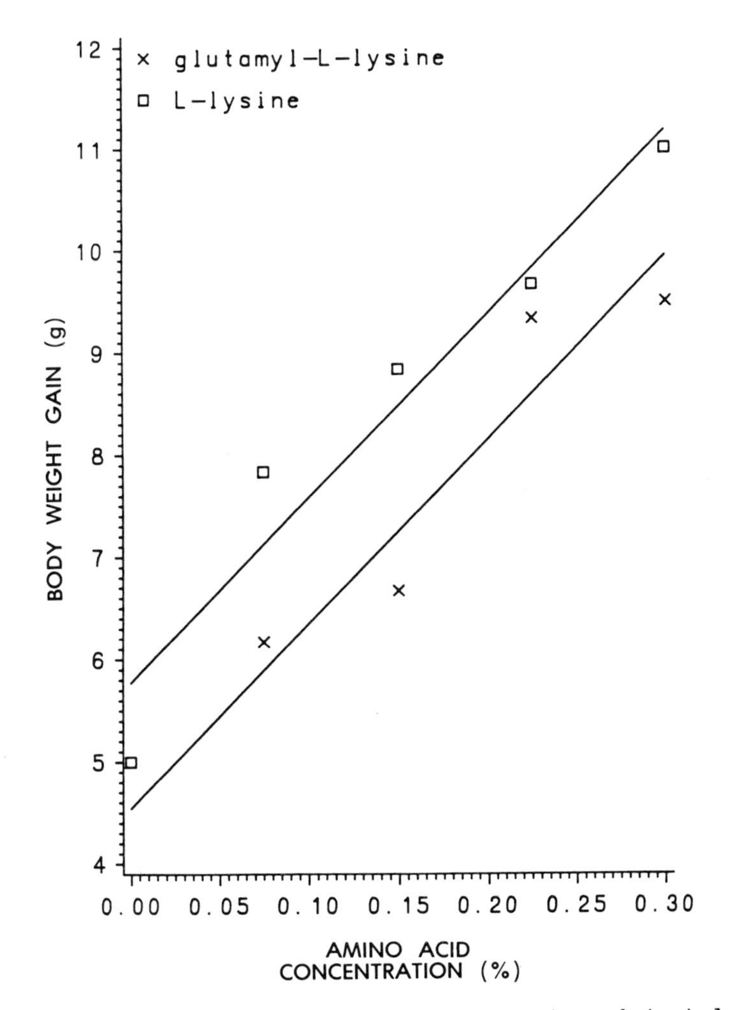

Figure 4. Weight gain in mice after 14 days fed whole bread supplemented with L-lysine or glutamyl-lysine in the diet (not co-baked, Table 5).

gluten. This appears to be an adaptive response conserving body lysine. At the same time catabolism of body proteins increases producing endogenous lysine needed for survival. These considerations imply that as the level of wheat gluten in the diet decreases, lysine is no longer the limiting amino acid. Total protein or some other amino acid then becomes limiting.

In contrast to the apparent mechanism of lysine catabolism, Chu and Hegsted (1976) report that threonine dehydratase does not appear to be substrate-induced. It appears, therefore, that when lysine is the limiting amino acid, the catabolic enzyme falls to low levels and lysine is conserved at the expense of body proteins. Loss of tissue proteins is much less when a diet low in threonine is fed, since the level of threonine dehydratase does not seem to be significantly affected by the protein or threonine content of the diet. Additional studies are needed to establish whether the catabolic enzyme patterns in mice parallel those of rats.

Fortification of Wheat Bread with Lysine and Glutamyl-Lysine After Baking. Tables 5, 6 compare stimulation of growth and effect on PER by lysine and glutamyl-lysine added after baking (not co-baked). The results suggest that glutamyl-lysine was slightly less effective than lysine, although differences were not statistically significant. Plotting the data (Figure 4) reveals that mice gained somewhat more weight from eating whole bread supplemented by lysine (not co-baked) than by glutamyl-lysine. The relative potencies of the two compounds in improving the diet were the same, as shown by weight gain per unit added amino acid. That is, the slopes of the two growth curves were equal after initial supplementation.

Table 6 shows that crumb produced slightly greater weight gain than did whole bread at each lysine level added. The differences were not statistically significant. The weight gain with crumb at the highest level of lysine was 82% that of casein. Crust, as expected, was associated with reduced growth. This reduction could be overcome to a large degree by lysine supplementation. Whether greater levels of lysine than those tested increase growth for crust to that produced by crumb is not known. The corresponding experiment with glutamyl-lysine was not done.

Figures 5-7 reveal that the protein quality of crust was less than that of whole bread, but differences were not significant mainly due to the increased variation associated with feed consumption data used in calculating weight gain. Thus, the PER calculation in mice has greater error than that of weight gain. Lysine supplementation improved protein quality of both crust and whole bread. At the highest level added, 0.3%, the PER of crumb was significantly greater than that of crust and whole bread. This may indicate greater digestibility and availability of other limiting amino acids in crumb and, thus, the greater potential for improvement in crumb when the first-limiting one, lysine, is supplemented.

Fortification of Wheat Flour with Lysine Before Baking. To establish the effect of baking on lysine utilization, wheat flour was fortified with lysine at the following levels: 0, 0.75, 1.50, 2.25, and 3.0% of flour protein. This flour was then baked into bread. Part of the bread was separated into crumb and crust, freeze-dried, and milled into a baked flour. The whole bread, crust, and crumb were then fed to mice in a 14-day growth assay in which diets contained 10% protein provided by the bread samples.

As with lysine supplemented diets (not co-baked), addition of lysine to flour before baking (co-baked) also greatly improved nutritional quality of the wheat protein. Table 6 show that both weight gain and protein quality (PER) were significantly increased. Supplementation had similar effects on weight gain whether co-baked or not. Maximum growth with crumb could be achieved with approximately 0.3% lysine, whereas crust might possibly be improved further with even greater lysine additions.

Figure 5 shows that improvement of protein quality of crumb through flour supplementation (co-baked) seemed to level off at less than 0.3% lysine. Lysine at 0.3% produced no further improvement, if not actual loss in protein quality at the higher level. Perhaps accelerated browning at higher lysine concentrations was responsible for this limit on the beneficial effect of lysine addition before baking.

Crust, as expected, had distinctly lower nutritive value than either crumb or whole bread. This suggests that crust relative to crumb is additionally deficient in lysine as a result of its greater exposure to heating during baking and, with sufficient lysine supplemention, may be improved close to the nutritional level of crumb. As already mentioned, further improvement of crust might be expected with greater lysine supplementation than included in this assay, whereas, that of the crumb and whole bread appeared to reach maximum.

Fortification of Wheat Flour with Glutamyl-Lysine Before Baking. The protocol of this assay was the same as that described in which lysine was used to fortify wheat flour before bread baking. The fortification with glutamyl-lysine was on a percent equivalent to lysine on a molar basis relative to protein content.

Table 7 reports body weight gain and PER. Glutamyl-lysine co-baked with wheat flour significantly stimulated growth of mice fed the bread preparations as the sole source of dietary protein and improved the quality of wheat protein. The relative nutritive value of crumb, crust, and whole bread are similar to those observed with lysine, crust being the most heat damaged. As with co-baked lysine, maximum improvement occurrred near 0.3% glutamyl-lysine.

Comparison of Lysine and Glutamyl-lysine Added Before Baking. Table 6 and Figures 5 and 6 compare the effect of lysine to that of glutamyl-lysine on weight gain and PER. When these amino acids were not co-baked with the bread diet, lysine was slightly more potent than glutamyl-lysine in stimulating growth and improving PER. This is still the case for PER (Figure 6) when the amino acids were co-baked with the flour in the simple bread formulation. With regard to weight gain (Figure 5), the overall effect of the two amino acids co-baked with flour was similar. Except for crust, both resulted in maximum growth at the 0.3% lysine-equivalent level.

Protein quality (PER) improvement with lysine supplementation showed an apparent maximum at levels below 0.3% for both crumb and whole bread, whereas continued improvement of crust at levels greater than 0.3% seemed possible For glutamyl-lysine, however, a maximum effect for crust at these levels was not apparent.

A statistical analysis of variance was run for weight gain and PER. Significant effects on weight gain and PER were associated with type of bread preparation (crumb, crust, and whole bread) and level of amino acid supplementation ($P < 0.0001$). Parallel line model analyses of variance were used for the entire sets of data represented by the

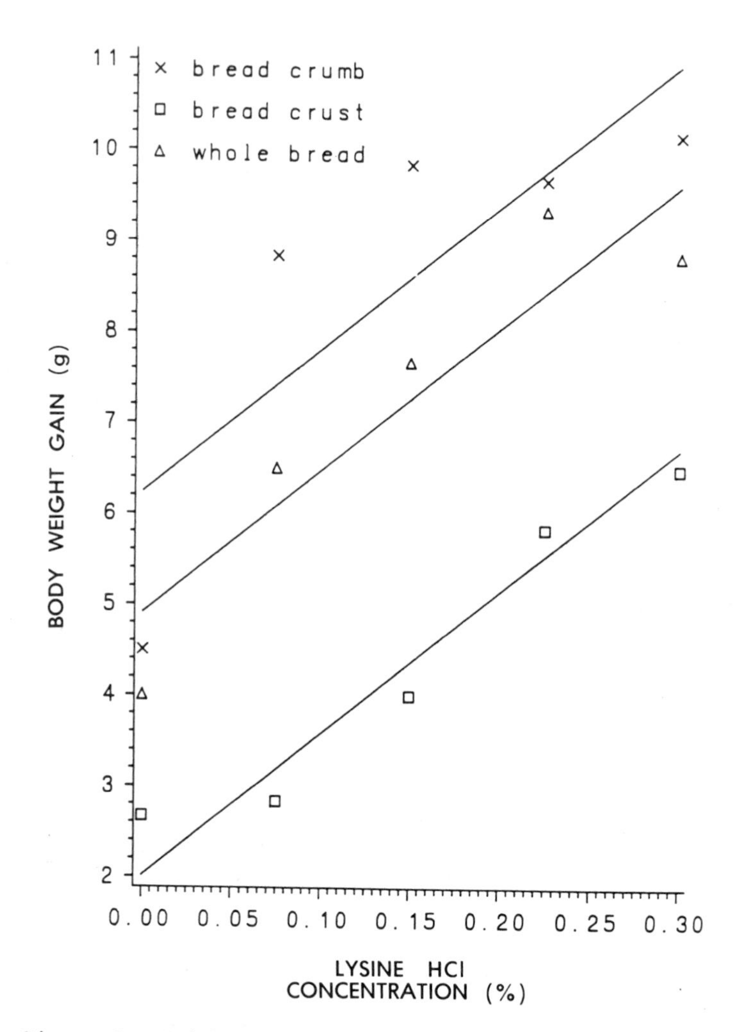

Figure 5. Weight gain in mice after 14 days fed whole bread supplemented with L-lysine or glutamyl-lysine in the diet (not co-baked, Table 6).

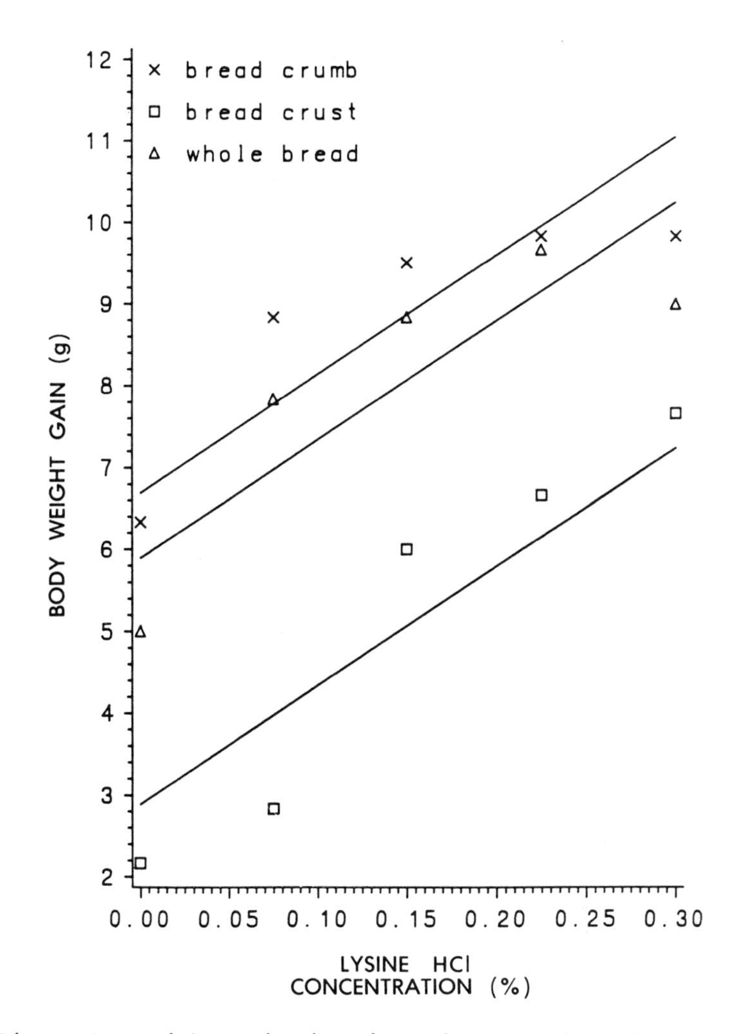

Figure 6. Weight gain in mice after 14 days fed whole bread, bread crumb, and bread crust supplemented with lysine (not co-baked, Table 6).

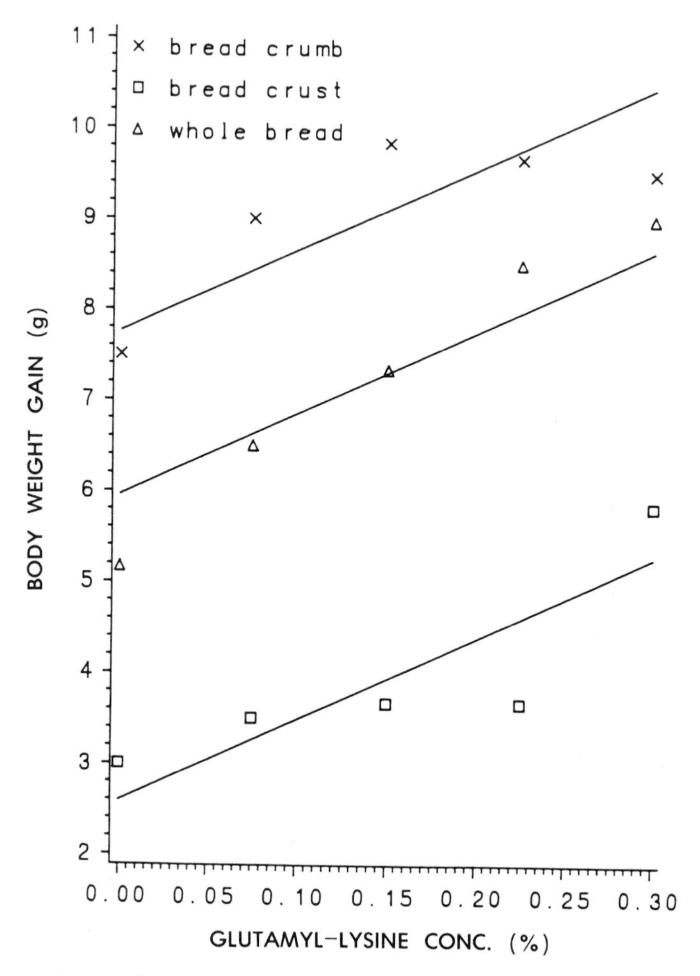

Figure 7. Weight gain in mice fed whole bread, bread crust, and bread crumb co-baked with glutamyl-lysine (Table 6).

figures. Overall, the two forms of lysine did not produce significantly different weight gains. Protein quality (PER), however, was improved to a greater extent by lysine ($P < 0.005$). Since Figures 5 and 7 represent separate experiments, the results can only be compared with regard to relative potencies among crumb, crust, and the whole bread.

In matching the two assays by analysis of variance, it was found that bread preparations with no amino acid supplementation (0% supplementation) resulted in overall PER's that were not significantly different. Thus, essentially the same protein quality for unsupplemented wheat proteins was obtained in both assays. In the two experiments, absolute weight gain was not the same, however. Weight gain was consistently greater for mice in the glutamyl-lysine run with each unsupplemented bread preparation (overall, $P < 0.0006$). One large assay with one lot of mice fed all diets simultaneously would be ideal to overcome this problem, but would be difficult to conduct.

Nutritional Significance. Several considerations relevant to the theme of this study will be briefly summarized to help place our findings into proper perspective.

(1). γ-Glutamyl-lysine crosslinks are also formed *in vivo* in various tissues by the catalytic action of transglutaminases (Fink et al., 1980). These authors also report on the partial purification and characterization of γ-glutamyl-cyclotransferase from rabbit kidneys, for which glutamyl-lysine is a substrate. Normal hydrolytic enzymes such as pepsin or trypsin do not cleave the isopeptide bond of glutamyl-lysine (Loewy, 1984).

(2). Raczynski et al. (1975) examined the metabolism of tritium-labeled glutamyl-lysine in the rat. They report that the compound is largely absorbed unchanged from the intestine and metabolized. This observation and the observed findings with tissue homogenates show that hydrolysis of the isopeptide bond of glutamyl-lysine largely takes place in the kidney.

(3). Waibel and Carpenter (1972) found that the growth-promoting activity of glutamyl-lysine supplementation to low-lysine diets was nearly equivalent to that of L-lysine for both young rats and chicks. In related studies, Hurrell and Carpenter (1977) and Otterburn et. al. (1977) report that heated food proteins containing isopeptide bonds are utilized less efficiently than unheated starting materials. This loss in protein quality may be due to reduced protein digestibility arising from steric hindrance of the introduced crosslinks to the action of proteolytic enzymes in the digestive tract.

(4). Other studies have successfully introduced lysine residues into low-lysine proteins such as gluten through (a) transglutaminase-catalyzed attachment of lysine residues to food proteins (Iwami and Yasumoto, 1986); (b) transglutaminase- catalyzed incorporation of lysyldipeptides into food proteins (Ikura et al., 1985); and, (c) carbodiimide-mediated covalent attachment of lysine, N-ε-acetyllysine, or N-ε-benzylidenelysine (Chan and Nakai, 1981). Although these approaches seem effective in improving the nutritional quality of food proteins, as well as their resistance to browning, they have disadvantages over the direct fortification described in the current study. These include (a) possible side reactions involving transglutaminase-catalyzed crosslinking of protein chains (Kurth and Rogers, 1984; Motoki and Nio, 1983); and (b) high cost of glutaminase and uncertainty about the safety of the chemical treatments.

Table 11. Plasma Cholesterol in Mice Fed Whole Bread, Bread Crumb, and Bread Crust with Lysine or Glutamyl-lysine Supplementation

Diet	Plasma cholesterol[1]			
	-L-lysine	+Lysine[2]	-Glutamyl-lysine	+Glutamyl-lysine[3]
	Not co-baked			
Whole bread	101[c]	140[ab]		152[a]
Bread crumb	112[bc]	127[abc]		--
Bread crust	111[bc]	123[abc]		--
Casein control		131[abc]		
Standard error		± 10		
	Co-baked			
Whole bread	89[c]	133[ab]	102[c]	131[ab]
Bread crumb	93[c]	135[a]	100[c]	136[ab]
Bread crust	94[c]	103[bc]	108[bc]	133[a]
Casein control	129[ab]	--	124[ab]	--
Standard error		± 10	± 7.2	

[1]Mean plasma cholesterol (mg/100 mL) at 14 days, 6 mice per group. Means without a superscript letter in common for a given baking procedure are significantly different, $P < 0.05$, Duncan's Multiple Range Test.

[2]L-lysine HCl concentration in the diet was 0.3%.

[3]Glutamyl-lysine was equimolar to 0.3% L-lysine HCl.

(5). In addition to catalyzing the formation of glutamyl-lysine, heat and alkali treatment of food proteins also induce the concurrent formation of N-ε-(DL-2-carboxyethyl)-L-lysine or lysinoalanine. In contrast to the described high utilization of glutamyl-lysine as a nutritional source of lysine, the biological utilization of lysinoalanine, determined in a growth assay in weanling male mice in which all L-lysine in a synthetic amino acid diet was replaced by a molar equivalent of lysinoalanine, produced a weight gain equivalent to that expected from diet containing 0.05% of L-lysine (Friedman et al., 1982; 1984). Evidently, acyl lysine derivatives such as glutamyl-lysine containing isopeptide bonds are hydrolyzed by kidney enzymes to lysine to a greater extent than alkyl lysine derivatives such as lysinoalanine (de Weck-Gaudard et al., 1988).

(6). In related studies we showed that (a) wheat gluten can be modified by vinyl compounds such as acrylonitrile, methyl acrylate, ethyl vinyl sulfone, and 4-vinylpyridine (Krull and Friedman, 1966; Cavins and Friedman, 1967, 1968; Eskins and Friedman, 1970; Friedman, 1973; Friedman and Finley, 1975; Friedman et al., 1970); (b) high pH and heat transform lysine residues in gluten to lysinoalanine side chains (Friedman, 1978, 1979; Friedman et al., 1984; Liardon et al., 1990); (c) significant amounts of mutagens are formed when gluten and gluten/carbohydrate blends are heated under simulated crust-baking conditions (Friedman et al., 1990; Ziderman et al., 1987, 1989); and (d) an improved ninhydrin assay may be useful to measure the available lysine content of gluten and other food proteins (Friedman et al., 1984; Pearce et al., 1988).

Plasma Cholesterol. Total plasma cholesterol sampled at the end of the 14-day feeding period was determined in mice fed bread and casein diets with and without supplementation by lysine and glutamyl-lysine (Table 11). Cholesterol levels were not significantly different between mice fed any of the bread diets. Increased cholesterol level in mice fed lysine- and glutamyl-lysine-fortified bread may be due to the nutritional improvement of the diet alone or to other factors. It is generally recognized that plasma cholesterol is depressed in mice fed diets poor in protein or amino acids (Sanchez and Hubbard, 1989), and that increases in the dietary lysine/arginine ratio increases cholesterol levels (Kritchevsky, 1989).

CONCLUSIONS

Our studies in mice suggest that glutamyl-lysine is hydrolyzed in the kidneys to lysine. The peptide is utilized more efficiently as a source of lysine when fed as part of a baking formulation than it is when fed as part of an all-amino-acid diet. Such lysine peptides merit further study as a potential source of lysine for animals and humans consuming low-lysine foods (Sarwar and Paquet, 1989; Sarwar et al., 1985). Our results also show that mice provide a good animal model to study protein quality of native, fortified, and processed wheat proteins (Cossak and Weber, 1983). Mouse bioassays have a major advantage. They require about one-fifth of the test material and about half the time to complete compared to the rat animal model. Mice feeding studies therefore merit wide adoption as a method of choice to obtain needed data on large numbers of samples to label foods and feeds for protein nutritional quality (Friedman, 1975, 1977, 1978, 1989).

ACKNOWLEDGEMENTS

We thank Carol E. Levin and Maura M. Bean for their help with the baking experiments, MacDoonald C. Calhoun and Michael R. Gumbmann for their help with the feeding studies and and interpretation of the results, Edith Magnenar for the metabolic experiments, and Bruce E. Mackey for the statistical analyses.

REFERENCES

Bell, J.M., and John, A.M. Amino acid requiremnts of growing mice: arginine, lysine, tryptophan and phenylalanine. J. Nutr., 111, 525-530. 1981.

Betschart, A.A. Improving protein quality of bread - nutritional benefits and realities. In "Nutritional Improvement of Food and Feed Proteins - Adv. Exp. Med. Biol. 105", M. Friedman, Ed., Plenum, New York, pp 703-734. 1978.

Bjorck, I., Noguchi, A., Asp, N.G., Cheftel, J., and Dahlqvist, A. Protein nutritional value of a biscuit processed by extrusion cooking: effects on available lysine. J. Agric. Food Chem., 31, 488-492. 1983.

Cavins, J.F., and Friedman, M. New amino acids derived from reactions of ε-amino groups in proteins with α, β-unsaturated compounds. Biochemistry 6, 3766-3770. 1967.

Cavins, J.F., and Friedman, M. Automatic integration and computation of amino acid analyses. Cereal Chem. 45, 172-176. 1968.

Chan, E.L., and Nakai, S. Comparison of browning in wheat glutens enriched by covalent attachment and addition of lysine. J. Agric. Food Chem., 29, 1200-1205. 1981.

Chan, E.L., Helbig, N., Holbek, E., Chan, S., and Nakai, S. Covalent attachment of lysine to wheat gluten for nutritional improvement. J. Agic. Food Chem., 27, 877-882. 1979.

Chu, S.H., and Hegsted, D.M. Adaptive response of lysine and threonine degrading enzymes in adult rats. J. Nutr., 106, 1089-1096. 1976.

Cossack, Z.T., and Weber, C.W. A proposed bioassay for evaluating protein quality using rats and mice. Nutr. Rep. Int., 28, 203-218. 1983

De Weck-Gaudard, D., Liardon, R., and Finot, P.A. Stereomeric composition of urinary lysinoalanine after ingestion of free or protein-bound lysinoalanine in rats. J. Agric. Food Chem., 36, 721-725. 1988.

Eskins, K., and Friedman, M. Solvent effects during the photochemistry of gluten-proteins. Photochem. Photobiol. 12, 245-247. 1970a.

Eskins, K., and Friedman, M. Graft photopolymerization of styrene to wheat gluten proteins in dimethyl sulfoxide. J. Macromol. Sci. A4, 947-956. 1970b.

Finley, J.W., and Friedman, M. Chemical methods for available lysine. Cereal Chemistry 50, 101-105. 1973.

Fink, M.L., Chung, S.I., and Folk, J.E. γ-Glutamylamine cyclotransferase: specificity toward ε-(L-γ-glutamyl)-L-lysine and related compounds. Proc. Natl. Acad Sci. USA, 77, 4564-4568. 1980.

Finney, D.J. Statistical Methods in Biological Assay, 3rd Ed., McMillan: New York, p 508. 1978.

Finot, P.A., Mottu, F., Bujard, E., and Mauron, J. N-substituted lysines as a source of lysine in nutrition. In "Nutritional Improvement of Food and Feed Proteins", M. Friedman, Ed., Plenum, New York, pp 549-569. 1978.

Friedman, M. Reactions of cereal proteins with vinyl compound. In "Industrial Uses of Cereal Grains", Y. Pomeranz, Editor, American Association of Cereal Chemists, Minneapolis, MN, PP. 237-251. 1973.

Friedman, M. (Editor). "Protein Nutritional Quality of Foods and Feeds", Marcel Dekker, New York. 2 Volumes. 1975.

Friedman, M. Effect of lysine modification on chemical, physical, nutritive,and functional properties of proteins. In "Food Proteins", J.R. Whitaker and S. Tannenbaum, Eds., Avi, Westport, CT, pp. 446-483. 1977a.

Friedman, M. Crosslinking amino acids - stereochemistry and nomenclature. In "Protein Crosslinking: Nutritional and Medical Consequences" - Adv. Exp. Med. Biol. 86B, M. Friedman, Ed., Plenum, New York, pp. 1-27. 1977b.

Friedman, M. (Editor). "Protein Crosslinking: Nutritional and Medical Consequences", Plenum Press, New York, 1977c.

Friedman, M. (Editor). "Protein Crosslinking: Biochemical and Molecular Aspects", Plenum Press, New York, 1977d.

Friedman, M. Wheat gluten-alkali reactions. In "Proceedings of the 10th National Conference on Wheat Utilization Research", U.S. Department of Agriculture, Science and Education Administration, Western Regional Research Center, Berkeley, CA, ARM-W-4, pp. 81-100. 1978a.

Friedman, M. (Editor). "Nutritional Improvement of Food and Feed Proteins", Plenum Press, New York, 1978b.

Friedman, M. Alkali-induced lysinoalanine formation in structurally different proteins. In "Protein Structure Related to Functionality", A-Pour-El, Ed., ACS Symposium Series, Washington, D.C. 92, 225-235. 1979.

Friedman, M. Chemically reactive and unreactive lysine as an index of browning. Diabetes, 31(3), 5-14. 1982.

Friedman, M. (Editor). "Absorption and Utilization of Amino Acids", CRC Press, Boca Raton, Florida. 3 Volumes. 1989.

Friedman, M., and Finley, J. W. Vinyl compounds as reagents for determining available lysine in proteins. In "Protein Nutritional Quality of Foods and Feeds", Marcel Dekker, New York, Part 1, pp. 503-520. 1975.

Friedman, M., and Gumbmann, M.R. Bioavailability of some lysine derivatives in mice. J. Nutr., 111, 1362-1369. 1981.

Friedman, M., and Gumbmann, M.R. Nutritional value and safety of methionine deriviatives, isomeric dipeptides and hydroxy analogs in mice. J. Nutr., 118, 388-397. 1988.

Friedman, M., Krull, L.H., and Cavins, J.F. The chromatographic determination of cysteine and half-cystine residues in proteins as S-ß-(4-pyridylethyl)-L-cysteine. J. Biol. Chem. 245, 3868-3871. 1970.

Friedman, M., Noma, A.T., and Wagner, J.R. Ion-exchange chromatography of sulfur amino acids on a single-column amino acid analyzer. Anal. Biochem., 98, 293-304. 1979.

Friedman, M., Gumbmann, M.R., and Savoie, L. The nutritional value of lysinoalanine as a source of lysine for mice. Nutr. Rep. Int., 26, 939-947. 1982.

Friedman, M., Gumbmann, M.R., and Masters, P.M. Protein-alkali reactions: chemistry, toxicology, and nutritional consequences. In "Nutritional and Toxicological Aspects of Food Safety - Adv. Exp. Med. Biol. 177", M. Friedman, Ed., Plenum, New York, pp. 367-412. 1984a.

Friedman, M., Pang, J., and Smith, G.A. Ninhydrin-reactive lysine in food proteins. J. Food Sci. 49, 10-13. 1984b

Friedman, M., Gumbmann, M.R., and Ziderman, I.I. Nutritional value and safety in mice of proteins and their admixtures with carbohydrates and vitamin C. J. Nutr., 117, 508-518. 1987.

Friedman, M., Gumbmann, M.R., and Brandon, D.L. Nutritonal, toxicological, and immunological consequences of food processing. Frontiers Gastsrointestinal Research 14, 79-90. 1988.

Friedman, M., Wilson, R.W., and Ziderman, I.I. Mutagen formation in heated wheat gluten, carbohydrates, and gluten/carbohydrate blends. J. Agric. Food Chem., 38, 1019-1028. 1990.

Geervani, P., and Devi, P.Y. Effect of different heat treatments on losses of lysine in processed products prepared from unfortified flour and flour fortified with lysine. Nutr. Rep. Int., 33, 961-966. 1986.

Gumbmann, M.R.. Friedman, M., and Smith, G.A. The nutritional values and digestibilities of heat damaged casein and casein-carbohydrate mixtures. Nutr. Rep. Int., 28, 355-361. 1983.

Hegsted, D.M. Protein quality and its determination. In "Food Proteins", J.R. Whitaker, S.R. Tannenbaum, Eds., AVI, Westport, CT, pp. 347-362. 1977.

Hurrell, R.R., and Carpenter, K.J. Nutritional significance of cross-link formation during food processing. Adv. Exp. Med. Biol., 86B, 225-237. 1977.

Ikura, K., Okumura, K., Yoshikawa, M., Sasaki, R., and Chiba, H. Incorporation of lysyldipeptides into food protein by transglutaminase. Agric. Biol. Chem., 49, 1877-1878. 1985.

Iwami, K., and Yasumoto, K. Amine-binding capacities of food proteins in transglutaminase reaction and digestibility of wheat gliadin with ε-attached lysine. J. Sci. Food Agric., 37, 495-503. 1986.

Jansen, R.G. Amino acid fortification. In "New Protein Foods", A.M. Altschul and H.L. Wilcke, Eds., Academic Press, New York, Vo. 4, pp. 161-204. 1981.

Johnson, V.A., and Mattern, P.J. Improvement of wheat protein quality and uantity by breeding. In "Nutritional Improvement of Food and Feed Proteins - Adv. Exp. Med. Biol. 105", M. Friedman, Ed., Plenum, New York, pp. 301-316. 1978.

Khan-Siddiqui, K. Lysine-carnitine conversion in rat and man. In "Absorption and Utilization of Amino Acids", M. Friedman, Ed., CRC, Boca Raton, FL, Vol. 2, pp 41-57. 1989.

Kritchevsky, D.A. Dietary protein in atherosclerosis. In "Absorption and Utilization of Amino Acids", M. Friedman, Ed., CRC, Boca Raton, FL, Vol. 2, pp 235-245. 1989.

Krull, L.H., and Friedman, M. Anionic polymerization of methyl acrylate to protein functional groups. J. Polym. Sci. A-1, 5, 2535-2546. 1967.

Kurth, L., and Rogers, P.J. Transglutaminase catalyzed cross-linking of myosin to soya protein, casein and gluten. J. Food Sci., 49, 573-576. 1984.

Liardon, R., Friedman, M., and Phillippossian, G. Racemization kinetics of free- and protein-bound lysinoalanine (LAL) in strong acid media. Isomeric composition of protein-bound LAL. J. Agric. Food Chem. 38, 1990, submitted.

Loewy, A.G. The N-ε-(γ-glutamic) lysine cross-link: method of analysis, occurrence in extracellular and cellular proteins. Methods Enzymol., 107, 241-257. 1984.

Lougnon, L., and Kiener, T. Biological utilization of basic amino acids and cations. In "Absorption and Utilization of Amino Acids", M. Friedman, Ed., CRC, Boca Raton, FL, Vol. 2, pp 1-24. 1989.

Menefee, M., and Friedman, M. Estimation of structural components of abnormal hair from amino acid analyses. J. Protein Chem. 4, 333-341. 1985.

Milner, J.A. Arginine: a dietary modifier of ammonia detoxification and pyrimidine biosynthesis. In "Absorption and Utilization of Amino Acids", M. Friedman, Ed., CRC, Boca Raton, FL, Vol. 2, pp. 26-40. 1989.

Motoki, M., and Nio, N. Crosslinking between different food proteins by transglutaminase. J. Food Sci., 48, 561-566. 1983.

Otterburn, M.S. Protein crosslinking. In "Protein Quality and the Effects of Processing", R.D. Phillips, J.W. Finley, Eds., Marcel Dekker, New York, pp. 247-261. 1989.

Otterburn, M., Healy, M., and Sinclair, W. The formation, isolation, and maintenance and growth. In "Absorption and Utilization of Amino Acids", M. Friedman, Ed., CRC, Boca Raton, FL., Vol. 1, pp. 15-30. 1989.

Owens, F.N., and Pettigrew, J.E. Subdividing amino acid requirements for maintenance and growth. In "Absorption and Utilization of Amino Acids", M. Friedman, Ed., CRC, Boca Raton, FL., Vol 1, pp. 15-30. 1989.

Pearce, K.N., Karahalios, D., and Friedman, M. Ninhydrin assay for proteolysis in ripening cheese. J. Food Sci. 52, 432-435. 1988.

Raczynski, G., Snochowski, M., and Buraczewski, S. Metabolism of ε-(γ-L-glutamyl)-L-lysine in the rat. Br. J. Nutr., 34, 291-296. 1975.

Reichl, J.R. Absorption and metabolism of amino acids studies in vitro, in vivo, and with computer simulations. In "Absorption and Utilization of Amino Acids", M. Friedman, Ed., CRC, Boca Raton, FL,, Vol. 1, pp. 93-156. 1989.

Sanchez, A., and Hubbard, R.W. Dietary protein modulation and serum cholesterol: the amino acid connection. In "Absorption and Utilization of Amino Acids", M. Friedman, Ed., CRC, Boca Raton, FL, Vol. 2, pp. 247-273. 1989

Sarwar, G., and Paquet, A. Availability of amino acids in some tripeptides and derivatives present in dietary proteins. In "Absorption and Utilization of Amino Acids", M. Friedman, CRC, Boca Raton, FL, Volume 2, pp. 147-154. 1989.

Sarwar, G., Christensen, D. A., Finlayson, A. J., Friedman, M., Hackler, L. R., Mackenzie, S. L.,, Pellet, P. L., and Tkachuk, R. Intra- and inter-laboratory variation in amino acid analysis. J. Food Sci., 48, 526-531. 1983.

Sarwar, G., Blair, R., Friedman, M., Gumbmann, M. R., Hackler, L. R., Pellett, P. L., and Smith, T. L. Inter-and intra laboratory variability in rat growth assays for estimating protein quality in foods. J. Assoc. Off. Anal. Chem. 67, 976-981. 1984.

Sarwar, G., Blair, R., Friedman, M., Gumbmann, M. R., Hackler, L. R., Pellett, P. L., and Smith, T. K. Comparison of interlaboratory variation in amino acid analysis and rat growth assays for evaluating protein quality. J. Assoc. Off. Anal. Chem., 68, 52-56. 1985.

SAS. SAS/STAT Guide for Personal Computers; 6th Edition, SAS Institute: Cary, NC, 1028 p. 1987.

Scharrer, E. Regulation of intestinal amino acid transport. In "Absorption and Utilization of Amino Acids", M. Friedman, Ed., CRC, Boca Raton, FL, Volume 1, pp. 57-68. 1989.

Sherr, B., Lee, C. M., and Jelesciewicz, C. Absorption and metabolism of lysine Maillard products in relation to the utilization of L-lysine. J. Agric. Food Chem., 37. 119-122 1989.

Tsen, C.C., and Reddy, P.R.K. Effect of toasting on the nutritive value of bread. J. Food Sci., 42. 1370-1372. 1977.

Tsen, C.C., Reddy, P.R.K., and Gehrke, C.W. Effects of conventional baking, microwave baking, and steaming on the nutritive value of regular and fortified breads. J. Food Sci., 42, 402-406. 1977.

Waibel, P.E., and Carpenter, K.J. Mechanism of heat damage in protein. 3. Studies with ε-(γ-L-glutamyl)-L-lysine. Br. J. Nutr., 27, 509-515. 1972.

Zerbe, G. On Fieller's theorem and General Linear Model. Am. Stat., 32, 103-105 1978.

Ziderman, I.I., and Friedman, M. Thermal and compositional changes of dry wheat gluten-carbohydrate mixtures during simulated crust baking. J. Agric. Food Chem., 33, 1096-1102. 1985.

Ziderman, I. I., Gregorski, K. S., and Friedman, M. Thermal analysis of protein-carbohydrate mixtures in oxygen. Thermochica Acta, 114, 109-114. 1987.

Ziderman, I.I., Gregorski, K.S., Lopez, S.V., and Friedman, M. Thermal interaction of ascorbic acid and sodium ascorbate with proteins in relation to nonenzymatic browning and Maillard reactions in foods. J. Agric. Food Chem. 37, 1480-1486. 1989.

31

FORMATION, NUTRITIONAL VALUE, AND SAFETY OF D-AMINO ACIDS

Mendel Friedman

Western Regional Resarch Center
Agricultural Research Service, U.S. Department of
Agriculture, 800 Buchanan, Albany, California 94710

ABSTRACT

The extent of racemization of L-amino acid residues to D-isomers in food proteins increases with pH, time, and temperature. The nutritional utilization of different D-amino acids vary widely, both in animals and humans. In addition, some D-amino acids may be deleterious. For example, although D-phenylalanine is nutritionally available as a source of L-phenylalanine, high concentrations of D-tyrosine inhibit the growth of mice. The antimetabolic effect of D-tyrosine can be minimized by increasing the L-phenylalanine content of the diet. Similarly, L-cysteine has a sparing effect on L-methionine when fed to mice; however, D-cysteine does not. The wide variation in the utilization of D-amino acids is exemplified by the fact that D-lysine is not utilized as a source of L-lysine, whereas the utilization of D-methionine as a source of the L-isomer for growth is dose-dependent, reaching 76% of the value obtained with L-methionine. Both D-serine and the mixture of L-L and L-D isomers of lysinoalanine induce histological changes in the rat kidneys. D-tyrosine, D-serine, and lysinoalanine are produced in significant amounts under the influence of even short periods of alkaline treatment. Unresolved is whether the biological effects of D-amino acids vary, depending on whether they are consumed in the free state or as part of a food protein. Possible, metabolic interaction, antagonism, or synergism among D-amino acids *in vivo* also merits further study. The described results with mice complement related studies with other species and contribute to the understanding of nutritional and toxicological consequences of ingesting D-amino acids. Such an understanding will make it possible to devise food processing conditions to minimize or prevent the formation of undesirable D-amino acids in food proteins and to prepare better and safer foods.

INTRODUCTION

Processed proteins are increasingly used to meet human dietary needs. Alkali treatment of casein, corn, and soy proteins brings about desirable changes in flavor, texture, and solubility. Such treatments which are used to prepare protein isolates also destroy toxins and trypsin inhibitors. Treating food protein with alkali and heat may, however, produce undesirable changes in the constituent amino acids. Such changes may include crosslinking, browning reactions, and racemization (1-92).

Racemization of L-amino acids to D-isomers in food proteins often impair biological value and safety of foods by (a) forming nonutilizable forms of amino acids; (b) creating L-D, D-L, and D-D peptide bonds that may resist hydrolysis by proteolytic enzymes; and (c) forming unnatural amino acids that may be nutritionally antagonistic or act as antimetabolites (14-17).

In this paper, I present a limited overview of our studies on the factors which influence racemization of L-amino acid residues in food proteins and on the biological utilization and safety of selected D-amino acids.

CHEMISTRY OF RACEMIZATION

Since the early part of this century (17), alkali and heat treatments have been known to racemize amino acids. As a result of food processing using these treatments, D-amino acids are continuously consumed by animals and man. Levels of D-aspartic acid detected in some commercial foods are shown in Table 1. Because all of the amino acid residues in a protein undergo racemization simultaneously, but at differing rates, assessment of the extent of racemization in a food protein requires quantitative measurement of at least 36 optical isomers, 18 L and 18 D. Analytically, this is a difficult problem not yet solved.

Racemization of an amino acid proceeds by removal of a proton from the α-carbon atom to form a carbanion intermediate. The trigonal carbon atom of the carbanion, having lost the original asymmetry of the α-carbon, recombines with a proton from the environment to regenerate a tetrahedral structure. The reaction is written as:

$$\text{L-amino acid} \underset{k'}{\overset{k}{\rightleftharpoons}} \text{D-amino acid} \tag{1}$$

where k_{rac} and k_{rac} are the first-order rate constants for the forward and reverse racemization of the stereoisomers.

The product is racemic if recombination can take place equally well on either side of the carbanion, giving an equimolar mixture of L- and D-isomers. Recombination may be biased, if the molecule has more than one asymmetric center, resulting in an equilibrium mixture slightly different from a 1:1 enantiomeric ratio.

The following equation was derived to describe the kinetic course of the racemization process (65-66):

$$L \underset{k'}{\overset{k}{\rightleftharpoons}} D$$

$$-dL/dt = kL - k'D$$

if $D_{t=o} \ll L_{t=o}$, then $L_{t=o} = L + D$ and $L_{t=o} - L = D$

$$-dL/dt = kL - k'(L_{t=o} - L) = kL - k'L_{t=o} + k'L$$

$$= (k + k')L - k'L_{t=o}$$

$$dL/dt = -(k + k')L + k'L_{t=o}$$

$$dL/dt + (k + k')L = k'L_{t=o}$$

$$dL/dt\, e^{(k + k')t} + (k + k')L\, e^{(k+k')t} = k'L_{t=o}\, e^{(k + k')t}$$

$$d/dt\, (Le^{(k + k')t}) = k'L_{t=o}\, e^{(k + k')t}$$

$$\int d\,(Le)\,(Le^{(k + k')t}) = \int (k'L_{t=o}\, e^{(k + k')t}) \cdot dt$$

$$Le^{(k + k')t} = [k/(k + k')]L_{t=o}\, e^{(k + k')t} + \text{constant}$$

$$L = [k'/(k + k')]L_{t=o} + \text{constant} \cdot e^{-(k + k')t}$$

At $t = 0$, $L = L_{t=o}$ and $e^{-(k + k')t} = 1$

$$L_{t=o} = [k'/(k + k')]L_{t=o} + \text{constant}$$

$$L_{t=o} - [k'/(k + k')]L_{t=o} = \text{constant}$$

$$L_{t=o}[(1 - k'/(k + k')] = \text{constant}$$

$$L = [k'/(k + k')]L_{t=o} + L_{t=o}\, k/(k + k')\, e^{-(k + k')t}$$

if $L_{t=o} = L + D$

$$L = [k'/(k + k')](L + D) + (L + D)\, k/(k + k')\, e^{-(k + k')t}$$

$$L/(L + D) = k'/(k + k') + [k/(k + k')]e^{-(k - k')t}$$

$$\frac{L(k + k') - k'(L + D)}{(L + D)(k + k')} = [k/(k + k')]e^{-(k + k')t}$$

$$= \frac{Lk - Dk'}{(L + D)(k + k')}$$

$$\left[\frac{L - D(k'/k)}{L + D}\right] = e^{-(k + k')t}$$

$$= \left[\frac{L[(1 - (D/L)(k'/k)]}{L[(1 + (D/L)]}\right]$$

$$= \left[\frac{1 - (D/L)(k'/k)}{1 + (D/L)}\right]$$

$$\ln\left[\frac{1 - (D/L)(k'/k)}{1 + (D/L)}\right] = -(k + k')t$$

$$\ln\left[\frac{1 + (D/L)}{1 - (D/L)(k'/k)}\right] = (k + k')t$$

if $k'/k = K'$

$$\ln\left[\frac{1 + (D/L)}{1 - K'(D/L)}\right] = (1 + K') \cdot k \cdot t \qquad \text{(eq. 2)}$$

The following operational equation is useful to relate the apparent racemization rate constant (k) to the extent of racemization (D/L) during a specific time period, e.g. 3 hr:

$$k = \frac{\ln\left[\frac{1 + D/L}{1 - D/L}\right]_{3\ hr} - \ln\left[\frac{1 + D/L}{1 - D/L}\right]_{0\ hr}}{2 \cdot k\ (3\ hr)} \qquad (2a)$$

An alternative mathematical analysis of the first-order kinetics of racemization gives equation 3. However, equation 2 is operationally more useful to measure racemization rates in food proteins (36, 37, 59, 60, 65, 66) than equation 3 because we can measure D/L ratios of amino acids with a 1-3% error compared to a 15% error when concentrations of D only are measured, as required by equation 3.

$$\ln (D_e/D_e - D_t) = (k + k')t = k_{(obs)}t \qquad (eq.\ 3)$$

where D_e equals equilibrium value of D

and D_t equals D at time t.

Because the structural and electronic factors which facilitate the formation and stabilization of the carbanion intermediate are unique for each amino acid, it follows that the reaction rate for the isomerization of each amino acid is also unique. Thus, the inductive strengths of the R-substituents have been invoked to explain differing racemization of amino acid residues in food proteins as influenced by pH, time (t), and temperature, as illustrated in Figures 1-5. Plotting racemization for individual amino acids for soybean proteins against the inductive parameters clearly demonstrates strong correlations (Figures 6,7). The reader should consult the cited papers for detailed interpretations.

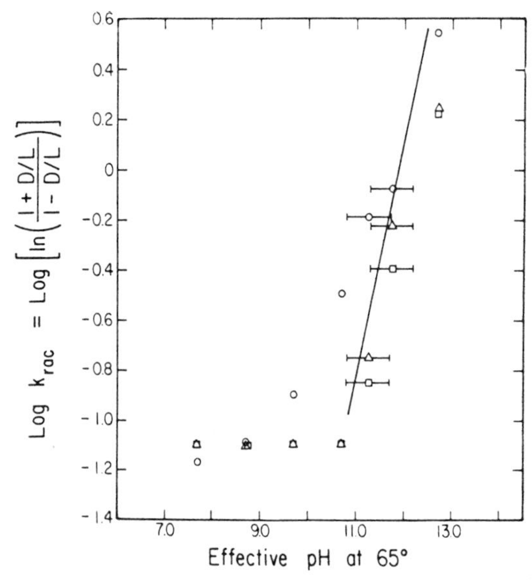

Figure 1. Effect of pH of aspartic acid (O), phenylalanine (Δ), and glutamic acid (□) racemization in casein. Ref. 37.

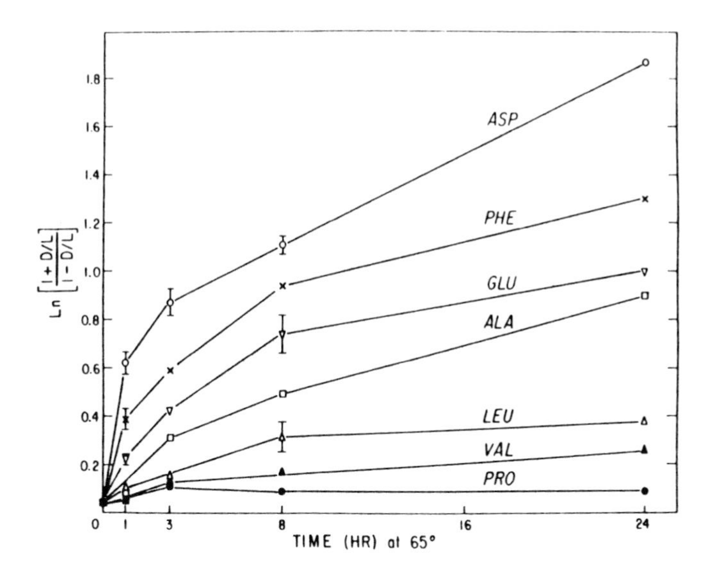

Figure 2. Time course of amino acid racemization of caseinn in 0.1 N NaOH at 65 C. Ref. 37.

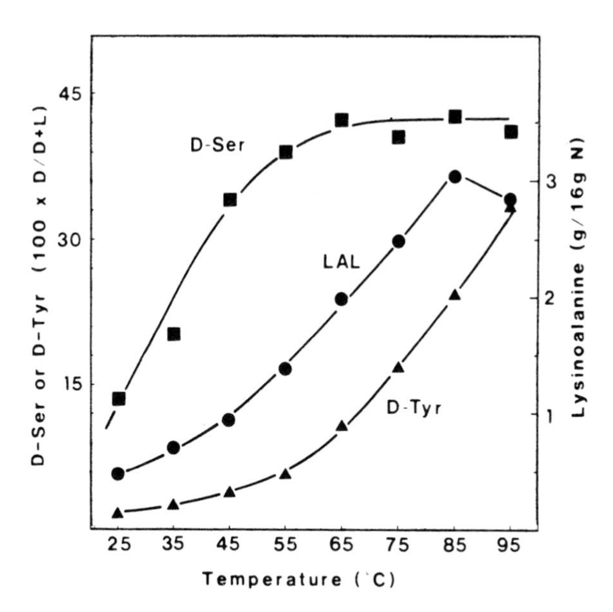

Figure 3. Effect of temperature on D-serine, D-tyrosine, and lysinoalanine content of alkali-treated soybean protein. Ref. 36.

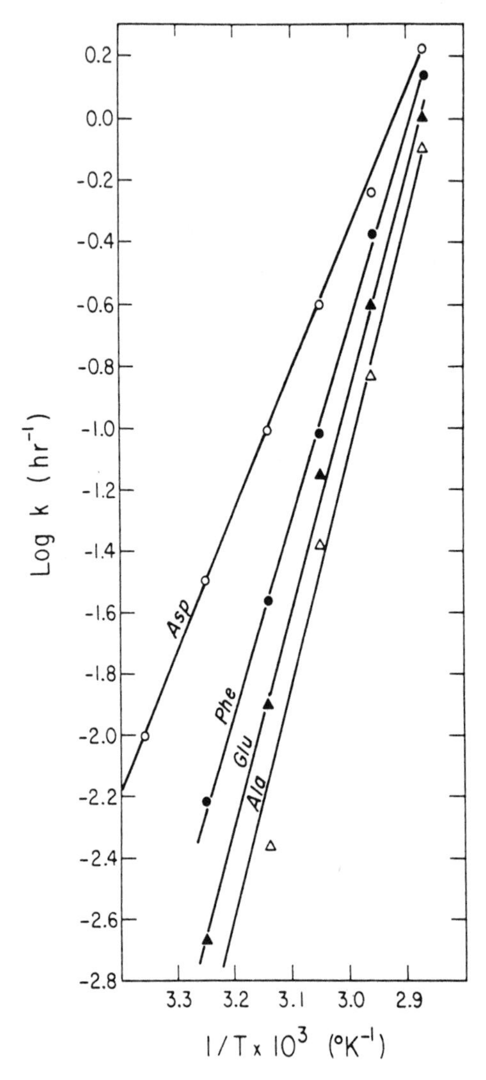

Figure 4. Temperature dependence (Arrhenius plots) for the racemization of amino acid residues in casein. Ref. 37.

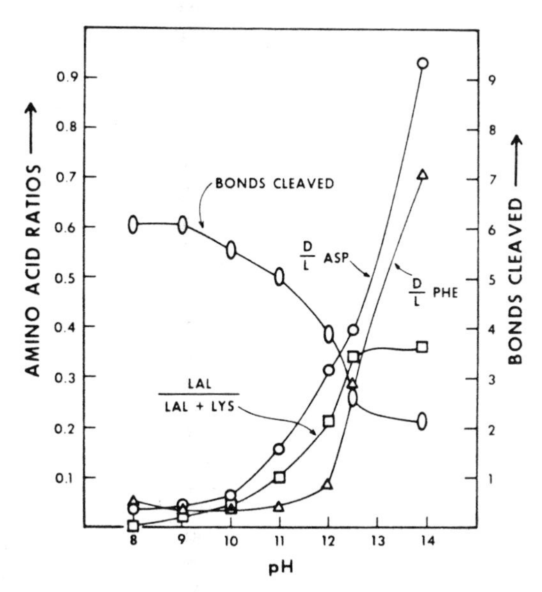

Figure 5. Effect of pH on the relationship between the extent of hydrolysis of peptide bonds by trypsin in casein and the content of lysinoalanine and D-amino acids. Ref. 47.

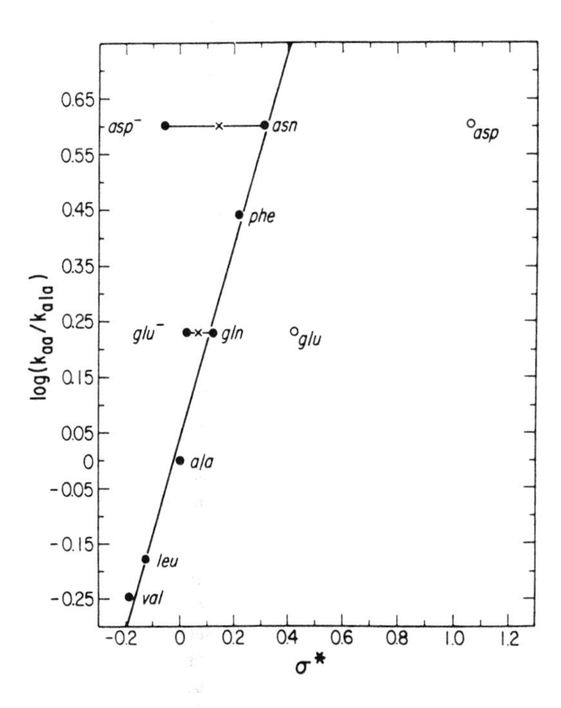

Figure 6. Relationship between the inductive constant (σ^*) of leucine, valine, phenylalanine, glutamic acid, and aspartic acid side chains of casein and the racemization rate constants relative to alanine in 0.1 N NaOH at 65°C for 3 hr. Ref. 37.

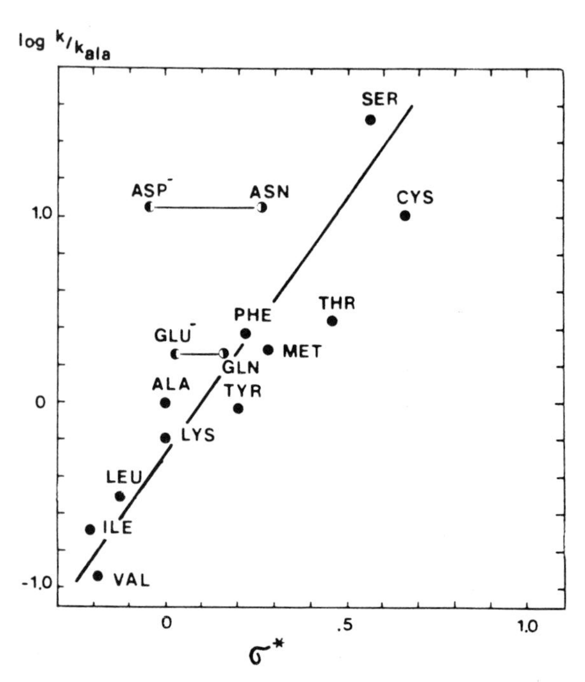

Figure 7. Relationship between the inductive constants (σ*) of the amino acid side chains of soybean protein and the logarithms of the racemization rate constants (k) relative to that of alanine (k_{ala}). Ref. 45.

BIOLOGICAL UTILIZATION OF D-AMINO ACIDS

Two pathways are available for the biological utilization of D-amino acids (55-56): (a) racemases or epimerases may convert D-amino acids directly to L-isomers or to (DL) mixtures; or (b) D-amino acid oxidases may catalyze oxidative deamination of the α-amino group to form α-keto acids, which can then be specifically reaminated to the L-form. Although both pathways may operate in microorganisms, only the latter activity has been demonstrated in mammals. The amounts and specificities of D-amino acid oxidase are known to vary in different animal species. In some, the oxidase system may be rate-limiting in the utilization of a D-amino acid as a source of the L-isomer. In this case, the kinetics of transamination of D-enantiomers would be too slow to support optimal growth. In addition, growth depression could result from nutritionally antagonistic or toxic manifestations of D-enantiomers exerting a metabolic burden on the organism.

The biological utilization of sulfur D-amino acids is conveniently discussed in terms of the transamination and transsulfuration pathways (Fig.8). D-methionine can be transformed to the L-isomer *via* oxidative deamination followed by reamination. L-Methionine may be incorporated into proteins. It acts as the methyl group donor *via* S-adenosylmethionine; is a precursor for cysteine, cystine, and taurine; and participates in transamination reactions in which 3-methylthiopropionate is an intermediate. Not all of the details in the cited pathways, and their possible relationship to inherited metabolic diseases such as cystathionuria and homocystinuria, have been completely elucidated (8).

D-Methionine

The reported wide variation in the biological utilization of sulfur D-amino acids for growth of various animal species mav be related to the relative activities of D-amino acid oxidases (12,32, 55-57, 91). However, the reported relative oxidation rates of D-methionine by D-amino acid oxidase in kidney homogenates from man, monkeys, chickens, frogs, rats, and mice do not support this hypothesis. For example, the reported rate of oxidation of D-methionine by the oxidase in mouse kidneys is about one third to one half of corresponding rates for rats or man. Nevertheless, we have observed that the nutritional utilization of D-methionine in mice is at least as good or better than reported bioavailabilities of D-methionine for either rats or man. In fact, D-methionine appears to be poorly utilized by humans (12,54,75,91). when consumed either orally or during total parenteral nutrition. One factor giving rise to inconsistencies regarding the utilization of D-methionine is the dose dependency (Figure 9) of the apparent potency of D-methionine relative to its L-isomer, *i.e.*, the dietary concentration of the D-form for any given growth response relative to that of the L-form which would produce the same growth response. This dose-dependency is a result of the non-linear nature of the dose-response curves. This complicates attempts to compare results from some earlier studies with other animal species, which often report data based on a single substitution of the D- for the L-isomer.

The extent of oxidation of D-methionine (and other D-amino acids) by D-amino acid oxidase may not be the limiting step governing

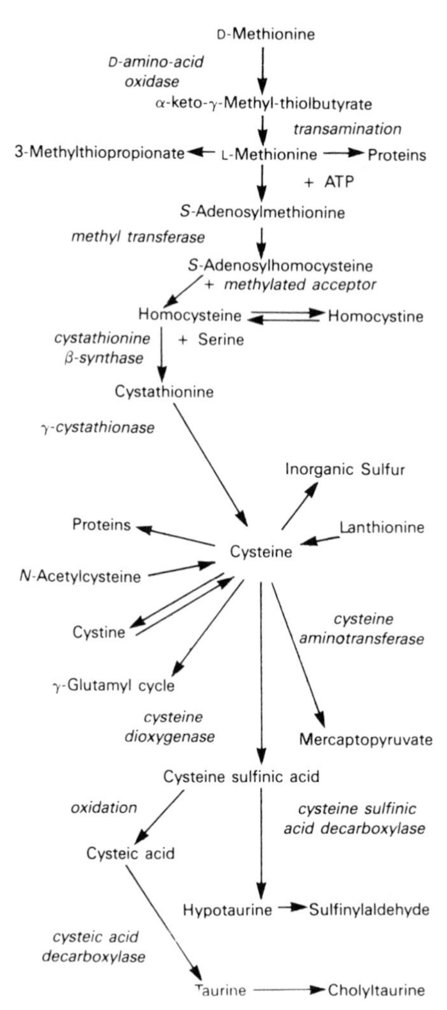

Figure 8. Transamination and transsulfuration pathways of D- and L-methionine. Ref. 32, 80a.

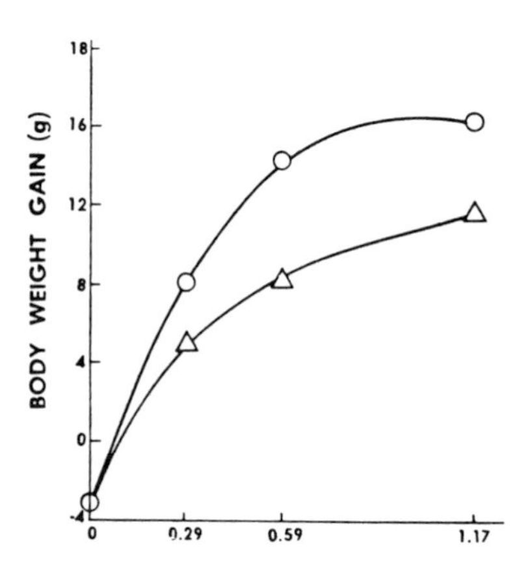

Figure 9. Relationship of weight gain to percent L- or D-methionine in amino acid diets fed to mice. Ref. 32.

utilization. Other factors that could influence utilization include: rates of transport, action of intestinal enzymes and bacteria, rates of absorption, renal clearance, and possible toxic effects.

D-Cystine

In discussing D-cystine, it is informative to review first the utilization of the L-isomer. Although L-cystine is not an essential amino acid for rodents, less L-methionine is needed for growth if the diet contains L-cystine (19, 32, 66, 82). The mechanism of this so-called sparing effect needs further clarification. One possibility is that L-cystine serves as a source of L-cysteine and other sulfur amino acids, thus minimizing the need for L-methionine to serve as a reservoir for these amino acids.

Our results show that L-cystine has a low level of apparent methionine- like activity in mice (32). Growth response was 6.5% when it was substituted for an optimum level of L-methionine (1.17% in an amino acid diet devoid of any other sulfur amino acids). This may reflect a limited sparing effect for depletion of endogenous L-methionine, which may then be reutilized in higher priority pathways.

L-cystine is somewhat more efficient in sparing D-methionine than L-methionine in amino acid diets containing a low level (0.29%) of L- or D-methionine (32). Supplementation of the D-isomer with an equal sulfur equivalent of L-cystine nearly doubled growth, bringing the overall response essentially equal to that produced by L-methionine in the presence of L-cystine. Since D-methionine must first be converted to the L-form before it can be utilized, these results illustrate a marked lessening in demand for L-methionine when L-cystine is available. Thus, even though D-methionine alone is somewhat less efficiently utilized in mice than is the L-isomer, the combination of L-cystine with either L- or D-methionine provides essentially equivalent nutritive value.

In contrast, supplementation of an amino acid diet containing 25% of the optimal level (0.29%) of L-methionine with the same concentration of D-cystine increased growth from 58.7 to only 70.7% of the weight gain with optimal (1.17%) L-methionine. Increasing D-cystine to 1.17% reduced relative growth from 58.7 to 43.5%, which may imply that excess D-cystine in the diet is toxic.

D-Cysteine

Table 2 shows that L-cysteine had a sparing effect on L-methionine, but that D-cysteine did not. In fact, D-cysteine imposed a metabolic burden indicated by depressed growth when fed with less than optimal levels of L-methionine. The 24% decrease in weight gain of the D-cysteine-plus- L-methionine compared to L-methionine alone could mean that D-cysteine was nutritionally antagonistic or toxic. The mechanism(s) of the growth-depressing effects of D-cysteine and D-cystine remains to be elucidated, as do the reported toxic manifestations resulting from excess consumption of L-methionine.

Cavallini et al. (17) and Krijgsheld et al. (57) report that D-cysteine appears to be a more efficient precursor of inorganic sulfate in the blood of rats than L-cysteine. These findings prompted Krijgsheld et al. to suggest that therapeutic administration of D-cysteine could be beneficial in cases of sulfate depletion. However, some caution may be in order, in view of our finding of a growth-depressing effect of D-cysteine in mice.

Lanthionine Isomers

Lanthionine ($S(CH_2CH(NH_2)COOH)_2$) isomers are formed during exposure of food and other proteins to alkali and heat (23-26, 41, 64), 72, 79). Such treatments generate dehydroalanine side chains from half-cystine and serine. Reaction of the SH group of cysteine and the double bond of dehydroalanine gives rise to one pair of optically active isomers (enantiomers) and one diastereoisomeric (meso) form. These forms are correctly named as follows:

(R)-L-lanthionine or S-[(2R)-2-amino-2-carboxyethyl] L-cysteine
(S)-D-Lanthionine or S-[(2S)-2-amino-2-carboxyethyl]-D-cysteine
(S)-L-lanthionine or S-[(2S)-2-amino-2-carboxyethyl]-L-cysteine
(meso-lanthionine).

Although food processing produces a mixture of all three isomeric lanthionines, it would be of interest to know the biological utilization of the individual isomers as a source of L-methionine. However, only the mixture of all three was available to us. Table 2 shows that DL + meso-Lanthionine has a moderate sparing effect on L-methionine, as evidenced by 27% greater weight gain when the two amino acids were fed together, than from suboptimal L-methionine alone.

The metabolic pathway for the utilization of lanthionine probably involves cystathionase-catalyzed transformation to form cysteine and pyruvic acid, analogous to the observed transformation of cystathionine to cysteine and α-ketobutyric acid (32, 49).

Stereochemical analysis of the cleavage shows that L-lanthionine will produce one molecule of L-cysteine; D-lanthionine, one molecule of D-cysteine; and meso-lanthionine, an equal mixture of both isomers. Complete cleavage of an equimolar mixture of all three isomers will, therefore, form an equimolar mixture of D-and L-cysteine, and only the latter, as our results have shown, will be utilized nutritionally in the presence of L-methionine.

D-Phenylalanine

Table 3 and Figure 10 show that the relative growth of animals fed L-and D-Phe ranged from 28.3 to 81.3%, depending on the concentrations selected for comparison. Inspection of Figure 10 reveals that near maximum growth may be expected with sufficient D-Phe in the diet. The data suggest the absence of any marked antinutritional effects or toxicity from feeding either Phe isomer at twice the optimum dietary level.

The dose dependence of the relative response of the two isomers in a range beyond that producing maximum growth with L-Phe complicates attempts to compare results from earlier studies with other species, which often report data based on a single substitution of the D- for the L-isomer. Nevertheless, the 62.1% relative response shown in Table 3 with a D-Phe concentration equal to the level of L-Phe in the complete amino acid diet formulation (1.51%) is similar to the reported relative efficacy of 68% in rats and 75% in poultry (9,33).

Although the semi-logarithmic plot shown in Figure 10 offers a useful graphic presentation of the relative potencies of the phenylalanine isomers, it does not limit correctly, *i.e.*, the point corresponding to 0% cannot be an extrapolation of the other data. An alternative way to represent the data would be to use the following simple exponential function which fits the data and limits correctly at both ends.

Table 1. D-Aspartic acid content in commercial food products

Commercial Product	D/L ASP	D-ASP / D-ASP+L-ASP
Texturized sov protein	0.095	0.09
Baby formula (sov protein)	0.108	0.10
Simulated bacon (soy protein)	0.143	0.13
Corn chips	0.164	0.14
Dairy creamer (sodium caseinate)	0.208	0.17

Ref. 66.

Table 2. Bioavailability of amino acid derivatives of L-cysteine with and without L-methionine in mice

Test Substance	Relative Percent in Diet[a]	Mean Weight Gain[b] (g)		Gain Relative to Methionine alone (%)
		L-Methionine		
		0%	25%	
None		4.0	7.4	100
L-Cysteine	100	-3.6	13.2	178
D-Cysteine	100	-4.2	5.6	76
DL + Meso-Lanthionine	100	-3.2	9.4	127

[a] Relative 100% is the molar equivalent of 1.17g L-methionine per 100g diet.
[b] Weight gain after 14 days. Ref. 32, 34.

Table 3. Utilization of L and D-phenylalanine for body weight gain in mice

Percent in diet	Relative percent in diet	Body Weight Gain[a] (g)		Relative Response for for D-Phenylalanine[b] (%)
		L-Phenylalanine	D-Phenylalanine	
0	0	-4.6	-4.6	-
0.38	25	7.4	-1.2	28.3
0.76	50	13.4	1.8	35.6
1.51	100	12.8	6.2	62.1
3.02	200	13.6	10.2	81.3

[a] Weight gain after 14 days.
[b] Net weight gain for D-form divided by net weight gain for L-form, times 100. Ref. 33.

Table 4. Weight gain of mice fed an amino acid diet supplemented with D-tyrosine

Diet		
Percent D-Tyr in diet	D-Tyr:L-Phe	Wt.gain[a]
0.76% L-Phe in diet		g
0	0	13.5
0.42	1:2	13.5
0.83	1:1	8.8
1.55	2:1	1.5
1.51% L-Phe in diet		
0	0	13.8
0.42	1:4	14.3
0.83	1:2	15.0
1.55	1:1	10.0

[a] Weight gain after 14 days. Ref. 33.

Table 5 Weight gain of mice fed a casein diet supplemented with D-tyrosine

Diet		
D-Tyr in diet (%)	D-Tyr:L-Phe	Mean wt gain[a]
10% protein		(g)
0	0	8.8
0.19	1:4	9.7
0.37	1:2	9.0
0.74	1:1	10.2
1.48	2:1	2.2
20% protein		
0	0	14.7
0.37	1:4	13.8
0.74	1:2	13.8
1.48	1:1	10.8
2.97	2:1	7.0

[a] Weight gain after 14 days. Ref. 33.

Assume that,

$$\frac{13.6 - \text{gain}}{13.6 - (-4.6)} = e^{-KP} \quad (3); \qquad \ln \frac{13.6 - \text{gain}}{13.6 - (-4.6)} = -KP \quad (4)$$

where: 13.6 is the maximum weight gain with L-Phe (Table 3); gain is the weight gain with 0 and 25% L-Phe and with all concentrations of D-Phe; -4.6 is the weight gain (loss) with 0% Phe; K is a constant; and P is the relative % of either Phe isomer (35).

Equation 4 predicts that a plot of the left-hand part against P should give a straight line with a slope of -K. Such plots (Figure 11) based on the data in Table 3 yield a value of K =0.0428 for L-Phe and 0.00844 for D-Phe. Thus, a measure of the relative potency of D-Phe compared to that of L-Phe, within the range of the growth response observed for the D-isomer, is 0.00844/0.0428 or approximately 20%.

Other workers have reported the following related observations. Healthly young adults can invert about one-third of D-Phe to its L-isomer (86). When a young human female ingested deuterium-labeled D- and L-Phe (8 mg/kg body weight), the amino acid appeared more rapidly and abundantly and disappeared more slowly from plasma with the D- than with the L-isomer (58). Although D- and L-isomers may be absorbed at the same rate from the intestine of rats, tissue accumulation of the isomers varies (88). The D- and L-Phe isomers differ in their ability to stimulate pancreatic secretions in dogs (67). The presence of other D-amino acids decreases the utilization of D-Phe, presumably because of the increased competition for available D amino-acid oxidase needed to transform D- to L-isomers (56). Finally, D-Phe produces analgesia in mice and humans by potentiating the action of enkephalins in the brain through inhibition of the enzyme that degrades brain opioids (6).

Metabolic studies of Phe, including the data shown in Figure 10, indicate that the D-isomer, because it must first be inverted, is utilized more slowly. This leads to the speculation that D-Phe might not have the same effect on phenylketonuria in children as the L-isomer (1). Should this be the case, the slow and uniform release of L-Phe from the D-isomer could possibly provide a basis for the utilization of D-Phe in the treatment of phenylketonuria. Animal and human studies are needed to demonstrate this possibility.

D-Tyrosine

The amino acid L-Tyr is an *in vivo* precursor for brain catecholamine, dopamine and norepinephrine; for the biogenic amine tyramine, and for the ubiquitous pigment melanin (7). *In vitro*, the phenolic hydroxyl group activates the benzene ring of tyrosine, making it more susceptible to chemical and radiation-induced modification than the corresponding ring in phenylalanine (40). The inductive nature of the phenolic group enhances such food processing factors as high pH and temperature to bring about the rapid racemization of L-Tyr residues in proteins (36, 59).

D-Tyr has been shown to prevent and reverse stress-induced chronic hypertension, which may indicate possible effects on the central nervous system (80).

Nutritionally, L-Tyr is classified as a semi-essential amino acid since it is synthesized in vivo from L-Phe (72). Combinations of L-Tyr and L-Phe are complementary in supporting the growth of mice (33) Thus, under conditions where L-Phe may be limiting, L-Tyr can supply

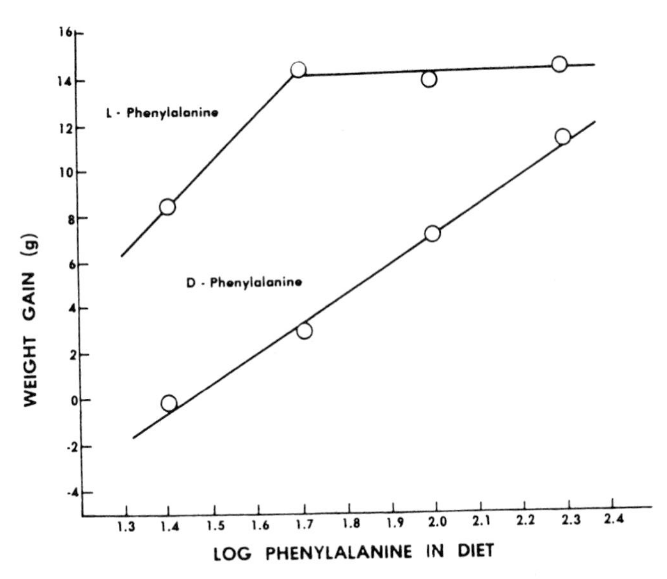

Figure 10. Growth response of mice fed D- and L-phenylalanine. Ref. 33.

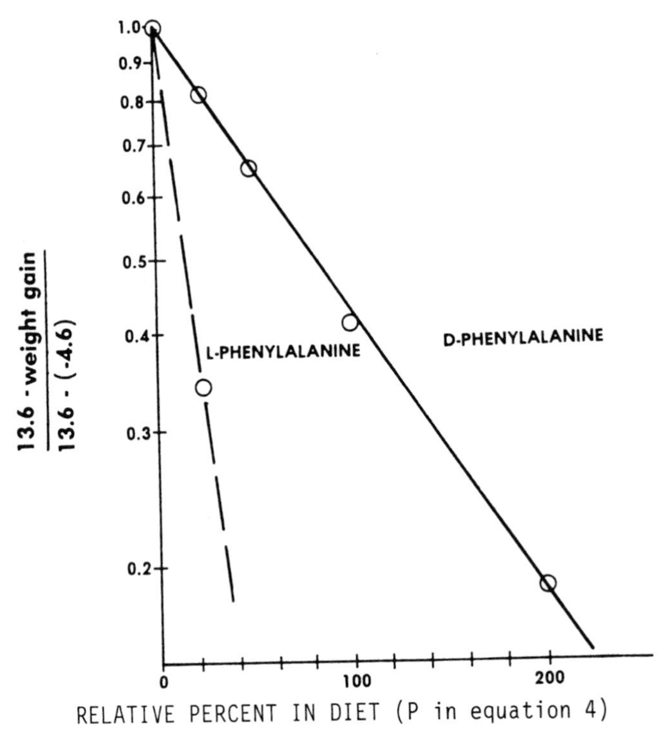

Figure 11 Growth reponse of mice fed D- and L-phenylalanine. Ref. 35..

about half the requirement of L-Phe alone, a value similar to those previously reported for humans, chicks, and rats.

Table 4 shows that with D-Tyr in an amino acid diet, growth inhibition was severe at a D-Tyr:L-Phe ratio of 2:1, but was much more moderate when the ratio was 1:1. Similar results were obtained with a casein based diet supplemented with D-Tyr (Table 5). One or more of the following six mechanisms may explain this inhibition of weight gain by dietary D-Tyr:

Tyrosyl ribonucleic acid synthethase was found to catalyze the incorporation of D-Tyr into tyrosyl-tRNA, an aminoacyl adenylate derivative similar to that formed by L-Tyr (16, 90). Since the structural features of an amino acid after combining with tRNA do not seem to be significant in the specificity of its incorporation into protein, D-Tyr ingestion should lead to the formation of faulty peptides and proteins. These, in turn, could interfere with normal metabolic processes and be responsible for the observed growth inhibition.

A related possibility is the suppression of normal protein synthesis by D-Tyr through competitive inhibition of L-Tyr or L-Phe incorporation into aminoacyl-tRNA, and thus, into proteins.

A third possibility is an interference by D-Tyr in the biosynthesis or biological action of vital neurotransmitters such as dopamine. The observed hypotensive effect of D-Tyr is postulated to involve this effect (80).

A fourth possibility is an interference by D-Tyr in the hydroxylation of L-Phe to L-Tyr.

A fifth possibility is that D-Tyr at concentrations provided in the present study could overload metabolic pathways needed to eliminate or detoxify excess D-amino acids. However, when the dietary level of L-Phe was increased, these same concentrations of D-Tyr failed to bring about marked growth inhibition.

A sixth possibility is competition of D-tyrosine with L-tyrosine or L-phenylalanine for membrane transport (4). A discussion of membrane transport of amino acids, which appeared in an article in Nutrition Reviews (4) is relevant to the theme of this paper, and accordingly it is quoted here:

"The catalytic transport of amino acids across the plasma membrane tends to have lower structural specificity than is usually characteristic of enzymatic reactions. Furthermore, the stereospecificity varies among amino acid transport systems and among membranes. In some cases, stereospecificity is high, as appears to be characteristic of the inner mitochondrial membrane. A study with the Ehrlich ascites tumor cell demonstrated differences in selectivity. In that study, there was only about threefold preference for L-methionine over its D-antipode. Furthermore, another study showed that a transport system that was able to discriminate between glutamic acid antipodes had little ability to discriminate between aspartic acid antipodes.

"Associated with this low stereoselectivity of transport is the familiar circumstance in higher animals that the transport of amino acids is mainly carried out by 'public' systems, able to handle a wide range of amino acids, rather than by the occasional 'private' systems restricted to a single amino acid. The probability is very high that not only phenylalanine and tyrosine, but also tryptophan, the branched-chain amino acids and several others have a large proportion of their inter-organ flows mediated by a common transport system designated L. This possibility occasions the currently re-emphasized suspicion that in untreated phenylketonuria, the high circulating L-phenylalanine levels interfere to critical degrees with the passage of several important amino acids across the blood-brain barrier. Tews et al. (85) have given special attention to analogue inhibition of amino acid transport across this barrier, arising from dietary amino acid imbalances.

"In connection with competition for membrane transport, Friedman and Gumbmann (33) are correct in focusing attention on the interference of D-tyrosine with L-phenylalanine as well as with L-tyrosine metabolism. They cite a study pertinent to possible competition by D-tyrosine with transport of L-amino acids. When a young woman ingested deuterium-labeled D-and L-phenylalanine (8 mg/kg body weight), the D-isomer appeared more rapidly and disappeared more slowly from the plasma than the L-isomer (58). The higher tolerance curve for the D- than for the L-tyrosine brings attention to the important differences in the rates at which D- and L-tyrosine are withdrawn from the circulation. In the case of D-phenylalanine, but not for D-tyrosine, inversion to the L-form limits and compensates for the competitive effects arising from elevated levels of the D- form".

"Although the specific mechanism by which D-tyrosine becomes an antimetabolite for the mouse is still not clear, these findings warn investigators to avoid the use of DL-tyrosine in experimental and clinical nutrition. They also dramatize the competition between enantiomorphs of some amino acids. In the case of D-tyrosine, the precise mechanismas can be elucidated only by further study."

In conclusion, the cited data demonstrate that D-Tyr, unlike L-Tyr, has no sparing effect for L-Phe. In fact, a metabolic stress, in the form of growth inhibition in mice, may become evident when D-Tyr is present in the diet at equal or greater molar concentrations to L-Phe. This acute effect remains to be defined toxicologically, and the potential for sub-chronic and chronic toxicity following exposure to lower levels of D-Tyr remains unknown.

D-Tryptophan

The relative potency of D-tryptophan (as defined in Tables 6 and 7) compared to the L-isomer was strongly dose-dependent, being inversely related to dietary concentration and ranging from 29% to 64%. A plot of this data showed that growth was linearly related to D-tryptophan. The maximum growth obtainable for L-tryptophan occurred at or slightly less than 0.174% in the diet.

By increasing the dietary concentration of D-tryptophan up to 0.52% (three times the highest level shown in Table 6), it was possible to demonstrate for D-tryptophan that growth also passed through a maximum, one which equalled 82% of that achieved with the L-isomer. This occurred at approximately 0.44% D-tryptophan for a relative potency of 25% (data not shown). This level in the short term assay appeared to be well tolerated. However, additional studies are needed, both in mice and other species, to define the effects of more prolonged dietary exposure to D-trytophan.

Table 6. Growth of mice fed D- and L-tryptophan

Amino acid tested	Percent in diet	Mean weight gain (g)[a]	Percent equivalent in diet[b]	Relative potency (%)[c]
L-tryptophan	0	-3.5 ± 0.29		
	0.022	-1.8 ± 0.37		
	0.044	2.2 ± 0.49		
	0.087	12.4 ± 1.25		
	0.174	15.6 ± 0.68		
D-tryptophan	0	-3.5 ± 0.29		
	0.022	-2.8 ± 0.49	0.014	64
	0.044	-1.8 ± 0.37	0.024	55
	0.087	-0.4 ± 0.81	0.033	38
	0.174	3.8 ± 0.58	0.051	29

[a] Weight gain after 14 days. N = 5. Means ± S.E. Duncan's Multiple Range test.

[b] Percent equivalent in the diet is that concentration of L-amino acid which would produce the same growth observed for the D-amino acid (determined graphically).

[c] Relative potency is the percent equivalent in the diet divided by the percent actually fed X 100.

Table 7. Growth of Mice fed D- and L-Tryptophan for 14 days (N = 5).

Relative percent in diet[a]	Body weight gain (g)[b]		Relative weight gain (D/L)
	D-tryptophan	L-tryptophan	%
0	-2.8 ± 0.4	-2.8 ± 0.4	
50	0.2 ± 0.6		17.3
100	4.4 ± 1.5	14.5 ± 0.4	41.6
150	10.6 ± 1.3		77.5
200	11.4 ± 1.0		82.1
250	12.2 ± 1.2		86.7
300	11.4 ± 1.1		82.1

[a] 100% relative equals 0.174% in the diet of either D or L-tryptophan.

[b] Means ± S.E. Duncan's Multiple Range test.

Considerable species variation is known to exist for the nutritive value of D-tryptophan. In chicks fed amino acid diets, the relative biological activity of the D- to L-isomer has been reported to be 20%, which is similar to the relative potency we find for mice (61,70,71) This is in marked contrast to rats also fed amino acid diets where the stimulation of growth was equivalent to that produced by L-tryptophan, for a biological activity of 100%. It should be noted that D-tryptophan supplementation of a protein based diet containing 0.085% protein-bound L-tryptophan resulted in growth stimulation in mice nearly equal to that produced by L-trytophan supplementation. Plasma analysis indicated that the rate of conversion of D to L closely reflected the difference in response between chicks and rats. Essentially no conversion of D-tryptophan to the L-isomer could be detected in chick plasma, and most of the D-isomer administered was excreted unchanged. In the rat, plasma levels of the L-isomer rapidly increased upon administration of D-tryptophan, and only 1% of the dose was excreted in the urine.

Lysinoalanine Isomers

Lysinoalanine ($HOOCCH(NH_2)CH_2CH_2CH_2CH_2NHCH_2CH(NH_2)COOH$,(LAL)), is an unnatural amino acid (formed as shown in Figures 12-14) has been identified in hyrolyzates of processed food proteins, in particular those subjected to alkali (19, 21, 25, 27, 31, 38, 39, 41, 42, 44, 46, 47, 53, 60, 68, 74, 81, 83, 89).

Lysinoalanine has two asymmetric carbon atoms. Four stereoisomers are therefore possible: LL, LD, DL, and DD. The following systematic name is now for the isomer formed in the greatest quantity during food processing (25):

$$N^{\varepsilon}-(\underline{DL}-2\text{-amino-}2\text{-carboxyethyl})-\underline{L}\text{-lysine.}$$

A more correct name with R and S designation for the substituents of lysine would be:

$$N^{6}-\left[(\underline{2RS})-2\text{-amino-}2\text{-carboxyethyl}\right]-\underline{L}\text{-lysine.}$$

In analogy with the two isomers derived from L-lysine, D-lysine would give rise to the following correctly named diastereoisomers:

$$N^{6}-\left[(\underline{2RS})-2\text{-amino-}2\text{-carboxyethyl}\right]-\underline{D}\text{-lysine.}$$

It is also worth noting that the reaction between the ε-NH_2 group of lysine and the double of dehydrothreonine can, in principle, give rise to five symmetric centers and 32 stereoisomers (25).

In rats, histological changes in the kidneys have been identified which are related to dietary exposure to this substance, either isolated or as part of intact proteins. The lesions are located in the epithelial cells of the straight portion of the proximal renal tubules and are characterized by enlargement of the nucleus and cytoplasm, increased nucleoprotein content, and disturbances in DNA synthesis and mitosis.

Because of these observations, concern has arisen about the safety of foods which may contain LAL and related dehydroalanine-derived amino acids which are known to produce similar lesions. However, since the mechanism by which these compounds damage the rat kidney is unknown, it is difficult to assess the risk to human health caused by their presence in the diet.

LAL has two asymmetric carbon atoms making possible four separate diastereoisomeric forms, LL, LD, DL, and DD. Its structure suggests that it should have excellent chelating potenial for metal ions, a property which may have relevance to its toxic action. Accordingly, we have examined LAL (a mixture of the LL and LD isomers) for its affinity towards a series of metal ions, of which copper (II) was chelated the most strongly (39, 42,74).On this basis, we have suggested a possible mechanism for kidney damage in the rat involving LAL's interaction with copper within the epithelial cells of the proximal tubules.

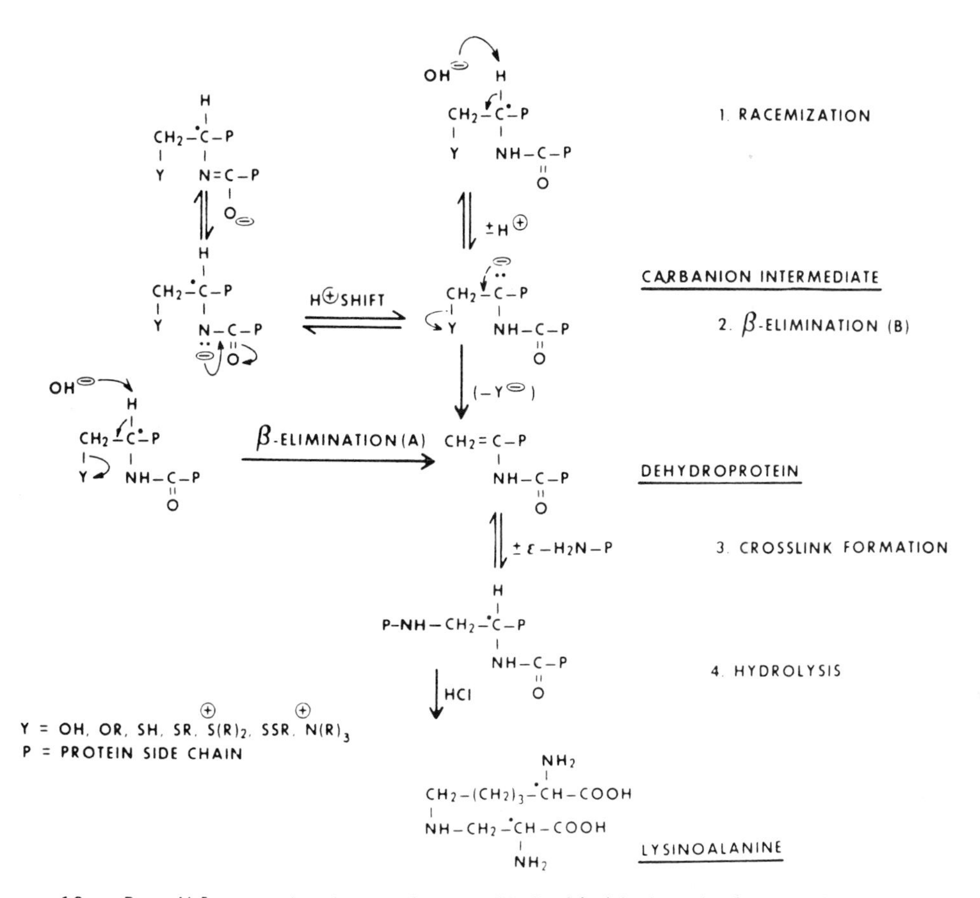

Figure 12. Possible mechanisms for alkali li iinduced lysinoalanine formation and concurrent racemization in a protien via a common carbanion intermediate. Ref. 46.

Of the four isomers of LAL, LL and LD are derived from L-lysine; the other two from D-lysine. Since L-lysine is the natural amino acid present in proteins, most of the LAL formed during food processing can be expected to be a mixture of LL and LD. However, since exposure of food proteins to heat and alkali may racemize a small fraction of L-lysine to the D-isomer, treated food proteins may also contain small amounts of DL- and DD-LAL.

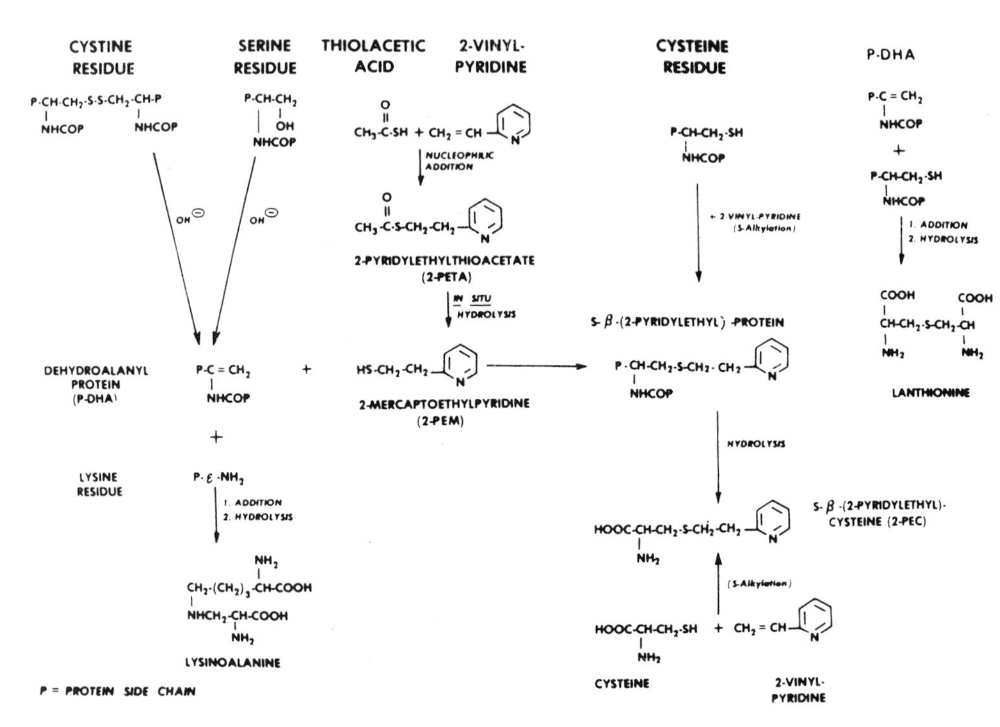

Figure 13. Transformation of dehydroalanyl (DHA) to lysinoalanyl (LAL), S-β-(2-pyridylethyl)-L-cysteine (2-PEC), and lanthionine (LAN) residues in a protein. Hydroxide ions induce the elimination reations in cystine and serine to produce dehydroalanine. The double bond of dehydroalanine can then interact with ε-NH_2 group of lysine to form LAL, with the SH group of cysteine to give LAN, and the SH group of added 2-mercaptoethylpyridine to form 2-PEC, as illustrated. The latter is identical with the compound obtained from cysteine and 2-vinylpyridine. All products have newly created asymmetric centers. Ref. 48, 64.

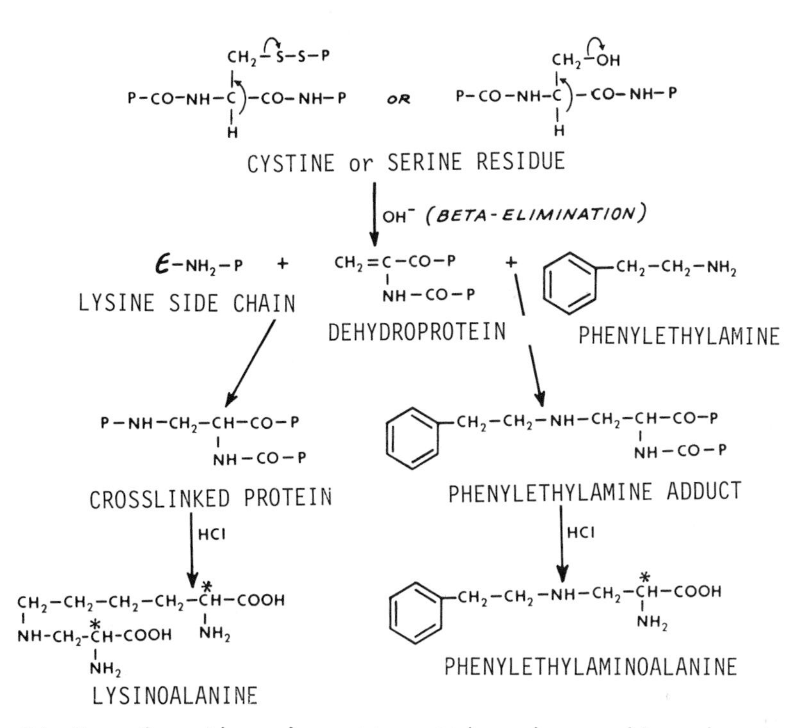

Figure 14. Transformation of cystine and serine residues in a protein to lysinoalanine and phenylethylaminoalanine. Newly created asymmetric centers are indicated by asterisks. Ref. 38.

As described in detail elsewhere (39, 74), it is possible to predict the equilibria in vivo between histidine, the major low-molecular weight copper carrier in plasma, and competing chelating agents such as LAL (eq. 5).

$$CuHis_2 \quad + \quad H_2LAL \rightleftharpoons CuLAL \quad + \quad 2HHis \qquad (5)$$

A mathematical analysis of the equilbrium shown in equation 5 was used to calculate LAL plasma levels needed to displace histidine as the major copper carrier in vivo. The calculated values are: 27 μM for LD-LAL; 100 μM for LL-LAL, and 49 μM for the mixture of the two.

The above considerations suggest that LD-LAL would be a better competitor for copper (II) in vivo than the LL-isomer, i.e., it should take about one fourth as much LD-LAL as LL-LAL to displace the same amount of histidine from copper-histidine. This difference could explain the greater observed toxicity of the LD-LAL. The apparent direct relationship between the observed affinities of the two LAL isomers for copper (II) ions in vitro and their relative toxic manifestation in the rat kidney is consistent with our hypothesis that LAL exerts its biological effect through chelation to copper in body fluids and tissues. Limited studies on the binding of LL- and LD-lysinoalanines to cobalt (II) ions imply that lysinoalanine could also influence cobalt utilization in vivo. Animal studies are needed to confirm the predicted role of lysinoalanine in metal ion transport, utilization, and histopathology

Dehydroamino acids

Dehydroalanine (CH_2=$CH(NH_2)COOH$) is the postulated reactive intermediates of a large series of novel amino acids, including lysinoalanine, formed during food processing (24, 25, 38, 39, 41, 50, 64). Since these novel amino acids all contain asymmetric carbon atoms, we will briefly examine the possible chemical and nutritional significance of dehydroalanine (derived from serine and cystine).

Direct evidence for dehydroalanine reactivity was obtained by Friedman et al (41). They showed that dehydroalanine derivatives convert lysine side chains in casein, bovine serum albumin, lysozyme, wool, or polylysine to lysinoalanine residues at pH 9 to 10. Related studies showed that protein SH groups generated by reduction of disulfide bonds are completely alkylated at pH 7.6 to form lanthionine. These studies demonstrate that lysinoalanine and lanthionine residues can be introduced into a protein under relatively mild conditions, without strong alkaline treatment. They also imply that it should be possible to explore nutritional and toxicological consequences of lysinoalanine and lanthionine consumption in the absence of racemization.

In a related study, Masri and Friedman (64) developed a procedure for detecting dehydroalanine in alkali-treated proteins based on addition of the SH group of 2-mercaptoethylpyridine to the double bond of dehydroalanline to form S-ß-(2-pyridylethyl) cysteine. The cysteine derivative can be assayed. after acid protein hydrolysis, by standard analytical techniques (48). This method revealed significant amounts of dehydroalanine in alkali- treated casein.

Because dehydroalanine residues occur naturally in peptide antibiotics and hormones (50), are present in alkali-treated proteins (64), and act as alkylating agents in vitro, a need exists to: (a) establish whether they act as biological alkylating agents in vivo;

and (b) assess their nutritional and toxicological effects, both in pure form and in foods.

It is also worth noting that in a series of classic kinetic and synthetic studies we previously examined the chemical reactivity of amino acids, peptides, and proteins with the double bonds of structurally different conjugated vinyl compounds related to dehydroalanine such as acrylonitrile, methyl acrylate, methyl vinyl ketone, vinylpyridine, and vinylquinoline (93-138). The chemical reactivities were correlated by linear free energy relationships and structural and electronic parameters associated with the reactants. These relationships permit calculating numerous predicted rate constants for many reactions and are also invaluable tools for designing experiments for selective and nonselective modification of functional groups in structural proteins and enzymes. Limited studies revealed that dehydroalanine derivatives are also influenced by similar electronic and inductive effects (122, 123, 125).

This review of our findings complements other reports and reviews on the nutritional utilization of D-amino acids (9, 12, 13, 34, 35, 43, 62, 66, 73, 87). These studies suggest tht a better understanding of the factors which induce racemization in food proteins and the resulting nutritional and toxicological consequences will make it possible to minimize adverse consequences of high pH during on processing and to improve the quality and safety of our diet. Finally, our studies show that mice are a useful animal model to study nutritional consequences of processing-induced changes in amino acids and proteins (90). Mouse feeding studies require less material and less time than corresponding studies with rats. They may therefore fulfill a need for an inexpensive bioassay to label foods for protein nutritional quality.

ACKNOWLEDGEMENT

I thank my colleagues whose names appear on the cited referenes for excelent scientific collaboration.

REFERENCES

1. Acosta, P. B., Trahms, D., Willman, N. S., and Williamson, M. Phenylalanine intakes of 1-and 6-year old children with phenylketonuria undergoing therapy. Am. J. Clin. Nutr. 38, 694. 1983.

2. Albanese, A.A., The utilization of D-amino acids by man. I. Tryptophan, methionine and phenylalanine. Bull. Johns Hopkins Hospital 75, 175. 1944.

3. Allenmark, S. and Andersson, S. Direct liquid chromatographic separation of enantiomers on immobilized protein statinary phases. J. Chromatogr. 351, 231. 1986.

4. Anonymous. Dietary D-tyrosine as an antimetabolite in mice. Nutr. Rev. 43, 156. 1985.

5. Baker, D. H. Utilization of isomers and analogs of amino acids and other sulfur-containing compounds. Progr. Food Nutr. Sci. 10, 133. 1986.

6. Balagot, R. C., Ehrenpreis, S., Kubota, K., and Greenberg, J. Analgesia in mice and humans by D-phenylalanine: relation to inhibition of enkephalin degradation and enkephalin levels. Adv. Pain Res. and Therapy 5, 289. 1983.

7. Benedict, C. R., Anderson, G.H., and Sole, M. J. The influence of oral tyrosine on plasma catecholamines in man. Am. J. Clin. Nutr. 38, 429. 1983.

8. Benevenga, N. U. Betaine in the treatment of homocystinuria. New Engl. J. Med. 310, 265. 1983.

9. Boebel, K. and Baker, D. H. Comparative utilization of the isomers of phenylalanine and phenyllactic acid by chicks and rats. J. Nutr. 112, 367. 1982.

10. Boehm, M. F. and Bada, J. L. Racemization of aspartic acid and phenylalanine in the sweetener aspartame. Proc. Natl. Acad Sci. USA 81, 5263. 1984.

11. Boehm, M. F. and Bada, J. L. The racemization rate of selenomethionine and methionine in yeast at 100°C and neutral pH. Analyt. Biochem. 145, 273. 1985.

12. Borg, B. S. and Wahlstrom, R.C. Species and isomeric variation in the utilization of amino acids. In "Absorption and Utilization of Amino Acids", Friedman, M., Ed., CRC Press, Boca Raton, Florida, Vol. 2, 1989. pp. 155-172.

13. Bruckner, H., and Hausch, M. D-amino acids in food: detection and nutritional aspects. In "Chirality and Biological Activity ", Helmstedt, B., Frank, H. and Testa, B., Eds., Alan R. Liss, Inc., New York, 1990, pp. 129-136.

14. Bruckner, H. and Hausch, M. Gas chromatographic detection of D-amino acids as common constituents of fermented foods. Chromatographia 18, 487. 1989.

15. Bruckner, H . and Hausch, M. D-amino acids in dairy products: detection, origin and nutritional aspects. II. Ripened cheeses. Milchwissenschaft 45, 7. 1990.

16. Calendar, R. and Berg, P. D-tyrosyl RNA: formation, hydrolysis and utilization for protein synthesis. J. Mol. Biol. 26, 39. 1967.

17. Cavallini, D., DeMarco, D., and Mondovi, B. Experiments with D-cysteine in the rat. J. Biol. Chem. 230, 25. 1958.

18. Dakin, H. D. The racemization of proteins and their derivatives resulting from tautomeric change. J. Biol. Chem. 14, 271. 1912.

19. De Weck-Guadard, D., Liardon, R., and Finot, P.A. Stereomeric composition of urinary lysinoalanine after ingestion of free or protein-bound lysinoalanine in rats. J. Agric. Food Chem. 36, 717. 1988.

20. Drake, M. R., Del La Rosa, J., and Stipanuk, M. H. Metabolism of cysteine in rat hepatocytes. Biochem. J. 244, 279. 1987.

21. Feron, V. J., van Beek, L., Slump, and Beems, R. B. Toxicological aspects of alkali-treatment of food proteins. In "Biological Aspects of New Protein Foods", Adler Nissen J., Ed., Pergamon, Oxford, England. 1977, p. 139.

22. Frank, H., Rettenmeier, A., Weicker, H., Nicholson, G.J., and Bayer, E. Determination of enantiomer-labeled amino acids in small volumes by gas chromatography. Analyt. Chem, 54, 715. 1982.

23. Friedman, M. "The Chemistry and Biochemistry of the Sulfhydryl Group in Amino Acids, Peptides, and Proteins", Pergamon, Oxford, England, 1973, Chapter 3.

24. Friedman, M. Chemical basis for pharmacological and therapeutic action penicillamine. Proc. Royal Soc. Med. 70,(Supplement 3), 50. 1977.

25. Friedman, M. Crosslinking amino acids -- stereochemistry and nomenclature. In "Protein Crosslinking: Nutritional and Medical Consequences", Friedman, M., Ed., Plenum, New York 1977, p. 1-27.

26. Friedman, M. (Editor). Protein Crosslinking, 2 Volumes, Plenum Press, New York, 1977.

27. Friedman, M. Inhibition of lysinoalanine formation by protein acylation. In "Nutritional Improvement of Food and Feed Proteins", M. Friedman, Ed., Plenum, New York, 1978, 613-648.

27a. Friedman, M., and Cuq, J.L. Chemistry, analysis, nutritional value, and toxicology of tryptophan in food. A Review. J. Agric. Food Chem. 36, 1079-1093. 1988.

28. Friedman, M. Chemically reactive and unreactive lysine as an index of browning. Diabetes 31(6), (Supplement 3), 5. 1982.

29. Friedman, M. and Gumbmann, M. R. Biological availability of epsilon-N-methyl-L-Lysine, 1-N-methyl-L-histidine, and 3-N-methyl-L-histidine in mice. Nutr. Repts. Int. 19,437.1979.

30. Friedman, M. and Gumbmann, M. R. Bioavailability of some lysine derivatives in mice. J. Nutr. 111, 1362. 1981.

31. Friedman, M., Gumbmann, M.R. and Savoie, L. The nutritional value of lysinoalanine as a source of lysine for mice. Nutr. Repts. Int. 26, 939. 1982.

32. Friedman, M. and Gumbmann, M. R. The utilization and safety of isomeric sulfur-containing amino acids in mice. J. Nutr. 114, 2301. 1984.

33. Friedman, M. and Gumbmann, M. R. The nutritive value and safety of D-phenylalanine and D-tyrosine in mice. J. Nutr. 114, 2089. 1984.

34. Friedman, M. and Gumbmann, M.R. Methionine derivatives as a nutritional source of methionine. In "Absorption and Utilization of Amino Acids", M. Friedman, Ed., CRC Press, Boca Raton, Florida, 1989, Vol. 2, pp. 117-132.

35. Friedman, M. and Gumbmann, M.R. Dietary significance of D-amino acids. In "Absorption and Utilization of Amino Acids", M. Friedman, Ed., CRC Press, Boca Raton, Florida, 1989, Vol. 2, pp. 173-190.

36. Friedman, M. and Liardon, R. Racemization kinetics of amino acid residues in alkali-treated soybean proteins. J. Agric. Food Chem. 33, 666. 1985.

37. Friedman, M. and Masters, P.M. Kinetics of racemization of amino acid residues in casein. J. Food Sci. 47, 760. 1982.

38. Friedman, M. and Noma, A. T. Formation and analysis of phenylethylaminoalanine in food proteins. J. Agric. Food Chem. 34, 497. 1986.

39. Friedman, M. and Pearce, K. N. Spectroscopic properties and copper (II) affinities of IL- and LD-lysinoalanine isomers. J. Agric. Food Chem. 37, 123. 1989.

40. Friedman, M., Diamond, M. J. and Broderick, G. A. Dimethylolurea as a tyrosine reagent and protein protectant against ruminal degration. J. Agric. Food Chem. 30, 72, 1982.

41. Friedman, M., Finley, J. W., and Yeh, I. S. Reactions of proteins with dehydroalanine. In "Protein Crosslinking: Nutritional and Medical Consequences", M. Friedman, Ed., Plenum, New York. 1977, p. 213.

42. Friedman, M., Grosjean, O.K., and Zahnley, J.C. Inactivation of metalloenzymes by lysinoalanine, phenylethylaminoalanine, alkali-treated food proteins, and sulfur amino acids. In "Nutritional and Toxicological Significance of Enzyme Inhibitors in Foods", M. Friedman, Ed., Adv. Exp. Med. Biiol. 199, 531. 1986.

43. Friedman, M., Gumbmann, M.R., and Brandon, D. I. Nutritional, toxicological, and immunological consequences of food processing. Frontiers Gastrointest. Res. 14, 79. 1988.

44. Friedman, M., Gumbmann, M.R., and Master, P.M. Protein-alkali reactions: chemistry, toxicology, and nutritional consequences. In "Nutritional and Toxicological Aspects of Food Safety", M. Friedman, Ed., Plenum Press, New York, 1984, p. 367.

45. Friedman, M., Gumbmann, M. R. , and Ziderman, I. R. Nutritional value and safety in mice of proteins and their admixtures with carbohydrates and vitamin C. J. Nutr. 117, 508. 1987.

46. Friedman, M., Levin, C. E., and Noma, A.T. Factors governing lysinoalanine formation in soy proteins. J. Food Sci. 49, 1282. 1984.

47. Friedman, M., Zahnley, J. C., and Masters, P.M. Relationship between in vitro digestibility of casein and its content of lysinoalanine and D amino acids. J. Food Sci. 46, 127. 1981.

48. Friedman, M., Zahnley, J. C., and Wagner, J. R. Estimation of the disulfide content of trypsin inhibitors as S β (2-pyridylethyl) L-cysteine. Analyt. Biochem. 106, 27. 1980.

49. Funk, M.A., Lowry, K.R., and Baker, D.H. Utilization of the L- and DL-isomers of α-keto-β-methylvaleric acid by rats and comparative efficacy of of the keto analogs branched-chain amino acids provided as ornithine, lysine, and histidine salts. J. Nutr. 117, 1550. 1987.

50. Gross, E. α, β-Unsaturated and related amino acids in peptides and proteins, Adv. Exp. Med. Biol. 86A, 131. 1977.

51. Gumbmann, R. R., Friedman, M., and Smith, G.A. The nutritional values and digestibilities of heat damaged casein and casein carbohydrate mixtures. Nutr. Repts. Int. 28, 355. 1983.

52. Herrera, M., Zyman, J., Pena, J. G., Segurajauregui, J. S., and Vernon, J. Kinetic studies on the alkaline treatment of corn for tortilla preparation. J. Food Sci. 51, 1486. 1986.

53. Karayiannis, N. I., MacGregor, J. T., and Bjeldanes, L. F. Biological effects of alkali-treated soy protien and lactalbumin in the rat and mouse. Food Cosm. Toxicol. 17, 509. 1979.

54. Kies, C., Fox, H., and Aprahamian, S. Comparative value of L-, DL-, and D-methionine supplementation of an oat-based diet for humans. J. Nut. 105, 809. 1975.

55. Konno, R., Uchiyama, S., and Yasumura, Y. Intraspecies and interspecies variations in the substrate specificity of D-amino acid oxidase. Comp. Biochem. Physiol. 71B, 735. 1982.

56. Konno, R. and Yasumura, Y. Involvement of D-amino acid oxidase in D-amino acid utilization in the mouse. J. Nutr. 114, 1617. 1984.

57. Krijgsheld, K. R., Glazenburg, E. J., Scholtlens, E., and Mulder, G. J. The oxidation of L and D-cysteine to inorganic sulfate and taurine in the rat. Biochim. Biophys. Acta 677, 7. 1981.

58. Lehmann, W.D., Theobald, N., Fischer, R., and Heinrich, H. C. Stereospecificity of phenylalanine plasma kinetics and hydroxylation in man following oral application of stable isotope labeled pseudo racemic mixture of L- and D-phenylalanine, Clin. Chim. Acta 128, 181. 1983.

59. Liardon, R. and Friedman, M. Effect of peptide bond cleavage on the racemization of amino acid residues in soybean proteins. J. Agric. Food Chem. 35. 1987.

60. Liardon, R., Friedman, M., and Philippossian, G. Racemization kinetics of protein-bound lysinoalanine in strong acid media. Isomeric composition of protein-bound ly sinoalanine. J. Agric. Food Chem. 39, in press. 1991.

61. MacEwan, K. L. and Carpenter, K. J. The nutritional value of supplementary D-tryptophan for growing mice. Nutr. Repts. Int. 21, 279. 1980.

62. Mann, E. H. and Bada, J. L. Dietary D-amino acids. Ann. Rev. Nutr. 7. 209. 1987.

63. Mann, E.H., Fisher, G.H., Payan, I.L., Cadilla Perezrios, R., Garcia, N.M., Chemburkar, R., Arends, G., and Frey II, W.H. D-aspartate in human brain. J. Neurochem. 48, 5110. 1987.

64. Masri, M. S. and Friedman, M. Transformation of dehydroalanine to S-beta-(2-pyridylethyl)-L-cysteine Biochem. Biophys. Res. Commun. 104, 321. 1982.

65. Masters, P. M. and Friedman, M. Racemization of amino acids in alkali treated food proteins.. J. Agric. Food Chem. 27, 507. 1979.

66. Masters, P.M. and Friedman, M. Amino acid racemization in alkali treated food proteins - chemistry, toxicology, and nutritional consequences. In "Chemical Deterioration of Proteins", J. R. Whitaker and M. Fujimaki, Eds. American Chemical Society Symposium Series, 123, 165. 1980.

67. Meyer, J. H. and Grossman, M. I. Comparison of D- and L-phenylalanine as pancreatic stimulants. Am. J. Physiol. 222, 1058. 1972.

68. Miller, S.P.F. and Thompson, J. Biosynthesis stereochemical configuration of N^5 (1-carboxyethyl) ornithine. J. Biol. Chem. 262 No. 33, 16109. 1987.

69. Ogino, T., Ogino, H., and Nagy, B. Application of aspartic acid racemization to forensic odontology: post-mortem designation of age of death. Forensic Sci. Int. 29, 259. 1985.

70. Ghara, I., Otsuka, S., Yugari, Y., and Ariyoshi, S. Comparison of the nutritive values of L , DL and D-tryptophan in the rat and chick. J. Nutr. 110, 634. 1980.

71. Ohara, I., Otsuka, S., Yugari, Y., and Ariyoshi, S. Inversion of D-tryptophan to L-tryptophan and excretory patterns in the rat and chick. J. Nutr. 110, 641. 1980.

72. Owens, F.M., and Pettigrew, J.F. Subdividing amino acid requirements into portions for maintenance and growth. In "Absorption and Utilization of Amino Acids", M. Friedman, Ed., CRC Press, Boca Raton, Florida, 1989. Vol. 2, pp. 15-30.

73. Pall, G., Marchelli, R., Dossena, A., and Casnati, G. Occurrence of D-amino acids in foods. J. Chromatogr. 475, 45. 1989.

74. Pearce, K. N. and Friedman, M. The binding of copper (II) and other metal ions by lysinoalanine and related compounds and its significance for food safety. J. Agric. Food Chem. 36, 707. 1988.

75. Printen, K. J., Brummel, M. C., Ericson, M. S., and Stegink, L. D. Utilization of D methionine during total parenteral nutrition in surgical patients. Am. J. Clin. Nutr. 32, 1200. 1979.

76. Sarwar, G., Blair, R., Friedman, M., Gumbmann, M. R., Hackler, L.R., Pellett, P. L., and Smith, T. L. Inter- and intra laboratory variability in rat growth assays for estimating protein quality in foods. J. Assoc. Off. Anal. Chem. 67, 976. 1984.

77. Sarwar, G., Blair, R., Friedman, M., Gumbmann, M. R., Hackler, L. R. Pellett, P. L., and Smith, T. K. Comparison of interlaboratory variation in amino acid analysis and rat growth assays for evaluating protein quality. J. Assoc. Off. Anal. Chem. 68, 52. 1985.

78. Sarwar, G., Christensen, D. A., Finlayson, A. J., Friedman, M., Hackler, L. R., Mackenzie, S. L., Pellet, P. L., and Tkachuk, R. Intra- and inter-laboratory variation in amino acid analysis. J. Food Sci. 48, 526. 1983.

79. Schormuller, J. and Weder, J. The presence of D amino acids in milk. Nahrung 6, 622. 1962.

80. Segal, M. Effect of D-tyrosine on chronic stress-induced hypertension in the rat Res. Commun. Psychol. Psychiatry Behav. 6, 285. 1982.

80a. Smolin, L.A., and Benevenga, N.J. Methionine, homocyst(e)ine, cyst(e)ine - metabolic interrelationships. In "Absorption and Utilization of Amino Acids", Friedman, M., Ed., CRC, Boca Raton, FL, Vol. 1, pp. 158-187.

81. Snow, J. T., Finley, J. W., and Friedman, M. Relative reactivities of sulfhydryl groups with N-acetyldehydroalanine. Int. J. Peptide Protein Res. 7, 461. 1976.

82. Stipanuk, M. H. Effects of excess dietary methionine on the catabolism of cysteine in rats. J. Nutr. 109, 2126. 1979.

83. Tas, A.C. and Kleipool, R. J. C. The stereoisomers of lysinoalanine. Lebensm. Wiss. Technol. 9, 360. 1979.

84. Tateishi, N., Hirasawa, M. Higashi, T., and Sakamoto, Y. The L-methionine sparing effect of dietary glutathione in rats. J. Nutr. 112, 2217. 1982.

85. Tews, J. K., Kim, Y. W. L., and Harper, A. E. Induction of threonine imbalance by dispensable amino acids; relation to competition for transport into brain J. Nutr. 109, 304. 1979.

86. Tokuhisa, S., Saisu, K., Yoshikawa, H., and Baba, S. Biotransformation of D- and L-phenylalanine in man. Chem. Pharm. Bull. 29, 514. 1981.

87. Tovar, L. and Schwass, D.E. D-amino acids in processed proteins: their nutritional consequences. ACS Symp. Series 234, 169. 1983.

88. Winne, D. The influence of blood flow on the absorption of L- and D-phenylalanine from the jejunum of the rat. Arch. Pharmacol. 277. 113. 1977.

89. Woodard, J. C., Short, D. D., Alvarez, M. R., and Reyniers, J. Biologic effects of lysinoalanine. In "Protein Nutritional Quality of Foods and Feeds", M. Friedman, Ed., Marcel Dekker, New York, 1975, p. 595.

90. Yamane, T., Miller, D.L., and Hopfield, J.J. Discrimination between D- and L-tyrosyl transfer ribonucleic acids in peptide chain elongation. Biochemistry 20, 7059. 1981.

91. Zezulka, A. Y., and Calloway, D. H. Nitrogen retention in men fed isolated soybean protein supplemented with L-methionine, D-methionine, N-acetyl-L-methionine, or inorganic sulfate. J. Nutr. 106, 1286. 1976.

92. Friedman, M. and Finot, P.A. Nutritional improvement of bread. 1991. This volume.

93. Friedman, M., and Wall, J.S. Application of a Hammett-Taft relation to kinetics of alkylation of amino acid and peptide model compounds with acrylonitrile. J. Amer. Chem. Soc. 86, 3735-3741. 1964.

94. Friedman, M., Cavins, J.F., and Wall, J.S. Relative nucleophilic reactivities of amino groups and mercaptide ions in additional reactions with α, β-unsaturated compounds. J. Amer. Chem. Soc. 87, 3572-3582. 1965.

95. Krull, H., and Friedman, M. Anionic polymerization of methyl acrylate to protein functional groups. J. Poly. Sci. A-1 5, 2535-2546. 1967.

96. Friedman, M., and Wall, J.S. Additive linear free energy relationships in reaction kinetics of amino groups with α, β-unsaturated compounds. J. Org. Chem. 31, 2888-2894. 1966.

97. Friedman, M., and Sigel, C.W. A kinetic study of the ninhydrin reaction. Biochemistry 5, 478-484. 1966.

98. Friedman, M. A novel differential titration to determine pK values of phenolic groups in tyrosine and related aminophenols. Biochem. Biophys. Res. Commun. 23, 626-32. 1966.

99. Krull, L.H., and Friedman, M. Reduction of protein disulfide bonds by sodium hydride in dimethyl sulfoxide. Biochem. Biophys. Res. Commun. 29, 373-377. 1967.

100. Friedman, M. Solvent effects in reaction of amino groups in amino acids, peptides, and proteins with α, β-unsaturated compounds. J. Amer. Chem. Soc. 89, 4709-4713. 1967.

101. Cavins, J.F., and Friedman, M. New amino acids derived from reactions of ε-amino groups in proteins with α, β-unsaturated compounds. Biochemistry 6, 3766-3770. 1967.

102. Cavins, J.F., and Friedman, M. Specific modification of protein sulfhydryl groups with α, β-unsaturated compounds. J. Biol. Chem. 243, 3357-3360. 1968.

103. Friedman, M., and Romersberger, J.A. Relative influence of electron-withdrawing functional groups on basicities of amino acid derivatives. J. Org. Chem. 33, 154-157. 1968.

104. Friedman, M., and Krull, L.H. A novel spectrophotometric procedure for the determination of half-cystine residues in proteins. Biochem. Biophys. Res. Commun. 37, 630-633. 1969.

105. Friedman, M. and Krull, L.H. N- and C-Alkylation of proteins in dimethyl sulfoxide. Biochem. Biophys. Acta 20, 301-363. 1970.

106. Cavins, J.F., and Friedman, M. Preparation and evaluation of S-ß -(4-pyridylethyl)-L-cysteine as an internal standard for amino acid analyses. Analyt. Biochem. 35, 389-493. 1970.

107. Eskins, K., and Friedman, M. Solvent effects during the photochemistry of gluten proteins. Photochem. Photobiol. 12, 245-247. 1970.

108. Friedman, M., Krull, L.H., and Cavins, J.F. The chromatographic determination of cysteine and half-cysteine residues in proteins as S-ß-(4-pyridylethyl)-L-cysteine. J. Biol. Chem. 245, 3868-3871. 1970.

109. Friedman, M., and Tillin, S. Flame-resistant wool. Tex. Res. J. 40, 1045-1047. 1970.

110. Friedman, M., and Noma, A.T. Cystine content of wool. Tex. Res. J. 40, 1073-1078. 1970.

111. Krull, L.H., Gibbs, D.E., and Friedman, M. 2-Vinylquinoline, a reagent to determine protein sulfhydryl groups spectrophotometrically. Anal. Biochem. 40, 8085. 1971.

112. Wu, Y.V., Cluskey, J.E., Krull, L.H., and Friedman, M. Some optical properties of S-ß-(4-pyridylethyl)-L-cysteine and its wheat gluten and serum albumin derivatives. Canad. J. Biochem. 49, 1042-1049. 1971.

113. Masri, M.S., Windle, J.J., and Friedman, M. p-Nitrostyrene: New alkylating agent for sulfhydryl groups in soluble proteins and keratins. Biochem. Biophys. Res. Commun. 47, 1408-1413. 1972.

114. Finley, J.W., and Friedman, M. Chemical methods for available lysine. Cereal Chemistry 50, 101-105. 1973.

115. Friedman, M., Noma, A.T., and Masri, M.S. New internal standards for basic amino acid analyses. Analyt. Biochem. 51, 280-287. 1973.

116. Koenig, N.H., Muir, M.W., and Friedman, M. Properties of wool modified by activated vinyl compounds. Text. Res. J. 43, 682-688. 1973.

117. Friedman, M. Reactions of cereal proteins with vinyl compounds. In "Industrial Uses of Cereal Grains", Y. Pomeranz, Editor, American Association of Cereal Chemists, Minneapolis, MN, p. 237-251. 1973.

118. Friedman, M., and Tillin S. Partly-reduced-alkylated wool. Tex. Res. J. 44, 578-580. 1974.

119. Friedman, M., Williams, L.D., and Masri, M.S. Reductive alkylation of proteins with aromatic aldehydes and sodium cyanoborohydride. Int. J. Peptide Protein Res. 6, 183-185. 1974.

120. Friedman, M., and Finley, J.W. Reactions of proteins with ethyl vinyl sulfone. Int. J. Peptide and Protein Res. 7, 481-486. 1975.

121. Friedman, M., and Finley, J.W. Vinyl compounds as reagents for determining available lysine in proteins. In "Protein Nutritional Quality of Foods and Feeds", Friedman, M., Ed., Marcel Dekker, New York, Part 1, 503-520. 1975.

122. Snow, J.T., Finley, J.W., and Friedman, M. A kinetic study of the hydrolysis of N-acetyldehydroalanine methyl ester. Int. J. Peptide and Protein Res. 7, 461-466. 1975.

123. Snow, J.T., Finley, J.W., and Friedman, M. Relative reactivities of sulfhydryl groups with N-acetyldehydroalanine and N-acetyldehydro-alanine methyl ester. Int. J. Peptide Protein Res. 8, 57-64. 1976.

124. Chauffe, L., and Friedman, M. Factors affecting the cyano-borhydride reduction of aromatic Schiff's bases of proteins. In "Protein Crosslinking: Biochemical and Molecular Aspects", M. Friedman, Ed., Plenum Press, New York. Adv. Exp. Med. Biol. 86A, 415-424. 1977.

125. Finley, J.W., Snow, J.T., Johnston, P.H., and Friedman, M. Inhibitory effect of mercaptoamino acids on lysinoalanine formation during alkali treatment of proteins. In "Protein Crosslinking: Nutritional and Medical Consequences", Friedman, M., Ed., Plenum Press, New York. Adv. Exp. Med. Biol. 86B, 85-92. 1977.

126. Friedman, M., and Williams, L.D. A mathematical analysis of kinetics of consecutive, competitive reactions of protein amino groups. Adv. Ex. Med. Biol. 86B, 299-319. 1977.

127. Friedman, M., and Broderick, G.A. Protected proteins in ruminant nutrition. In vitro evaluation of casein derivatives. Adv. Exp. Med. Biol. 86B, 545-558. 1977.

128. Friedman, M. Chemical basis for pharmacological and therapeutic action of penicillamine. Proceedings of the Royal Society of Medicine 70(3), 50-60. 1977.

129. Friedman, M. Effect of lysine modification on chemical, physical, nutritive, and functional properties of proteins. In "Food Proteins", J.R. Whitaker and S. Tannenbaum, Eds., Avi, Westport, CT., pp. 446-483. 1977.

130. Friedman, M., and Orraca-Tetteh, R. Hair as an index of protein malnutrition. Adv. Exp. Med. Biol. 105, 131-154. 1978.

131. Friedman, M., Zahnley, J.C., and Wagner, J.R. Estimation of the disulfide content of trypsin inhibitors as S-ß-(2- pyridyl-ethyl)-L- cysteine. Analytical Biochem. 106, 27-34. 1980.

132. Friedman, M. Inhibition of lanthionine formation during alkaline treatment of keratinous fibers. U.S. Patent 4,212,800. July 15, 1980.

133. Carter, E.G.A., Carpenter, K.J. and Friedman, M. The nutritional value of some niacin analogs for rats. Nutr. Repts. International 25, 389-397. 1982.

134. Friedman, M. Chemically reactive and unreactive lysine as an index of browning. Diabetes 31, 5-14. 1982.

135. Zahnley, J.C. and Friedman, M. Absorption and fluorescence spectra of S-quinolylethylated Kunitz soybean trypsin inhibitor. J. Protein Chemistry 1, 225-240. 1982.

136. Friedman, M. Sulfhydryl groups and food safety. Adv. Exp. Med. Biol. 177, 31-63. 1984.

137. Menefee, E., and Friedman, M. Estimation of structural components of abnormal hair from amino acid analyses. J. Protein Chem. 4, 333-341. 1985.

138. Masri, M.S., and Friedman, M. Protein reactions with methyl and ethyl vinyl sulfones. J. Protein Chem. 7, 49-54. 1988.

32

EFFECT OF FOOD PROCESSING AND PREPARATION ON MINERAL UTILIZATION

Phyllis E. Johnson

USDA, ARS, Grand Forks Human Nutrition Research Center
PO Box 7166, University Station, Grand Forks, ND 58202

ABSTRACT

While effects of various nutrients and certain non-nutrient components of food on mineral utilization have been intensively studied, less is known about the effects of food processing and preparation procedures. Fermentation during the production of beer, wine, yogurt, and African tribal foods affects bioavailability of Zn and Fe. Baking affects the chemical form of Fe in fortified bread products and these changes can affect its bioavailability. Availability of Fe in milk-based infant formula depends on whether Fe is added before or after heat processing. Food packaging (e.g., tin cans) can alter food composition and thus potentially affects mineral bioavailability. Maillard browning has been reported to cause slight decreases in Zn availability both *in vitro* and in humans. However, we found that feeding of highly browned casein-glucose products to rats as 5% of diet produced no effect on Zn absorption (59.5 $\pm$ 8.2% vs 54.1 $\pm$7.3%) or Fe absorption (45.6 $\pm$ 7.7% vs 46.9 $\pm$ 12.6%) for browned vs control, respectively; nor did we find any of the adverse health effects reported by others. We found no effect on stable Zn or Cu absorption in seven men when browned foods were fed, compared to the same diets without browning. Zinc absorption was 34 $\pm$ 13% (browned) vs 24 $\pm$ 15% (unbrowned), and Cu absorption was 55 $\pm$ 5% vs 55 $\pm$ 8% ($p>0.05$).

INTRODUCTION

Many factors relating to food composition have been intensively studied with regards to their effects on mineral absorption and utilization. For the most part, the factors that have received the most attention are variables in food composition, whether they are nutrients or non-nutrients such as phytates, fiber, tannins, etc. Food processing and preparation can also have marked effects on the composition or nature of a food, yet the effects of these processes on mineral bioavailability have been studied to a much lesser extent. Indeed, one review of the effects of food production and processing methods on nutritive value of foods did not even mention effects on mineral content or bioavailability (Hollingsworth & Martin, 1972). Processes such as fermentation, heat processing, and extrusion, contamination from packaging, or formation of Maillard products during cooking all have the potential of affecting mineral utilization. This paper will briefly review the effects of some

Nutritional and Toxicological Consequences of Food Processing
Edited by M. Friedman, Plenum Press, New York, 1991

of these processes on mineral bioavailability from food and will also present some original data from my laboratory concerning the effects of Maillard products on Zn, Cu, and Fe absorption in animals and humans.

FERMENTATION

Fermentation occurs during the production of alcoholic beverages, of yogurt and similar foods, and during the leavening of bread by yeast. Although the specific chemical transformations occurring during production of these foods differ, all seem to result in effects on the availability of minerals.

When bread is leavened with yeast, the phytate content may be decreased as much as 60 to 80% (Nävert et al., 1985). Reduction of the phytate content by leavening resulted in increased Zn absorption (Nävert et al., 1985). Fermentation of sorghum during preparation of *aceda*, an African tribal food, resulted in higher ^{65}Zn absorption by rats than from unfermented sorghum foods (Stuart et al., 1987). Fermentation of the sorghum to produce *aceda* also resulted in a reduction of the phytate content by approximately one-third, compared to non-fermented sorghum foods. There was no effect of fermentation of sorghum on absorption of Fe by rats in the same experiment. Fermentation of pearl millet, another staple food in Asia and Africa, improved extractability of Ca, Zn, Fe, Mn, and Cu (Mahajan & Chauhan, 1988), and presumably improved availability of these minerals.

Fermentation of milk to produce yogurt affects the bioavailability of different minerals to varying degrees. Calcium is absorbed equally well from milk and yogurt (Smith et al., 1985; Schaafsma et al., 1988). However, the apparent absorption ([intake - fecal excretion]/intake) of phosphorus by rats was higher from yogurt than from milk or lactase-treated milk (Schaafsma et al., 1988); apparent absorption of Mg and Fe was less from yogurt than from milk, and apparent of absorption of Zn was the same from yogurt and milk. Although the differences between yogurt and milk in their effects on Fe and Mg availability were statistically significant, they were very small and may not be biologically important. In rats fed high phytate diets, growth and food efficiency were greater when either yogurt or inactivated yogurt were added to the diet; bone, plasma, and tissue levels of Zn were not increased by dietary yogurt, however (Toleman, 1987).

Fermentation of soybean meal by lactic acid-producing bacteria or by *Rhizopus oligosporus* in tempeh fermentation significantly increased bioavailability of Zn to rats (Moeljopawiro et al., 1988). Both destruction of phytate and production of Zn-binding ligands during fermentation were suggested as the reason for the increased Zn availability.

Beer made in South Africa from maize and sorghum undergoes two sequential fermentations. The first is a lactic acid fermentation; this is followed by the alcoholic fermentation (Derman et al., 1980). Iron absorption from beer prepared by this process was more than twelve-fold greater than from a gruel made from the ingredients used to prepare the beer (Derman et al., 1980). Three factors were responsible for the enhanced Fe absorption; these were removal of solids during fermentation and the presence of ethanol and lactic acid in the beer.

Wine contains various kinds of organic Fe(II) complexes, and wine has often been said to have a hematopoeic effect. Tabata & Tanaka (1986) studied the Fe complexes produced in wine when grape juice was fermented by *Saaaccharomyces cerevisiae*. They found that Fe absorption and incorporation into hemoglobin were higher in rats given wine than in rats given $FeCl_3$ or $FeSO_4$.

LIQUID MILK PRODUCTS

Heat processing of milk-based products or milk substitutes during sterilization and canning can significantly affect the chemical form of Fe in the product and its bioavailability. This is of particular interest in the production of infant formulas, which are normally fortified with Fe. Theuer et al. (1973) found that heat sterilization of milk-based infant formulas containing ferric pyrophosphate and sodium iron pyrophosphate increased Fe availability substantially. Heat processing of soy isolate-based infant formulas increased the bioavailability of Fe from ferric pyrophosphate and sodium pyrophosphate two- to four-fold, but did not increase bioavailability from ferrous sulphate (Theuer et al., 1971). When cow's milk was fortified with a citrate phosphate iron complex before pasteurization, bioavailability of the Fe to rats was unaffected (Ranhotra et al., 1981). When slurried chick diets were processed with heat and pressure, bioavailability of Fe from sodium ferric pyrophosphate and ferric pyrophosphate was increased, but processing did not change the bioavailability from ferrous sulfate or ferric orthophosphate (Wood et al., 1978). Clemens and Mercurio (1981) found that retort processing of liquid milk-based products significantly increased the bioavailability of carbonyl Fe and electrolytic Fe, but not ferric orthophosphate. The improvement in bioavailability was related to oxidation and solubilization of the elemental Fe. Differences in heat treatment of infant formula (spray-dried, UHT [ultra-high temperature], or sterilized) did not affect Ca absorption in infant monkeys (Rudloff & Lönnerdal, 1990) or rats (Weeks & King, 1985).

Changes in the form of Fe can also occur during storage of liquid milk-based products. Iron added to liquid formulas marketed for weight control as ferric orthophosphate is largely insoluble and suspended rather than dissolved at the time of addition (Hodson, 1970). Hodson found that after two to five months of storage, all or most of the ferric Fe was converted to more soluble ferrous Fe, which is presumably more available for absorption (Hodson, 1970). In contrast, Clemens (1981) found that none of the Fe from ferric orthophosphate in a liquid milk-based product was soluble after 6 or 12 months, and the relative biological value of this Fe was low. There were no changes in the solubility or valency of Fe added to a liquid milk-based product between between 6 and 12 months of storage (Clemens, 1981). Evidently, most of the change in bioavailability of elemental Fe added to liquid milk-based products takes place during the thermal processing and during the first six months of storage. It is clear that measuring the bioavailability of Fe from the salt used for food fortification is not enough to determine the bioavailability of a mineral in the food to which it is added. Changes in bioavailability depend on both the chemical form of the added fortification Fe and on the type of processing. Storage after processing may also affect Fe bioavailability.

HEAT PROCESSING OR COOKING OF OTHER FOODS

Many non-liquid foods are processed by methods that also involve heat. These include processes such as extrusion, puffing, or flaking. Carlson and Miller (1983) studied the effects of processing on the dialyzable Fe in various breakfast cereals. The dialyzable Fe in several Fe-fortified wheat and corn cereals ranked as puffed > extruded > flaked. Differences in the conditions of heat treatments such as time, temperature, moisture, and the final degree of browning may influence Fe availability. Maillard browning products have not been found to have much effect on Fe availability in *in vivo* studies (see below). However, processing conditions may also affect the relative amounts of ferrous and ferric Fe in the products and thus alter Fe availability. Extrusion processing did not affect Zn bioavailability from egg white- or soy-based diets fed to rats (Hess et al., 1984). Although total Fe concentration decreased slightly, Fe diffusibility *in vitro* increased during the manufacture of a maize-based snack food, from 2.3% (whole maize) to 5.0% (maize grits) to 14.7% (pre-extrusion mix) to 18.4% (extruded mix) to 21.8% (final product) (Hazell & Johnson, 1989). *In vivo* studies showed no significant influence of extrusion processing on Fe availability in rats (Fairweather-Tait et al., 1987) or humans (Kivisto et al., 1986). Heat treatment of egg white did not affect *in vitro* Fe bioavailability until the heating was sufficient to cause pronounced Maillard browning (Leahey & Thompson, 1989).

Various food preparation procedures that involve heat can affect mineral bioavailability in ways other than the formation of Maillard products, which will be discussed later. When enriched flour is used to prepare baked goods, a marked effect of the baking process is the formation of insoluble forms of Fe (Lee & Clydesdale, 1980). Bioavailability of Fe is generally lower from poorly soluble forms of Fe than from soluble forms (Pla et al., 1976) so that these changes in the solubility of Fe after cooking might reduce the bioavailability of the Fe in the food. However, *in vitro* estimation of Fe availability in breads fortified with various forms of Fe showed no effect of baking on Fe bioavailability (Schricker & Miller, 1982). Baking did not affect bioavailability of the endogenous Fe in wheat bran muffins (Buchowski et al., 1988).

Frying of a soy-hamburger patty resulted in marked reductions in the percent soluble Fe at simulated intestinal pH values (Rizk & Clydesdale, 1985). In contrast, compared to raw beef, boiling or baking of beef did not result in any change in Fe bioavailability to rats (Jansuittivechakul et al., 1985). Copper in cooked beef is more available for absorption than Cu in raw beef (Moore et al., 1964).

Food cooked in iron utensils contains more Fe than food cooked in non-iron utensils. Cooking in iron utensils can more than double the amount of Fe in foods (Burroughs & Chan, 1972; Brittin & Nossaman, 1986; Mistry et al., 1988). The amount of increase in Fe depends on the type of food (Mistry et al., 1988) and the amount of prior use of the iron utensil (Brittin & Nossaman, 1986). Both *in vivo* studies in rats (Martinez & Vannucchi, 1986) and *in vitro* studies (Mistry et al., 1988) showed that Fe added to food by cooking in an iron pot was as available as native food Fe.

EFFECT OF CURING AND COOKING OF MEAT ON FE BIOAVAILABILITY

Although bioavailability to rats of Fe in meat was not changed by boiling or baking (Jansuittivechakul et al., 1985), rats do not discriminate between heme and non-heme Fe as humans do. There is evidence that both cooking and curing of meat affect the proportion of heme and non-heme Fe in the meat. Because heme Fe is more available for absorption by humans than is non-heme Fe (Monsen et al., 1978), these changes may affect the bioavailability of Fe from meat. Slow heating of muscle extracts resulted in the release of more non-heme Fe than did fast heating (Chen et al., 1984), but nitrite stabilized the heme Fe and prevented the increase in non-heme Fe levels. Both baking and microwave cooking increased non-heme Fe in ground beef, and a linear relationship was observed between non-heme Fe in meat and the time of exposure to heat treatment (Schricker & Miller, 1983). Changes in the amount of heme Fe were felt to be large enough to cause significant changes in absorbable Fe in the meat. Again, nitrite appeared to protect against heat-induced changes in non-heme Fe. In the same study (Schricker & Miller, 1983), ascorbic acid was found to significantly increase non-heme Fe levels in ground beef even when heating was mild. Another study, however, showed only a very small decrease in hemoglobin Fe (not total Fe) absorption with cooking; hemoglobin Fe absorption by three human subjects was 15.7% from uncooked hemoglobin and 14.4% from cooked hemoglobin (Turnbull et al., 1962).

Heme Fe levels in sausages were unchanged after curing with sodium nitrite or sodium erythorbate (Lee & Shimaoka, 1984). More of the non-heme Fe was in an ionic form in erythorbate-cured sausage than in sausage cured with nitrite. More of the Fe in the erythorbate-cured sausage was in a soluble ferrous form. There were no changes in vacuum-packed meats after 14 days at 5^0C, but ferrous Fe was depleted when sausages were exposed to air at 5^0. Atmospheric oxidation of beef decreased Fe bioavailability to rats (Mahoney et al., 1979).

GERMINATION

The consumption of bean sprouts, alfalfa sprouts, and other sprouted seeds has become more prevalent in recent years. Germination of pea seeds increased the bioavailability of Zn to rats (Beal et al., 1984); the increased Zn bioavailability was attributed to a twelve-fold increase in phytase enzyme content which resulted from germination and a concomitant reduction in phytate content of the germinated peas. Germination of peas reduced phytate to a much greater extent (75%) than did cooking (25%); 10-15% of Zn in peas was lost to cooking water (Beal & Mehta, 1985).

PROCESSING OF SOY PRODUCTS

Conditions involved in the processing of soy protein have been thought to be partially accountable for the poor bioavailability of minerals from soy protein products. Rackis (1979) suggested that Zn bioavailability in isoelectric (acid-precipitated) soy isolates is high, while isolates processed with alkali exhibit low Zn bioavailability. Erdman et al. (1980) found poorer growth in rats fed neutral soy than rats fed acid soy, but there was no significant difference in tibia Zn or Mg between rats fed acid or neutral soy. In a subsequent study based on measurement of ^{65}Zn retention, bioavailability of Zn was significantly higher from acid-precipitated soy products than from neutralized products (Ketelsen et al., 1984).

The bioavailability of Fe was also found to be affected by the form of soy product in the diet (Picciano et al., 1984). Rats were able to discriminate among soy products, and Fe bioavailability was ranked as soy concentrate > soy flour > soy beverage, although all three products were considered good plant sources of Fe. Thompson & Erdman (1988) found that when various radiolabeled test meals were fed in either acid or neutralized form to rats, neutralization of soy protein isolate, but not of casein, resulted in less retention of ^{59}Fe from the meal. However, the difference between the acid and neutralized soy products was not present after the products were subjected to heat treatment. In addition to pH and heat treatment, other differences among soy products may also have an effect on Fe bioavailability. Thompson & Erdman (1988) also compared ^{59}Fe retention from casein test meals in rats fed diets based on soy flour, acid-precipitated soy concentrate, ethanol-washed soy concentrate, water-washed soy concentrate, tofu analogue concentrate, or soy protein isolate. For two of the soy products, the acid precipitated concentrate and the ethanol-washed concentrate, there was no significant depression in ^{59}Fe retention relative to casein. They suggested that the lack of effect by the ethanol-washed product might have been caused by removal of some non-protein factors during ethanol extraction.

Differences in bioavailability from various soy products are probably unimportant in a practical sense if the soy products make up a small portion of the total diet. On the other hand, if soy products comprise a major portion of the diet, as infant formula or as meat substitutes or extenders, for example, processing technology may have an impact on the nutritional quality of the diet.

FOOD PACKAGING

Composition of food may be affected by its packaging. Migration of nutrients or other substances into or out of food can result in changes in food composition that have the potential of affecting mineral utilization from the food. Elkins (1979) found that the canning process may cause loss of minerals from green beans to blanching water and canning brine. After 12 months of storage, however, beans showed a 131% increase in Fe, a 94% increase in Zn, and a 25% decrease in Cu. Peaches also decreased in Cu content during storage. Henriksen et al. (1985) found significant changes in trace element content of canned tomatoes and green beans, but not peaches, during storage. Changes ranged from a 14% increase in Zn content to a 630% increase in Fe in tomatoes and from a 307% increase in Mg to a 555% increase in Cu in green beans. Schmitt and Weaver (1982) also found losses of Zn and Cr from kale and bush beans after canning or blanching and freezing. Canning, but not blanching, reduced Mg in peas by 18%; Fe and Ca were not affected by either process (Lee et al., 1982).

Storage of food in tin cans, especially storage after opening the can, was found to increase concentrations of Sn and Fe in foods (Greger & Baier, 1981). Storage or canning of foods in glass does not affect trace element content (Greger & Baier, 1981; Theriault & Fellers, 1942). Tin has been reported to decrease Zn retention significantly (Johnson et al., 1982). Food Fe which originates from tin cans is soluble and seems to be nearly 100% available for absorption (Theriault & Fellers, 1942). Because trace element interactions can affect bioavailability, changes in mineral content of foods may affect bioavailability of minerals in addition to those whose concentrations are changed.

MAILLARD BROWNING

It is well known that Maillard browning can affect the protein quality of foods (Dworschák, 1980). The effect of Maillard products on mineral nutrition is unclear. Copper can form complexes with Maillard products (Petit, 1956). On the other hand, Rendleman (1987) found that melanoidins in model systems, toasted bread, or coffee did not bind Ca; toasted bread and coffee did bind Ca, but binding was to various organic acids in these foods. Maillard products in heat-sterilized solutions for parenteral nutrition were found to increase urinary excretion of Zn, Cu, and Fe when administered intravenously, but not when administered nasogastrically (Freeman et al., 1975; Stegink et al., 1981). This stimulated research into the effects of Maillard products on mineral availability from foods.

In my laboratory, we found that corn flakes (browned) bound more Cu and less Fe than did corn grits (unbrowned) (Johnson et al., 1983). Camire and Clydesdale (1981) also found an effect of toasting on binding of Zn,

Mg, and ferrous Fe by wheat bran. We found that adult men fed a single meal containing either corn grits or corn flakes at breakfast absorbed significantly less Zn from corn flakes than from corn grits (corn grits, 48%; cornflakes, 36%) (Lykken et al., 1986). When subjects were fed diets with cornflakes or corn grits for 42 days, there were no significant differences in absorption of Fe, Zn, or Cu measured with stable isotopes (Johnson, 1983), suggesting that perhaps some adaptation to the diets occurred. Subjects fed the corn flakes diet excreted higher molecular weight Zn-binding substances in urine, while low molecular weight Zn-binding substances predominated in urine of subjects fed corn grits (Johnson et al., 1983).

In a subsequent study, we fed "browned" and "unbrowned" diets to adult men for four months in a crossover design. The "browned" diets contained cookies, crackers and quick breads cooked in a conventional oven, toasted bread, meats browned with flour, tortillas, and browned spaghetti. The unbrowned diets contained the same foods, but foods were untoasted and bread was served without crust. Foods were prepared with moist heat processes or in a microwave oven to minimize browning. Zinc and Cu absorption were measured by giving oral doses of the stable isotopes ^{67}Zn and ^{65}Cu. Neither Cu nor Zn absorption was significantly affected by the dietary treatment in this study, but mean Zn absorption was about one-third higher when browned diets were fed (Table 1) (PE Johnson, unpublished data).

Table 1. Absorption of Zinc and Copper from Browned and Unbrowned Diets by Adult Men

Vol. #	% Zn Absorption		% Cu Absorption	
	Browned	Unbrowned	Browned	Unbrowned
2104	49.2	43.0	58.5	63.5
2109	35.8	11.8	49.5	55.0
2110	9.3	39.2	47.5	65.5
2118	40.4	20.2	63.1	43.9
2119	24.0	5.5	52.5	44.4
2122	38.5	13.1	54.6	56.6
2123	41.3	34.8	57.7	59.2
mean	34.1	23.9	54.8	55.4
± SD	13.3	14.9	5.4	8.5

There were no significant differences between browned and unbrowned diets for either Zn or Cu absorption ($p>0.05$).

Recent *in vitro* studies showed that Maillard browning products formed from amino acid/glucose solutions bind Zn and decrease apparent Zn availability compared to an unheated control (Whitelaw & Weaver, 1988). Toasting of corn meal also reduced apparent Zn availability compared to unbrowned corn meal (Whitelaw & Weaver, 1988), and the authors commented that the degree of Zn binding observed seemed to be related to the extent of browning.

In contrast, we found that Zn binding seemed to depend not only on the degree of protein browning but also on the nature of the protein (PE Johnson and TC Lee, unpublished data). Browned proteins (casein, albumin, gluten) were prepared by mixing three parts protein with two parts glucose and adjusting the moisture to 15%. The samples were then stored for 0, 5,

10, 20, 30, or 40 days at 37^0 in a sealed glass chamber. Relative humidity was maintained at 68% in the chamber. At the end of the treatment, 5- and 10-day browned proteins were light tan, 20- and 30-day proteins were darker tan, and 40-day proteins were dark brown. Subsequent to removal from the chamber, samples were stored at -20^0. Adsorption isotherms for Zn with each browned protein were determined by a published method (Sarazinni, 1983) and the apparent stability constants for Zn were calculated (Table 2). Apparent stability constants for Zn and Cu with browned albumin using modified gel filtration chromatography (Evans et al., 1979) gave the results in Table 3.

When we fed browned casein/glucose products (browned 40 days by the above procedure) to rats as 5% of a protein-adequate diet containing 12 ppm Zn and 35 ppm Fe, we found no effect on Zn or Fe absorption compared to feeding unbrowned casein and glucose (PE Johnson, TC Lee, TL Starks, unpublished data) (Table 4). Hemoglobin, hematocrit, plasma, liver, and femur mineral concentrations were also unaffected by feeding the browned casein (Table 4). This is consistent with the data shown in Table 2, which indicate that Zn binding by 40-day browned casein is *less* than Zn binding by unbrowned casein.

Furniss et al. (1989) fed casein-lactose Maillard reaction products to rats and found little effect except a slight increase in urinary Zn excretion compared to rats fed unheated casein-lactose mixtures. Increasing the degree of browning increased the amount of urinary Zn excretion. Free fructose-lysine had no effect on urinary Zn excretion. In contrast, O'Brien et al. (1989) fed Maillard reaction products prepared

Table 2. Apparent Stability Constants for Zn and Browned Casein and Gluten

Days Browned	$K \times 10^{-4}$ Casein	Gluten
0	1.40 ± 0.29	1.10 ± 0.14
5	---	0.93 ± 0.18
10	1.14 ± 0.11	2.86 ± 1.06
20	0.78 ± 0.31	3.03 ± 1.73
30	0.46 ± 0.05	1.53
40	0.48 ± 0.11	---

Table 3. Binding of Zn and Cu by Browned Albumin

Days	Zinc		Copper	
Browned	Atoms Zn/mole alb	log $ß_n$	Atoms Cu/mole alb	log $ß_n$
0	4.3	3.74	0.62	4.18
10	5.7	3.62	2.62	3.12
20	4.5	4.69	1.82	3.19
30	5.8	4.17	1.88	3.21
40	4.2	3.77	1.54	3.15

from glucose and monosodium glutamate to rats and found increases in urinary excretion of Ca, Mg, Zn, and Cu. Retention of Zn, but not Ca, Mg, Fe, or Cu, was significantly less (62% vs 41%) in rats fed Maillard reaction products than in control rats. In a similar study, rats fed fructose-tryptophan reflux products absorbed slightly, but significantly less, ^{59}Fe than rats fed diets with free fructose and tryptophan (Mahalko et al., 1984).

It seems that *in vivo* effects of Maillard products prepared from whole proteins on mineral bioavailability are minimal or non-existent. Thus, the nutritional impact of Maillard products on mineral availability from real diets is minimal. In situations where individual amino acids may react to form Maillard products, as in parenteral nutrition solutions heated after amino acids and sugar are mixed, Maillard products may increase urinary losses of minerals.

Table 4. Effect of 5% Casein-Glucose Maillard Product in the Diet on Zinc and Iron Absorption in Rats

Diet	Maillard	Control
% ^{65}Zn Absorption	59.5 ± 8.2	54.1 ± 7.3
% ^{59}Fe Absorption	45.6 ± 7.7	46.9 ± 12.6
Femur Fe (μg/g)	99.3 ± 20.4	106.3 ± 16.5
Femur Zn (μg/g)	253 ± 21	263 ± 14
Liver Zn (μg/g)	68.3 ± 5.7	72.2 ± 7.8
Liver Fe (μg/g)	289 ± 68	279 ± 59
Hemoglobin (g/dL)	15.2 ± 1.1	15.4 ± 1.2

FOOD ADDITIVES

A variety of chemicals are added to foods during processing for several purposes: to maintain or improve nutritional value, to maintain freshness, to help in processing or preparation, and to make food more appealing. The addition of nutrients to enrich or fortify food obviously can affect bioavailability of minerals; discussion of these effects is outside the scope of this review.

Preservatives and antioxidants include substances such as calcium lactate, calcium sorbate, citric acid, EDTA, lactic acid, potassium sorbate, sodium erythorbate, sodium nitrate, sodium nitrite, TBHQ (tertiary butyl hydroquinone), ascorbic acid, and tocopherols. Many of these substances are known to affect mineral metabolism. The effects of sodium erythorbate, sodium nitrate, and sodium nitrite have been discussed above in the section on curing meats. Calcium interacts with Fe and inhibits its absorption (Latunde-Dada & Neale, 1986; Dawson-Hughes et al., 1986). Citric acid and ascorbic acid improve Fe absorption (Gillooly et al., 1983; Monsen et al., 1978). Iron is easily complexed by EDTA, and Fe(III)EDTA is well absorbed (MacPhail et al., 1981); EDTA also improves Zn availability (Oberleas et al., 1966). TBHQ was reported to protect against some of the manifestations of Cu deficiency in rats (Johnson & Saari, 1989).

Table 5. Effect of Some Food Additives on Zn Absorption in Women

	% Zn Absorption	n
Zn only (control)	25.9 ± 12.0	7
PEG 4000	26.4 ± 18.5	7
FD & C Blue No. 1	30.5 ± 10.6	5
FD & C Red No. 40	19.9 ± 9.3	5

Other food additives are used as emulsifiers, stabilizers, leavening agents, pH control agents, humectants, dough conditioners, and anti-caking agents. These include substances such as calcium alginate, calcium bromate, calcium phosphate, calcium silicate, cellulose, lactic acid, lecithin, mannitol, pectin, polyethylene glycols (PEG), sorbitol, tartaric acid, and various types of gums. The effects of Ca on availability of other minerals were mentioned above. Cellulose and pectin improved Fe absorption in humans (Gillooly et al., 1984), but cellulose had no effect on Zn absorption (Turnlund et al., 1984). Sorbitol improved Fe absorption in humans (Loría et al., 1962) as did tartaric acid (Gillooly et al., 1983). Adjustment of the pH of bread dough with lactic acid resulted in improved Zn availability from whole-meal bread (Harmuth-Hoene & Meuser, 1988). We found no effect of PEG 4000 on Zn absorption in humans (PE Johnson, unpublished data) (Table 5). *In vitro* tests of Zn and Fe availability showed that locust bean gum and guar gum reduced Zn availability, but sodium alginate had no effect on Zn or Fe availability (Zemel & Zemel, 1985).

Food additives which affect appeal characteristics of foods include flavor enhancers, flavors, colors, and sweeteners. Food dyes are complex organic molecules which have the potential of forming complexes with trace minerals. However, we found that neither FD & C Blue No. 1 nor Red No. 40 affected Zn absorption in humans (PE Johnson, unpublished data) (Table 5). Addition of flavoring (beef, cheese, or pickled onion flavors) increased *in vitro* Fe diffusibility from maize grits and a commercial maize product (Hazell & Johnson, 1989).

OTHER TYPES OF PROCESSING

Physical processing such as the milling of grain can also affect mineral bioavailability. Pearling of sorghum to remove the outer layers of the grain reduced the polyphenol and phytate contents by 96% and 2%, respectively (Gillooly et al., 1984). This resulted in an increase of Fe absorption from 1.7% to 3.5%. It is well known that removal of bran during milling of wheat results in improved bioavailability of Fe and Zn; this may be caused by the removal of fiber or by the reduction in phytate content of the flour (Davies, 1978; Elwood et al., 1968; Andersson et al., 1983).

ACKNOWLEDGEMENTS

The author wishes to thank Cheryl Stjern for mass spectrometric analyses, James Normandin for sample preparation, Lisa Hesse and Thad Bowman for stability constant determinations, Rodger Sims for mineral analyses, and LoAnne Mullen for metabolic diet preparation.

REFERENCES

Andersson, H., Nävert, B., Bingham, S. A., Englyst, H. N., and Cummings, J. H., 1983, The effects of breads containing similar amounts of phytate but different amounts of wheat bran on calcium, zinc, and iron balance in man, Br. J. Nutr. 50:503-510.

Mention of a trademark or proprietary product does not constitute a guarantee or warranty of the product by the USDA and does not imply its approval to the exclusion of other products that may be suitable.

Beal, L., Finney, P. L., and Mehta, T., 1984, Effects of germination and dietary calcium on zinc bioavailability from peas, J. Food Sci. 49:637-641.

Beal, L., and Mehta, T., 1985, Zinc and phytate distribution in peas. Influence of heat treatment, germination, pH, substrate, and phosphorus on pea phytate and phytase, J. Food Sci. 50:96-100, 115.

Brittin, H. C., and Nossaman, C.E., 1986, Iron content of food cooked in iron utensils, J. Am. Diet. Assoc. 86:897-901.

Buchowski, M., Vanderstoep, J., and Kitts, D.D., 1988, Effect of heat treatment and organic acids on bioavailability of endogenous iron from wheat bran in rats, Can. Inst. Food Sci. Technol. J. 21:161-166.

Burroughs, A. L., and Chan, J. J., 1972, Iron content of some Mexican-American foods, J. Am. Diet. Assoc. 60:123.

Camire, A. L., and Clydesdale, F.M., 1981, Effect of pH and heat treatment on the binding of calcium, magnesium, zinc, and iron to wheat bran and fractions of dietary fiber, J. Food Sci. 46:548-551.

Carlson, B. L., and Miller, D. D., 1983, Effects of product formulation, processing, and meal composition on in vitro estimated iron availability from cereal-containing breakfast meals, J. Food Sci. 48:1211-1216.

Chen, C. C., Pearson, A. M., Gray, J. I., Fooladi, M. H., and Ku, P. K., 1984, Some factors influencing the nonheme iron content of meat and its implications in oxidation, J.Food Sci. 49:581-584.

Clemens, R. A., 1981, Effects of storage on the bioavailability and chemistry of iron powders in a heat-processed liquid milk-based product, J. Food Sci. 47:228-230.

Clemens, R. A., and Mercurio, K. C., 1981, Effects of processing on the bioavailability and chemistry of iron powders in a liquid milk-based product, J. Food Sci. 46:930-935.

Davies, N. T., 1978, The effects of dietary fibre on mineral availability, J. Plant Foods 3:113-123.

Dawson-Hughes, B., Seligson, F. H., and Hughes, V. A., 1986, Effects of calcium carbonate and hydroxyapatite on zinc and iron retention in postmenopausal women, Am. J. Clin. Nutr. 44:83-88.

Derman, D. P., Bothwell, T. H., Torrance, J. D., Bezwoda, MacPhail, A. P., Kew, M. C., Sayers, M. H., Disler, P. B., and Charlton, R. W., 1980, Iron absorption from maize (*Zea mays*) and sorghum (*Sorghum vulgare*) beer, Brit. J. Nutr. 43:271-279.

Dworschák, E., 1980, Nonenzyme Browning and its effect on protein nutrition, CRC Crit. Rev. Food & Nutr. 13:1-40.

Elkin, E. R., 1979, Nutrient content of raw and canned green beans, peaches, and sweet potatoes, Food Technol. 33(2):66-70.

Elwood, P. C., Newton, D., Eakins, J. D., and Brown, D. A., 1968, Absorption of iron from bread, Am. J. Clin. Nutr. 21:1162-1169.

Erdman, J. W. Jr., Weingartner, K. E., Mustakas, G. C., Schmutz, R. D., Parker, H. M., and Forbes, R. M., 1980, Zinc and magnesium bioavailability from acid-precipitated and neutralized soybean protein products, J. Food. Sci. 45:1193-1199.

Evans, G. W., Johnson, P. E., Brushmiller, J. G., and Ames, R. W., 1979, Detection of labile zinc-binding ligands in biological fluids by modified gel filtration chromatography, Anal. Chem. 51:839-843.

Fairweather-Tait, S. J., Symss, L. L., Smith, A. C., and Johnson, I. T., 1987, The effect of extrusion cooking on iron absorption from maize and potato, J. Sci. Food Agric. 39:341-348.

Freeman, J. B., Stegink, L. D., Meyer, P. D., Fry, L. K., and Denbesten, L., 1975, Excessive urinary zinc losses during parenteral alimentation, J. Surg. Res. 18:463-469.

Furniss, D. E., Vuichoud, J., Finot, P. A., and Hurrell, R. F., 1989, The effect of Maillard reaction products on zinc metabolism in the rat, Br. J. Nutr. 62:739-749.

Gillooly, M., Bothwell, T. H., Torrance, J. D., MacPhail, A. P., Derman, D. P., Bezwoda, W. R., Mills, W., Charlton, R. W., and Mayet, F., 1983, The effects of organic acids, phytates and polyphenols on the absorption of iron from vegetables, Br. J. Nutr. 49:331-342.

Gillooly, M., Bothwell, T. H., Charlton, R. W., Torrance, J. D., Bezwoda, W. R., MacPhail, A. P., Derman, D. P., Novelli, L., Morrall, P., and Mayet, F., 1984, Factors affecting the absorption of iron from cereals, Br. J. Nutr. 51:37-46.

Greger, J. L., and Baier, M., 1981, Tin and iron content of canned and bottled foods, J Food Sci. 46:1751-1754, 1765.

Harmuth-Hoene, A. E., and Meuser, F., 1988, Verbesserung der biologischen Verfügbarkeit von Zink in Schrot- un Knäckebrot, Z. Ernährungwiss. 27:244-251.

Hazell, T., and Johnson, I. T., 1989, Influence of food processing on iron availability *in vitro* from extruded maize-based snack foods, J. Sci. Food Agric. 46:365-374.

Henriksen, L. K., Mahalko, J. R., and Johnson, L. K., 1985, Canned foods: appropriate in trace element studies? J. Am. Diet. Assoc. 85:563-568.

Hess, R, L., Gordon, D. T., Hanna, M., and Satterlee, L., 1984, Influence of extrusion processing on zinc bioavailability, Abstracts, 44th Annual IFT Meeting, pp. 155.

Hodson, A. Z., 1970, Conversion of ferric to ferrous iron in weight control dietaries J. Agr. Food Chem. 18:946-947.

Hollingsworth, D. F., and Martin, P. E., 1972, Some aspects of the effects of different methods of production and of processing on the nutritive value of food, in: World Rev. Nutr. Diet. Vol. 15, G. H. Bourne, ed., S. Karger, New York, pp. 2-36.

Jansuittivechakul, O., Mahoney, A. W., Cornforth, D. P., Hendricks, D. G., Kangsadalampai, 1985, Effect of heat treatment on bioavailability of meat and hemoglobin iron fed to anemic rats, J. Food Sci. 50:407-409.

Johnson, M. A., Baier, M. J., and Greger, J. L., 1982, Effects of dietary tin on zinc, copper, iron, manganese, and magnesium metabolism of adult males, Am. J. Clin Nutr. 35:1332-1338.

Johnson, P. E., Lykken, G. I., Mahalko, J. R., Milne, D. B., Inman, L., Sandstead, H. H., Garcia, W. J., and Inglett, G. E., 1983, The effect of browned and unbrowned corn products on absorption of zinc, iron, and copper in humans, in: The Maillard Reaction in Foods and Nutrition, ACS Symposium Series 215, G. R. Waller, M. S. Feather, eds., American Chemical Society, Washington, DC, pp. 349-360.

Johnson, W. T., and Saari, J. T., 1989, Dietary supplementation with *t*-butylhydroquinone reduces cardiac hypertrophy and anemia associated with copper deficiency in rats, Nutr. Res. 119:1404-1410.

Ketelsen, S. M., Stuart, M. A., Weaver, C. M., Forbes, R. M., and Erdman, J. W. Jr., 1984, Bioavailability of zinc to rats from defatted soy flour, acid-precipitated soy concentrate and neutralized soy concentrate as determined by intrinsic and extrinsic labeling techniques, J. Nutr. 114:536-542.

Kivisto, B., Andersson, H., Cederblad, Å., Sandberg, A-S., and Sandström, B., 1986, Extrusion cooking of a high-fibre cereal product. 2. Effects on apparent absorption of zinc, iron, calcium, magnesium, and phosphorus in humans, Br. J. Nutr. 55:255-260.

Latunde-Dada, G. O., and Neale, R. J., 1986, Review: Availability of iron from foods, J. Food Technol. 21:255-268.

Leahey, J. M., and Thompson, D. B., 1989, Effect of heat processing of dried egg white on *in vitro* iron bioavailability, J. Food Sci. 54:154-158.

Lee, C. Y., Parson, G. F., and Downing, D. L., 1982, Effects of processing on amino acid and mineral contents of peas, J. Food Sci. 47:1034-1035.

Lee, K., and Shimaoka, J. E., 1984, Forms of iron in meats cured with nitrite and erythorbate, J. Food Sci. 49:284-287.

Lee, K., and Clydesdale, F. M., 1980, Effect of baking on the forms of iron in iron-enriched flour, J. Food Sci. 45:1500-1504.

Loría, A., Sánchez Medal, L., and Elizoondo, J., 1962, Effect of sorbitol on iron absorption in man, Am. J. Clin. Nutr. 10:124-127.

Lykken, G. I., Mahalko, J. R., Johnson, P. E., Milne, D. B., Sandstead, H. H., Garcia, W. J., Dintzis, and Inglett, G. E., 1986, Effect of browned and unbrowned corn products intrinsically labelled with ^{65}Zn on absorption of ^{65}Zn in humans, J. Nutr. 116:795-801.

MacPhail, A. P., Bothwell, T. H., Torrance, J. D., Derman, D. P., Bezwoda, W. R., Charlton, R. W., and Mayet, F., 1981, Factors affecting the absorption of iron from Fe(III)EDTA, Br. J. Nutr. 45:215-227.

Mahajan, S., and Chauhan, B. M., 1988, Effect of natural fermentation on the extractability of minerals from pearl millet flour, J. Food Sci. 53:1576-1577.

Mahalko, J. R., Johnson, P. E., and Lykken, G. I., 1984, Effect of fructose-tryptophan reflux product on the absorption and retention of iron in the rat, Fed. Proc. 43:1050 (Abstract).

Mahoney, A. W., Hendricks, D. G., Gillett, T. A., Buck, D. R., and Miller, C. G., 1979, Effect of sodium nitrite on the bioavailability of meat iron for the anemic rat, J. Nutr. 109:2182-2189.

Martinez, F. E., and Vanucchi, H., 1986, Bioavailability of iron added to the diet by cooking food in an iron pot, Nutr. Res. 6:421-428.

Mistry, A. N., Brittin, H. C., and Stoecker, B. J., 1988, Availability of iron from food cooked in an iron utensil determined by an *in vitro* method, J. Food Sci. 53:1546-1573.

Moeljopawiro, S., Fields, M. L., and Gordon, D. D., 1988, Bioavailability of zinc in fermented soybeans, J. Food Sci. 53:460-463.

Monsen, E. R., Halberg, L., Layrisse, M., Hegsted, D. M., Cok, J. D., Mertz, W., and Finch, C. A., 1978, Estimation of available dietary iron, Am. J. Clin. Nutr. 31:134-141.

Moore, T., Constable, B. J., Day, K. C., Impey, S. G., and Symonds, K. R., 1964, Copper deficiency in rats fed upon raw meat, Br. J. Nutr. 18:135-146.

Nävert, B., Sandström, B., and Cederblad, Å., 1985, Reduction of the phytate content of bran by leavening in bread and its effect on zinc absorption in man, Brit. J. Nutr. 53:47-53.

Oberleas, D., Muhrer, M. E., and O'Dell, B. L., 1966, Dietary Metal-complexing agents and zinc availability in the rat, J. Nutr. 90:56-62.

O'Brien, J. M., Morrissey, P. A., and Flynn, A., 1989, Mineral balance study of rats fed Maillard reaction products, in: Trace Elements in Man and Animals-6, L. S. Hurley, C. L. Keen, B. Lönnerdal, and R. B. Rucker, eds., Plenum, New York, pp. 563-564.

Petit, L., 1956, Complexion du cuivre par les produits de la réaction de Maillard: dosage d'un intermédiare de la formation des mélanoïdines. Compte Rend. Acad. Sci. Paris 242:54-829-831.

Picciano, M. F., Weingartner, K. E., and Erdman, J. W. Jr., 1984, Relative bioavailability of dietary iron from three processed soy products, J Food Sci. 49:1558-1561.

Pla, G. W., Fritz, J. C., and Rollinson, C. L., 1976, Relationship between the biological availability and solubility rate of reduced iron, J. Assoc. Off. Anal. Chem. 59:582-583.

Rackis, J. J., 1979, Comments following paper by O'Dell (see O'Dell, 1979), in: Soy Protein and Human Nutrition, H. L. Wilcke, D. T. Hopkins, and D. H. Waggle, eds., Academic Press, New York, p. 205.

Ranhotra, G. S., Gelroth, J. J. A., Torrence, F. A., Bock, M. A, and Winterringer, G. L., 1981, Bioavailability of iron in iron-fortified fluid milk, J. Food. Sci. 46:1342-1344.

Rendleman, J. A. Jr., 1987, Complexation of calcium by melanoidin and its role in determining bioavailability, J. Food Sci. 52:1699-1705.

Rizk, S. W., Clydesdale, F. M., 1985, Effect of organic acids on the *in vitro* solubilization of iron from a soy-extended meat patty, J. Food Sci. 50:577-581.

Rudloff, S., and Lönnerdal, B., 1990, Effects of processing of infant formulas on calcium retention in suckling rhesus monkeys, FASEB J. 4: A521(Abstract).

Sarzanini, C., 1983, Evaluation of stability constants in the association between activated sludge and Cu(II), Zn(II), and Cr(III) ions. Separ. Sci. & Technol. 18:1-14.

Schaafsma, G., Dekker, P. R., and de Waard, H., 1988, Nutritional aspects of yogurt. 2. Bioavailability of essential minerals and trace elements, Neth. Milk Dairy J. 42:135-146.

Schmittt, H. A., and Weaver, C. M., 1982, Effects of laboratory scale processing on chromium and zinc in vegetables, J. Food Sci. 47:1693-1694.

Schricker, B. R., and Miller. D. D., 1982, In vitro estimation of relative iron availability in breads and meals containing different forms of fortification iron, J. Food Sci. 47:723-727.

Schricker, B. R., and Miller, D. D., 1983, Effects of cooking and chemical treatment on heme and nonheme iron in meat, J Food Sci. 48:1340-1349.

Smith, T. M., Kolars, J. C., Saviano, D. A., and Levitt, M. D., 1985, Absorption of calcium from milk and yogurt, Am. J. Clin. 42:1197-1200.

Stegink, L. D., Freeman, J. B., Denbesten, L., and Filer, L. J. Jr., 1981, Maillard reaction products in parenteral nutrition. Prog. Food Nutr. Sci. 5:265-278.

Stuart, M. A., Johnson, P. E., Hamaker, B., and Kirleis, A., 1987, Absorption of zinc and iron by rats fed meals containing sorghum food products, J. Cereal Sci. 6:81-90.

Tabata, S., and Tanaka, K., 1986, Studies on Fe complexes produced by yeast. I. Separation of Fe complexes from wine and their incorporation into hemoglobin in rats, Chem. Pharm. Bull. 34:5045-5055.

Theriault, F. R., and Fellers, C. R., 1942, Effect of freezing and of canning in glass and in tin on available iron content of foods. Food Res. 7:503-508.

Theuer, R. C., Kemmerer, K. S., Martin, W. H., Zoumas, B. L., and Sarett, H. P., 1971, Effect of processing on availability of iron salts in liquid infant formula products. Experimental soy isolate formulas. J. Agric. Food Chem. 19:555-558.

Theuer, R. C., Martin, W. H., Wallander, J. F., and Sarett, H. P., 1973, Effect of processing on availability of iron salts in liquid infant formula products. Experimental milk-based formulas, J. Agric. Food Chem. 21: 482-485.

Thompson, D. B., and Erdman, J. W. Jr., 1988, Effect of various soy protein products on retention of nonheme iron from a casein test meal or from soy-based test meals, J. Food Sci. 53:1460-1469.

Toleman, C. J., 1987, The effect of yogurt on the growth and zinc status of rats fed diets high in phytic acid. MS Thesis, University of Kentucky, Lexington, KY.

Turnbull, A., Cleton, F., Finch, C. A., Thompson, L., and Martin, J., 1962, Iron absorption IV. The absorption of hemoglobin iron, J. Clin. Invest. 41:1897-1907.

Turnlund, J. R., King, J. C., Keyes, W. R., Gong, B., and Michel, M. C., 1984, A stable isotope study of zinc absorption in young men: effects of phytate and α-cellulose, Am. J. Clin. Nutr. 40:1071-1077.

Weeks, C. E., and King, R. L., 1985, Bioavailability of calcium in heat-processed milk, J. Food Sci. 50:1101-1105.

Whitelaw, M. L., and Weaver, M. L., 1988, Maillard browning effects on *in vitro* availability of zinc, J. Food Sci. 53:1508-1510.

Wood, R. J., Stake, P. E., Eisman, J. H., Shippee, R. L., Wolski, K. E., and Koehn, U., 1978, Effects of heat and pressure processing on the relative biological value of selected dietary supplemental inorganic iron salts as determined by chick hemoglobin repletion assay, J. Nutr. 108:1477-1484.

Zemel, P. C., and Zemel, M. B., 1985, Effects of food gums on zinc and iron solubility following *in vitro* digestion, J. Food Sci. 50:547-550.

33

THE EFFECT OF FOOD PROCESSING ON PHYTATE HYDROLYSIS AND AVAILABILITY OF IRON AND ZINC

Ann-Sofie Sandberg

Department of Food Science
Chalmers University of Technology
Gothenburg, Sweden

ABSTRACT

Phytate is one of the major inhibiting factor for zinc and iron absorption. When phytate is hydrolyzed during the food process the mineral availability is increased. By activation of the endogenous enzyme phytase which is present in plant foods, or addition of phytase, phytate is degraded to various inositolphosphates containing 1 - 5 phosphate groups per an inositol molecule. The effects of degradation products of phytate on availability of zinc, calcium and iron have to be further investigated. Food processes including soaking, germination and fermentation were under optimal conditions demonstrated to completely reduce the phytate content of cereals and vegetables. The results were related to in vitro measurements of iron availability and human iron and zinc absorption studies.

INTRODUCTION

Bioavailability of minerals is usually defined as the portion of the total intake absorbed and utilized by the organism. It depends mainly on the presence of factors in the diet promoting or depressing absorption. The inhibiting factors of importance for iron absorption are phytates, polyphenols and calcium, whereas stimulating factors include the "meat factor" ascorbic acid and certain other organic acids. These factors affect the absorption of non-heme iron which is present in plant foods. The meat factor and calcium also affect the absorption of heme iron which is present in meat and blood products (Hallberg 1984, Hallberg et al 1990). For zinc phytate seems to be the major inhibiting factor or combined effects of calcium, phytate and protein, while animal protein has a stimulating effect (Sandström 1987).

Food processing can lead to both positive and negative effects on mineral availability: Heat treatment often leads to inactivation of the enzyme phytase which hydrolises phytate (McCance and Widdowson 1944), vitamine C can be destroyed during cooking and baking (Sayers et al 1973), and the structure of polyphenols altered. An example of effect of heat treatment is extrusion cooking of a bran product causing lost phytase activity and a decrease in zinc, magnesium and phosphorous absorption compared to a meal containing raw bran (Kivistö et al 1986, Kivistö et al 1989). The decreased absorption was explained by lack of phytate hydrolysis of the heat treated product in the stomach and small intestine of humans (Sandberg et al 1986, Sandberg et al 1987).

Nutritional and Toxicological Consequences of Food Processing
Edited by M. Friedman, Plenum Press, New York, 1991

Other food processes, e.g. germination, fermentation, activate the endogenous phytase and polyphenol-oxidase of plant foods, and during fermentation stimulating factors for iron absorption such as organic acids are produced. The purpose of the present investigation was to use the food process to increase the availability of iron and zinc by activating the endogenous enzymes or by adding phytase during the food process.

Phytate

Phytate is the storage form of phosphorus in plants. Phytate constitutes 1 -3 % of cereals, nuts, legumes, but occurs also in low concentrations in roots, tubers and vegetables. Whole grain cereals contain high levels of minerals (Fe, Zn, Mg) but also of phytate, which - due to formation of insoluble complexes - makes the minerals unavailable for absorption. Some cereals and legumes also contain polyphenols that bind iron, and interfer with iron absorption (Brune 1989).

Degradation Products of Phytate

By activation of the endogenous phytase of plant foods the phytate is hydrolyzed to myo-inositol and inorganic phosphate via intermediate myo-inositolphosphates (penta- to monophosphates). When the phosphate groups are removed from the phytate the mineral binding capacity is decreased. Recent development of an HPLC method makes it possible to analyse phytate and its degradation products in foods, diets and intestinal contents also when present in low concentrations (Sandberg and Ahderinne 1986, Sandberg et al 1987).

By preparing pure fractions of inositolphosphates (via non-enzymatic and enzymatic hydrolysis) and addition to a diet with no phytate content we have investigated the effect of inositol tri-, tetra-, penta- and hexaphosphates on availability of iron, zinc and calcium in vitro and in vivo. Inositol hexa- and pentaphosphate added to a white wheat roll reduced the iron availability estimated in vitro (Sandberg et al 1989) and in vivo in humans (Hallberg et al 1989), while inositol tri- and tetraphosphate had no such effect. A similar result was obtained in suckling rats for zinc and calcium absorption (Lönnerdal et al 1989), and in humans for zinc absorption (Sandström and Sandberg 1989). In a mixed diet or a food containing various hydrolysis products of phytate which possibly can interact with each other and with minerals, the situation is more complicated.

Trials to correlate the content of inositolphosphates in a meal with iron (in cooperation with Hallberg, L. et al) and zinc absorption showed that the highest correlation was obtained when the sum of mg phosphorus from IP_3 - IP_6 was correlated to zinc and iron absorption, especially in the diets containing high amounts of IP_3 and IP_4 compared to IP_5 + IP_6 . The curve (Figure 1) shows the zinc absorption related to inositolphosphate content in different bread and porridge meals containing cereals (wheat, rye, oat, barley, triticale) and soy flour (Sandström, B. and Sandberg, A.-S. submitted for publ.). In all cases the meal was served with 200 g milk or fermented milk (Sandström et al 1987 a, Sandström et al 1987 b). The bread meals served with 200 g fermented milk contained bran baked in bread with different fermentation times. The fibre content of these breads were thus similar. The reduction in phytate content during fermentation resulted in increased zinc absorption. A meal with a wheat roll fermented to contain no detectable amounts of phytate gave a high zinc absorption and shows that it is possible to reach 40 % zinc absorption when the phytate was removed. Thus, a great improvement in absorption is possible if inhibiting factors are removed from the diet.

Food Process Activating the Endogenous Phytase

Examples of food processes activating the endogenous phytases are soaking, malting, sour-dough leavening of bread or fermentation of porridges, infant formulas or legumes and vegetables. During such food process hydrolysis of phytate occurs and various inositol phosphates are formed. The alternative to activate the endogenous enzymes is adding phytase in the food process.

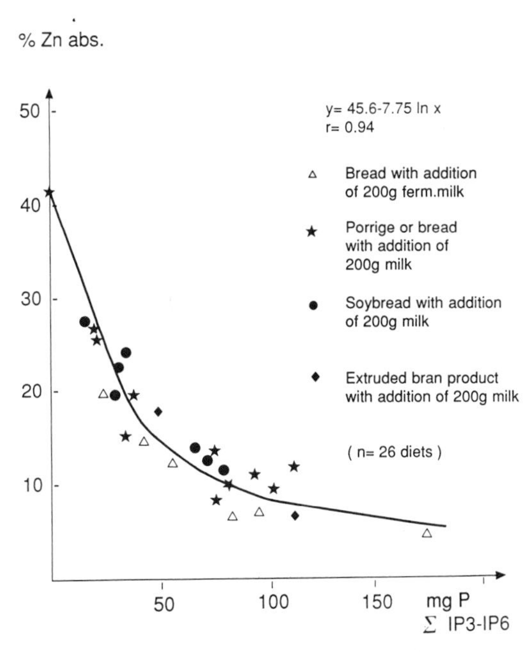

Figure 1. Zinc absorption from composite meals was related to content of inositolphosphate-P (sum of IP_3, IP_4, IP_5 and IP_6) (Sandström, B. and Sandberg, A.-S., submitted for publ.). In all cases the meals included 200 g of milk or fermented milk. The bread meal (Δ) with addition of 200 g fermented milk contained bran baked in bread with different fermentation times. The fibre content of these breads was thus similar (Nävert et al 1985).

Soaking

The naturally occurring phytases in cereals is activated by soaking under optimal conditions (for wheat phytase pH 4.5 - 5, 55°C). Soaking of wheat bran resulted in hydrolysis of 95 % of the phytate within one hour and a complete degradation within two hours. The amount of inositol hexa- and pentaphosphate, which in a previous study was found to decrease iron availability (Sandberg et al 1989), was related to the iron availability estimated in vitro under simulated physiological conditions (Sandberg, A.-S. and Svanberg, U., submitted for publ.) (Figure 2). It was found that the phytate content must be reduced to levels beyond 0.5 µmol/g to give the strong increase in iron solubility, if no promoting factors were present. The complete reduction of inositol hexa- and pentaphosphate increased the iron solubility by a factor six times.

For whole-meal rye flour soaking resulted in complete hydrolysis of phytate within 30 minutes, while the phytate hydrolysis of oat whole-meal flour under similar conditions after 4 h soaking were only 17 % and the iron solubility did not exceed 10 %, despite soaking for 17 h. This is in agreement with previous studies which show that phytase activity varies between species and that oats have a low phytase activity compared to wheat and rye (McCance and Widdowson 1944, Bartnik and Szafranska 1987).

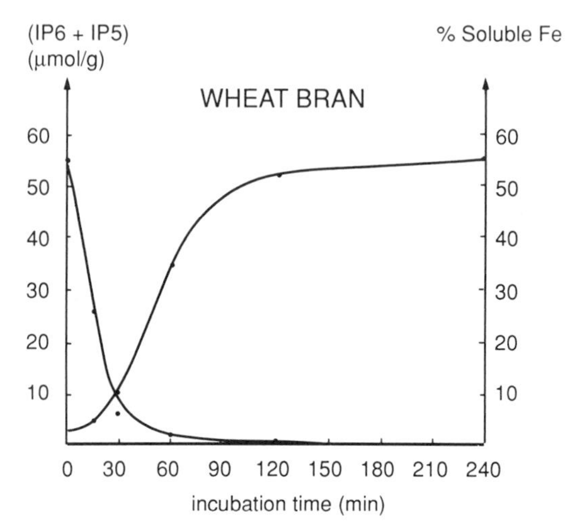

Figure 2. Soaking of wheat bran under optimal conditions for phytase (pH 4.5 - 5, 55°C) (Sandberg, A.-S. and Svanberg, U., submitted for publ.). The degradation of inositol hexa- and pentaphosphates were analysed by HPLC according to Sandberg et al 1988 and the iron solubility under simulated physiological conditions.

Malting

Malting is a process during which the whole grain is soaked and then germinated. During this process enzymes, e.g. phytase, are activated and possibly synthesized. We found that malting of wheat, barley, rye and oats for 30 - 44 h at 15°C only resulted in slightly reduced amount of phytate (Table 1). The grains were germinated until the sprout was 1 mm, furhter germination would make the grain impossible to use for production of flour. If the malted cereals were ground and soaked at pH 4.5 - 5 there was a complete degradation of phytate, except for oats which under the conditions studied had a low phytase activity.

Table 1. Reduction of phytate content in whole grain cereals after malting 30 - 44 h at 15°C and then soaking at pH 4.5 - 5, 4 h.

Cereal	Raw material	After malting	After malting and soaking
	Inositol hexa- and pentaphosphate (μmol/g)		
Wheat	10.8	10.8[a]	0
Rye	10.0	9.3[a]	0
Barley	9.8	6.0[b]	0
Oats	12.5	11.6[c]	10

[a] Germination time 30 h
[b] Germination time 4o h
[c] Germination time 44 h

Fermentation

Fermentation is an old method for food processing and preservation of food. Due to production of lactic acid and other organic acids the pH is lowered and the phytase activated. Fermentation of maize, soy beans and sorghum has been demonstrated to reduce the phytate content in the food (Sudarmadji, S. and Markakis, T. 1977, Lopez, Y. et al 1983, Moeljopawiro, S. et al 1987, Svanberg and Sandberg 1989). Our studies of sour-dough fermentation of rye-bread demonstrate the importance of the pH-level in dough and bread for phytate hydrolysis (Table 2). When 20 or 30 % sour-dough was added to the dough a decreased phytate hydrolysis was observed compared to 10 % sour-dough (Larsson, M. and Sandberg, A.-S., submitted for publ.). The remaining phytate in the breads containing 20 - 30 % sour-dough were quite low amounts but probably high enough to negatively affect iron absorption.

Table 2. Phytate content in rye-bread after sour-dough fermentation, 10 %, 20 % or 30 % sour-dough was added. pH was measured in doughs and breads.

Rye-bread	pH Dough	pH Bread	Percentage phytate hydrolyzed	Amount of $IP_6 + IP_5$ in breads (μmol/g)
10 % sour-dough	4.6	4.7	98	0.07
20 % sour-dough	4.1	4.4	89	0.5
30 % sour-dough	3.9	4.2	86	0.8

Soaking, germination and fermentation of white sorghum was used to increase iron availability (Svanberg and Sandberg 1989). Dehulling to 85 % extraction rate removed a minor part of the phytate, and germination caused a decrease of 35 % of the inositol hexa- and pentaphosphate ($IP_6 + IP_5$) content. However, no effect was observed on the iron solubility. Soaking the flours in water at pH 5 for 12 hrs caused a further degradation - over 90 % in the flour of germinated grains - which also resulted in an improved iron solubility, doubled compared to the untreated whole flour (Figure 3).

In the fermented samples the amount of $IP_6 + IP_5$ was reduced by more than 90 % and the fermented flour of germinated grains contained no detectable amounts. The lactic acid fermentation increased the amount of soluble iron two to six times. The combination of germination and fermentation gave the highest increase in soluble iron. The explanation to this seems to be that the germination process activates or synthesizes phytase, which during the fermentation process is active under optimal pH conditions, due to bacterial production of organic acids, mainly lactic acid. It is also interesting to notice that despite the high fibre content in the fermented flour, the iron availability was comparable to that of a white wheat roll with no detectable phytate content (Sandberg et al 1989). Sour-dough fermentation of whole-meal rye bread to a very low phytate level also lead to an increased iron solubility of the same level as the white wheat roll. These results indicate that phytate rather than fibre is responsible for the negative effect on iron availability. The in vitro results from sour-dough fermentation of rye bread has now been confirmed in human absorption studies (Brune 1989), showing 18,2 % iron absorption from a rye-roll with almost no phytate content and containing 6 g of dietary fibre compared to 18,2 % absorption from a white wheat roll with 1,4 g dietary fibre content.

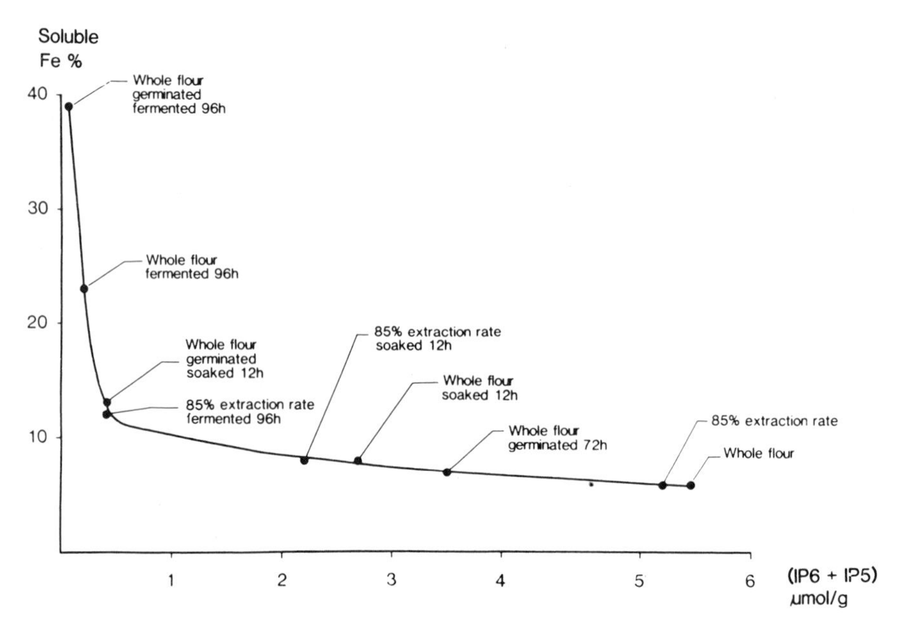

Figure 3. In vitro availability (% soluble iron after digestion) related to the amount of inositol hexa- and pentaphosphates in processed sorghum porridges (Svanberg and Sandberg 1989)

Vegetables generally contain low amounts of phytate but sufficient amounts to have an inhibiting effect on iron absorption. Lactic acid fermentation of vegetables was performed using Lactobacillus plantarium as a starter culture. (The study was performed in cooperation with U. Svanberg and R. Andersson.) The fresh vegetables; carrots/turnips/onion contained small amounts of phytate (1 μ mol/g dry weight).

During lactic acid fermentation the amount of phytate was reduced to a level under the detection limit and the percentage soluble iron increased three times. The amount of soluble iron was also increased in phytate-rich meals (whole meal rye and wheat rolls) when fermented vegetables were added (Table 3). These effects indicate the formation of iron promoting factors in lactic acid fermented vegetables. Different organic acids (Derman et al 1980, Gillooly et al 1983) have previously been demonstrated to improve iron absorption from rice and maize meals in studies in humans.

Table 3. Soluble iron (%) in different diets mixed with fresh or fermented vegetables (Svanberg, U., Andersson, R. and Sandberg, A.-S.1990).

Diet/vegetable	With vegetables	
	Fresh	Fermented
Rye roll, wholemeal/carrot juice	13	32
Wheat roll, 60 % extraction/ mixed vegetables[a]	27	42
Wheat roll, whole flour/mixed vegetables	13	18
Hamburger, potatoes and french beans/mixed vegetables	24	33

[a]carrot/turnip/onion

Addition of Phytase

During the making of whole-meal bread phytate is usually only reduced to about half except during sour-dough fermentation. The use of phytase addition to the dough was, therefore, studied.

Bread was baked from whole wheat flour and 60 % extraction wheat flour. Phytase preparations from A. niger var. were added to the doughs. The effect of addition of standard milk and fermented milk on phytate hydrolysis was studied with and without addition of phytase (Figure 4). The phytase addition resulted in an increased phytate reduction in all doughs and breads. Addition of standard milk inhibited phytate hydrolysis almost completely, while the same amount of fermented milk not significantly decreased phytate hydrolysis, probably due to the content of lactic acid. When phytase was added to the dough, the phytate hydrolysis was considerably lower than in the corresponding dough without milk (Lingaas, M. and Sandberg, A.-S., submitted for publ.). Recent human absorption studies demonstrate that these interactions between calcium and phytate have nutritional implications, as low levels of phytate has a strong inhibiting effect on iron absorption (Hallberg et al 1990).

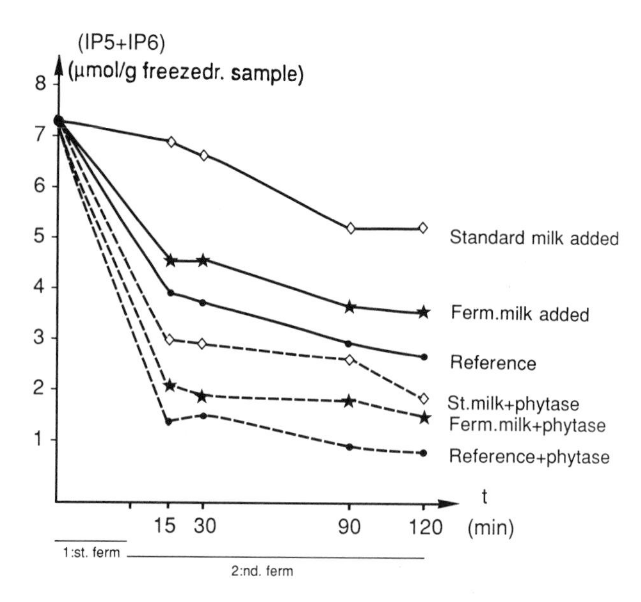

Figure 4. Phytate hydrolysis during bread-making with and without addition of phytase to the dough. Effect of addition of fermented milk and standard milk to the dough was also investigated (Lingaas, M. and Sandberg, A.-S., submitted for publ.)

Addition of phytase does not always result in complete phytate hydrolysis. When phytase was added to oat-meal flour, the hydrolysis was incomplete. Pre-cooking of the oat-meal flour during 10 minutes before enzyme addition significantly increased the phytate hydrolysis (Sandberg and Svanberg, submitted for publ.). The increased phytate digestibility after cooking cannot at this state be fully explained, but will be further investigated.

Table 4. Effect of adding phytase to oat whole meal flour on phytate hydrolysis and iron solubility (Sandberg, A.-S. and Svanberg, U., submitted for publ.). The samples were incubated for 4 h at 55°C and pH 5.

	Inositolphosphates (μmol/g)				Iron solubility (%)
	IP_6	IP_5	IP_4	IP_3	
Raw oat-meal not incubated	10.8	0.2	-	-	4
Raw oat-meal incubated without phytase addition	7.3	2.0	1.0	-	10
Raw oat meal incubated with 10 mg phytase	2.9	0.1	traces	-	12
Pre-cooked oat meal incubated with 10 mg phytase	1.2	0.09	0.03	0.1	20

CONCLUSIONS

Increased knowledge of phytate hydrolysis and formation of promoting factors during food processing makes it possible to optimize the food process from a nutritional point of view. Further studies of possible interactions between various inositol-phosphates and minerals and the effect on mineral availability are warranted as well as identification of promoting factors for mineral absorption. The results presented in this paper show that it is possible to reduce the content of inhibiting factors (phytate) and increase the mineral (iron and zinc) availability by soaking, malting, fermentation as well as with addition of phytase in the food process. This means that a diet of "low-bioavailability" via food processing can be changed into an "intermediate to high bioavailability diet", which otherwise only can be achieved by including generous quantities of meat or foods containing high amounts of ascorbic acid.

REFERENCES

M. Bartnik, and I. Szafranska, Changes in phytate content and phytase activity during the germination of some cereals, J. Cer. Sci. 5:23-28 (1987).

M. Brune, Food iron absorption in man. The inhibitory effect of phytate, calcium and phenolic compounds on non-heme iron absorption. Ph.D.Thesis, University of Gothenburg, Sweden, ISBN 91-7900-723-6 (1989).

D.P. Derman, T.H. Bothwell, J.D. Torrance, W.R. Bezwoda, A.P. MacPhail, M.C. Kew, M.H. Sayers, P.B. Disler, and R.W. Charlton, Iron absorption from maize and sorghum beer, Br. J. Nutr. 43:271-279 (1980).

M. Gillooly, T.H. Bothwell, J.D. Torrance, A.P. MacPhail, D.P. Derman, W.R. Bezwoda, W.R. Mills, and R.W. Charlton, The effects of organic acids, phytates and polyphenols on the absorption of iron from vegetables, Br. J. Nutr. 49:331-342 (1983).

L. Hallberg, Iron, in Present Knowledge in Nutrition. 5th ed., R.E. Olson, ed., The Nutrition Foundation Inc., Washington D.C. 459-478 (1984).

L. Hallberg, M. Brune, L. Rossander, A.-S. Sandberg, Inhibitory effect of phytates containing different number of phosphate groups on iron absorption in man (Abstr.), FASEB J. 3:A759 (1989).

L. Hallberg, M. Brune, M. Erlandsson, A.-S. Sandberg, L. Rossander-Hultén, Calcium: effect of different amounts on nonheme and heme iron absorption in man, Am. J. Clin. Nutr., (accepted for publication) (1990).

B. Kivistö, H. Andersson, G. Cederblad, A.-S. Sandberg, and B. Sandström, Extrusion cooking of a high-fibre cereal product. 2. Effects on apparent absorption of zinc, calcium, magnesium and phosphorus in humans, Br. J. Nutr. 55:255-260 (1986).

B. Kivistö, Å. Cederblad, L. Davidsson, A.-S. Sandberg, and B. Sandström, Effect of meal composition and phytate content on zinc absorption in humans from an extruded bran product, J. Cer. Sci. 10:189-197 (1989).

Y. Lopez, D.T. Gordon, L. Fields, Release of phosphorus from phytate by natural lactic acid fermentation, J. Food Sci. 48:953-954 (1983).

B. Lönnerdal, A.-S. Sandberg, B. Sandström, and C. Kunz, Inhibitory effects of phytic acid and other inositol phosphates on zinc and calcium absorption in suckling rats, J. Nutr. 119:211-214 (1989).

R.A. McCance, and E.M. Widdowson, Activity of the phytase in different cereals and its resistance to dry heat, Nature 153:650-651 (1944).

S. Moeljopawiro, D.T. Gordon, M.L. Fields, Bioavailability of iron in fermented soybeans, J. Food Sci 52:102-105 (1987).

B. Nävert, B. Sandström, and Å. Cederblad, Reduction of the phytate content of bran by leavening in bread and its effect on zinc absorption in man, Br. J. Nutr. 53:47-53 (1985).

M.H. Sayers, S.R. Lynch, P. Jacobs, R.W. Charlton, T.H. Bothwell, R.B. Walker, F. Mayet, The effect of ascorbic acid supplementation on the absorption of iron in maize, wheat and soya, Br. J. Haematol. 24:209-218 (1973).

A.-S. Sandberg, H. Andersson, B. Kivistö, and B. Sandström, Extrusion cooking of a high-fibre cereal product. 1. Effects on digestibility and absorption of protein, fat, starch, dietary fibre and phytate in the small intestine, Br. J. Nutr. 55:245-254 (1986).

A.-S. Sandberg, H. Andersson, N.-G. Carlsson, and B. Sandström, Degradation products of bran phytate formed during digestion in the human small intestine: effect of extrusion cooking on digestibility, J. Nutr, 117:2061-2065 (1987).

A.-S. Sandberg, N.-G. Carlsson, and U. Svanberg, Effects of inositol tri-, tetra-, penta- and hexaphosphates on in vitro estimation of iron availability, J. Food Sci. 54(1):159-161 (1989).

A.-S. Sandberg, and R. Ahderinne, HPLC-method for determination of inositol tri-, tetra-, penta- and hexaphosphates in foods and intestinal contents, J. Food Sci. 51:547-550 (1986).

B. Sandström, Cereals as a source of minerals in human nutrition, in Cereals in a European context. I. Morton, ed., Ellis Horwood, Ltd., Chichester, England, 241-247 (1987).

B. Sandström, B. Kivistö, and Å. Cederblad, Absorption of zinc from soy protein meals in humans, J. Nutr. 117:321-327 (1987 b).

B. Sandström, A. Almgren, B. Kivistö, and Å. Cederblad, Zinc absorption in humans from meals based on rye, barley, oatmeal, triticale and whole wheat, J. Nutr. 117:1898-1902 (1987 a).

B. Sandström, and A.-S. Sandberg, The effect of penta- and tetra inositol phosphate on zinc absorption in humans (Abstract), FASEB J. 3:A759 (1989).

S. Sudarmadji, P. Markakis, The phytate and phytase of soybean Tempeh, J. Sci. Fd. Agric., 28:381-383 (1977).

U. Svanberg, and A.-S. Sandberg, Improved iron availability in weaning foods using germination and fermentation, in Nutrient availability: Chemical and biological aspects, D.A.T. Southgate, I.T. Johnson, and G.R. Fenwick, eds., The Royal Society of Chemistry, Cambridge, CB4 4WF, 179-181 (1989).

U. Svanberg, R. Andersson, and A.-S. Sandberg, Bioavailability of iron in lactic acid fermented foods (Abstract), Proceedings of COST 91, Gothenburg (1990).

34

ANTI-NUTRITIVE EFFECTS OF DIETARY TIN

Jeanne I. Rader

Division of Nutrition, Food and Drug Administration

200 C Street, S.W., Washington, DC 20204

ABSTRACT

Tin is usually present in foods at levels of less than 4 µg/g. Higher levels may be found in some processed foods due to the addition of tin-based preservatives and stabilizers or to corrosion and leaching of the metal from unlacquered cans or from tin foils used in packaging. Estimates of dietary intake range from about 0.2 to $>$ 5 mg Sn/day. Diets including a high proportion of canned vegetables and fish could supply $>$ 30 mg Sn/day. Although intakes from dietary sources are generally considered to be harmless, a variety of adverse effects of tin have been reported, including effects on serum and bone alkaline phosphatase, lactic dehydrogenase, heme oxygenase, and 5-aminolevulinic acid dehydratase. Perturbations in glutathione metabolism have been reported, as have adverse effects on metabolism of essential trace minerals such as copper, zinc, and iron. Specific effects on calcium content of bone, serum, and kidney have also been described. Reported effects vary with the chemical form, dose of tin, and route and frequency of administration. Effects of tin in animal systems and on essential trace mineral absorption and excretion in human volunteers are reviewed. A summary of recent investigations on dietary tin-copper interactions and effects of tin on rat hepatocellular antioxidant protection are also presented.

INTRODUCTION

Exposure to tin occurs primarily through foods in which it is present as a result of processing and packaging. Much of the research interest in tin has been related to human exposure to tin from canned foods and to various organotin compounds such as plasticizers and fungicides. Nielsen (1986) and Greger and Lane (1987) have reviewed the toxicity of a variety of tin compounds. The high level of toxicity of specific organotin compounds has been the subject of several reviews (Barnes and Stoner, 1959; Magos, 1986; Boyer, 1989) and will not be discussed further.

DIETARY EXPOSURE TO TIN

Many early estimates of tin content of biological materials are unreliable because of loss of volatile tin compounds or formation of

Nutritional and Toxicological Consequences of Food Processing
Edited by M. Friedman, Plenum Press, New York, 1991

insoluble tins during sample preparation or because of the insensitivity of the analytical methods. Relatively few data have been obtained by reliable methods about the tin content of foods. An average English total diet has been reported to supply 0.187 mg Sn/day (mean value), while intakes in the United States have been estimated to be 1.5 to 3.5 mg/day (Nielsen, 1986). Tin content of canned foods may be increased by the absence of an inside resin or lacquer coating, by the presence of oxidants, such as nitrates, by low pH, increased storage temperature, and by failure to use canned foods promptly after opening (WHO/UNEP, 1980). Foods that have been contaminated by processing and packaging provide the highest levels of tin. Tin levels as high as 0.669 mg/g have been reported in some canned foods versus < 10 ng Sn/g in bottled foods (Greger and Baier, 1981). Highest levels have been reported in foods stored at refrigerator temperatures in their opened containers (Capar and Boyer, 1980; Greger and Baier, 1981). Levels of tin may exceed 250 µg/g under such conditions. Acute intoxications, including nausea, abdominal cramps, vomiting, and diarrhea, have been reported in individuals following consumption of acidic foods or juices prepared in tinned vessels (Barker and Runte, 1972; Benoy et al., 1971; Warburton et al., 1962). Levels of tin in such contaminated products have ranged from 500 to 1,000 µg Sn/ml.

ABSORPTION, RETENTION, AND EXCRETION OF INORGANIC TIN

The toxicity of ingested tin has generally been considered to be low because of its poor absorption and low retention in tissues. Hiles (1974) reported that 2.8% of Sn (II) and $< 1\%$ of Sn (IV) was absorbed by rats following oral dosing. Furchner and Drake (1976) estimated absorption at about 5% in mice, rats, dogs, and monkeys. In humans, apparent absorption of about 3% from a diet supplemented with 50 mg Sn/day has been reported, with apparent absorption increasing significantly when intake was reduced to 0.11 mg Sn/day (Johnson and Greger, 1982). Following acute or chronic exposure, tin is distributed primarily to the liver and kidneys. When fed to humans at relatively high levels, tin is excreted primarily in the feces. Significant fractions of tin doses are excreted in the urine following intravenous administration to rats. The appearance of tin in feces following intravenous administration indicates that the biliary system can contribute significantly to the clearance of tin (Hiles, 1974). The body burden of tin is carried primarily in bones, muscle, and liver (Furchner and Drake, 1976). Tin content of bone increases during prolonged oral exposure and a half-life of 34-40 days has been estimated (Hiles, 1974).

EFFECTS OF INORGANIC TIN

Chronic ingestion by rats of > 500 µg Sn/g is usually associated with growth depression, anemia, and adverse effects on the metabolism of specific trace minerals such as zinc, copper, and iron (De Groot, 1973; DeGroot et al., 1973; Fritsch et al., 1977).

Studies of effects of dietary tin on zinc metabolism have shown that levels of >500 µg Sn/g result in depressed levels of zinc in bone and soft tissues (Greger and Johnson, 1981; Johnson and Greger, 1984). Some of these changes may be due to effects of dietary tin on absorption of zinc (Johnson and Greger, 1984). Human subjects lost an additional 2 mg of zinc daily in the feces when fed 50 mg Sn/day (Johnson et al., 1982). This observation was confirmed by Valberg et al. (1984), who found that inorganic tin depressed the absorption of 65-Zn from zinc chloride and from a turkey test meal.

The mechanism by which tin affects zinc absorption is not known but appears to be dose-dependent. Rats fed levels of tin >2000 µg/g had hypertrophied gastrointestinal tracts and significantly increased endogenous losses of zinc in the feces (Johnson and Greger, 1984). With more moderate doses (200-500 µg Sn/g), endogenous losses of zinc in the feces were constant but true absorption of zinc tended to be depressed. Human subjects fed 50 mg versus 0.11 mg Sn/day apparently absorb significantly less selenium (Greger et al., 1982).

Ingestion of >200 µg Sn/g generally depresses copper levels in soft tissues (Greger and Johnson, 1981; Johnson and Greger, 1985). DeGroot (1973) demonstrated that the signs of anemia associated with ingestion of high levels of tin (1,500 µg/g) could be prevented or reduced by addition of copper or iron to the diets; however, tin-associated depression in growth was not prevented by these additions.

Ingestion of tin has also been reported to reduce the calcium content of bone, reduce serum calcium, and increase kidney calcium levels (Yamaguchi et al., 1980, 1981; Yamamoto et al., 1976). Yamaguchi et al. (1982) reported that collagen synthesis was depressed in rats orally dosed with tin. Other workers have reported that low levels of dietary tin (about 100 µg Sn/g diet) depressed calcium content of bone, but they did not observe changes in plasma calcium concentrations (Johnson and Greger, 1985). Addition of 50 mg Sn/day to diets of human volunteers did not affect calcium or magnesium excretion or retention (Johnson and Greger, 1982; Johnson et al., 1982).

The activities of serum alkaline phosphatase and serum lactic dehydrogenase were depressed in rats administered 3.0 mg/kg stannous chloride by gavage at 12-hour intervals twice daily for 90 days (Yamaguchi et al., 1980). Following a single injection of stannous chloride, hepatic azo-reductase and aromatic hydroxylase activities were reported to be reduced (Burba, 1983). Injections of tin have been reported to induce heme oxygenase activity in rat kidney (Kappas and Maines, 1976). Adult male rabbits injected intravenously with stannous chloride (5 µmol/kg) daily for 13 days had decreased activity of erythrocyte 5-aminolevulinic dehydratase (5-ALAD) and increased coproporphyrin concentrations in blood and urine (Chiba et al., 1980). In rats, levels of dietary tin >2000 µg Sn/g diet inhibited blood 5-ALAD activity. Ingestion of additional zinc did not counteract the effect of tin on 5-ALAD activity (Johnson and Greger, 1985).

EXPERIMENTAL SECTION

The studies described below were carried out to examine in more detail the effects of levels of 100-1,100 µg/g of dietary tin on tissue trace elements and bone minerals in weanling rats. Our initial studies showed that copper metabolism was particularly sensitive to dietary tin. We investigated the response of copper-depleted rats to tin as well as effects of tin on enzymes involved in hepatocellular antioxidant protection.

Diets

Diet composition was based upon diet AIN-76A (American Institute of Nutrition, 1977, 1980). Mineral contents of major dietary ingredients (casein, cornstarch, glucose, cellulose) were determined by inductively coupled argon plasma-atomic emission spectrometry (ICP-AES). The composition of the mineral mix was adjusted to compensate for copper, iron, zinc, and phosphorus in casein used in the diets (Rader et al., 1990).

Diets contained (%): protein (casein), 20; choline bitartrate, 0.2; DL-methionine, 0.3; cornstarch, 15; corn oil (Mazola), 5; fiber (cellulose), 5; mineral mix, 3.5; vitamin mix, 1; and glucose, 50. Minerals added to copper-adequate diets (Study 1) included (mg/kg diet): Ca, 5100; P, 3910; Na, 1090; K, 3600; Mg, 500; Mn, 54; Fe, 35.8; Zn, 32.2; Cu, 5.5; I, 0.21; Se, 0.11; and Cr, 2.0. Copper carbonate was omitted from mineral mixes used to prepare copper-deficient diets (Study 2). Vitamins were included in all diets at levels specified for diet AIN-76A (American Institute of Nutrition, 1977, 1980). Stannous chloride ($SnCl_2 \cdot 2H_2O$, Mallinkrodt, Inc., Paris, KY) was included in copper-adequate diets (Study 1) at levels of 100, 330, and 1,100 µg Sn/g and in copper-adequate and copper-deficient diets (Study 2) at 100 µg Sn/g.

Details of purified diet preparation are described in Rader et al. (1986). Salts used in mineral mix preparation were Baker-analyzed reagent grade (J.T. Baker Chemical Co., Phillipsburg, NJ). Mineral compositions of the diets were confirmed by ICP-AES analysis (see below). Vitamins and major diet ingredients (vitamin-free casein [90% protein], cornstarch, glucose monohydrate, non-nutritive fiber [cellulose], choline bitartrate, and DL-methionine) were obtained from Teklad, Madison, WI. Corn oil (Mazola, Best Foods, CPC International, Englewood Cliffs, NJ) was obtained locally. Retinyl acetate was purchased from Eastman Laboratory Chemicals, Rochester, NY.

Experimental Animals

Male Long-Evans rats (22 days old, 46 ± 3 g for Study 1 and 48 ± 4 g for Study 2, mean ± SD, 1 day post-weaning) were obtained from Blue Spruce Farms, Altamont, NY. A basal group of 10 animals was euthanized upon receipt, and the same analyses were performed as for the experimental groups. All other rats were randomly assigned to diet groups. Rats were housed singly in suspended stainless steel cages with wire-mesh flooring. Powdered diets were provided in ceramic food bowls equipped with stainless steel lids and food followers. Rats were weighed daily. Food consumption by rats fed diets containing 0, 100, and 330 µg Sn/g was measured weekly. Because reduced food intake was expected, food consumption by rats fed 1,100 µg Sn/g was measured daily. Food intake of rats in a pair-fed control group (group 1 PF) was reduced to the mean amount consumed by rats fed diets containing 1,100 µg Sn/g (group 4 AL).

Distilled deionized water was provided ad libitum in glass bottles equipped with polyethylene stoppers and stainless steel sipper tubes. Room temperature was maintained at 70-74° C. Room lighting was automatically controlled to provide alternating 12-hour periods of light and dark.

After 4 weeks, rats were fasted overnight and then euthanized by carbon dioxide asphyxiation. Blood was collected by cardiac puncture. Serum was prepared and stored at $< -70^{o}$ C until analysis. Soft tissues (liver, kidney, duodenum) and femurs were removed; duodena were flushed with ice-cold 0.9% NaCl. All tissues were stored in polyethylene bags at -17° C until analysis.

Determination of Minerals in Diets and Tissues

Portions of tissues and diets were weighed and wet-digested in mixtures of nitric and perchloric (femurs), or nitric, perchloric, and sulfuric acids (soft tissues, diets) (Rader et al., 1984) and analyzed by ICP-AES using a sequential Perkin-Elmer Plasma II system (Norwalk,

CT). Portions of National Institute of Standards and Technology (NIST; formerly National Bureau of Standards) Standard Reference Material Bovine Liver 1577 were digested and analyzed initially and then after each set of 20 samples for the same elements. Values fell within ±5% of certified values for all elements of interest. Tin in the diets was determined by hydride generation atomic absorption spectrometry (Alvarez and Capar, 1987).

Other Methods

Serum ceruloplasmin was measured by the method of Schosinsky et al. (1974). Cholesterol was measured enzymatically using Sigma Diagnostics Kit No. 352 (Sigma Chemical Co., St. Louis, MO). Sigma lipid controls for normal and elevated cholesterol levels were used as quality control standards. Liver reduced glutathione concentration (GSH) was determined according to an enzymatic recycling assay based on glutathione reductase and 5, 5'-dithiobis-(2-nitrobenzoic acid) (Griffith, 1980). Liver GSH-peroxidase (GSH-Px) activity was determined by the spectrophotometric method of Paglia and Valentine (1967) and liver superoxide dismutase (SOD) was assayed by a method utilizing inhibition of auto-oxidation of pyrogallol (Prohaska et al., 1983). Liver malondialdehyde (MDA) production was determined by the method of Levine (1982). Liver homogenate supernatant protein was assayed using the Bio-Rad protein reagent (Bio-Rad, Richmond, CA).

Statistical Methods

Means and standard deviations were estimated for the parameters measured. Statistical analysis included an analysis of variance and separation of means by Duncan's tests.

RESULTS

Study 1. Effects of Dietary Tin (Tables 1-6)

Food consumption by rats in group 1 PF was restricted to that amount consumed by rats fed 1,100 µg Sn/g. Comparison of group 1 PF with group 1 AL which was fed a tin-free diet ad libitum indicated that restricted feeding of the control diet reduced body weight, efficiency of nutrient utilization, relative liver weight, and femur percent ash and significantly increased relative kidney weight (Tables 1 and 5). Restricted food intake per se was also associated with significant alterations in tissue minerals. For example, significant decreases were measured for duodenal calcium, manganese, and zinc (Table 2) and femur zinc (Table 6). Significant increases were observed in liver copper, iron and zinc (Table 3) and kidney iron and zinc (Table 4).

Tin reduced weight gain in a dose-dependent manner (Table 1). Food consumption among groups of rats fed 0, 100, and 330 µg Sn/g diet did not differ significantly during the first 3 weeks of the study. However, rats fed 330 µg Sn/g consumed less food during week 4 with the result that cumulative (final) food intake by this group was significantly lower than that of control group 1 AL (Table 1). Food intake was not significantly different between groups 1 PF and 4 AL.

Significantly reduced copper levels were found in the duodena of all animals fed tin (Table 2). Effects of 330 µg Sn/g diet on duodenal calcium, manganese and zinc are probably related in part to reduced food consumption because decreases in these elements were also measured in group 1 PF.

TABLE 1. Responses of Weanling Rats to Purified Diets Containing Tin.

Group	Dietary Sn (μg/g)	Weight gain (g)	Food consumption (g)	Efficiency of nutrient utilization (g gain/ g food)	Liver weight (% bw)	Kidney weight (% bw)
Basal	-	-	-	-	4.6e	1.43a
1 AL	0	182a	399a	0.46a	5.9ab	0.96d
2 AL	100	168b	397a	0.42b	5.5bc	1.00cd
3 AL	330	140c	326b	0.43b	5.3c	1.01cd
1 PF	0	92d	237c	0.39c	4.9cd	1.04bc
4 AL	1,100	101d	243c	0.41bc	5.2c	1.10b

Values are means, 10 determinations/group. Means sharing the same letter are not significantly different ($P < 0.05$); bw = body weight; AL = ad libitum; PF = pair-fed.

TABLE 2. Duodenal Minerals in Weanling Rats Fed Diets Containing Tin.

Group	Dietary Sn (μg/g)	Ca	Mn	Fe	Zn	Cu
		(μg/g fresh weight)				
Basal		117±51cd	3.5±2.3bc	19.6±3.0a	23.1±2.7a	1.6±0.2a
1 AL	0	248±53a	9.5±3.4a	15.4±2.8b	23.7±1.2a	1.5±0.2a
2 AL	100	231±72a	8.8±3.3a	16.4±2.2ab	22.7±1.3ab	1.0±0.3b
3 AL	330	176±42b	5.8±2.4b	17.2±4.0ab	20.2±1.2c	0.7±0.2c
1 PF	0	70±19d	2.9±1.4c	16.3±3.8ab	21.7±1.0b	1.5±0.2a
4 AL	1,100	150±59bc	3.7±2.7bc	14.0±2.5b	20.1±1.7c	0.5±0.2d

Values are means ± SD; 10 determinations/group. Values sharing the same letter are not significantly different ($P < 0.05$); AL = ad libitum; PF = pair-fed.

Significant copper depletion in liver was observed when rats were fed diets containing 100 μg Sn/g and higher levels (Table 3). Food restriction increased liver iron by 75% and caused smaller increases in liver copper (22%) and zinc (13%). Rats fed 1,100 μg Sn/g diet had significantly lower liver iron than did pair-fed control rats.

Kidney copper and zinc were significantly reduced by 100 μg Sn/g diet (Table 4). Pair-fed rats in group 1 PF had significantly higher kidney iron (68%) than did rats fed ad libitum. Kidney iron for rats fed 1,100 μg Sn/g diet was significantly lower than that for rats in the pair-fed control group 1 PF.

Femur ash and femur magnesium were significantly reduced when 100 and 330 μg Sn/g were included in diets (Table 5). Reductions for both of these parameters in rats fed 1,100 μg Sn/g were significantly greater than those caused by reduced food intake per se. Restricted feeding also decreased zinc and increased iron in femur (Table 6). Femur copper

TABLE 3. Liver Minerals in Weanling Rats Fed Diets Containing Tin.

Group	Dietary Sn (µg/g)	Cu	Fe (µg/g fresh weight)	Zn
Basal	-	7.7 ± 1.6a	96.3 ± 29.1ab	35.6 ± 2.6a
1 AL	0	3.2 ± 0.4c	62.8 ± 20.9c	22.8 ± 2.2c
2 AL	100	1.9 ± 0.7d	83.0 ± 26.4bc	21.7 ± 2.6c
3 AL	330	1.2 ± 0.7e	83.5 ± 20.9bc	21.9 ± 2.1c
1 PF	0	3.9 ± 0.5b	110.1 ± 24.2a	25.8 ± 2.3b
4 AL	1,100	0.8 ± 0.2e	75.0 ± 4.1bc	21.6 ± 1.4c

Values are means ± SD; 10 determinations/group. Values sharing the sameletter are not significantly different ($P < 0.05$). AL = ad libitum; PF = pair-fed.

TABLE 4. Kidney Minerals in Weanling Rats Fed Diets Containing Tin.

Group	Dietary Sn (µg/g)	Cu	Fe (µg/g fresh weight)	Zn
Basal	-	4.0 ± 0.5b	38.5 ± 5.0b	25.6 ± 1.6a
1 AL	0	5.9 ± 1.4a	51.2 ± 13.9b	21.4 ± 1.1c
2 AL	100	3.0 ± 0.4c	50.6 ± 12.7b	19.7 ± 1.5d
3 AL	330	2.8 ± 0.3c	47.3 ± 11.9b	18.6 ± 0.9de
1 PF	0	5.7 ± 1.9a	86.2 ± 51.6a	22.6 ± 1.2b
4 AL	1,000	2.4 ± 0.2c	35.0 ± 8.1b	17.8 ± 0.6e

Values are means ± SD; 10 determinations/group. Values sharing the same letter are not significantly different ($P < 0.05$). AL = ad libitum; PF = pair-fed.

was not affected by restricted feeding. Animals fed tin at all levels tested had significantly reduced femur copper and zinc.

Study 2. Responses to Tin in Copper-Adequate and Copper-Deficient Diets (Tables 7-10)

Rats fed diets containing < 0.5 µg Cu/g did not become anemic during the 4-week experimental period (Table 7). Inclusion of 100 µg Sn/g in copper-deficient, but not copper-adequate, diets reduced hemoglobin significantly. Ceruloplasmin levels for 3 of 10 rats in group 3 were < 5 U/l and the mean value for the remaining rats was significantly lower than that for rats fed the copper-adequate diet (Group 1). When rats were fed 100 µg Sn/g in copper-adequate diets, ceruloplasmin values for 5 of 10 animals were below the limits of quantification, and the mean value for the remaining rats was significantly lower than that for rats fed the tin-free control diet. Ceruloplasmin was measurable in only 1 rat in 10 fed a copper-deficient diet containing 100 µg Sn/g (Group 4). Serum cholesterol was significantly elevated by dietary tin in both copper-adequate and copper-deficient diets.

TABLE 5. Major Minerals and Ash Content in Femurs of Weanling Rats Fed Dietary Tin.

Group	Dietary Sn (μg/g)	Ca	P	Mg	Femur ash (%)
		(mg/g fat-free dry weight)			
Basal		166 ± 6d	86 ± 3c	3.8 ± 0.1a	46.3d
1 AL	0	214 ± 16a	105 ± 8a	3.8 ± 0.3a	57.0a
2 AL	100	206 ± 10ab	100 ± 5ab	3.6 ± 0.2b	55.4b
3 AL	330	206 ± 8ab	101 ± 4b	3.6 ± 0.3b	54.9b
1 PF	0	202 ± 6bc	100 ± 4b	3.7 ± 0.2ab	54.5b
4 AL	1,100	196 ± 19c	98 ± 5c	3.3 ± 0.3c	50.3c

Values are means ± SD or means; 10 determinations/group. Values sharing the same letter are not significantly different ($P < 0.05$). AL = ad libitum. PF = pair-fed.

TABLE 6. Zinc, Copper and Iron in Femurs of Weanling Rats Fed Diets Containing Tin.

Group	Dietary Sn (μg/g)	Zn	Cu	Fe
		(μg/g fat-free dry weight)		
Basal	-	150 ± 11d	6.8 ± 1.9a	82.2 ± 8.0a
1 AL	0	194 ± 17a	2.5 ± 0.5bc	45.9 ± 9.5c
2 AL	100	165 ± 11c	1.6 ± 0.6d	41.7 ± 9.4c
3 AL	330	146 ± 7d	1.5 ± 0.5d	44.4 ± 7.1c
1 PF	0	183 ± 15b	3.0 ± 0.5b	63.4 ± 10.7b
4 AL	1,100	142 ± 18d	1.7 ± 0.7cd	34.3 ± 6.0d

Values are means ± SD; 10 determinations/group. Values sharing the same letter are not significantly different ($P < 0.05$). AL = ad libitum; PF = pair-fed.

Liver copper declined by 24% in rats fed a copper-deficient diet for 4 weeks (Table 8). Inclusion of 100 μg Sn/g in a copper-adequate diet (Group 2) reduced liver copper by 43% within 4 weeks. Liver copper in rats fed tin in a copper-deficient diet was < 20% of that for rats fed a tin-free copper-deficient control diet.

Liver GSH was not significantly different from control values in rats fed tin in copper-adequate diets, but decreased slightly for rats fed tin in copper-deficient diets (Table 9). Tin ingestion significantly increased GSH-Px activity and decreased SOD activity regardless of dietary copper content. Ingestion of tin by rats fed copper-deficient diets increased liver MDA production (Table 10). Liver non-heme iron was also significantly increased in this group.

TABLE 7. Responses of Weanling Rats to Dietary Tin in Copper-Adequate and Copper-Deficient Diets.

Group	Dietary Cu (μg/g)	Dietary Sn (μg/g)	Weight gain (g)	Food intake (g)	HGB (g/dl)	Serum Ceruloplasmin (U/l)	Serum Cholesterol (mg/dl)
Basal	-	-	-	-	10.5 ± 1.3a	-	-
1	5.5	0	146a	338a	9.5 ± 1.3a	83 ± 33a	81 ± 10a
2	5.5	100	155a	356a	9.5 ± 1.6a	25 ± 35bc	121 ± 33b
3	<0.5	0	136a	323a	9.1 ± 0.9a	45 ± 50b	62 ± 10a
4	<0.5	100	140a	339a	7.2 ± 1.1b	2 ± 3c	117 ± 30b

Values are means or means ± SD; 10 determinations/group except as follows: Ceruloplasmin: Values for 5 of 10 rats in group 2 were 5 U/l; values for 3 of 10 rats in group 3 were 5 U/l; values for 9 of 10 rats in group 4 were 5 U/l. Values sharing the same letter are not significantly different ($P < 0.05$). HGB = hemoglobin; Rats weighed 48 ± 4 g on receipt.

TABLE 8. Copper, Iron, and Zinc in Livers of Weanling Rats Fed Tin in Copper-Adequate and Copper-Deficient Diets.

Group	Dietary Cu (μg/g)	Dietary Sn (μg/g)	Cu (μg/g fresh weight)	Fe (μg/g fresh weight)	Zn (μg/g fresh weight)
Basal	-	-	8.8 ± 2.4a	36.3 ± 4.7b	29.9 ± 2.5a
1	5.5	0	4.2 ± 0.6b	25.9 ± 2.7b	26.3 ± 2.3b
2	5.5	100	2.4 ± 1.4c	32.9 ± 11.5b	24.8 ± 2.4b
3	<0.5	0	3.2 ± 0.8c	25.7 ± 2.0b	25.9 ± 2.2b
4	<0.5	100	0.6 ± 0.2d	60.2 ± 8.5a	21.8 ± 1.2c

Values are means ± SD, 10 determinations/group. Values sharing the same letter are not significantly different ($P < 0.05$).

TABLE 9. Hepatic GSH, GSH-Px and SOD in Weanling Rats Fed Tin in Copper- Adequate and Copper-Deficient Diets.

Group	Dietary Cu (μg/g)	Dietary Sn (μg/g)	GSH (μmol/g liver)	GSH-Px (Units/mg protein)	SOD (Units/mg protein)
1	5.5	0	4.52 ± 0.37ab	11.1 ± 4.5c	133.7 ± 15.7a
2	5.5	100	4.28 ± 0.45b	18.1 ± 3.8a	110.8 ± 30.9b
3	<0.5	0	4.82 ± 0.38a	8.5 ± 2.7c	121.6 ± 19.1ab
4	<0.5	100	4.18 ± 0.53b	14.6 ± 3.2b	38.6 ± 19.0c

GSH, reduced glutathione; PSH-Px, glutathione peroxidase; SOD, superoxide dismutase. Values are means ± SD, 9-10 determinations/group. Values sharing the same letter are not significantly different, $p < 0.05$.

TABLE 10. Malondialdehyde and Non-Heme Fe in Livers of Rats Fed Tin in Copper-Adequate and Copper-Deficient Diets.

Group	Dietary Cu (µg/g)	Dietary Sn (µg/g)	Malondiadehyde (mM)	Non-Heme Fe (µg/g)
1	5.5	0	11.2 ± 3.6bc	13.7 ± 2.9b
2	5.5	100	8.8 ± 1.0cd	20.7 ± 10.8b
3	<0.5	0	12.3 ± 2.9b	13.8 ± 2.8b
4	<0.5	100	15.8 ± 4.4a	59.8 ± 14.6a

Values are means ± SD, 8-10 determinations/group. Values sharing the same letter are not significantly different, $p < 0.05$.

DISCUSSION

Anemia and growth depression are common effects of chronic ingestion of inorganic tin. Growth depression generally occurs when dietary tin levels exceed 500 µg/g. DeGroot (1973) reported that diets high in copper and iron (50 and 250 µg/g, respectively) reduced signs of anemia in rats fed tin at levels up to 1,500 µg Sn/g diet. The growth-depressing effect of tin was only slightly diminished by this combination of trace minerals, however. Because DeGroot (1973) did not report tissue copper or iron levels in the tin-treated, copper- and/or iron-supplemented animals, direct assessment of trace mineral problems associated with tin ingestion is not possible. Johnson and Greger (1985) did report tissue copper levels for rats fed levels of tin ranging from 100 to 1,954 µg/g. They showed that ingestion of diets containing 200 µg Sn/g for 27 days reduced copper concentrations by 38% in liver and 19% in kidney. No significant changes in levels of iron in liver, kidney, or tibia were reported at this level of dietary tin. Tibia zinc was reduced to about 88% of the value measured for control animals. The variable effects of diets containing 100 µg Sn/g depended upon the specific tissue studied and the level of zinc in the diets.

The results of the present studies indicate that low levels of dietary tin primarily affect copper and zinc status. Pronounced copper depletion in all tissues examined was observed at tin levels as low as 100 µg/g diet. Subsequent studies have shown that this effect of tin on tissue copper concentrations occurs at tin levels as low as 30 µg Sn/g diet.

Feeding tin in copper-adequate and copper-deficient diets demonstrated that copper depletion was accentuated by low levels of dietary tin. Copper depletion, induced in weanling rats within 4 weeks by feeding a diet containing < 0.5 µg Cu/g, was characterized by significant declines in serum ceruloplasmin and reductions in liver copper. Indices of copper deficiency, such as increased serum cholesterol and increased liver iron, were not altered during the 4-week study unless diets also contained 100 µg Sn/g.

In normal animals, approximately 90% of copper in plasma is contained in ceruloplasmin. Our finding of decreased ceruloplasmin in tin-treated rats is consistent with observations of decreased copper in plasma of rats fed diets containing 200 to 500 µg Sn/g (Johnson and Greger, 1985).

Reduced growth is a common symptom of zinc deficiency. Our studies indicate that low dietary tin adversely affects growth and zinc status in rats fed zinc-adequate diets. This observation may explain the failure of copper and iron supplementation in earlier studies to correct the diminished growth observed in rats fed high-tin diets (DeGroot, 1973).

The mechanism(s) by which tin affects the metabolism of copper and zinc is not understood but could include effects on absorption or on fecal losses of endogenous or diet-derived minerals. Greger and Johnson (1981) demonstrated that rats fed 200-2,000 µg Sn/g diet had significantly increased fecal losses of zinc, but they did not observe correspondingly increased copper losses. Additional studies are needed to understand how tin influences the metabolism of these essential minerals.

Effects of high dietary tin on enzymes involved in hematopoiesis have been reviewed by Nielsen (1986). Our studies did not identify a specific effect of low dietary tin on iron status. Johnson and Greger (1985) observed no changes in iron content of liver, kidney, or tibia in rats fed levels of 100-500 µg Sn/g diet. However, kidney iron was significantly reduced in rats fed 2,000 µg Sn/g (Johnson and Greger, 1985). In the present study, iron concentrations in soft tissues and bone were more markedly affected by reduced food intake than were concentrations of copper and zinc. Addition of 1,100 µg Sn/g diet reduced iron in all tissues examined. The role, if any, of ceruloplasmin, which has an important function in iron utilization, is not known.

Tin would be expected to adversely affect bone metabolism as a result of its effects on copper metabolism. Copper plays a major role in bone metabolism through its involvement in collagen cross-linking. The activity of osteoblasts is depressed in copper deficiency in all species studied. However, effects of copper deficiency on other parameters of bone formation vary widely (Davis and Mertz, 1987). We did not observe the marked reductions in bone calcium that Yamaguchi et al. (1980, 1981, 1982) reported for their 90-day studies in which rats were provided with tin in food or by intubation. We did observe, however, that tin at low levels significantly reduced femur ash and femur magnesium, effects that have not been previously reported. The types of diets used, the calcium, phosphorus, and magnesium contents of diets, and the route of administration of tin may account for the differences observed between the present study and those of Yamaguchi et al. (1980, 1981, 1982).

A number of trace minerals play major roles in cellular antioxidant protection. SOD activity requires copper, zinc, and manganese (Paynter et al., 1979; DeRosa et al., 1980) and selenium is required for GSH-Px (Rotruck et al., 1973). Dietary copper deficiency depresses cytosolic SOD activity and decreases Se-dependent GSH-Px activity in lung and liver (Jenkinson et al., 1982). Allen et al., (1988) reported that feeding rats a copper-deficient diet for 51 days caused a 50-60% increase in liver GSH concentration and reductions in GSH-Px and SOD activities. These authors suggested that increased GSH was secondary to the reductions in GSH-Px and SOD activities. In our 28-day study, we observed a significant reduction in GSH-Px activity in rats fed copper-deficient diets but did not observe significant changes in hepatic GSH or SOD activity. The copper-depleted rather than copper-deficient status of our rats was probably responsible for the differences observed. Addition of tin to both copper-adequate and copper-deficient diets caused significant reductions in SOD and increases in GSH-Px.

The reported effects of heavy metals on cellular antioxidant protection mechanisms vary. For example, hepatic GSH levels have been reported to be increased by ingestion of nickel (Athar et al., 1987) and decreased by ingestion of cadmium or mercury (Shukla et al., 1987; Stacey and Klaassen, 1981). The administration of tin tartrate to sham-operated and partially hepatectomized rats resulted in hepatic GSH depletion and an increase in lipid peroxide formation (Dwivedi et al., 1984). In the present study, we observed hepatic GSH depletion in tin-fed copper-depleted rats. The addition of tin significantly reduced liver copper and SOD activity in rats fed copper-adequate and copper-deficient diets. Liver MDA concentration and total (and non-heme) liver iron were increased only in rats fed copper-deficient diets with tin. At the present time, it is not possible to estimate the contribution of liver iron accumulation or decreased antioxidant protection to the increased MDA production observed.

CONCLUSIONS

Adverse effects of feeding rats diets containing 100 µg Sn/g include 1) copper depletion in rats fed copper-adequate diets; 2) accelerated development of copper deficiency in copper-depleted rats; and 3) reduction in hepatocellular antioxidant protection. The level of tin used in these studies is within the range of values reported for tin present in foods from unlacquered cans. The sensitivity of copper status to tin indicates that effects of chronic ingestion of tin may need to be reexamined.

The studies reported herein were conducted according to the principles set forth in the Guide for the Care and Use of Laboratory Animals, Institute of Laboratory Animal Resources, NRC, NIH Publ. No 86-23, 1986.

ACKNOWLEDGMENTS

We gratefully acknowledge the capable technical assistance of Miss Catherine J. Paul, Miss Clarisse Jones, and Mrs. Edythe Smith in these studies.

REFERENCES

Allen, K. G. D., Arthur, J. R., Movica, P. C., Nichol, F., and Mills, C. F. (1988). Copper deficiency and tissue glutathionine concentration in the rat. Proc. Soc. Exp. Biol. Med. 187, 38-43.

Alvarez, G. H. and Capar, S. G. (1987). Determination of tin in foods by hydride generation atomic absorption spectrometry. Anal. Chem. 59 (3), 530-533.

American Institute of Nutrition. (1977). Report of the AIN ad hoc Committee on Standards for Nutritional Studies. J. Nutr. 107, 1340-1348.

American Institute of Nutrition. (1980). Second report of the ad hoc Committee on Standards for Nutritional Studies. J. Nutr. 110, 1726.

Athar, M., Hasan, S. K., and Srivastava, R.C. (1987). Role of glutathione metabolizing enzymes in nickel-mediated induction of hepatic glutathione. Res. Commun. Chem. Pathol. Pharmacol. 57, 421-424.

Barker, W. H., Jr. and Runte, V. (1972). Tomato juice-associated gastroenteritis, Washington and Oregon 1969. Am. J. Epidemiol. 96, 219-226.

Barnes, J. M. and Stoner, H. B. (1959). The toxicology of tin compounds. Pharmacol. Rev. 11, 211-231.

Benoy, C. J., Hooper, P. A., and Schneider, R. (1971). The toxicity of tin in canned fruit juices and solid foods. Food Cosmet. Toxicol. 9, 645-656.

Boyer, I. J. (1989). Toxicity of dibutyl tin, tributyl tin and other organotin compounds to humans and to experimental animals. Toxicol. 55, 253-298.

Burba, J. V. (1983). Inhibition of hepatic azo-reductase and aromatic hydrozylase by radiopharmaceuticals containing tin. Toxicol. Lett. 18, 269-272.

Capar, S.G. and Boyer, K.W. (1980). Multielement analysis of foods stored in their opened cans. J. Food Saf. 2, 105-118.

Chiba, M., Ogihara, K., and Kikuchi, M. (1980). Effect of tin on porphyrin biosynthesis. Arch. Toxicol. 45, 189-195.

Davis, G. K. and Mertz, W. (1987). Copper. In: "Trace Elements in Human and Animal Nutrition", W. Mertz, ed., Academic Press, New York, 5th edition, pp. 301-364.

DeGroot, A. P. (1973). Subacute toxicity of inorganic tin as influenced by dietary levels of iron and copper. Food Cosmet. Toxicol. 11, 955- 962.

DeGroot, A. P., Feron, V. J., and Til, H. P. (1973). Short-term toxicity studies on some salts and oxides of tin in rats. Food Cosmet. Toxicol. 11, 11-30.

DeRosa, G., Keen, C. L., Leach, R. M., and Hurley, L.S. (1980). Regulation of superoxide dismutase activity by dietary manganese. J. Nutr. 110: 895-804.

Dwivedi, R. S., Kaur, G., Srivastava, R. C., and Krishna Murti, C. R. (1984). Lipid peroxidation in tin-intoxicated partially-hepatectomized rats. Bull. Environ. Contam. Toxicol. 33, 200-209.

Fritsch, P., DeSaint-Blanquat, G., and Derache, R. (1977). Effect of various dietary components on absorption and tissue distribution of orally administered inorganic tin in rats. Food Cosmet. Toxicol. 15, 147-149.

Furchner, J. E. and Drake, G. A. (1976). Comparative metabolism of radionuclides in mammals - XI. Retention of 113-Sn in the mouse, rat, monkey, and dog. Health Phys. 31, 219-224.

Greger, J. L. and Baier, M. (1981). Tin and iron content of canned and bottled foods. J. Food Sci. 46, 1751-1753 and 1765.

Greger, J. L. and Johnson, M. A. (1981). Effect of dietary tin on zinc, copper and iron utilization by rats. Food Cosmet. Toxicol. 19, 163-166.

Greger, J. L. and Lane, H. W. (1987). The toxicology of dietary tin, aluminum, and selenium. In: "Nutritional Toxicology", J.N. Hathcock, ed., Academic Press, New York, Vol. 2, pp. 223-247.

Greger, J. L., Smith, S. A., Johnson, M. A., and Baier, M. J. (1982). Effects of dietary tin and aluminum on selenium utilization by adult males. Biol. Trace Elem. Res. 4, 269-278.

Griffith, O. W. (1980). Determination of glutathione and glutathione disulfide using glutathione reductase and 2-vinylpyridine. Anal. Biochem. 106, 207-212.

Hiles, R. A. (1974). Absorption, distribution, and excretion of inorganic tin in rats. Toxicol. Appl. Pharmacol. 27, 366-379.

Jenkinson, S. G., Lawrence, R.A., Burk, R. F., and Williams, D.M. (1982). Effects of copper deficiency on the activity of the selenoenzyme glutathione peroxidase and on the excretion and tissue retention of $^{75}SeO_3^{2-}$. J. Nutr. 112: 197-204.

Johnson, M. A., Baier, M. J., and Greger, J. L. (1982). Effects of dietary tin on zinc, copper, iron, manganese, and magnesium metabolism in adult males. Am. J. Clin. Nutr. 35, 1332-1338.

Johnson, M. A. and Greger, J. L. (1982). Effects of dietary tin on tin and calcium metabolism in adult males. Am. J. Clin. Nutr. 35, 655-660.

Johnson, M. A. and Greger, J. L. (1984). Absorption, distribution and endogenous excretion of zinc by rats fed various dietary levels of inorganic tin and zinc. J. Nutr. 114, 1843-1852.

Johnson, M. A. and Greger, J. L. (1985). Tin, copper, iron and calcium metabolism in rats fed various levels of inorganic tin and zinc. J. Nutr. 115, 615-624.

Kappas, A. and Maines, M. D. (1976). Tin: A potent inducer of heme oxygenase in kidney. Science 192, 60-62.

Levine, W.G. (1982). Glutathione, lipid peroxidation and regulation of cytochrome P-450 activity. Life Sci. 31, 779-784.

Magos, L. (1986). Tin. In: "Handbook on the Toxicology of Metals", L. Friberg, G. F. Nordberg, and V. B. Vouk, eds., Elsevier, New York, 2nd ed., Vol. II, pp. 568-593.

Nielsen, F. H. (1986). Other Elements: Sb, Ba, B, Br, Cs, Ge, Rb, Ag, Sr, Sn, Ti, Zr, Be, Bi, Ga, Au, In, Nb, Sc, Te, Tl, W. In: "Trace Elements in Human and Animal Nutrition", W. Mertz, ed., Academic Press, New York, 5th ed., Vol. 2, pp. 415-463.

Paglia, D. E. and Valentine, W. N. (1967). Studies on the quantitative and qualitative characterization of erythrocyte glutathione peroxidase. J. Lab. Clin. Med. 70, 158-169.

Paynter, D. I., Moir, R. J., and Underwood, E. J. (1979). Changes in activity of Cu-Zn superoxide dismutase enzyme in tissues of the rat with changes in dietary copper. J. Nutr. 109, 1570-1576.

Prohaska, J.R., Downing, S. W., and Lukasewycz, O. A. (1983). Chronic dietary deficiency alters biochemical and morphological properties of mouse lymphoid tissue. J. Nutr. 113: 1583-1590.

Rader, J. I., Hight, S. C., and Capar, S. G. (1990). Effects of dietary sugars and cellulose on tissue minerals in weanling rats fed purified diets of adequate or marginal nutrient content. J. Trace Elem. Exptl. Med., in press

Rader, J. I., Wolnik, K. A., Gaston, C. M., Celesk, E. M., Peeler, J. T., Fox, M. R. S., and Fricke, F. L. (1984). Trace element studies in weanling rats: Maternal diets and baseline tissue mineral values. J. Nutr. 114, 1946-1954.

Rader, J. I., Wolnik, K. A., Gaston, C. M., Fricke, F. L., and Fox, M. R. S. (1986). Purified reference diets for weanling rats: Effects of biotin and cellulose. J. Nutr. 116, 1777-1788.

Rotruck, J. T., Pope, A. L., Ganther, H.E., Hafeman, D. G., and Hoekstra, W. G. (1973). Selenium: Biochemical role as a component of glutathione peroxidase. Science 179, 588-590.

Schosinsky, K. H., Lehmann, H. P., and Beeler, M. F. (1974). Measurement of ceruloplasmin from its oxidase activity in serum by use of o-dianisidine dihydrochloride. Clin. Chem. 20, 1556-1563.

Shukla, G. S., Srivasta, R. S., and Chandra, S. V. (1987). Glutathione metabolism in liver, kidney, and testis of rats exposed to cadmium. Ind. Health 25, 139-146.

Stacey, N. H. and Klaassen, D. C. (1981). Comparison of the effects of metals on cellular injury and lipid peroxidation in isolated rat hepatocytes. Toxicol. Environ. Health 7, 139-147.

Valberg, L. S., Flanagan, P. R., and Chamberlain, M. J. (1984), Effects of iron, tin, and copper on zinc absorption in humans. Am. J. Clin. Nutr. 40, 536-541.

Warburton, S., Udler, W., Ewert, R. M., and Haynes, W. S. (1962). Outbreak of foodborne illness attributed to tin. Public Health Rep. 77, 798-800.

World Health Organization United Nations Environmental Program. (1980). Tin and Organotin Compounds, Environmental and Health Criteria, 15. World Health Organization, Geneva.

Yamaguchi, M., Saito, R., and Okada, S. (1980). Dose-effect of inorganic tin on biochemical indices in rats. Toxicol. 16, 267-273.

Yamaguchi, M., Sugii, K., and Okada, S. (1981). Inorganic tin in the diet affects the femur in rats. Toxicol. Lett. 9, 207-209.

Yamaguchi, M., Sugii, K., and Okada, S. (1982). Tin decreases femoral calcium independently of calcium homeostasis in rats. Toxicol. Lett. 10, 7-10.

Yamamoto, T., Yamaguchi, M., and Sato, H. (1976). Accumulation of calcium in kidney and decrease in calcium in serum of rats treated with tin chloride. J. Toxicol. Environ. Health 1, 749-756.

CONTRIBUTORS

Aabin, B.
Department of Biochemistry and Nutrition
The Technical University of Denmark
DK-2800 Lyngby, DENMARK

Adrian, Jean
Chaire de Biochimie Industrielle et Agroalimentaire
Conservatoire National des Arts et Metiers,
292, rue Saint-Martin, 75003 Paris, FRANCE

Aeschbacher, H. U.
Nestle Research Centre, Nestec Ltd.
Vers-chez-les-Blanc
CH 1000 Lausanne 26, SWITZERLAND

Albright, K
Food Research Institute
University of Wisconsin
Madison, Wisconsin 53706

Barkholt, V
Dept. of Biochemistry and Nutrition
The Technical Unversity of Denmark
DK-2800 Lyngby, DENMARK

Bates, Anne H.
Western Regional Research Center
ARS, USDA, Albany, California 94710

Becker, M. Ines
P. Universidad Catolica de Chile
Casilla 6177, Santiago, CHILE

Bellion, I.
Procter Department of Food Science
University of Leeds
Leeds, LS2 9JT, GREAT BRITAIN

Benjamin, H.
Food Research Institute
University of Wisconsin
Madison, Wisconsin 53706

Bjeldanes, Leonard F.
Dept. of Nutritional Sciences
University of California
Berkeley, California 94720

Boncristiani, Guido
Dipartmento di Scienze dell Ambiente
Universita di Pisa, 56126 Pisa, ITALY

Bradfield, Christopher
Dept. of Pharmacology and Toxicology
Northwestern University Medical School, Chicago, Illinois

Brandon, David L.
Western Regional Research Center
ARS, USDA, Albany, California 94710

Brooks, James R.
Ross Laboratories
Columbus, Ohio

Buhl, Karin
Institute of Human Nutrition
Christian-Albrechts University
2300 Kiel 1, GERMANY

Buonorati, M. H.
Lawrence Livermore National Laboratory, Livermore, California 94550

Burks, Wesley A.
Arkansas Children's Hospital
800 Marshall St.,
Little Rock, Arkansas 72202

Chin, J.
Food Research Institute
University of Wisconsin
Madison, Wisconsin 53706

De Ioannes, Alfred E.
P. Universidad Catolica de Chile
Casilla 6177, Santiago, CHILE

Djurtoft, R.
Dept. of Biochemistry and Nutrition
The Technical University of Denmark
DK-2800 Lyngby, DENMARK

Edwards, Ana M.
P. Universidad Catolica de Chile
Casilla 6177, Santiago, CHILE

Erbersdobler, Helmut F.
Institute of Human Nutrition
Chirstian-Albrechts University
2300 Kiel 1, GERMANY

Faith, N.
Food Research Institute
University of Wisconsin
Madison, Wisconsin 53706

Felton, J. S.
Lawrence Livermore Natl. Laboratory
Livermore, California 94550

Finot, Paul-Andre
Nestle Research Centre, Nestec Ltd.
Vers-chez-les-Blanc
1000 Lausanne 26, SWITZERLAND

Frangne, Regine
Conservatoire National des Arts et Metiers
292, rue Saint-Martin, 75003 Paris, FRANCE

Friedman, Mendel
Western Regional Research Center
ARS, USDA, Albany, California 94710

Fulz, E.
Lawrence Livermore Natl. Laboratory
Lovermore, California 94550

Goddard, S. J.
Procter Department of Food Science
University of Leeds
Leeds, LS2 9JT, GREAT BRITAIN

Gruter, A.
Food Research Institute
University of Wisconsin
Madison, Wisconsin 53706

Ha, Y. L.
Food Research Institute
University of Wisconsin
Madison, Wisconsin 53706

Hanna, Marta Salim
P. Universidad Catolica de Chile
Casilla 6177, Santiago, CHILE

Hathcock, John N.
Center for Foof Safety and Applied Nutrition, FDA
Washington, D. C. 20204

Helm, Rick, M.
Arkansas Children's Hospital
800 Marshall St.,
Little Rock, Arkansas 72202

Hymowitz, Theodore
Department of Agronomy
University of Illinois
Urbana, Illinois 61801

Jagerstad, Margaretha
Dept. of Food Chemistry
Chemical Center
University of Lund, P. O. Box 124
S-221 00 Lund, SWEDEN

Johnson, Phyllis E.
Grand Forks Human Nutrition Center
USDA, ARS, P. O. Box 7166
University Station,
Grand Forks, North Dakota 58202

Jonsson, Lena
SIK, Swedish Institute for Food Research, P. O. Box 5401
S-402 29 Goteborg, SWEDEN

Jost, R.
Nestle Research Centre, Nestec Ltd.
Vers-chez-les-Blanc
CH 1000 Lausanne 26, SWITZERLAND

Kanazawa, Kazuki
Dept. of Agricultural Chemistry
Kobe University
Nada-ku, Kobe 657, JAPAN

Kinsella, J.
Institute of Food Science
Cornell University
Ithaca, New York 14853

Knize, M. G.
Lawrence Livermore Natl. Laboratory
Livermore, California 94550

Lohmann, Michael
Institute of Human Nutrition
Christian-Albrecths University
2300 Kiel 1, GERMANY

Loprieno, Gregorio
Dipartmento di Scienze dell Ambiente
Universita di Pisa
56126 Pisa, ITALY

Loprieno, Nicola
Departmento di Scienze dell Ambiente
Universita di Pisa
56126 Pisa, ITALY

Monti, J. C.
Nestle Research Centre, Nestec LTD.
Vers-chez-les-Blanc
CH 1000 Lausanne 26, SWITZERLAND

Oste, Rickard E.
Department of Applied Nutrition
The Chemical Center
University of Lund
S-221 00 Lund, SWEDEN

Pahud, J. J.
Nestle Research Centre, Nestec LTD.
Vers-chez-les-Blanc
CH 100 Lausanne 26, SWITZERLAND

Paik, In-Kee
Department of Animal Science
Chung-Ang Unversity
Assung-Kun, Kyonggi-Do
SOUTH KOREA

Pariza, M. W.
Food Research Institute
University of Wisconsin
Madison, Wisconsin 53706

Pedersen, H. S.
Dept. of Biochemistry and Nutrition
The Technical University of Denmark
DK-2800 Lyngby, DENMARK

Poiffait, Annie
Conservatoire National des Arts et
Metiers
292, rue Saint-Martin, 75003 Paris,
FRANCE

Quattrucci, Enrica
Istituto Nazionale della Nutrizione
00179 Rome, ITALY

Rader, Jeanne I.
Division of Nutrition, FDA
200 C. Street, S. W.
Washington, D. C. 20204

Sampson, Hugh, H.
Johns Hopkins University Medical
School
Baltimore, Maryland

Sandberg, Anne-Soffie
Department of Food Science
Chalmers University of Technology
Gothenburg, SWEDEN

Sarwar, G.
Bureau of Nutritional Science
Food Directorate, National Health
and Welfare, Tunney's Pasture,
Ottawa, Ontario, CANADA K1A 0L2

Silva, Eduardo
P. Universidad Catolica de Chile
Casilla 6177, Santiago, CHILE

Skog, Kerstin
Department of Food Chemistry
University of Lund, P. O. Box 124
S-221 00 Lund, SWEDEN

Smith, T. K.
Department of Nutritional Sciences
University of Guelph
Guelph, Ontario, CANADA N1G 2W1

Storkson, J.
Food Research Institute
University of Wisconsin
Madison, Wisconsin 53706

Swallow A. John
Patterson Institute for Cancer
Research
Christie Hospital and Holt Radium
Institute
Manchester M20 9BX, ENGLAND

Sword, J. T.
Food Research Institute
University of Wisconsin
Madison, Wisconsin 53706

Thompson, L. H.
Lawrence Livermore Natl. Laboratory
Livermore, California 94550

Thresher, Wayne
Central Soya
Fort Wayne, Indiana

Tucker, J. D.
Lawrence Livermore Natl. Laboratory
Livermore, California 94550

Turteltaub, M. H.
Lawrence Livermore Natl. Laboratory
Livermore, California 94550

Vanderlaan, M.
Lawrence Livermore Natl. Laboratory
Livermore, California 94550

Walker, R.
Department of Biochemistry
University of Surrey
Guildford GU2 5XH
UNITED KINGDOM

Watkins, B. Y.
Lawrence Livermore Natl. Laboratory
Livermore, California 94550

Wedzicha, B. L.
Procter Department of Food Science
University of Leeds
Leeds, LS 2 9 JT, GREAT BRITAIN

Weisburger, J. H.
American Health Foundation
Valhalla, New York 10595

Williams, Larry
Arkansas Children's Hospital
800 Marshall St.,
Little Rock, Arkansas 72202

INDEX